Recent Titles in This Series

133 **P. I. Naumkin and I. A. Shishmarev,** Nonlinear nonlocal equations in the theory of waves, 1994
132 **Hajime Urakawa,** Calculus of variations and harmonic maps, 1993
131 **V. V. Sharko,** Functions on manifolds: Algebraic and topological aspects, 1993
130 **V. V. Vershinin,** Cobordisms and spectral sequences, 1993
129 **Mitsuo Morimoto,** An introduction to Sato's hyperfunctions, 1993
128 **V. P. Orevkov,** Complexity of proofs and their transformations in axiomatic theories, 1993
127 **F. L. Zak,** Tangents and secants of algebraic varieties, 1993
126 **M. L. Agranovskiĭ,** Invariant function spaces on homogeneous manifolds of Lie groups and applications, 1993
125 **Masayoshi Nagata,** Theory of commutative fields, 1993
124 **Masahisa Adachi,** Embeddings and immersions, 1993
123 **M. A. Akivis and B. A. Rosenfeld,** Élie Cartan (1869–1951), 1993
122 **Zhang Guan-Hou,** Theory of entire and meromorphic functions: Deficient and asymptotic values and singular directions, 1993
121 **I. B. Fesenko and S. V. Vostokov,** Local fields and their extensions: A constructive approach, 1993
120 **Takeyuki Hida and Masuyuki Hitsuda,** Gaussian processes, 1993
119 **M. V. Karasev and V. P. Maslov,** Nonlinear Poisson brackets. Geometry and quantization, 1993
118 **Kenkichi Iwasawa,** Algebraic functions, 1993
117 **Boris Zilber,** Uncountably categorical theories, 1993
116 **G. M. Fel′dman,** Arithmetic of probability distributions, and characterization problems on abelian groups, 1993
115 **Nikolai V. Ivanov,** Subgroups of Teichmüller modular groups, 1992
114 **Seizô Itô,** Diffusion equations, 1992
113 **Michail Zhitomirskiĭ,** Typical singularities of differential 1-forms and Pfaffian equations, 1992
112 **S. A. Lomov,** Introduction to the general theory of singular perturbations, 1992
111 **Simon Gindikin,** Tube domains and the Cauchy problem, 1992
110 **B. V. Shabat,** Introduction to complex analysis Part II. Functions of several variables, 1992
109 **Isao Miyadera,** Nonlinear semigroups, 1992
108 **Takeo Yokonuma,** Tensor spaces and exterior algebra, 1992
107 **B. M. Makarov, M. G. Goluzina, A. A. Lodkin, and A. N. Podkorytov,** Selected problems in real analysis, 1992
106 **G.-C. Wen,** Conformal mappings and boundary value problems, 1992
105 **D. R. Yafaev,** Mathematical scattering theory: General theory, 1992
104 **R. L. Dobrushin, R. Kotecký, and S. Shlosman,** Wulff construction: A global shape from local interaction, 1992
103 **A. K. Tsikh,** Multidimensional residues and their applications, 1992
102 **A. M. Il′in,** Matching of asymptotic expansions of solutions of boundary value problems, 1992
101 **Zhang Zhi-fen, Ding Tong-ren, Huang Wen-zao, and Dong Zhen-xi,** Qualitative theory of differential equations, 1992
100 **V. L. Popov,** Groups, generators, syzygies, and orbits in invariant theory, 1992
99 **Norio Shimakura,** Partial differential operators of elliptic type, 1992
98 **V. A. Vassiliev,** Complements of discriminants of smooth maps: Topology and applications, 1992

(*Continued in the back of this publication*)

Nonlinear Nonlocal Equations in the Theory of Waves

Translations of

MATHEMATICAL MONOGRAPHS

Volume 133

Nonlinear Nonlocal Equations in the Theory of Waves

P. I. Naumkin
I. A. Shishmarev

American Mathematical Society
Providence, Rhode Island

Наумкин П. И., Шишмарёв И. А.

НЕЛИНЕЙНЫЕ НЕЛОКАЛЬНЫЕ УРАВНЕНИЯ В ТЕОРИИ ВОЛН

Translated by Boris Gommerstadt from an original Russian manuscript
Translation edited by Simeon Ivanov

1991 *Mathematics Subject Classification.* Primary 35Lxx, 45K05;
Secondary 35Q35, 76L05.

ABSTRACT. Nonlinear evolutional equations of mathematical physics are studied. The major part of the book is devoted to the analysis of breaking and decay of solutions in finite time. The methods developed in the book can be applied to a wide class of conservative and dissipative nonlinear equations, both local and nonlocal. Among the important examples, the authors consider the Kolmogorov-Petrovskiĭ-Piskunov equation, the nonlinear nonlocal Schrödinger equation, the Kuramoto-Sivashinsky equation, the Korteweg-de Vries-Burgers equation, and several other important equations of mathematical physics.

Library of Congress Cataloging-in-Publication Data

Naumkin, P. I. (Pavel Ivanovich) 1961–
[Nelineĭnye nelokal′nye uravneniia v teorii voln. English]
Nonlinear nonlocal equations in the theory of waves / P. I. Naumkin, I. A. Shishmarev.
p. cm. — (Translations of mathematical monographs, ISSN 0065-9282 ; v. 133)
Includes bibliographical references.
ISBN 0-8218-4573-X (acid-free paper)
1. Waves–Mathematics. 2. Nonlinear wave equations–Numerical solutions. I. Shishmarev, Il′ia Andreevich. II. Title. III. Series.
QC157.N3813 1994
532′.593′01515353—dc20

93-8452
CIP
r93

The paper used in this book is acid-free and falls within the guidelines
established to ensure permanence and durability.
Printed on recycled paper.

Information on Copying and Reprinting can be found at the back of this volume.

This publication was typeset using AMS-TEX,
the American Mathematical Society's TEX macro system.

10 9 8 7 6 5 4 3 2 1 99 98 97 96 95 94

Contents

Introduction

§1. Physical problems leading to nonlinear nonlocal equations

Modern mathematical physics is almost exclusively a mathematical theory of nonlinear models—equations describing various physical processes. A large number of publications have been devoted to classical models of nonlinear hydrodynamics and nonlinear equations that are solvable exactly by inverse scattering techniques (see [**1, 18, 24, 34, 70, 79, 68, 111**]) This method has rapidly developed by the growing theory of nonlinear structures [**30**]. In addition to this literature, the attention of researchers has recently been focused on the theory of nonlinear nonlocal equations (NNE), which is the topic of the present monograph. Though nonlinear nonlocal equations had been encountered earlier in the physics literature (see, e.g. [**27**]), G. Whitham was the first (in 1967) to draw general attention to the mathematical aspects of these equations [**81, 82**]. He proposed the following very general nonlinear nonlocal equation:

$$u_t + uu_x + \int_{-\infty}^{\infty} k(x-s)\,U_s(s,t)\,ds = 0\,. \tag{1}$$

This equation combines two factors: the nonlinear term uu_x characteristic of water waves, and an integral term describing the complete dispersions of the linear theory.

The dispersion relation for the Whitham equation (1) is equal to the Fourier transform of the kernel $k(x)$:

$$c(p) = \frac{w}{p} = \int_{-\infty}^{\infty} k(x)e^{-ipx}\,dx \tag{2}$$

and can be arbitrary for a suitable choice of $k(x)$, so that equation (1) represents a wide class of various nonlinear equations that are of great interest for physical applications. The presentation of the Whitham equation in the form (1) and of the dispersion relation in the form (2) presupposes that the kernel $k(x)$ is symmetric ($k(x) = k(-x)$) and integrable.

Whitham's motivation for introducing equation (1) was to have a simple equation suitable for describing such typical water wave phenomena as sharp crests and breaking of waves, which the Burgers and Korteweg-de-Vries equations fail to do. Indeed, as we shall see further, the Whitham nonlinear nonlocal equation (1) with various kernels describes all the characteristic properties of waves in a finite time interval, existence of the wave solutions global in time, eventual smoothing of discontinuous initial perturbations, etc.

Whitham's original idea was to combine the shallow water equation with the integral term that produces arbitrary wave dispersion, so that the resulting equation would describe both phenomena of peaking and breaking, but it proves to be fruitful also in other aspects. First, if one eliminates the requirement that the kernel $k(x)$ be even, the equation includes not only the complete dispersion of the linear theory, but also dissipative processes and those related to energy pumping. This substantially enlarges the class of nonlinear equations under consideration. Second, the next natural step is to generalize (1) to include nonlocal equations with arbitrary singular kernels (including those with integrable singularity at the origin) as well as local equations (with δ-function singularities and their derivatives). This requires replacing Whitham's integral operator $\int_{-\infty}^{\infty} k(x-s)u_s(s,t)\,ds$ by arbitrary pseudodifferential operator, and that means that instead of equation (1) one should consider the following equation:

$$u_t + uu_x + \mathbb{K}(u) = 0, \tag{3}$$

where the operator $\mathbb{K}(u)$ is given by

$$\mathbb{K}(u) = \frac{1}{2\pi}\int_{-\infty}^{\infty} e^{ipx} K(p)\hat{u}(p,t)\,dp; \tag{4}$$

here $\hat{u}(p,t)$ is the Fourier transform of $u(x,t)$:

$$\hat{u}(p,t) = \int_{-\infty}^{\infty} e^{-ipx} u(x,t)\,dx\,.$$

The function $K(p)$, defining the operator $\mathbb{K}(u)$, is called the symbol of the operator. The dispersion relation $c(p)$ and the symbol $K(p)$ are related by equality $K(p) = ipc(p)$. With this approach, the Whitham equation represents, as particular cases, many equations that are of great interest for physical applications, e.g., the equations of Korteweg-de-Vries, Burgers, Kawahara, Kuramoto-Sivashinsky, and others.

Finally, it is reasonable to start not from the shallow water equation but from equations with other kinds of nonlinearities that are typical for problems of modern mathematical physics. In this way, we are led to the generalized nonlinear Schrödinger equation:

$$u_t + i|u|^2 u + \mathbb{K}(u) = 0, \tag{5}$$

to the generalized Kolmogorov-Petrovskiĭ-Piskunov equation (KPP):

$$u_t - u^2 + \mathbb{K}(u) = 0, \tag{6}$$

where the operator $\mathbb{K}(u)$ is defined by formula (4), and also to various systems of nonlinear nonlocal equations generalizing the Broer, Kaup, Dobrokhotov, and other systems of equations.

We outline a number of physical problems leading to nonlinear nonlocal equations. The Burgers equation **[101]** $u_t + uu_x - \nu u_{xx} = 0$ and the Korteweg-de Vries equation (KdV) **[137]** $u_t + uu_x + \alpha u_{xxx} = 0$ are well known in the theory of water surface waves. Both equations contain the nonlinear term uu_x of the shallow water theory. In addition, the Burgers equation contains the term $-\nu u_{xx}$ describing simple diffusion (dissipation), while the KdV equation contains the term αu_{xxx} describing simple dispersion. These equations find extensive application in various areas of physics, and have been studied in numerous papers **[76, 77, 82, 84, 105, 120]**. Both of these equations are particular cases of equation (3): for the Burgers equations the symbol is $K(p) = \nu p^2$, and for the KdV equation it is $K(p) = -i\alpha p^3$.

Also of great interest in physics is the combination of these two equations, the so-called Korteweg-de Vries-Burgers equation [**118, 123, 144**]:

$$u_t + uu_x + \alpha u_{xxx} - \nu u_{xx} = 0\,. \tag{7}$$

It appears, for example, in nonlinear acoustics for fluids with gas bubbles [**26, 75, 148**].

The Benjamin-Ono equation [**90, 94, 102, 109, 147, 151**]:

$$u_t + uu_x + \frac{1}{\pi} \fint_{-\infty}^{\infty} \frac{u_{ss}(s,t)}{x-s}\, ds = 0, \tag{8}$$

(the integral is understood in the sense of its principal value) describes long internal waves in a stratified fluid of infinite depth (its symbol is $K(p) = ip|p|$).

The intermediate or the Joseph equation [**103, 127, 138**]

$$u_t + uu_x + \fint_{-\infty}^{\infty} \left[\frac{1}{2\delta}\coth\frac{\pi(x-s)}{2\delta} - \frac{1}{2\delta}\operatorname{sgn}(x-s)\right] u_{ss}(s,t)\, ds = 0 \tag{9}$$

describes long internal waves in a stratified fluid of finite depth δ. For $\delta \to 0$, equation (9) transforms into the KdV equation, and for $\delta \to \infty$, it reduces to the Benjamin-Ono equation (this property explains the name of the equation). The symbol of equation (9) is $K(p) = -ip^2(\coth 2\delta p - \frac{1}{2\delta p})$. The inverse scattering technique is applicable to equations (8) and (9) and their theory has thus been substantially advanced (see [**102, 109, 136**]).

Leibovitz [**95, 142**] derived the equation:

$$u_t + \alpha u_x + uu_x + A\frac{\partial^2}{\partial x^2} \int_{-\infty}^{\infty} \frac{u_s(s,t)}{\sqrt{(x-s)^2 + B^2}}\, ds = 0 \tag{10}$$

to describe axially-symmetric long waves with a low amplitude in a nonviscous incompressible rotating fluid which is infinite in radial direction. Its kernel $k(x) = (\partial^2/\partial x^2)(1/\sqrt{x^2+B^2})$ has no singularities at zero ($B \neq 0$) i.e., it is regular, in contrast to the other equations quoted above).

Klimontovich [**27**] proposed the following magnetohydrodynamic equation for a nonisothermal collisionless plasma:

$$u_t + uu_x + c\int_{-\infty}^{\infty} K_0(|x-s|)u_s(s,t)\, ds = 0, \tag{11}$$

where u is the ion velocity, K_0 is the zero-order MacDonald function, and c is a constant.

Besides the above-mentioned KdV-Burgers equation for a mixture of a fluid with gas bubbles, other particular cases of the Whitham nonlinear nonlocal equation (3) are also found in nonlinear acoustics, for example, the equation

$$u_t + uu_t + \alpha u_{xxx} + \beta u + \gamma \int_0^x \frac{u_s(s,t)}{\sqrt{x-s}}\, ds = 0, \tag{12}$$

proposed by Kobelev and Ostrovskiĭ [**28**] and by Nakoryakov and Shreiber [**45**], where u is the acoustic pressure inside a bubble and α, β, and γ are constants.

When the joint action of nonlinearity, dissipation and relaxation due to molecular viscosity, and the presence of salts dissolved in the ocean are considered, we arrive at the following nonlinear nonlocal equation:

$$u_t + uu_x + \alpha u_{xx} + \beta \frac{\partial}{\partial x} \int_{-\infty}^{x} e^{-(x-s)/\gamma} u_s(s,t)\, ds = 0, \tag{13}$$

proposed by Rudenko and Soluyan **[75]**, where u is the velocity of particles in a wave, and α, β, and γ are constants. Engelbrecht **[12, 108, 122]** also arrived at this equation, while treating mathematical models of nonlinear elasticity for viscoelastic media. He also studied (13) in the case when the kernel $k(x)$ is the sum of exponents.

Ott, Sudan, and Ostrovskiĭ **[128, 152, 153, 155]** proposed the following generalizations of the KdV equation:

$$u_t + uu_x + \alpha u_{xxx} + \frac{1}{\pi} \fint_{-\infty}^{\infty} \frac{u_s(s,t)}{x-s}\, ds = 0, \tag{14}$$

$$u_t + uu_x + \alpha u_{xxx} + u + \int_{-\infty}^{\infty} \frac{\operatorname{sgn}(s-x)}{\sqrt{|s-x|}} u_s(s,t)\, ds = 0, \tag{15}$$

$$u_t + uu_x + \alpha u_{xxx} - \int_{-\infty}^{\infty} \frac{1-\operatorname{sgn}(s-x)}{\sqrt{|s-x|}} u_s(s,t)\, ds = 0, \tag{16}$$

describing the ion-acoustic waves in plasma with the Landau damping. In these examples, the symbols of the operators $\mathbb{K}(u)$ have the following form, respectively: $K(p) = -i\alpha p^3 + |p|$, $K(p) = -i\alpha p^3 + 1 + |p|^{1/2}$ and $K(p) = -i\alpha p^3 - ip|p|^{-1/2} + |p|^{1/2}$.

The Kawahara equation **[131]**

$$u_t + uu_x + \alpha u_{xxx} - u_{xxxxx} = 0, \tag{17}$$

where $K(p) = -i\alpha p^3 - ip^5$, describes propagation of signals in transmission lines, propagation of long waves under ice cover in liquids of finite depth **[21]**, and also gravity waves on the surface of a liquid with surface tension **[171]**.

The Kuramoto-Sivashinsky equation **[139, 149]**

$$u_t + \tfrac{1}{2}u_x^2 + u_{xx} + \alpha u_{xxxx} = 0 \tag{18}$$

is applied, for instance, in the theory of combustion to model a flame front and also in the study of two-dimensional turbulence.

Finally, we present two examples of the nonlocal equation (3), the first of which has the symbol

$$K(p) = ip\sqrt{g\frac{\tanh ph}{p}\left(1 + \frac{T}{\rho g}p^2\right)}$$

and is encountered in the theory of capillary waves **[82]**, and the second has the symbol $K(p) = ip\sqrt{1+|p|^2}$ and is used to describe shelf waves.

As was noted above, a different character of nonlinearity is manifested in the nonlinear nonlocal Schrödinger (NNS) equation (5) describing the effects of self-focusing for plane monochromatic beams in media with nonlinear refractive indexes

[**19, 117, 133, 150**]. A particular case of the NNS equation is the generalized Landau-Ginzburg equation [**107, 156**]:

$$u_t + iu^2u^* + iu_{xx} + \alpha u = 0, \tag{19}$$

where $\alpha > 0$. Another example is the equation

$$u_t + iu^2u^* + iu_{xx} + \int_{-\infty}^{\infty} q(x-s)u(s,t)\,ds = 0, \tag{20}$$

encountered in the theory of the Langmuir waves in plasma [**16, 156**].

Now we turn to nonlinear nonlocal systems of equations. We consider a model system describing surface waves in an incompressible liquid [**55, 60**]:

$$\begin{cases} \eta_t + (\nabla, \eta\nabla\varphi) + \mathbb{K}_1(\eta) + \mathbb{K}_2(\varphi) = 0, \\ \varphi_t + \frac{1}{2}(\nabla\varphi)^2 + \mathbb{K}_3(\eta) + \mathbb{K}_4(\varphi) = 0, \end{cases} \tag{21}$$

where $\eta(x,t)$ is the free surface of the liquid, $\varphi(x,t)$ is the velocity potential, $x \in R_2$, $t \geq 0$, and $\mathbb{K}_j$ are linear operators defined by the formula

$$\mathbb{K}_j(\psi) = \frac{1}{4\pi^2} \iint_{-\infty}^{\infty} e^{i(p,x)} K_j(p)\hat{\psi}(p,t)\,dp\,. \tag{22}$$

The symbols $K_j(p)$ of the operators $\mathbb{K}_j$ define the dispersion relation for the system (21). Different choices of kernels correspond to different processes of wave propagation: conservative, dissipative, or processes associated with energy pumping, so that system (21) is rather general.

Thus, in the particular case where $\mathbb{K}_3(\eta) = \eta$ and the remaining $\mathbb{K}_j \equiv 0$, system (21) is reduced to the shallow water system of equations [**35, 38, 76, 77, 82**]:

$$\begin{cases} \eta_t + (\nabla, \eta\nabla\varphi) = 0, \\ \varphi_t + \frac{1}{2}(\nabla\varphi)^2 + \eta = 0, \end{cases} \tag{23}$$

which contains nonlinear terms but disregards dispersion.

If we confine ourselves only to the simplest dispersion term, then the system of equations (21) goes over into the familiar system of Boussinesq equations [**98, 143**]:

$$\begin{cases} \eta_t + (\nabla, \eta\nabla\varphi) + \beta\nabla^4\varphi = 0, \\ \varphi_t + \frac{1}{2}(\nabla\varphi)^2 + \eta = 0, \qquad \beta = \text{const}\,. \end{cases} \tag{24}$$

System (21) also includes equations recently derived by Dobrokhotov [**13**], which, in the first approximation of the nonlinearity, take the form:

$$\begin{cases} \eta_t + (\nabla, \eta\nabla\varphi) + \mathbb{B}(\varphi) = 0, \\ \varphi_t + \frac{1}{2}(\nabla\varphi)^2 + \eta = 0\,. \end{cases} \tag{25}$$

The symbol of the operator $\mathbb{B}$ is $B(p) = |p|\tanh|p|$ and corresponds to the exact potential theory of waves on water. An important special case of system (21) is connected with the choice of $K_3 = 0$. Then (21) becomes

$$\eta_t + (\nabla, \eta\nabla\varphi) + \mathbb{K}_1(\eta) + \mathbb{K}_2(\varphi) = 0, \tag{26}$$

$$\varphi_t + \frac{1}{2}(\nabla\varphi)^2 + \mathbb{K}_4(\varphi) = 0\,. \tag{27}$$

The second equation in this system can be solved independently of the first. The solution thus obtained, if substituted in (26), converts (26) into a linear equation with respect to η. In its turn, in the case of one spatial dimension, (27) reduces to the Whitham equation and thus includes in itself many important one-dimensional equations, such as the KdV, Burgers, Benjamin-Ono, and other equations (see (7)–(18)).

Thus, the system (21) of equations for surface waves on water includes many equations and systems, both one-dimensional and multidimensional, important from a physical viewpoint.

A more general than (21) is the following system **[55, 60]**:

$$\begin{cases} \eta_t + (\nabla, \eta u) + \mathbb{K}_1(\eta) + \mathbb{K}_2(u) = 0, \\ u_t + (u, \nabla)u + \mathbb{K}_3(\eta) + \mathbb{K}_4(u) = 0, \end{cases} \tag{28}$$

where $\eta = \eta(x, t)$ is the free surface and $u = u(x, t)$ is the two-dimensional velocity vector. For physics this system is just as interesting as (21). The systems (23)–(27) are particular cases of it. The familiar Khokhlov-Zabolotskaya **[14]** and Kadomtsev-Petviashvili **[23]** equations also follow from (28) for a suitable choice of the operator $\mathbb{K}_4$, provided η and one of the components of the velocity u vanish. For instance, the Kadomtsev-Petviashvili equation

$$(u_t + \alpha u_{xxx} - uu_x)_x \pm u_{yy} = 0$$

is derived from (28) if $\eta \equiv 0$, $u^{(2)} \equiv 0$, and the symbol of the operator $\mathbb{K}_4$ is

$$K_4(p) = \begin{pmatrix} (ip_1)^3 + c\frac{(ip_2)^2}{ip_1}, & 0 \\ 0, & 0 \end{pmatrix}.$$

The one-dimensional in the spatial variable x system of equations, proposed by Kaup **[130]** and Broer **[100]**,

$$\begin{cases} \eta_t = (u\eta + \alpha u_{xx} - \beta \eta_x)_x, \\ u_t = (\frac{1}{2}u^2 + \eta + \beta u_x)_x \end{cases} \tag{29}$$

is also a special case of (28). The systems (21) and (28) are interconnected in a natural way: by differentiating the second equation in (21) with respect to x_1 and x_2, and putting $u = \nabla\varphi$ and $\mathbb{K}_2(\varphi) = \mathbb{K}_2(\nabla\varphi)$, we obtain system (28).

In conclusion, we emphasize that the description of all the above-mentioned numerous and various examples of physical problems are described in a unified way by the nonlinear nonlocal equation (3) or by the system of equations (21). Thus, the study of nonlinear nonlocal equations and their systems enables one to proceed from the analysis of individual equations to the investigation of wide classes of nonlinear equations that are of great interest for physical application.

§2. Brief review of the content of this book

The success achieved in the last twenty years in the study of nonlinear equations as a result of development and application of the inverse scattering problem technique is well known (see **[1, 18, 19, 24, 34, 41, 68, 70, 79, 88, 111, 119, 140]**). While providing astonishingly faithful solutions of nonlinear equations and, in some cases, permitting the determination of the asymptotic behavior of solutions to the Cauchy problem for long time scales, this method has certain substantial limitations: it is applicable only to conservative equations, there are no general criteria for its applicability, so that each equation has to be analyzed individually, and, finally, this method is unstable with respect to a small perturbation of the equation. The methods developed in this book are applicable to a wide class of conservative and dissipative nonlinear equations, both local and nonlocal.

Let us outline the circle of problems considered in this monograph. A substantial part of this book is devoted to the traditionally important problem of the theory of nonlinear evolutional equations, i.e., breaking and decaying of solutions in finite time. A great number of mathematical publications have dealt with this problem in the last thirty years. We do not even attempt to provide a complete review of these publications; nevertheless, we give a list of some of the well-known results: [**4, 33, 35, 38, 74, 82, 92, 96, 110, 112, 113, 114, 125, 126, 129, 132, 154, 161, 163, 164, 169**]. We note, however, that these publications deal with local nonlinear equations, and, as a rule, do not consider breaking of waves, i.e., situations when the first derivative of a smooth solution to the corresponding Cauchy problem becomes infinite in finite time. Whitham in his monograph [**82**, p. 457] emphasized that "This breaking phenomenon is one of the most intriguing long-standing problems of water wave theory." As pointed out above, Whitham's motivation for suggesting equation (1) was to derive a relatively simple model equation describing the effect of breaking of waves (as opposed to the KdV equation). In his book he raised the following question. Do waves, described by (1) with the kernel $k_g(x)$ arising in the exact potential theory of water waves and corresponding to the dispersion $c(p) = ((g/p)\tanh ph)^{1/2}$, break or not?

Chapter 2 gives an affirmative answer to this question for the problem on the entire line. Chapter 3 does that for the periodic problem. Moreover, the corresponding theorems provide sufficiently precise two-sided estimates of the time of breaking. The breaking of waves is also discussed in Chapter 4, which is devoted to the theory of surface waves. Here we find conditions guaranteeing breaking of the solutions for the system of equations of surface waves, and give two-sided estimates for the time of breaking.

The existence of classical and generalized solutions of the Cauchy problem is closely related to the phenomenon of breaking of waves. The most general results on local existence of classical solutions are derived for conservative and dissipative solutions. We introduce the notions of a conservative and a dissipative operator on the example of equation (3). We define two functions $K^1(p)$ and $K^2(p)$ related to the symbol $K(p)$ of the operator as follows: $K^1(p) = \operatorname{Re} K(p)$ and $K^2(p) = \operatorname{Im} K(p)$. In the case when the solution $u(x, t)$ of (3) is real, the symbol $K(p)$ must satisfy the condition $K(p) = K^*(-p)$, so that $K^1(p)$ is an even and $K^2(p)$ is an odd function. As shown in Chapter 1, if $K^1(p) = 0$ equation (3) has a number of conservation laws, so that the function $K^2(p)$ is naturally called the conservative part of the symbol $K(p)$. Conversely, if $K^1(p) > 0$ (< 0), then the norm $\|u\|_{L_2}$ of the solution decreases (increases) with time; therefore, $K^1(p)$ reflects dissipation (pumping) of energy in the system. We also define the order of the nonlocal operator $\mathbb{K}(u)$. We assume everywhere that the symbol $K(p)$ is locally bounded: $K(p) \in L_\infty^{\mathrm{loc}}(R_1)$, and that the minimum α_0 of the numbers $\alpha > 0$ such that $|K(p)| \le c(1 + |p|)^\alpha$, $p \in R_1$ exists. We say that α_0 is the order of the operator $\mathbb{K}(u)$ and of its symbol $K(p)$. Various theorems on local existence of classical solutions of the Cauchy problem for equation (3) are presented in Chapter 2, and for the periodic problem with respect to the spatial coordinate in Chapter 3. In these results the operator $\mathbb{K}$ is assumed to be either bounded (the operator $\mathbb{K}$ has order zero, $\alpha_0 = 0$), or conservative ($K^1(p) = 0$), or dissipative ($K^1(p) \ge 0$). Analogous results are obtained in Chapter 4 for the system of equations of surface waves (21) and (28).

Of great interest is the problem of finding conditions ensuring existence in the large of classical solutions. In the theory of nonlinear equations this problem is traditional, as is the problem of breaking of waves. Again, with no claim of completeness, we list a number of recent publications in this direction: [**33, 36, 74, 83, 104, 115, 116,**

121, 135, 146, 168]. Mainly, all these publications are devoted to local nonlinear equations of evolution. As for nonlocal nonlinear equations, we single out the work [**87**], where existence in the large of classical solutions for the Benjamin-Ono, Joseph, and Korteweg-de Vries equations is proved.

In Chapters 2 and 3, for a dissipative operator $\mathbb{K}(u)$, i.e., such that $K^1(p) > 0$ and $K^1(p) \geq c > 0$ for $|p| \geq p_0 > 0$, we establish the existence of global in time solutions of the Cauchy problem for (3) in the case when the steepness of the initial perturbation is sufficiently small. Comparison of this result with the theorem on breaking of waves shows that for the dissipative operator $\mathbb{K}(u)$ the condition of sufficiently large steepness of the initial perturbation is necessary and sufficient for breaking of the solution in finite time.

We introduce the notion of a strongly dissipative operator $\mathbb{K}$. Such is an operator whose symbol satisfies the condition

$$K^1(p) \geq c|p|^\alpha \quad \text{for } |p| \geq p_0 > 0, \tag{30}$$

where $c > 0$, $\alpha > 0$. The requirement that the steepness of the initial perturbation be small for the global existence of a solution turns out to be redundant in the case of a strongly dissipative operator $\mathbb{K}$ with $\alpha > 3/2$. Moreover, in Chapters 2 and 3 we show that strong dissipation provides not only the global existence of classical solutions for any smoothed initial perturbation, but also guarantees the smoothing out in time of a discontinuous initial perturbation in the class(*) $H^\rho(R_1)$, $\rho \geq 0$.

Similar results regarding global existence of classical solutions and smoothing of discontinuous initial perturbations for the system of equations of surface waves are established in Chapter 4.

Chapter 5 is devoted to various generalized solutions of the Cauchy problem for the nonlinear nonlocal equations and systems of equations.

The above-mentioned results concerning local existence of solutions to the Cauchy problem, the global in time existence, smoothing of discontinuous initial perturbations, and generalized solutions, are presented in this book for the nonlinear nonlocal equation (3) and system (28). However, by the methods developed in this monograph, these results can easily be extended to nonlinear nonlocal equations with different types of nonlinearity, in particular, to equations (5) and (6).

The final circle of problems considered in the book concerns the asymptotic behavior of solutions of nonlinear nonlocal equations and systems. The asymptotics of solutions of partial differential equations with a small parameter multiplying the derivatives has been discussed by many mathematicians in the last four decades: [**7, 20, 31, 32, 42, 43, 71, 80, 120, 141, 170**]. In our book we study the rather narrow problem of preservation of the property of breaking of solutions for weak nonlocal interaction. Similar problems were considered in [**2, 3, 44**] in connection with nonlinear equations that can be solved exactly by the inverse scattering problem method, and equations that are small perturbations of exactly solvable equations.

In §5 of Chapter 3 and §8 of Chapter 4 we analyze the asymptotics of solutions to the Cauchy problem as the parameter of smallness μ of nonlocal interaction approaches zero (equation (3) and system (28) with nonlocal operators $\mathbb{K}$ depending on a small parameter μ are considered).

(*) $H^\rho(R_1)$ is the Sobolev space [**37, 67**] in which the norm given as follows: $\|u\|_\rho^2 = \int_{-\infty}^{\infty}(1+|p|)^{2\rho}|\hat{u}(p,t)|^2\,dp$, $H^\infty(R_1) = \bigcap_{\rho=1}^{\infty} H^\rho(R_1)$.

Chapters 6–10 are devoted to a very important question concerning the asymptotics, as $t \to \infty$, of solutions to the Cauchy problem for nonlinear nonlocal equations and systems of equations. Great progress has been achieved in this area lately, mainly in connection with the possibility of applying the inverse scattering problem method to conservative nonlinear equations: **[5, 17, 22, 25, 39, 69, 78, 89, 145, 159, 166]**. A very important (for physics) class of dissipative nonlinear equations remains outside the scope of applicability of this method. The theory of these equations is indeed the main topic of this monograph. Widely known examples of dissipative equations are the nonlinear Schrödinger equation with linear dissipation (also called the generalized Landau-Ginzburg equation), the KdV-Burgers equation, the Kuramoto-Sivashinsky equation, and others.

The asymptotics of various integrals of solutions to the KdV-Burgers equation and to the regularized equation of long waves was found in **[91]**. Another approach to studying the asymptotic behavior, as $t \to \infty$, of solutions to nonlinear equations was proposed by Kolmogorov, Petrovskiĭ, and Piskunov in **[29]**. In this work the so-called step-decaying problem was considered for the KPP equation (6). The method elaborated in **[29]** is mainly founded on application of the explicit form of the Green function for the equation of thermal conductivity. In this monograph we use a different approach based on a detailed study of the nonlinear nonlocal equation in the momentum representation and on incorporating of a special diagram technique of the perturbation theory.

Asymptotic formulas, as $t \to \infty$, uniform with respect to the spatial variables and with bounds for the remainder term, are found in Chapters 6 through 9 for the solutions of the Cauchy problem for the following dissipative equations of evolution: the generalized Kolmogorov-Petrovskiĭ-Piskunov equation (and, in particular, for its classical version), the Whitham equation (and, in particular, for the Korteweg-de Vries-Burgers, the Kuramoto-Sivashinsky, the Ott-Sudan-Ostrovskiĭ, and other equations), the nonlinear nonlocal Schrödinger equation (and, in particular, for the Landau Ginzburg equation), and also for the solutions of the system of equations of surface waves (and, in particular, for the Broer, Dobrokhotov, and Kaup systems of equations with viscosity).

In the last chapter we consider the step-decaying problem for the Korteweg-de Vries-Burgers equation and give an affirmative answer to a problem of Bona and Schonbek **[97]** concerning the stability of solutions of the travelling wave type. We also consider the analogous problem for the Kuramoto-Sivashinsky equation.

A more detailed description of the contents is given in the introduction to each chapter. We have strived to keep the presentation in each chapter as independent as possible, with some unavoidable repetitions as a consequence.

CHAPTER 1

Simplest Properties of Solutions of Nonlinear Nonlocal Equations

Whitham **[81, 82]** suggested his model equation by simply combining two factors: the nonlinearity typical for hydrodynamics and the general law of linear dispersion. This "symbiosis" turned out to be rather fruitful: on one hand, many classical nonlinear equations, such as those of Korteweg-de Vries, Burgers, and Benjamin-Ono, are particular cases of the Whitham equation:

$$u_t + uu_x + \int_{-\infty}^{\infty} k(x-s)u_s(s,t)\,ds = 0\,.$$

On the other hand, it is a relatively simple model equation describing practically all characteristic effects of the nonlinear wave theory: peaking and breaking of waves, the existence of solitons, the appearance of waves of limited amplitude, etc. In this chapter, we shall consider the simplest examples of these effects.

§1. Conservation laws. Solitary waves

1.1. Conservation laws, or integrals of motion, are quantities which conserve their values with time during the evolution of the wave process described by the equation examined here. Consider the main conservation laws corresponding to the solution of the Whitham equation. It should be noted that the latter is invariant under the Galilean transformations:

$$t' = t, \quad x' = x + ct, \quad u'(x', t') = u(x,t) + c, \quad c = \text{const}\,.$$

To derive the conservation laws, we shall assume that $u(x,t)$—the classical solution of Whitham's equation—decreases sufficiently rapidly at infinity. We shall also assume that the kernel $k(x)$ is symmetric: $k(x) = -k(-x)$ (this condition means that the phase velocity $c(p)$ [see Introduction] is an even function of the wave number p), and that it and its first derivative are absolutely integrable over the whole real line. For brevity, we denote

$$\widetilde{K}v = \int_{-\infty}^{\infty} k(x-s)v(s,t)\,ds\,. \tag{1.1}$$

The operator $\widetilde{K}$ is symmetric in $L_2(R_1)$, $(\widetilde{K}u, v) = (u, \widetilde{K}v)$, where (u, v) is the scalar product in $L_2(R_1)$, and the operator $\partial_x = \frac{\partial}{\partial x}$ of differentiation with respect to x

commutes with $\widetilde{K}$, i.e. $\partial_x \widetilde{K} = \widetilde{K}\partial_x$. Integrating the Whitham equation with respect to $x \in R_1$, we find:

$$\begin{aligned}
\dot{I}_1(u) &\equiv \frac{d}{dt}\int_{-\infty}^{\infty} u(x,t)\,dx = \int_{-\infty}^{\infty} u_t\,dx \\
&= -\tfrac{1}{2}\int_{-\infty}^{\infty} \frac{d}{dx}u^2(x,t)\,dx - \int_{-\infty}^{\infty} dx \int_{-\infty}^{\infty} k(x-s)u_s\,ds \\
&= -\tfrac{1}{2}u^2\big|_{x=-\infty}^{x=+\infty} - \int_{-\infty}^{\infty} u_s(s,t)\,ds \int_{-\infty}^{\infty} k(x-s)\,dx \\
&= -c(0)u\big|_{x=-\infty}^{x=+\infty}, \quad \text{where } c(0) = \int_{-\infty}^{\infty} k(y)\,dy < \infty\,.
\end{aligned}$$

Thus, the quantity $I_1(u) = \int_{-\infty}^{\infty} u(x,t)\,dx$ is conserved in time: $I_1(u) = I_1(\overline{u}) = \text{const}$, where $\overline{u}(x) = u(x,0)$.

Consider the quantity

$$I_2(u) = \int_{-\infty}^{\infty} u^2(x,t)\,dx\,.$$

In order to prove that $I_2(u)$ is the integral of the motion, we multiply the Whitham equation by $u(x,t)$ and integrate with respect to $x \in R_1$. Then, integrating by parts we obtain:

$$\begin{aligned}
\frac{1}{2}\dot{I}_2(u) &\equiv \frac{1}{2}\frac{d}{dt}\int_{-\infty}^{\infty} u^2(x,t)\,dx = -\int_{-\infty}^{\infty} \partial_x\left(\frac{u^3}{3}\right)dx - \int_{-\infty}^{\infty} u\partial_x\widetilde{K}u\,dx \\
&= -(u,\partial_x\widetilde{K}u) = (\partial_x u, \widetilde{K}u) = (u, \widetilde{K}\partial_x u) = (u, \partial_x\widetilde{K}u),
\end{aligned}$$

i.e., $(u,\partial_x\widetilde{K}u) = 0$, so that $\dot{I}_2(u) = 0$. Thus, $I_2(u) = I_2(\overline{u}) = \text{const}$. A third conserved quantity is the integral

$$I_3(u) = \int_{-\infty}^{\infty}\left[\frac{u^3}{3} + u\widetilde{K}u\right]dx\,. \tag{1.2}$$

Indeed:

$$\begin{aligned}
\dot{I}_3(u) &= \int_{-\infty}^{\infty} u^2u_t\,dx + (u_t,\widetilde{K}u) + (u,\widetilde{K}u_t) \\
&= 2(u_t,\widetilde{K}u) - \int_{-\infty}^{\infty} u^2[uu_x + \widetilde{K}\partial_x u]\,dx \\
&= -2\left(\partial_x\left(\frac{u^2}{2} + \widetilde{K}u\right), \widetilde{K}u\right) - \tfrac{1}{4}u^4\big|_{x=-\infty}^{x=+\infty} - (u^2, \widetilde{K}\partial_x u) \\
&= (u^2,\partial_x\widetilde{K}u) + 2(\widetilde{K}u,\partial_x\widetilde{K}u) - (u^2,\widetilde{K}\partial_x u) = \int_{-\infty}^{\infty}\frac{\partial}{\partial x}(\widetilde{K}u)^2\,dx = 0\,.
\end{aligned}$$

The integral $I_3(u)$ represents the energy. Its time-independence enables one to write the Whitham equation in the Hamiltonian form:

$$\frac{\partial u}{\partial t} = -\frac{\partial}{\partial x}\frac{\delta\mathscr{H}(u)}{\delta u},$$

where the Hamiltonian $\mathscr{H}(u)$ is equal to $\frac{1}{2}I_3(u)$.

In addition to the polynomial integrals of motion $\dot{I}_i$, $i = 1, 2, 3$, there exist other time-invariant quantities that are explicitly containing the coordinate x or time t. Consider the integral

$$I_4(u, x) = \int_{-\infty}^{\infty} xu(x, t)\, dx\,.$$

We have

$$\dot{I}_4 = \int_{-\infty}^{\infty} xu_t\, dx = -\int_{-\infty}^{\infty} x\partial_x \left[\frac{u^2}{2} + \widetilde{K}u\right] dx = \int_{-\infty}^{\infty} \left(\frac{u^2}{2} + \widetilde{K}u\right) dx$$

$$= \tfrac{1}{2}I_2(u) + \int_{-\infty}^{\infty} dx \int_{-\infty}^{\infty} k(x-s)u(s,t)\, ds = \tfrac{1}{2}I_2(u) + c(0)I_1(u)\,. \tag{1.3}$$

Integrating relation (1.3) with respect to t and taking into consideration the time-invariance of $I_1(u)$ and $I_2(u)$, we obtain

$$I_4(u, x) - \frac{t}{2}I_2(u) - c(0)tI_1(u) = \text{const}\,. \tag{1.4}$$

The equality (1.4) can be interpreted kinematically. Let us introduce the "center of gravity" coordinate of the solution of Whitham's equation ($I_1(u) \neq 0$)

$$R(t) = \frac{1}{I_1(u)} \int_{-\infty}^{\infty} xu(x, t)\, dx\,.$$

Then, in view of (1.4) we obtain

$$\dot{R}(t) = \frac{I_2(u)}{2I_1(u)} + c(0) = \text{const};$$

i.e., the center of gravity of the solution moves at constant velocity.

There are infinitely many other polynomial conservation laws **[1, 18, 136]** for the following important particular cases of the Whitham equation: the shallow water equation (i.e., $u_t + uu_x = 0$), the Korteweg-de Vries equation, the Benjamin-Ono equation, and the intermediate equation (see Introduction). It was recently shown in **[93]** that these equations are the only representatives of the Whitham equation that possess this property.

1.2. We now show that the Whitham equation has solutions which are solitary traveling waves. A simple example of such a wave was given by Whitham in **[82]**. As the kernel $k(x)$ we choose the function $k_0(x) = \frac{\pi}{4}e^{-\nu|x|}$, $\nu = \frac{\pi}{2}$. It is easy to verify the formula

$$(\partial_x^2 - \nu^2)\widetilde{K}_0 v = -\nu^2 v, \tag{1.5}$$

for any function $v \in L_2(R_1) \cap C^0(R_1)$.

We look for a solution of the Whitham equation in the form of a traveling wave $u(x, t) = \varphi(x - \lambda t)$, where λ is the velocity of the wave. Upon substitution of this function into the Whitham equation, we obtain $-\lambda\varphi' + \varphi\varphi' + \widetilde{K}_0\varphi' = 0$, whence, integrating with respect to x, we find

$$-\lambda\varphi + \tfrac{1}{2}\varphi^2 + \widetilde{K}_0\varphi - A = 0, \tag{1.6}$$

where A is the constant of integration. Applying the operator $\partial_x^2 - \nu^2$ to (1.6) and using (1.5), we find that

$$(\partial_x^2 - \nu^2)\left(\lambda\varphi - \frac{\varphi^2}{2}\right) - \nu^2 A + \nu^2\varphi = 0\,. \tag{1.7}$$

Multiplying (1.7) by $\partial_x(\lambda\varphi - \varphi^2/2) = (\lambda - \varphi)\varphi'$ and integrating with respect to x, we obtain

$$(\lambda-\varphi)^2\varphi'^2 - \nu^2\left[\left(\lambda\varphi - \frac{\varphi^2}{2}\right)^2 - \lambda\varphi^2 + \frac{2}{3}\varphi^3\right] - 2\nu^2 A\left(\lambda\varphi - \frac{\varphi^2}{2}\right) - B = 0, \tag{1.8}$$

where B is the constant of integration. We set $A = B = 0$ and rewrite (1.8) in the form

$$(\lambda-\varphi)^2\varphi'^2 = \nu^2\varphi^2\left[\frac{\varphi^2}{4} - \left(\lambda - \frac{2}{3}\right)\varphi + \lambda^2 - \lambda\right]. \tag{1.9}$$

If λ satisfies the condition $1 < \lambda < \frac{4}{3}$, the roots of the quadratic trinomial in square brackets on the right-hand side of (1.9) are

$$\varphi_0 = 2\left(\lambda - \frac{2}{3}\right) - 2\sqrt{\frac{4}{9} - \frac{\lambda}{3}} \quad\text{and}\quad \varphi_1 = 2\left(\lambda - \frac{2}{3}\right) + 2\sqrt{\frac{4}{9} - \frac{\lambda}{3}},$$

whereby

$$0 < \varphi_0 < \lambda < \varphi_1\,. \tag{1.10}$$

We rewrite (1.9) in the form

$$\gamma^2\left(\frac{dx}{d\varphi}\right)^2 = \frac{(\lambda-\varphi)^2}{\varphi^2\left[\frac{\varphi^2}{4} - \left(\lambda - \frac{2}{3}\right)\varphi + \lambda^2 - \lambda\right]} \equiv f^2(u) \tag{1.11}$$

and study it in the interval $\varphi \in [0, \varphi_0]$. After extracting the square root, we can integrate (1.11) explicitly and obtain the function $(x - x_0)^2 = h^2(\varphi)$, the inverse of the function sought, where $h(\varphi) = \int_{\varphi_0}^{\varphi} f(\xi)\,d\xi$ and x_0 is a constant of integration; here $\varphi = \varphi_0$ corresponds to $x = x_0$. It is obvious from (1.11) that the sign of the derivative $\frac{dx}{d\varphi}$ is constant under conditions (1.10) and approaches $\pm\infty$ as $\varphi \to +0$ and $\varphi \to \varphi_0 - 0$. Thus, the desired function, $\varphi(x)$, turns out to be a solitary wave.(*) The theory of solitary waves for the Whitham equation was also developed in **[9, 15]**.

§2. Wave peaking

In this section we consider peaking of waves described by the Whitham equation. Whitham **[82]** showed that if the kernel $k(x)$ is the exponential $k(x) = \alpha e^{-\beta|x|}$, where $\alpha, \beta > 0$ are constants, the crests of the waves of limiting amplitude have an angle of 110°, which is very close to the Stokes angle.

The question arises as to the time behavior of the solution to the Whitham equation if the wave was peaking at the initial instant. Will the wave remain peaked or will it

(*) The material of this subsection is borrowed from the monograph **[11]**.

smooth out in time? In this section we prove a theorem asserting that the wave remains peaked for all time as long as a solution exists, provided the kernel $k(x)$ in the Whitham equation is continuous and absolutely integrable.

Consider the Cauchy problem for the Whitham equation:

$$\begin{cases} u_t + uu_x + \int_{-\infty}^{\infty} k(x-s)u_s(s,t)\,ds = 0, \\ u|_{t=0} = \overline{u}(x), \end{cases} \tag{2.1}$$

where the kernel is $k(x) \in C^0(R_1) \cap L_1(R_1)$, and the initial perturbation is $\overline{u}(x) \in C^0(R_1)$ and $\overline{u}'(x)$ has a jump discontinuity at $x = 0$, i.e., the wave "peaks"at the initial time.

We examine a generalized solution of problem (2.1).

DEFINITION. A function $u(x,t)$ will be called a generalized solution of the Cauchy problem (2.1) if it satisfies the following conditions:

1) $u(x,t)$ is continuous and its first and second x-derivatives are piecewise smooth in any rectangle

$$\Pi = \{(x,t) \colon -\infty < a < x < b < \infty, 0 < t < T\},$$

where $T > 0$ is the time for which the solution exists;

2) there exists a constant $M > 0$ such that

$$|u_x(x,t)| \leq M, \quad |u_{xx}(x,t)| \leq M, \qquad (x,t) \in \Pi;$$

3) $u(x,t)$ satisfies the integral identity:

$$\int_\Pi \left[u\psi_t + \left(\frac{u^2}{2} + \widetilde{K}u \right) \psi_x \right] dx\,dt = 0 \tag{2.2}$$

for any Π and any $\psi(x,t) \in C_0^\infty(\Pi)$; $\widetilde{K}u$ is defined in (1.1);

4) $u(x,0) = \overline{u}(x)$.

Without loss of generality, we assume that the function $u_x(x,t)$ has a single discontinuity along a smooth line $s = \{(x,t) \colon x = x_0(t)\}$ on the plane (x,t).

THEOREM 1. *Let the kernel $k(x)$ of the Whitham equation be continuous and absolutely integrable on the entire line $x \in R_1$, let the initial perturbation $\overline{u}(x)$ be continuous, and let its first derivative have a discontinuity at $x = 0$: $\overline{u}'(+0) = c_1$, $\overline{u}'(-0) = c_2$, $c_1, c_2 < \infty$, and $c_1 \neq c_2$. Then the jump of the first x-derivative of the generalized solution remains as long as the solution exists.*

PROOF. It is evident that everywhere outside the line of discontinuity of the first derivatives, the generalized solution to (2.1) satisfies the Whitham equation in the usual classical sense: it is sufficient to integrate (2.2) by parts, taking into account that $u(x,t)$ is continuously differentiable outside the discontinuity line. We now show that on the discontinuity line $x = x_0(t)$, where the derivatives of the generalized solution (2.1) are discontinuous, the Hugoniot relation holds:

$$\dot{x}_0(t) = u(x,t)\,. \tag{2.3}$$

In doing so, we choose $\psi(x,t) = \varphi_x(x,t)$ in identity (2.2), where $\varphi(x,t) \in C_0^\infty(\Pi)$, and integrate it by parts with respect to x. In view of the continuity of $u(x,t)$ in Π we obtain

$$\int_\Pi [v\varphi_t + uv\varphi_x + (\widetilde{K}v)\varphi_x]\,dx\,dt = 0, \tag{2.4}$$

where the notation $v(x,t) \equiv u_x(x,t)$ is introduced. It should be noted that in (2.4) Π is any rectangle, possibly containing the discontinuity line s.

Due to the arbitrariness in the choice of the function $\varphi(x, t)$, we conclude from (2.4), as above, that, outside the discontinuity line s, $v(x, t)$ satisfies the equation

$$v_t + \frac{\partial}{\partial x}(uv + \widetilde{K}v) = 0\,. \tag{2.5}$$

We examine identity (2.4) in a rectangle Π containing the discontinuity line s; s divides it into two domains Π_1 and Π_2. Transforming separately the integrals over the domains Π_1 and Π_2, and using (2.5), we obtain:

$$\begin{aligned}\int_{\Pi_1} &= \int_s \varphi[v^+\cos(n,t) + (uv^+ + F^+)\cos(n,x)]\,ds\\ &\quad - \int_{\Pi_1}\varphi\left[v_t + \left(uv + \int_{-\infty}^{\infty} k(s)v(x-s,t)\,ds\right)_x\right]dx\,dt\\ &= \int_s \varphi[v^+\cos(n,t) + (uv^+ + F^+)\cos(n,x)]\,ds,\\ \int_{\Pi_2} &= -\int_s \varphi[v^-\cos(n,t) + (uv^- + F^-)\cos(n,x)]\,ds,\end{aligned} \tag{2.6}$$

where v^+ and v^- are the limiting values of $v(x, t)$ on the curve s from within Π_1 and Π_2, respectively, and $F^\pm \equiv \lim_{\Delta\to+0}\int_{-\infty}^{\infty} k(s)v(x_0(t)\pm\Delta - s, t)\,ds$. Adding the equalities (2.6) and taking (2.4) into account we obtain

$$0 = \int_\Pi = \int_{\Pi_1} + \int_{\Pi_2} = \int_s \varphi\{[v]\cos(n,t) + (u[v] + [F])\cos(n,x)\}\,ds,$$

where $[v] = v^+ - v^- \neq 0$, $[F] = F^+ - F^-$. Hence, due to the arbitrariness of $\varphi(x, t)$, it follows that

$$[v]\cos(n,t) + (u[v] + [F])\cos(n,x) = 0\,. \tag{2.7}$$

Since

$$\cos(n,t) = \frac{\dot{x}_0(t)}{\sqrt{1+\dot{x}_0^2(t)}}, \qquad \cos(n,x) = \frac{-1}{\sqrt{1+\dot{x}_0^2(t)}},$$

from (2.7) we find

$$\dot{x}_0(t) = u(x,t) + \frac{[F]}{[v]}\,. \tag{2.8}$$

Consider the integral

$$[F] = \lim_{\Delta\to+0}\int_{-\infty}^{\infty} k(s)(v(x_0(t)+\Delta-s,t) - v(x_0(t)-\Delta-s,t))\,ds\,.$$

Since $|u_s(s,t)| \le M$, and $\int_{-\infty}^{\infty}|k(s)|\,ds < \infty$, the integrand has an integrable majorant and, according to the Lebesgue convergence theorem, the limit can be taken under the integral sign. As a result we obtain $[F] = 0$. Equality (2.8) now leads to the Hugoniot relation (2.3), so that the discontinuity line of the first derivatives of the generalized solution to the Cauchy problem (2.1) coincides with the characteristic of the Whitham equation.

We differentiate the Whitham equation with respect to x outside the discontinuity line of the derivatives

$$u_{xt} + u_x^2 + uu_{xx} + \frac{\partial}{\partial x}\int_{-\infty}^{\infty} k(x-s)u_s(s,t)\,ds = 0 \tag{2.9}$$

and consider in more detail the derivative of the integral. We have

$$\begin{aligned}\frac{\partial}{\partial x}&\int_{-\infty}^{\infty} k(x-s)u_s(s,t)\,ds\\ &= \frac{\partial}{\partial x}\left(\int_{x-x_0(t)}^{\infty} k(y)u_x(x-y,t)\,dy + \int_{-\infty}^{x-x_0(t)} k(y)u_x(x-y,t)\,dy\right)\\ &= \int_{-\infty}^{\infty} k(y)u_{xx}(x-y,t)\,dy + I_0(x,t),\end{aligned}$$

where

$$I_0(x,t) = k(x-x_0(t))[u_x(x_0(t)+0,t) - u_x(x_0(t)-0,t)]\,. \tag{2.10}$$

We denote by $x = x(y,t)$ the characteristic through an arbitrary point $y \in R_1$,

$$\dot{x}(t) = u(x,t), \qquad x|_{t=0} = y\,.$$

Introducing the notation $w(y,t) = u_x(x(y,t),t)$ for the function $u_x(x,t)$ on the characteristic, we note that $w_t(y,t) = u_{xt} + u_{xx}u$. Then equation (2.9) takes the form

$$w_t(y,t) + w^2(y,t) + \int_{-\infty}^{\infty} k(s)u_{ss}(x(y,t)-s,t)\,ds + I_0(x(y,t),t) = 0 \tag{2.11}$$

on the characteristic.

We consider the family of characteristics emanating from the points $y \in R_1$ and pass to the limit in (2.11) as $y \to \pm 0$; i.e., we approach from two sides the characteristic coinciding with the line of possible discontinuity of the derivatives of the generalized solution of the Cauchy problem (2.1). This limit exists because the solution of the Whitham problem is classical outside of the discontinuity line.

We obtain

$$w_t^{\pm} + (w^{\pm})^2 + I^{\pm} = 0, \tag{2.12}$$

where $w^{\pm}(t) \equiv w(\pm 0, t)$,

$$I^{\pm} \equiv \lim_{\Delta \to +0}\int_{-\infty}^{\infty} k(s)u_{ss}(x(0\pm\Delta,t)-s,t)\,ds + I_0(x_0(t),t)\,. \tag{2.13}$$

We denote $g(t) \equiv w^+(t) - w^-(t)$, $f(t) \equiv w^+(t) + w^-(t)$ and subtract one of the equations (2.12) from the other. Noting that

$$\lim_{\Delta\to+0}\int_{-\infty}^{\infty} k(s)[u_{ss}(x(0+\Delta,t)-s,t) - u_{ss}(x(0-\Delta,t)-s,t)]\,ds = 0$$

(which can be proved as the equality $[F] = 0$ above), we arrive at the problem

$$\begin{cases} g_t + fg = 0,\\ g(0) = g_0,\end{cases}$$

where $g_0 = c_2 - c_1 \neq 0$. It follows from this that $g(t) = g_0 e^{-\int_0^t f(\tau)\,d\tau} \neq 0$, i.e., the function $g(t) = u_x(+0,t) - u_x(-0,t)$ does not become zero as long as the solution exists, i.e., the wave peaking holds for $t \in [0, T]$. Theorem 1 is proved. □

§3. Breaking of waves in the case of a monotone kernel

Whitham's main goal in deriving equation (2.1) was to obtain a relatively simple equation suitable for describing the effect of breaking of waves as opposed, for example, to the Korteweg-de Vries equation. Whitham used the fact that the shallow water equation

$$u_t + uu_x = 0 \tag{3.1}$$

with any compactly supported initial condition $\overline{u}(x) \not\equiv 0$ always giving rise to the wave breaking phenomenon in finite time $T_0 = |\min_{x\in R_1} \overline{u}'(x)|^{-1}$, i.e., $\max_{x\in R_1} |u_x(x,t)| \to +\infty$ for $t \to T_0 < \infty$; T_0 is called the time of breaking. It is natural to expect that not too large a perturbation of the shallow water equation (3.1) retains the effect of wave-breaking. The question of exact quantitative evaluation of such a perturbation is very subtle.

In his monograph **[82]** Whitham formulated a conjecture that equation (2.1) with a special kernel $k_g(x)$, corresponding to the dispersion $c(p) = \sqrt{\frac{\tanh p}{p}}$ of the exact potential theory of water waves, has solutions breaking in finite time. It should be noted that the kernel $k_g(x)$ is symmetric, singular at zero, i.e., it has a singularity of order $|x|^{-1/2}$, $x \to 0$, and is exponentially decaying at infinity.

When the kernel $k(x)$ in (2.1) is regular, i.e., it has an absolutely integrable derivative, Seliger **[160]** and Gabov **[10]** showed that a wave described by (2.1) breaks in a finite time T_0 if the initial perturbation has an asymmetric profile and its steepness is sufficiently large. We show below that the condition of asymmetry for the initial perturbation is redundant.

The first study of the Cauchy problem (2.1) for the Whitham equation with a singular kernel was undertaken in **[47–50]**. It was proved that a wave that is sufficiently steep at the initial moment $t = 0$ is breaking in finite time, if the kernel $k(x)$ is monotonically increasing in the vicinity of the singular point $x = 0$ and it has a singularity of order $|x|^{-\alpha}$, $0 < \alpha < 1$, as $x \to 0$, and also if the kernel is monotonically decreasing in the vicinity of $x = 0$ and has a singularity of order $|x|^{-\alpha}$, $0 < \alpha < \frac{2}{3}$.

In the case of a monotonically decreasing kernel, the integral term in equation (2.1) describes dissipative effects; as opposed to monotonic increasing of the kernel corresponding to processes with energy pumping.

The conclusive answer to the problem posed by Whitham was obtained in publication **[52]** and is given in Chapter 2.

Consider the Cauchy problem (2.1) with the initial condition $u(x,0) = \overline{u}(x)$, where the function $\overline{u}(x)$ sufficiently rapidly approaches zero as $|x| \to \infty$. We assume that the kernel $k(x)$ is a smooth function of x everywhere except at the origin $x = 0$, where $k(x)$ has a power singularity and sufficiently fast approaches zero together with its derivatives as $|x| \to \infty$. Until the time of breaking, T_0, the solution $u(x,t)$ of the Cauchy problem is a smooth function of x and t, vanishing together with its x-derivatives as $|x| \to \infty$. This statement is justified in Chapter 2.

The following is a simple assertion concerning the breaking of waves for the Whitham equation with a singular kernel.

THEOREM 2. Let the kernel $k(x)$ satisfy the following conditions:

1) $k(x) \in C^1(R_1 - 0)$, $\int_{|x|\geq a}^{\infty} |\mathcal{D}^j k(x)|\, dx < \infty$, $a > 0$, $j = 1, 2$;

2) $|k(x)| \leq c_0|x|^{-\alpha}$, $|k'(x)| \leq c_0|x|^{-\alpha-1}$, $x \to 0$, $\alpha < 1$, $c_0 \geq 0$,

3) $k'(x) \geq 0$, $x \in R_1 - 0$.

Then, if the condition $|m(0)| \equiv |\min_{x\in R_1} u_x(x,0)| > 0$ is satisfied at the initial moment $t = 0$, there exists a positive constant T_0 such that the wave $u(x,t)$ will be breaking up to the moment of $T_0 < \infty$.

PROOF. Differentiating (2.1) with respect to x (the kernel in the integral term is regularized by adding and subtracting $u_x(x,t)$):

$$u_{xt} + u_x^2 + uu_{xx} + \int_{-\infty}^{\infty} k'(x-s)[u_s(s,t) - u_x(x,t)]\, ds = 0\,. \tag{3.2}$$

Setting $m(t) = \min_{x\in R_1} u_x(x,t)$, $x_1 = x_1(t)$, $u_x(x_1(t),t) = m(t)$, $m(t) < 0$, where $x_1(t)$ is a smooth function of t, and using the fact that $u_{xx}(x_1(t),t) = 0$ and, consequently, $\frac{du_x}{dt}|_{x=x_1(t)} = \frac{\partial u_x}{\partial t}|_{x=x_1(t)}$, taking $x = x_1(t)$, from equality (3.2) we get

$$\frac{dm}{dt} = -m^2(t) - \int_{-\infty}^{\infty} k'(\xi)[u_x(x_1(t)-\xi,t) - u_x(x_1(t),t)]\, d\xi\,. \tag{3.3}$$

Since $k'(\xi) \geq 0$ and $u_x(x_1(t)-\xi,t) - u_x(x_1(t),t) \geq 0$ by virtue of the definition of $m(t)$, the integral in (3.3) is nonnegative. Therefore, $dm(t)/dt \leq -m^2(t)$. Upon integration of this inequality, we find that $|m(t)| \geq |m(0)|/(1 - |m(0)|t)$. Hence, if $|m(0)| \neq 0$ at the initial moment $t = 0$, i.e.,$|m(0)| = c > 0$, then $m(t) \to -\infty$ as $t \to T_0$, where $T_0 = 1/c < \infty$. Theorem 2 is proved. □

In this theorem condition 3) can be replaced by the condition $k'(x) \geq 0$, $x \in [-\delta, \delta]$, $\delta > 0$, $x \neq 0$ (under the additional conditions that $k(x)$ is regular for $|x| > \delta$).

We note that the quantity $I_1(t) = \int_{-\infty}^{\infty} u(x,t)\, dx$ is an integral of motion for any integrable kernel $k(x)$. Indeed,

$$\begin{aligned}\frac{dI_1}{dt} &= \int_{-\infty}^{\infty} u_t\, dx = -\frac{1}{2}\int_{-\infty}^{\infty} \frac{\partial u^2}{\partial x}\, dx - \int_{-\infty}^{\infty} dx \int_{-\infty}^{\infty} k(x-s)u_s\, ds \\ &= -\int_{-\infty}^{\infty} u_s\, ds \int_{-\infty}^{\infty} k(x-s)\, dx = 0\,.\end{aligned}$$

As for the quantity $I_2(t) = \int_{-\infty}^{\infty} u^2(x,t)\, dx$ the analogous statement is false. Indeed,

$$\frac{dI_2}{dt} = 2\int_{-\infty}^{\infty} uu_t\, dx = -\frac{2}{3}\int_{-\infty}^{\infty} \frac{\partial u^3}{\partial x}\, dx - 2\int_{-\infty}^{\infty} u(x,t)\, dx \int_{-\infty}^{\infty} k(x-s)u_s(s,t)\, ds\,. \tag{3.4}$$

Here, the first integral on the right-hand side is equal to zero, and the second one can be transformed to the form

$$\iint_{-\infty}^{\infty} k'(x-s)[u(x,t) - u(s,t)]^2\, dx\, ds\,.$$

It follows that a sufficient condition for the conservation law $I_2(t) = I_2(0)$ to hold is the requirement that the kernel be symmetric: $k(x) = k(-x)$. Under the conditions of Theorem 2 with $k'(x) > 0$, $x \in R_1\backslash\{0\}$, the function $I_2(t)$ increases with t.

We now turn to the question of breaking of waves for equation (2.1) with a monotonically decreasing kernel ($k'(x) \leq 0$ for $x \in [-a, a]$, $a > 0$). This case is considerably more complicated, although the basic idea of eliminating the integral over the monotonicity domain of the kernel $k(x)$ is also applicable here. This difficulty is explained by the fact that in the case of a decreasing kernel the integral term in the Whitham equation (2.1) prevents the breaking of waves, and rather careful estimates are required to demonstrate that the nonlinear term in (2.1) prevails over the integral term.

THEOREM 3. Let the kernel $k(x)$ satisfy the following conditions:

1) $k(x) \in C^2(R_1\backslash 0)$, $\int_{|x|\geq a}^{\infty} |\mathscr{D}^j k(x)|^p\, dx < \infty$, $a > 0$, $j, p = 1, 2$;

2) $|k(x)| \leq c_0|x|^{-\alpha}$, $|k'(x)| \leq c_0|x|^{-\alpha-1}$, $x \neq 0$, $\alpha = \frac{2}{3} - \gamma$, $0 < \gamma < \frac{2}{3}$, $c_0 \geq 0$;

3) there exists an interval $(-a_1, a_2)$, $a_1, a_2 > 0$, such that $k'(x) \leq 0$ for $x \in (-a_1, a_2)$, $x \neq 0$.

Then, if $|m(0)|$ is sufficiently large, the wave $u(x, t)$ breaks at the moment of time

$$T_0 = \frac{1}{|m(0)|(1 - \gamma/8)} < \infty .$$

We introduce the notation

$$A(t) = -\frac{1}{m^2(t)} \int_{-\infty}^{\infty} k(\xi) u_{xx}(x_1(t) - \xi, t)\, d\xi,$$

$$Z(t) = 1 + m(0)\left(t - \int_0^t A(t')\, dt'\right)$$

with $m(t)$ and $x_1(t)$ as in Theorem 2. The proof of Theorem 3 is based on the following lemma.

LEMMA. *Suppose conditions* 1)–3) *of Theorem* 3 *are satisfied, and* $A(t) < \gamma/4$ *for* $t \in [0, t_0]$, $t_0 > 0$. *If* $|m(0)|$ *is sufficiently large, then for all* $t \in [0, t_0]$ *the following estimates are valid*:

1) $n(t) \equiv \max_{x\in R_1} u_x(x, t) \leq n(0) + \overline{k}t$;

2) $\tilde{l}(t) \equiv \max_{x\in R_1} |u_{xx}(x, t)| \leq l(0)Z^{-4}(t)$;

3) $r(t) \equiv \max_{x\in R_1} u_{xxx}(x, t) \leq (r(0) + t\overline{\overline{k}}l(0))Z^{-\beta}(t)$, *where* $\beta = 4 + 24\gamma/(20 - 5\gamma)$, $\overline{k}, \overline{\overline{k}} > 0$ *are constants*.

PROOF. We first estimate $n(t) = \max_{x\in R_1} u_x(x, t)$. Let $x_2(t)$ be determined by the equality $u_x(x_2(t), t) = n(t)$, where $n(t) > 0$, and $x_2(t)$ is a smooth function of t. Differentiating (2.1), as in Theorem 2, and putting $x = x_2(t)$ we obtain

$$\frac{dn}{dt} + n^2(t) + \int_{-\infty}^{\infty} k'(\xi)[u_x(x_2(t) - \xi, t) - n(t)]\, d\xi = 0 . \tag{3.5}$$

We split the integral in (3.5) into three parts: from $-\infty$ to $-a_1$, from $-a_1$ to a_2, and from a_2 to ∞. For the integral over the region $[-a_1, a_2]$ we have

$$\int_{-a_1}^{a_2} k'(\xi)[u_x(x_2(t) - \xi, t) - n(t)]\, d\xi \geq 0, \tag{3.6}$$

because $k'(\xi) \leq 0$ and $u_x(x_2(t) - \xi, t) - n(t) \leq 0$ by the definition of $n(t)$. The integrals over the regions $(-\infty, a_1)$ and (a_2, ∞) are estimated in the same way. Without loss of generality, we assume that $k'(-a_1) = k'(a_2) = 0$; therefore,

$$\begin{aligned}
&\left| \int_{-\infty}^{-a_1} k'(\xi)[u_x(x_2(t) - \xi, t) - n(t)]\, d\xi \right| \\
&= \left| \int_{-\infty}^{-a_1} k''(\xi) u(x_2(t) - \xi, t)\, d\xi - n(t) \int_{-\infty}^{-a_1} k'(\xi)\, d\xi \right| \\
&\leq n(t)|k(-a_1)| + \left\{ \int_{-\infty}^{-a_1} |k''(\xi)|^2\, d\xi \int_{-\infty}^{\infty} u^2(\xi, t)\, d\xi \right\}^{1/2} .
\end{aligned} \tag{3.7}$$

To estimate $\int_{-\infty}^{\infty} u^2(x, t)\, dx$ we apply (3.4):

$$\begin{aligned}
\frac{dI_2}{dt} &= \iint_{-\infty}^{\infty} k'(x - s)[u(x, t) - u(s, t)]^2\, dx\, ds \\
&\leq \int_{-\infty}^{\infty} dx \int_{R_1 \setminus [-a_1, a_2]} k'(y)[u(x, t) - u(x - y, t)]^2\, dy \\
&\leq 4 I_2(t) \int_{R_1 \setminus [-a_1, a_2]} |k'(y)|\, dy .
\end{aligned} \tag{3.8}$$

We set

$$\begin{aligned}
&|k(-a_1)| + |k(a_2)| - 2k_1, \qquad \int_{R_1 \setminus [-a_1, a_2]} |k''(x)|^2\, dx = k_2^2, \\
&\int_{R_1 \setminus [-a_1, a_2]} |k'(x)|\, dx = k_5, \quad I_2(0) e^{4k_5 t_0} = k_3^2, \quad 2k_2 k_3 + k_1^2 = k_4^2 .
\end{aligned} \tag{3.9}$$

We conclude from (3.8) that

$$\int_{\infty}^{\infty} u^2(x, t)\, dx \equiv I_2(t) \leq I_2(0) e^{4k_5 t} . \tag{3.10}$$

Using (3.6), (3.7), (3.9), (3.10), instead of (3.5) we obtain the inequality

$$\frac{dn}{dt} \leq -n^2(t) + 2k_1 n(t) + 2k_2 k_3 = -(n(t) - k_1)^2 + k_4^2 .$$

Hence, $n(t) \leq n(0) + k_4^2 t$ for $t \in [0, t_0]$, and this proves 1).

We next establish estimate 2). Differentiating (2.1) with respect to x we obtain

$$u_{xt} + u_x^2 + u u_{xx} + \int_{-\infty}^{\infty} k(\xi) u_{xx}(x - \xi, t)\, d\xi = 0. \tag{3.11}$$

Since $m(0) < 0$, we have $m(t) < 0$ on some interval. Setting $x = x_1(t)$ in (3.11) we find that

$$\frac{dm}{dt} + m^2(t)(1 - A(t)) = 0, \tag{3.12}$$

where

$$A(t) = -\frac{1}{m^2(t)} \int_{-\infty}^{\infty} k(\xi) u_{xx}(x_1(t) - \xi, t)\, d\xi .$$

Integration of (3.12) yields

$$m(t) = \frac{m(0)}{Z(t)}, \tag{3.13}$$

$$Z(t) = 1 + m(0)\left(t - \int_0^t A(t')\,dt'\right). \tag{3.14}$$

According to (3.14) and the condition of the lemma,

$$Z(0) = 1, \quad \frac{dz}{dt} = m(0)(1 - A(t)) < m(0)\left(1 - \frac{\gamma}{4}\right) < 0, \qquad t \in [0, t_0], \tag{3.15}$$

because $A(t) < \frac{\gamma}{4}$ and $\gamma < \frac{2}{3}$. Hence, $Z(t)$ is decreasing for $t \in [0, t_0]$, and (3.13) implies that $m(t) < 0$ for all $t \in [0, t_0]$. Substituting $x - s = \xi$ in the integral in (3.2) and taking the derivative of the resulting equation with respect to x, we obtain

$$u_{xxt} + 3u_x u_{xx} + uu_{xxx} + \int_{-\infty}^{\infty} k'(\xi)[u_{xx}(x - \xi, t) - u_{xx}\,(x, t)]\,d\xi = 0\,. \tag{3.16}$$

Putting

$$l_1(t) = \min_{x \in R_1} u_{xx}(x, t), \quad \overline{x}_1(t)\colon u_{xx}(\overline{x}_1(t), t) = l_1(t), \quad l_1(t) < 0,$$

$$l_2(t) = \max_{x \in R_1} u_{xx}(x, t), \quad \overline{x}_2(t)\colon u_{xx}(\overline{x}_2(t), t) = l_2(t), \quad l_2(t) > 0,$$

and writing equation (3.16) for $x = \overline{x}_i(t)$, $i = 1, 2$, we obtain

$$\frac{dli}{dt} = -3u_x l_i - \int_{-\infty}^{\infty} k'(\xi)[u_{xx}(\overline{x}_i(t) - \xi, t) - l_i(t)]\,d\xi\,. \tag{3.17}$$

We split the integral in (3.17) into three parts:

$$\begin{aligned} &-\int_{-\infty}^{\infty} k'(\xi)[u_{xx}\overline{x}_i(t) - \xi, t) - l_i(t)]\,d\xi \\ &= -\int_{-\infty}^{-a_1} - \int_{-a_1}^{a_2} - \int_{a_2}^{\infty} \equiv j_1^{(i)} + j_2^{(i)} + j_3^{(i)}, \qquad i = 1, 2\,. \end{aligned} \tag{3.18}$$

The integrals over infinite intervals are estimated identically:

$$\begin{aligned} |j_1^{(i)} + j_3^{(1)}| \le &\int_{-\infty}^{-a_1} |k'(\xi)|\,|u_{xx}(\overline{x}_i(t) - \xi, t) - l_i(t)|\,d\xi \\ &+ \int_{a_2}^{\infty} |k'(\xi)|\,|u_{xx}(\overline{x}_i(t) - \xi, t) - l_i(t)|\,d\xi \\ \le &(l_2(t) - l_1(t))k_5, \qquad i = 1, 2, \end{aligned} \tag{3.19}$$

where

$$k_5 = \int_{-\infty}^{-a_1} |k'(\xi)|\,d\xi + \int_{a_2}^{\infty} |k'(\xi)|\,d\xi\,. \tag{3.20}$$

By virtue of the monotonicity of the kernel k and by the definition of the $l_i(t)$ the integrals over $[-a_1, a_2]$ satisfy the following inequalities:

$$j_2^{(1)} = -\int_{-a_1}^{a_2} k'(\xi)[u_{xx}(\overline{x}_1(t) - \xi, t) - l_1(t)]\, d\xi \geq 0,$$
$$j_2^{(2)} = -\int_{-a_1}^{a_2} k'(\xi)[u_{xx}(\overline{x}_2(t) - \xi, t) - l_2(t)]\, d\xi \leq 0\,. \tag{3.21}$$

Combining (3.17) through (3.21), we get

$$\frac{dl_1}{dt} \geq -3ml_1 - k_5(l_2 - l_1), \qquad \frac{dl_2}{dt} \leq -3ml_2 + k_5(l_2 - l_1)\,. \tag{3.22}$$

Introducing the notation $l(t) = l_2(t) - l_1(t)$ and subtracting the first inequality in (3.22) from the second, we obtain $\frac{dl}{dt} \leq l(t)(-3m(t) + 2k_5)$. Hence, by virtue of (3.13) and the inequality $Z(t) \leq 1$, we conclude that

$$\frac{dl}{dt} \leq l(t)\left(\frac{3|m(0)| + 2k_5 Z(t)}{Z(t)}\right) \leq \frac{l(t)}{Z(t)}(3|m(0)| + 2k_5)\,. \tag{3.23}$$

If $|m(0)| > 2k_5/(1-\gamma)$, the inequality (3.23) becomes

$$\frac{dl}{dt} \leq \frac{l(t)}{Z(t)}|m(0)|(4-\gamma)\,. \tag{3.24}$$

Inequality (3.15) can be written as follows:

$$-4\frac{dZ}{dt} > |m(0)|(4-\gamma), \tag{3.25}$$

and (3.24) and (3.25) imply that

$$\frac{dl}{l} < -4\frac{dz}{Z}\,. \tag{3.26}$$

Integrating (3.26), we find that for $t \in [0, t_0]$

$$l(t) \leq l(0)Z^{-4}(t)\,. \tag{3.27}$$

This proves estimate 2).

We next prove estimate 3). Taking the x-derivative of (3.16), writing the result as an equation for $r(t) = \max_{x\in R_1} u_{xxx}(x,t)$, and denoting by $x_0(t)$ the curve for which $u_{xxx}(x_0(t), t) = r(t)$, we obtain

$$\frac{dr}{dt} = -4u_x r(t) - 3u_{xx}^2 - \int_{-\infty}^{\infty} k'(\xi)[u_{xxx}(x_0(t) - \xi, t) - r(t)]\, d\xi\,. \tag{3.28}$$

We estimate the integral in (3.28):

(3.29)
$$-\int_{-\infty}^{\infty} k'(\xi)[u_{xxx}(x_0(t) - \xi, t) - r(t)]\, d\xi = -\int_{-\infty}^{-a_1} - \int_{-a_1}^{a_2} - \int_{a_2}^{\infty} \equiv j_1 + j_2 + j_3\,.$$

Integrating by parts we get (assuming, as before, that $k'(-a_1) = k'(a_2) = 0$)

$$|j_1| = \left| -\int_{-\infty}^{-a_1} k''(\xi) u_{xx}(x_0(t) - \xi, t)\, d\xi + r(t)k(-a_1) \right|$$
$$\leq r(t)|k(-a_1)| + l(t)\int_{-\infty}^{-a_1} |k''(\xi)|\, d\xi\,.$$

An analogous estimate is valid for j_3 as well. Therefore, denoting

$$|k(-a_1)| + |k(a_2)| = 2k_1, \qquad \int_{-\infty}^{-a_1} |k''(\xi)|\, d\xi + \int_{a_2}^{\infty} |k''(\xi)|\, d\xi = k_6, \tag{3.30}$$

we arrive at the inequality

$$|j_1 + j_3| \le 2k_1 r(t) + k_6 l(t)\,. \tag{3.31}$$

It follows from the monotonicity of the kernel k and the definition of $r(t)$ that

$$j_2 = -\int_{-a_1}^{a_2} k'(\xi)[u_{xxx}(x_0(t) - \xi, t) - r(t)]\, d\xi \le 0\,. \tag{3.32}$$

Substituting (3.29) through (3.32) into (3.28), we obtain

$$\begin{aligned} \frac{dr}{dt} &\le -4u_x r(t) - 3u_{xx}^2 + 2k_1 r(t) + k_6 l(t) \\ &\le -4m(t)r(t) + 2k_1 r(t) + k_6 l(t)\,. \end{aligned} \tag{3.33}$$

We now integrate inequality (3.33), first considering the homogeneous inequality

$$\frac{dr}{dt} \le -4mr(t) + 2k_1 r(t)\,. \tag{3.34}$$

Using (3.13) and the inequality $Z(t) \le 1$ and assuming that $|m(0)| > \frac{10}{\gamma} k_1$, we transform the right-hand side in (3.34) to obtain

$$\frac{dr}{dt} \le r(t)\frac{|m(0)|}{Z(t)}(4 + \gamma/5)\,.$$

This and inequality (3.25) yield

$$\frac{dr}{dt} \le -\beta \frac{r(t)}{Z(t)} \frac{dZ}{dt}, \tag{3.35}$$

where $\beta = 4 + 24\gamma/(20 - 5\gamma)$.

Integrating inequality (3.35) we find

$$r(t) \le r(0) Z^{-\beta}(t)\,. \tag{3.36}$$

By varying the constant and using the previously established inequality (3.27), we then integrate the inhomogeneous inequality (3.33):

$$\frac{d\tilde{r}}{dt} \le k_6 \frac{l(0)}{Z^4(t)} Z^{\beta}(t) = k_6 l(0) Z^{\sigma}(t) \le k_6 l(0), \qquad \sigma = \frac{24\gamma}{20 - 5\gamma} > 0,$$

because $Z(t) \le 1$. Calculating $\tilde{r}(t)$ and substituting it into (3.36) we obtain $r(t) \le (r(0) + k_6 l(0) t) Z^{-\beta}(t)$.

Thus, if the initial value satisfies the inequality

$$|m(0)| > \max\left\{ \frac{2k_5}{1 - \gamma}, \frac{10}{\gamma} k_1 \right\}, \tag{3.37}$$

then all three estimates of the lemma are proved. □

We next prove Theorem 3, first establishing that, if $A(0) < \frac{\gamma}{8}$, then $A(t) < \frac{\gamma}{8}$ for all $t \in [0, T_0)$, where $T_0 = 1/|m(0)|(1 - \frac{\gamma}{8})$. Assuming the contrary, we let $t_1 > 0$, $t_1 \in [0, T_0]$, be the least time for which $A(t_1) = \frac{\gamma}{8}$. The continuity of $A(t)$ up to the time when the wave breaks implies that there is some t_0, $t_1 < t_0 < T_0$, such that $A(t) < \frac{\gamma}{4}$

for $t \in [0, t_0]$. Hence our lemma holds for the time interval $[0, t_0]$, we use the lemma to estimate $A(t)$ on $[0, t_0]$:

$$A(t) = -\frac{1}{m^2(t)} \int_{-\infty}^{\infty} k(\xi) u_{xx}(x_1(t) - \xi, t)\, d\xi \equiv \frac{j_0}{m^2(t)}\,. \tag{3.38}$$

We now introduce a function $P(x)$, $x \in [-a_1, a_2]$, by means of the equality

$$P(x) = \frac{c_0}{\frac{2}{3} - \gamma} |x|^{-2/3+\gamma} \operatorname{sgn} x, \qquad x \in [-a_1, a_2], x \neq 0\,. \tag{3.39}$$

It follows from conditions 2) and 3) of the theorem that, for $x \in [-a_1, a_2]$, the following inequalities hold:

$$0 \geq k'(x) \geq -\frac{c_0}{|x|^{5/3-\gamma}} = P'(x), \quad x \neq 0, \quad |P(x)| \geq |k(x)|\,. \tag{3.40}$$

If $\mathscr{G}(x) = k(x) - P(x)$, then $\mathscr{G}'(x) \geq 0$ for $x \in [-a_1, a_2]$, by virtue of inequalities (3.40). We next estimate the integral

$$\begin{aligned} j_0 &= -\int_{-\infty}^{\infty} k(\xi) u_{xx}(x_1(t) - \xi, t)\, d\xi \\ &= -\int_{-\infty}^{-a_1} - \int_{-a_1}^{-\delta(t)} - \int_{-\delta(t)}^{\delta(t)} - \int_{\delta(t)}^{a_2} - \int_{a_2}^{\infty} \equiv j_1 + \cdots + j_5, \end{aligned} \tag{3.41}$$

where $\delta(t) \leq a = \min\{a_1, a_2\}$ will be determined below. We have

(3.42)
$$\begin{aligned} |j_1 + j_5| &= |-k(\xi) u_x(x_1(t) - \xi, t)|_{-\infty}^{-a_1} - k(\xi) u_x(x_1(t) - \xi, t)|_{a_2}^{\infty} \\ &+ \int_{-\infty}^{-a_1} k'(\xi) u_x(x_1(t) - \xi, t)\, d\xi + \int_{a_2}^{\infty} k'(\xi) u_x(x_1(t) - \xi, t)\, d\xi| \leq k_7 h(t), \end{aligned}$$

where $k_7 = |k(-a_1)| + |k(a_2)| + \int_{-\infty}^{-a_1} |k'(\xi)|\, d\xi + \int_{a_2}^{\infty} |k'(\xi)|\, d\xi$, $h(t) = \max_{x \in R_1} |u_x(x, t)|$. Since the kernel $k(x)$ is monotonically nonincreasing on $[-a_1, a_2]$, then for an estimate of the integrals j_2 and j_4 we apply Bonnet's form of the mean-value theorem to obtain

(3.43)
$$\begin{aligned} |j_2 + j_4| &= \left| k(-\delta) \int_{-\xi_1}^{-\delta(t)} u_{xx}(x_1(t) - \xi, t)\, d\xi + k(\delta) \int_{\delta(t)}^{\xi_2} u_{xx}(x_1(t) - \xi, t)\, d\xi \right| \\ &\leq 2(|k(-\delta(t))| + |k(\delta(t))|) h(t) \leq 4 c_0 \delta(t)^{-2/3+\gamma} h(t)\,. \end{aligned}$$

We now estimate the integral j_3:

(3.44)
$$j_3 = -\int_{-\delta(t)}^{\delta(t)} P(\xi) u_{xx}(x_1(t) - \xi, t)\, d\xi - \int_{-\delta(t)}^{\delta(t)} \mathscr{G}(\xi) u_{xx}(x_1(t) - \xi, t)\, d\xi \equiv \bar{j}_3 + \bar{\bar{j}}_3.$$

Denote by $F(x)$ the primitive of $P(x)$:

$$F(x) = \frac{c_0}{(\frac{2}{3} - \gamma)(\frac{1}{3} + \gamma)} [|x|^{1/3+\gamma} - \delta(t)^{1/3+\gamma}], \qquad x \in [-\delta(t), \delta(t)],$$

this primitive possesses the properties $F(\pm\delta(t)) = 0$, $-F(x) \geq 0$, $x \in [-\delta(t), \delta(t)]$. Integrating by parts and using the properties of the function $F(x)$, we transform $\overline{j}_3$:

$$\begin{aligned}\overline{j}_3 &= \int_{-\delta(t)}^{\delta(t)} (-F(\xi)) u_{xxx}(x_1(t) - \xi, t)\, d\xi \\ &\leq \max_{x \in R_1} u_{xxx}(x, t) \int_{-\delta(t)}^{\delta(t)} (-F(\xi))\, d\xi \\ &= \frac{2r(t) c_0 \delta(t)^{4/3+\gamma}}{(\frac{2}{3} - \gamma)(\frac{1}{3} + \gamma)} \left[1 - \frac{1}{\frac{4}{3} + \gamma} \right].\end{aligned} \tag{3.45}$$

Finally, we estimate the integral $\overline{\overline{j}}_3$:

$$\begin{aligned}-\overline{\overline{j}}_3 &= \int_{-\delta(t)}^{\delta(t)} \mathscr{G}(\xi) u_{xx}(x_1(t) - \xi, t)\, d\xi = \frac{\partial}{\partial x} \int_{-\delta(t)}^{\delta(t)} \mathscr{G}(\xi) u_x(x - \xi, t)\, d\xi \Bigg|_{x = x_1(t)} \\ &= \frac{\partial}{\partial x} \int_{-\delta(t)}^{\delta(t)} \mathscr{G}(\xi) [u_x(x - \xi, t) - u_x(x, t)]\, d\xi |_{x = x_1(t)} \\ &\quad + u_{xx}(x_1(t), t) \int_{-\delta(t)}^{\delta(t)} \mathscr{G}(\xi)\, d\xi \\ &= \frac{\partial}{\partial x} \int_{x-\delta(t)}^{x+\delta(t)} \mathscr{G}(x - z) [u_x(z, t) - u_x(x, t)]\, dz |_{x = x_1(t)} \\ &= \mathscr{G}(x_1(t) - z) [u_x(z, t) - u_x(x_1(t), t)] |_{z = x_1(t) - \delta(t)}^{z = x_1(t) + \delta(t)} \\ &\quad + \int_{-\delta(t)}^{\delta(t)} \mathscr{G}'(\xi) [u_x(x_1(t) - \xi, t) - u_x(x_1(t), t)]\, d\xi\, .\end{aligned}$$

The last integral is nonnegative because $\mathscr{G}'(\xi) \geq 0$, and $u_x(x_1(t) - \xi, t) - u_x(x_1(t), t) \geq 0$ by virtue of the definition of $x_1(t)$. Thus, for $\overline{\overline{j}}_3$ we have the estimate $\overline{\overline{j}}_3 \leq 2h(t)(|\mathscr{G}(\delta(t))| + |\mathscr{G}(-\delta(t))|)$. But $|\mathscr{G}(x)| = |k(x) - P(x)| \leq 2|P(x)|$; hence, (3.39) implies that this estimate takes the form

$$\overline{\overline{j}}_3 \leq \frac{8 c_0 h(t)}{(\frac{2}{3} - \gamma)} \delta(t)^{-2/3+\gamma}\, . \tag{3.46}$$

Combining (3.41) through (3.46) we obtain the following estimate for the integral j_0:

$$j_0 \leq k_7 h(t) + k_8 \delta(t)^{-2/3+\gamma} h(t) + k_g r(t) \delta(t)^{4/3+\gamma}, \tag{3.47}$$

where

$$k_8 = 4c_0 + \frac{8c_0}{\frac{2}{3} - \gamma}, \qquad k_g = \frac{2c_0}{(\frac{2}{3} - \gamma)(\frac{1}{3} + \gamma)} \left(1 - \frac{1}{\frac{4}{3} + \gamma} \right).$$

A further estimate of the integral j_0 is obtained by applying the lemma for the interval $[0, t_0]$,$t_0 < T_0 = 1/|m(0)|(1 - \frac{\gamma}{8})$. We have $h(t) \leq n(t) - m(t)$. Since $Z(t)$ is decreasing for $t \in [0, t_0]$, we conclude from (3.13) that $|m(t)| \geq |m(0)|$. According to the Lemma, $n(t) \leq n(0) + \bar{k}t \leq n(0) + \bar{k}T_0$, and since we can always assume that $2|m(0)| \geq n(0) + \bar{k}/|m(0)|(1 - \frac{\gamma}{8})$, for $t \in [0, t_0]$ we have

$$h(t) \leq 3|m(t)| \ . \tag{3.48}$$

Inequality 3) of the Lemma implies that for $t \in [0, t_0]$

$$r(t) \leq k_{10} Z^{-\beta}(t), \tag{3.49}$$

where $k_{10} = r(0) + \bar{\bar{k}} l(0) \cdot 1/|m(0)|(1 - \frac{\gamma}{8})$. We now choose

$$\delta(t) = Z(t)^{3/2} a, \qquad a = \min\{a_1, a_2\} \ . \tag{3.50}$$

Since $Z(t) \leq 1$, we have $\delta(t) \leq a$. Substituting of (3.48) through (3.50) into (3.47), with account taken of (3.13), we obtain

$$j_0 \leq \frac{|m(0)|}{Z(t)} (k_{11} + k_{12} Z(t)^{-1+3/2\gamma}) + k_{13} Z(t)^{-\beta+2+3\gamma/2}, \tag{3.51}$$

where $k_{11} = 3k_7$, $k_{12} = 3k_8 a^{-2/3+\gamma}$, $k_{13} = k_9 k_{10} a^{4/3+\gamma}$.

Finally, recalling the expression (3.38) for $A(t)$ and using the equality (3.13) and the estimate (3.51), we find that

$$A(t) = \frac{j_0}{m^2(t)} = \frac{j_0 Z^2(t)}{m^2(0)} \leq \frac{Z^\sigma(t)}{|m(0)|} \left(k_{11} + k_{12} + \frac{k_{13}}{|m(0)|} \right), \tag{3.52}$$

where $\sigma = \gamma(\frac{3}{2} - \frac{24}{20-5\gamma}) > 0$, since $0 < \gamma < \frac{2}{3}$. Since $z(t) \leq 1$ for $t \in [0, t_0]$, from (3.52) we conclude that $A(t) < \frac{\gamma}{8}$ for all $t \in [0, t_0]$, provided

$$\frac{1}{|m(0)|} \left(k_{11} + k_{12} + \frac{k_3}{|m(0)|} \right) < \frac{\gamma}{8}. \tag{3.53}$$

In particular, $A(t_1) < \frac{\gamma}{8}$, i.e., we have arrived at a contradiction.

Thus, we have proved that if $A(0) < \frac{\gamma}{8}$, then $A(t) < \frac{\gamma}{8}$ for all $t \in [0, T_0)$, where $T_0 = 1/|m(0)|(1 - \frac{\gamma}{8})$.

We now turn to formula (3.14). In view of the inequality $A(t) < \frac{\gamma}{8}$ just proved for $t \in [0, T_0]$, we have $Z(t) \leq 1 + m(0)t(1 - \frac{\gamma}{8})$. Hence, $Z(t) \to +0$ as $t \to \widetilde{T}_0 \leq 1/|m(0)|(1 - \frac{\gamma}{8})$. Since $m(t) = m(0)/Z(t)$, $m(t) \to -\infty$ as $t \to T_0$, i.e., the wave breaks after a finite time interval not exceeding $T_0 = 1/|m(0)|(1 - \frac{\gamma}{8})$.

To complete the proof of Theorem 3 it remains to verify that the various conditions imposed on the initial function $\bar{u}(x)$ are not contradictory. In condition (3.37) the constants k_1 and k_5 depend only on the kernel $k(x)$, and condition (3.37) is satisfied if $|m(0)|$ is sufficiently large. In condition (3.53) the constants k_{11} and k_{12} are determined only by the kernel $k(x)$, and k_{13} is a constant multiple of k_{10}, so that to ensure that (3.53) is satisfied it is sufficient to verify that $k_{10}/|m(0)|$ is bounded when $|m(0)|$ is sufficiently large. Since the function $\bar{u}(x)$ decreases at infinity together with its derivatives, the maxima of the moduli of all derivatives of $\bar{u}(x)$ are attained in a certain finite domain $\mathscr{D} \subset R_1$. Hence, subtracting (if necessary) a function $cx\varphi(x)$ from $\bar{u}(x)$, where $\varphi \in C_0^\infty(\mathscr{D}_1)$, $\mathscr{D} \subset \mathscr{D}_1$, $\varphi(x) \equiv 1$ on $\mathscr{D}$, $c > 0$, and $\mathscr{D}_1$ is a compact set, we can increase $|m(0)|$ by a constant c, while leaving $r(0)$ and $l(0)$ unchanged. Thus, $k_{10}/|m(0)|$ is bounded and (3.53) holds if $|m(0)|$ is sufficiently large. The condition $A(0) < \frac{\gamma}{8}$ is satisfied by the same token.

Thus, all the required conditions are satisfied if $|m(0)|$ is sufficiently large. Theorem 3 is completely proved.

CHAPTER 2

The Cauchy Problem for the Whitham Equation

§1. Introduction

In the present chapter we study the Cauchy problem for the nonlinear, nonlocal Whitham equation on the line $x \in R_1$, $t \geq 0$:

$$u_t + uu_x + \mathbb{K}(u) = 0, \qquad u|_{t=0} = \overline{u}(x) \tag{1.1}$$

where $\mathbb{K}$ is a linear pseudodifferential operator of the following form:

$$\mathbb{K}(u) = \frac{1}{2\pi}\int_{-\infty}^{\infty} e^{ipx}K(p)\hat{u}(p,t)\,dp;$$

$\hat{u}(p,t)$ is the Fourier transform: $\hat{u}(p,t) = \int_{-\infty}^{\infty} e^{-ipx}u(x,t)\,dx$.

$K(p)$ is called the symbol of the operator of $\mathbb{K}(u)$; if the kernel $k(x)$ of the Whitham equation (1.2.1) belongs to $L_1(R_1)$, then $K(p) = ip\hat{k}(p)$.

We consider real solutions to this equation. Therefore, the symbol $K(p)$ of the operator $K(u)$ satisfies the condition $K(p) = K^*(-p)$ so that its real part $K^1(p) = \operatorname{Re} K(p)$ is an even, and the imaginary part $K^2(p) = \operatorname{Im} K(p)$ is an odd function of p.

We describe briefly the content of this chapter. In §2 we study local in time existence of a classic solution to the Cauchy problem (1.1). In subsection 2.1. the existence and uniqueness of a solution $u(x,t) \in C^\infty([0,T_0]; H^\infty(R_1))$ for initial perturbations $u(x) \in H^\infty(R_1)$ in the case of a regular operator $\mathbb{K}$ is proved; here the time T_0 is either infinite or is equal to the time of breaking. The time of breaking T_0 is defined as the moment when $\max_{x\in R_1}|u_x(x,t)| \to \infty$ as $t \to T_0$, $T_0 < \infty$. In subsection 2.2, this same result is proved for dissipative operators $\mathbb{K}$, i.e., for operators whose symbol $K(p)$ satisfies the following condition:

$$\operatorname{Re} K(p) \geq 0 \quad \text{for } |p| \geq h > 0\,.$$

Subsection 2.3 is devoted to the proof of the existence of a classical solution $u(x,t) \in C^\infty([0,T]; H^\infty(R_1))$ in the case of an antidissipative operator $\mathbb{K}$, i.e., an operator whose symbol $K(p)$ satisfies the condition

$$\operatorname{Re} K(p) \geq -b(1+|p|^\alpha),$$

where $b > 0$, $0 < \alpha < 1$. Therein the initial condition $\overline{u}(x)$ cannot be an arbitrary function from the class $H^\infty(R_1)$ anymore, but must belong to some Gevrey class and be sufficiently small. The time T of existence of a solution $u(x,t)$ in the case of an antidissipative operator is not necessarily the time of its breaking.

In §3, we study general conditions that provide the existence of a global in time solution for the problem (1.1). Subsection 3.1 deals with the proof of the existence of a global solution in the presence of sufficiently small initial perturbations $\overline{u}(x) \in H^\infty(R_1)$

for the strongly dissipative operator $\mathbb{K}$, i.e., such that $\operatorname{Re} K(p) \geq 0$ for $|p| \leq h$; $\operatorname{Re} K(p) \geq \varepsilon > 0$ for $|p| \geq h > 0$.

The smallness of the initial perturbation $\overline{u}(x)$ is understood as the boundedness of the norm $\|\overline{u}\|_{H^2(R_1)}$, which ensures a small steepness of the wave at the initial moment, and this, in turn, prevents breaking of the waves as time passes. A global solution of problem (1.1) thus exists.

It turns out that in the case of a strongly dissipative operator $\mathbb{K}$

$$\operatorname{Re} K(p) \geq \varepsilon |p|^{\alpha} \quad \text{for } |p| \geq h > 0 \tag{1.2}$$

where $\varepsilon > 0$, $\alpha > \frac{3}{2}$, and the smallness of the initial perturbation is redundant. Indeed, in subsection 3.2, we establish the existence of a unique solution $u(x,t) \in C^{\infty}([0,\infty]; H^{\infty}(R_1))$ of the Cauchy problem (1.1) for a strongly dissipative operator and arbitrary initial perturbations $\overline{u}(x) \in H^{\infty}(R_1)$.

In §4 smoothing of discontinuous initial perturbations is analyzed. In subsection 4.1, it is proved that, in the case of nonsmooth initial conditions $\overline{u}(x) \in H^{\rho}(R_1)$, $0 \leq \rho \leq \frac{1}{2}$, for some $T > 0$ there exists a unique solution $u(x,t) \in C^{\infty}([0,T]; H^{\infty}(R_1)) \cap C^0([0,T]; H^{\rho}(R_1))$ for the Cauchy problem (1.1), if the operator $\mathbb{K}$ is strongly dissipative (1.2) with $\alpha > \max(1, \frac{3}{2} - \rho)$. The existence of a smoothed global in time solution is established in subsection 4.2. It is shown that if the condition of dissipation (1.2) is satisfied with $\alpha > \frac{3}{2}$ and the initial perturbation $\overline{u}(x) \in L_2(R_1)$, then the problem (1.1) has a unique solution

$$u(x,t) \in C^{\infty}((0,\infty); H^{\infty}(R_1)) \cap C^0([0,\infty); L_2(R_1)) .$$

EXAMPLE. The theorem on smoothing and global existence holds, in particular, for the Burgers equation $u_t + uu_x - \nu u_{xx} = 0$, the Korteweg-de Vries-Burgers equation $u_t + uu_x + \gamma u_{xxx} - \nu u_{xx} = 0$, and the Kuramoto-Sivashinsky equation $u_t + uu_x - \nu u_{xx} + \beta u_{xxxx} = 0$.

In the second part of this chapter, §§5 through 7, we consider the problem of breaking of the classical solution to the Cauchy problem for the Whitham equation in a finite time-interval. In §5 we prove Theorem 9 concerning the breaking of waves in the case of conservative and dissipative operators $\mathbb{K}(u)$ of order less than $\frac{3}{5}$. In particular, here we completely solve the Whitham problem of breaking of waves described by equation (1.2.I) with a kernel $k_g(x)$ occurring in exact potential theory of water waves. We give sufficiently precise two-sided estimates for the time of breaking. In §6, we formulate Theorem 10 on the breaking of waves in the case of an arbitrary operator $\mathbb{K}(u)$ of order less than $\frac{2}{3}$. In this theorem, the absence of requirement on the type of dissipation, or at least conservation, is compensated by strengthening the requirements on the initial perturbation, which now must belong to some Gevrey class. In the same §6, we prove several preparatory lemmas for the proof of Theorem 10. The proof itself is given in §7.

In §3, the existence of a global solution of the Cauchy problem (1.I) for the Whitham equation is proved under the condition that the initial steepness of the wave is sufficiently small. Thus, condition 3) in Theorems 9 and 10 concerning the large steepness of the initial perturbation is not only sufficient, but in some sense, also necessary for the breaking of waves. The results of this chapter were published in [**46, 52, 53, 63, 64**].

§2. The existence of a classical solution for the Cauchy problem on a finite time-interval

2.1. Regular operator. We begin our study of the Cauchy problem (1.I) with the case of a regular operator $\mathbb{K}$, which in the x-representation has the following form:

$$\mathbb{K}(\varphi) = \int_{-\infty}^{\infty} k(x-s)\varphi(s)\,ds + \xi\varphi(x), \tag{2.1}$$

for any function $\varphi(x) \in H^\infty(R_1)$. Here the kernel $k(x)$ is absolutely integrable on the line R_1, and ξ is a real constant.

We consider the Cauchy problem for an ordinary differential equation:

$$\dot{y} = u(y,t), \quad t > 0; \qquad y|_{t=0} = \xi, \quad \xi \in R_1\,. \tag{2.2}$$

The solution $y(\xi, t)$ of problem (2.2), called the characteristic of (1.I), depends on ξ as a parameter. The following result is well known [**73**].

ASSERTION. *Suppose that for some $T > 0$ the right-hand side $u(y,t)$ of (2.2) belongs to $C^\infty([0,T]; H^\infty(R_1))$. Then*:

1) *There exists a unique solution $y(\xi,t) \in C^\infty(R_1 \times [0,T])$ of problem (2.2).*

2) *For each $t \in [0,T]$ the function $y(\xi,t)$ realizes a diffeomorphism $\xi \in R_1 \leftrightarrow y \in R_1$.*

THEOREM I. *Let*:

1) *The operator $\mathbb{K}$ in the Whitham equation be regular*;

2) *The initial perturbation $\overline{u}(x)$ belong to $H^\infty(R_1)$.*

Then for some $T_0 > 0$ there exists a unique solution $u(x,t)$ of the Cauchy problem (1.1) *in the class $C^\infty([0,T_0]; H^\infty(R_1))$. This solution can destruct only as a result of breaking of waves.*

PROOF. We apply the method of successive approximations. We set $w^0(x,t) = \overline{u}(x)$, and define the functions $w^{(s)}(x,t)$, $s \geq 1$, as the solutions of the following linear problem:

$$w_t^{(s)} + w^{(s-1)} w_x^{(s)} + \mathbb{K}(w^{(s-1)}) = 0, \qquad w^{(s)}|_{t=0} = \overline{u}(x),$$

whence, temporarily omitting the index s, we obtain

$$w_t + g w_x + f = 0, \qquad w|_{t=0} = \overline{u}(x), \tag{2.3}$$

where $g = w^{(s-1)}$, $f = \mathbb{K}(w^{(s-1)})$. Assume by induction that the function $w^{(s-1)} \in C^\infty([0,\infty); H^\infty(R_1))$ has already been constructed. Then, by formula (2.1), we have $g, f \in C^\infty([0,\infty); H^\infty(R_1))$, so that (2.3) is the Cauchy problem for a linear hyperbolic equation with smooth coefficients and right-hand side. The solution of problem (2.3) is given by

$$w(y(\xi,t),t) = \overline{u}(\xi) + \int_0^t f(g(\xi,\tau),\tau)\,d\tau, \tag{2.4}$$

where the function $y(\xi,t)$ is a solution of problem (2.2) with right-hand side $g(x,t)$. Formula (2.4) immediately implies that $w(x,t) \in C^\infty([0,\infty); H^\infty(R_1))$. We show that for some $T > 0$ the sequence $\{w^{(s)}\}$ is uniformly bounded with respect to $s \geq 0$ in the class $C^\infty(R_1 \times [0,T])$. To this end, we differentiate the Whitham equation n times with respect to x:

$$\frac{\partial}{\partial t} w^{(s)}_{(n)} + \sum_{j=0}^{n} C^j_n w^{(s-1)}_j w^{(s)}_{(n+1-j)} + \mathbb{K}(w^{(s-1)}_{(n)}) = 0 \,. \tag{2.5}$$

Here and below $\varphi_{(n)}(x) = \partial^n \varphi/\partial x^n$, $\varphi_n = \max_{x\in R_1} |\varphi_{(n)}(x)|$ for each function $\varphi(x) \in H^\infty(R_1)$; C^j_n are binomial coefficients. We define the characteristic $y^{(s)}(\xi, t)$ as the solution of problem (2.2) with right-hand side $w^{(s-1)}(s, t)$ and introduce new functions $v^{(s)}_{(n)}(\xi, t) = w^{(s)}_{(n)}(y^{(s)}(\xi, t), t)$ for each $n \geq 0$ and $s \geq 1$. In view of (2.2) this allows us to remove the term with the highest-order derivative $w^{(s-1)}_{(0)} w^{(s)}_{(n+1)}$ in (2.5). From the result formulated in the assertion it follows that $v^{(s)}_n = w^{(s)}_n$; therefore, taking into account (2.1), from (2.5), we obtain

$$|\dot v^{(s)}_{(n)}| \leq \sum_{j=1}^{n} C^n_j v^{(s-1)}_j v^{(s)}_{n+1-j} + H v^{(s-1)}_n, \qquad v^{(s)}_{(n)}|_{t=0} = \overline{u}_{(n)}, \qquad n \geq 0; \tag{2.6}$$

here $H = |\xi| + \int_{-\infty}^{\infty} |k(x)|\, dx$. For $n = 0$ the sum in (2.6) is identically zero.

Now we consider one more system on $[0, T]$, $n \geq 0$:

$$\dot p_n = \sum_{j=1}^{n} C^n_j p_j p_{n+1-j} + H p_n, \qquad p_n(0) = 2\overline{u}_n > 0 \,. \tag{2.7}$$

Here $T > 0$ determines the time of existence of a classical solution for the system (2.7). We note that in (2.7) only the equation with $n = 1$ is nonlinear, while all other equations are linear with respect to the unknown function $p_n(t)$. Integration of (2.7) with $n = 1$ with respect to t yields:

$$p_1(t) = H p_1(0) e^{Ht} (H + p_1(0)(1 - e^{Ht}))^{-1} \,.$$

It follows that if T is chosen so that $0 < T < H^{-1} \ln(1 + H/p_1(0))$ the solution $p_1(t)$ is bounded by some constant $\mathscr{D}_1$ on the entire closed interval $[0, T]$. Upon integration of the remaining linear equations (2.7) ($n \neq 1$) on the interval $[0, T]$, we see that for some set of numbers $\mathscr{D}_n > 0$ the following estimates are valid:

$$p_n(t) \leq \mathscr{D}_n \quad \text{for } t \in [0, T],\ n \geq 0 \,. \tag{2.8}$$

We show that on $[0, T]$ a solution $p_n(t)$ of the system (2.7) majorizes the functions $v^{(s)}_{(n)}$, i.e.

$$v^{(s)}_n < p_n \quad \text{for all } n \geq 0,\ s \geq 0 \,. \tag{2.9}$$

Indeed, inequalities (2.9) hold for $t = 0$; by the definition of the successive approximations, they also hold for all $n \geq 0$ and $t \in [0, T]$, when $s = 0$. We apply induction on $s \geq 0$. For $s \geq 1$ we prove (2.9) by contradiction. Let n_1 be the least number for which (2.9) does not hold on the entire interval $[0, T]$. Further, suppose that T_1 is the first moment of time when (2.9) does not hold for this number, $n = n_1$. Since the functions in (2.9) are continuous, for $n \le n_1$ and $t \in [0, T_1]$ we have

$$v_n^{(s)}(t) \le p_n(t)\,. \tag{2.10}$$

Substituting (2.9) and (2.10) in (2.6) we see that

$$|\dot{v}_{(n)}^{(s)}(\xi, t)| \le \dot{p}_n(t), \qquad v_n^{(s)}(0) < p_n(0),$$

for $n \le n_1, \xi \in R_1, t \in [0, T_1]$. Hence, (2.9) follows for $n = n_1$ and $t = T_1$. The contradiction obtained proves (2.9). The induction is completed.

Estimates (2.8) and (2.9) imply that $v_n^{(s)}(t)$ are uniformly bounded with respect to s for $n \ge 0$ and $t \in [0, T]$. Similar arguments show that the expressions

$$\max_{0 \le t \le T} \max_{x \in R_1} \left| \frac{\partial^{j+n} w^{(s)}}{\partial x^n \partial t^j} \right|$$

are uniformly bounded with respect to s for all $n, j \ge 0$.

We now prove that the sequence $\{w^{(s)}\}$ is uniformly bounded with respect to s in the space $C^\infty([0, T]; H^\infty(R_1))$. To this end, we multiply (2.5) by $w_{(n)}^{(s)}$ and integrate it over x from $-\infty$ to $+\infty$:

$$\begin{aligned} &\dot{W}_n^{(s)} + 2\sum_{j=0}^{n} C_j^n \int_{-\infty}^{\infty} w_{(n)}^{(s)} w_{(n+1-j)}^{(s)} w_{(j)}^{(s-1)}\, dx \\ &\quad + 2\int_{-\infty}^{\infty} w_{(n)}^{(s)} \mathbb{K}(w_{(n)}^{(s-1)})\, dx = 0, \end{aligned} \tag{2.11}$$

where $W_n^{(s)} = \int_{-\infty}^{\infty} (w_{(n)}^{(s)})^2\, dx$. Applying (2.1) and the Cauchy inequality, we bound the absolute value of the last integral in (2.11) by $H\sqrt{W_n^{(s)} W_n^{(s-1)}}$. The functions $w_{(n)}^{(s)}$ belong to $H^\infty(R_1)$ for all $t \in [0, T]$; therefore $w_{(n)}^{(s)}(x, t) \to 0$ as $|x| \to \infty$. Hence, integration by parts yields

$$\int_{-\infty}^{\infty} w_{(n)}^{(s)} w_{(n+1)}^{(s)} w_{(0)}^{(s-1)}\, dx = -\frac{1}{2}\int_{-\infty}^{\infty} (w_{(n)}^{(s)})^2 w_{(1)}^{(s-1)}\, dx\,.$$

Applying (2.8) and (2.9), from (2.11) we obtain, $n \ge 0$,

$$\dot{W}_n^{(s)} \le c_n \sum_{j=1}^{n} \sqrt{W_n^{(s)} W_{n+1-j}^{(s)}} + 2H\sqrt{W_n^{(s)} W_n^{(s-1)}} \tag{2.12}$$

where $c_n > 0$ are some constants. By induction on s, (2.12) easily implies $\max_{0 \le t \le T} W_n^{(s)} \le \widetilde{\mathscr{D}}_n$ for $n \ge 0$, $s \ge 0$, where $\widetilde{\mathscr{D}}_n$ are numbers independent of s. Similar results hold for

$$W_{n,j}^{(s)} = \int_{-\infty}^{\infty} \left(\frac{\partial^j w_{(n)}^{(s)}(x, t)}{\partial t^j} \right)^2 dx\,.$$

Thus, the sequence $\{w^{(s)}\}$ is bounded in $C^\infty([0, T]; H^\infty(R_1))$ and, therefore, some subsequence of it converges to a function $u(x, t) \in C^\infty([0, T]; H^\infty(R_1))$, which is, obviously, a solution to the Cauchy problem (1.1) with a regular operator $\mathbb{K}$.

We show that the derived solution is unique. Assuming the contrary, suppose there are two solutions $u_1(x,t)$ and $u_2(x,t)$ from the class $C^\infty([0,T]; H^\infty(R_1))$. For the difference $\psi = u_1 - u_2$ we have

$$\dot\psi + (u_1 - \psi)\psi_x + \psi u_{1x} + \mathbb{K}(\psi) = 0, \qquad \psi|_{t=0} = 0\,. \tag{2.13}$$

We define the characteristic $y(\xi,t)$ as a solution of problem (2.2) with the right-hand side equal to $u_1 - \psi$. Applying (2.13), we conclude that the new function $\varphi(\xi,t) = \psi(y(\xi,t),t)$ satisfies

$$\dot\varphi + \varphi u_{1x} + \mathbb{K}(\psi) = 0, \qquad \varphi|_{t=0} = 0\,.$$

Since u_{1x} is bounded, we have

$$|\dot\varphi| \le c|\varphi|, \qquad \varphi|_{t=0} = 0, \tag{2.14}$$

for some $c > 0$. It follows from (2.14) that $\varphi \equiv 0$. The uniqueness is proved. We note that the time T of existence of a classical solution is determined by the fact that the function $p_1(t)$ is bounded. Indeed, if $p_1(t)$ is bounded, then all functions $p_n(t)$ and with them, by (2.9), the functions $v_n^{(s)}(t)$ are also bounded. That means that a classical solution $u(x,t)$ of the Cauchy problem (1.1) with a regular operator $\mathbb{K}$ can be extended in time as long as the first derivative with respect to x of the solution $u(x,t)$ remains bounded. Thus, destruction of the classical solution can occur only as a result of breaking. The theorem is proved. □

2.2. Dissipative operator. We establish now the local in time existence of a classical solution for the Cauchy problem (1.1) in the case of a dissipative operator $\mathbb{K}$. The operator $\mathbb{K}$ here can be nonregular (i.e., the kernel $k(x)$ in (2.1) does not belong to $L_1(R_1)$ and is, for instance, a distribution). Hence, it is more convenient to use p-representation of the operator $\mathbb{K}$ in the form of a pseudodifferential operator

$$\mathbb{K}(u) = \frac{1}{2\pi}\int_{-\infty}^{\infty} e^{ipx} K(p)\hat u(p,t)\,dp,$$

where $\hat u(p,t) = \int_{-\infty}^{\infty} e^{-ipx}u(x,t)\,dx$ is the Fourier transform. We assume that $K(p) \in L_\infty^{\mathrm{loc}}(R_1)$, and that there exists infimum $\overline{\alpha} < \infty$ of the numbers α such that

$$|K(p)| \le c(1+|p|)^\alpha \quad \text{for } p \in R_1, \text{ where } c > 0\,.$$

The number $\overline{\alpha}$ is called the order of the operator $\mathbb{K}$.

We call the operator $\mathbb{K}$ dissipative if its symbol $K(p)$ satisfies the following condition:

$$K^1(p) = \operatorname{Re} K(p) \ge 0 \quad \text{for } |p| \ge h > 0\,. \tag{2.15}$$

We remark that the case of conservative operators, i.e., $K^1(p) \equiv 0$, is included in (2.15).

THEOREM 2. *Assume that*

1) *Condition* (2.15) *is satisfied;*

2) *The initial perturbation* $\overline{u}(x)$ *belongs to* $H^\infty(R_1)$.

Then, on some closed interval $[0,T]$ *there exists a unique solution* $u(x,t)$ *of problem* (1.1) *in the class* $C^\infty([0,T]; H^\infty(R_1))$. *The time of existence* T *is determined by the norm* $\|\overline{u}\|_{H^2(R_1)}$.

PROOF. To regularize the operator $\mathbb{K}$ we define a sequence of symbols $K^{(l)}(p)$, $l = 1, 2, \ldots$, as follows:

$$K^{(l)}(p) = (1 + |p|)^{\beta} \int_{-l}^{l} p_l(p - q)(1 + |q|)^{-\beta} K(q)\, dq,$$

where $\beta = 2 + [\overline{\alpha}]$, and $\overline{\alpha}$ is the order of the operator $\mathbb{K}$, $p_l(p) = lp_1(pl)$, $p_1(p) \in C_0^{\infty}(R_1)$ is such that

$$\rho_1(p) \geq 0,\ \rho_1(p) \equiv 0 \quad \text{for } |p| \geq 1,\ \int_{-\infty}^{\infty} \rho_1(p)\, dp = 1\,.$$

We note that the operator $\mathbb{K}^{(l)}$ with the symbol $K^{(l)}(p)$ can be represented as the convolution (2.1) with a kernel $k^{(l)}(x)$ in the Schwartz space. From the definition and condition (2.15) it follows that for all $l = 1, 2, \ldots$,

$$K^{(l),1}(p) \geq 0 \quad \text{for } |p| \geq h + 2, \tag{2.16}$$

$$K^{(l)}(p)(1 + |p|)^{-\beta} \to K(p)(1 + |p|)^{-\beta} \tag{2.17}$$

in the norm of the space $L_2(R_1)$ as $l \to \infty$, and that there exists a number $b \geq 0$ such that

$$-K^{(l),1}(p) \leq b \quad \text{for } |p| \leq h + 2. \tag{2.18}$$

Theorem 1 yields a solution $u^{(l)}(x, t) \in C^{\infty}([0, T^{(l)}]; H^{\infty}(R_1))$ of problem (1.1) with the operator $\mathbb{K}^{(l)}$. We establish the boundedness of the sequence $\{u^{(l)}(x, t)\}$. Temporarily omitting the index l, multiplying the Whitham equation and its derivatives by $u_{(n)}(x, t)$, and integrating over x from $-\infty$ to $+\infty$, we find

$$\dot{J}_n + 2\sum_{j=0}^{n} C_j^n \int_{-\infty}^{\infty} u_{(n)}u_{(n+1-j)}u_{(j)}\, dx + 2I_n = 0, \tag{2.19}$$

where

$$J_n = \int_{-\infty}^{\infty} u_{(n)}^2\, dx = \frac{1}{2\pi}\int_{-\infty}^{\infty} |\hat{u}(p, t)|^2 |p|^{2n}\, dp,$$

$$I_n = \int_{-\infty}^{\infty} u_{(n)}(x, t)\mathbb{K}(u_{(n)})\, dx.$$

Using the identity

$$2\int_{-\infty}^{\infty} u_{(n)}u_{(n+1)}u_{(0)}\, dx = -\int_{-\infty}^{\infty} u_{(n)}^2 u_{(1)}\, dx$$

and the inequality

$$u_{(n)}(t) \leq (4J_nJ_{n+1})^{1/4},$$

we transform (2.19) to the following form:

$$\dot{J}_0 \leq -2I_0, \tag{2.20}$$

$$\dot{J}_2 \leq 5J_2(4J_1J_2)^{1/4} - 2I_2, \tag{2.21}$$

$$\dot{J}_n \leq (2n + 1)J_n(4J_1J_2)^{1/4} + 2\sum_{j=2}^{n-1} C_j^n (J_nJ_{n+1-j})^{1/2}(4J_jJ_{j+1})^{1/4} - 2I_n, \qquad n \geq 3\,. \tag{2.22}$$

It follows from the Plancherel theorem and (2.16), (2.18) that

$$
\begin{aligned}
I_n(t) &= \int_{-\infty}^{\infty} u_{(n)}\mathbb{K}^{(l)}(u_{(n)})\,dx = \frac{1}{2\pi}\int_{-\infty}^{\infty} K^{(l),1}(p)|p^n\hat{u}(p,t)|^2\,dp \\
&\geq \frac{1}{2\pi}\int_{|p|\leq h+2} K^{(l),1}(p)|p^n\hat{u}(p,t)|^2\,dp \geq -bJ_n\,.
\end{aligned}
\tag{2.23}
$$

Substituting (2.23) in (2.20)–(2.22), we obtain

$$
\dot{J}_0 \leq 2bJ_0, \qquad \dot{J}_2 \leq 10J_2^{5/4}J_1^{1/4} + 2bJ_2, \tag{2.24}
$$

$$
\dot{J}_n \leq (2n+1)(4J_1J_2)^{1/4} + 2\sum_{j=2}^{n-1} C_j^n\sqrt{J_nJ_{n+1-j}}(4J_jJ_{j+1})^{1/4} + 2bJ_n, \qquad n\geq 3\,. \tag{2.25}
$$

Put $J = J_0 + J_2$; then, since $J_1 \leq \sqrt{J_0J_2} \leq J$, we derive from (2.24) the inequality

$$
\dot{J} \leq 10J^{3/2} + 2bJ\,.
$$

Integrating it with respect to t we get

$$
J(t) \leq J(0)e^{2bt}\left(1 - \frac{5}{b}\sqrt{J(0)}(e^{bt}-1)\right)^{-2}. \tag{2.26}
$$

It follows from (2.26) that if T is chosen from the condition $0 < T < \frac{1}{b}\ln(1 + b/5\sqrt{J(0)})$, then one can find a number $c_1 > 0$ independent of l such that $J(t) \leq c_1$ for all $t \in [0,T]$. Integration of the linear inequalities (2.25) with respect to t yields

$$
J_n(t) \leq c_n, \tag{2.27}
$$

for $t \in [0,T]$, $n \geq 0$, where $c_n > 0$ is independent of l. The boundedness of $\int_{-\infty}^{\infty}(\partial^{n+m}u/\partial x^n\partial t^m)^2\,dx$ is established similarly. Thus, the sequence $\{u^{(l)}(x,t)\}$ is bounded in $C^\infty([0,T];H^\infty(R_1))$; therefore a certain subsequence of it converges to a function $u(x,t) \in C^\infty([0,T];H^\infty(R_1))$ in the norm of the space $C^m([0,T];H^n(R_1))$, where n,m are arbitrary. Without loss of generality, we suppose that the sequence $\{u^{(l)}(x,t)\}$ itself converges to the function $u(x,t)$. To prove that $u(x,t)$ is a classical solution of the problem (1.1), it is sufficient to establish that the sequence $\mathbb{K}^{(l)}(u^{(l)})$ converges to $\mathbb{K}(u)$ uniformly with respect to $x \in R_1$ and $t \in [0,T]$. Indeed, we have

$$
\begin{aligned}
&\frac{1}{2\pi}\int_{-\infty}^{\infty} e^{ipx}K^{(l)}(p)\hat{u}^{(l)}(p,t)\,dp \\
&\quad = \mathbb{K}(u) + \frac{1}{2\pi}\int_{-\infty}^{\infty}\frac{e^{ipx}(K^{(l)}(p)-K(p))\hat{u}^{(l)}(p,t)\,dp}{(1+|p|)^\beta(1+|p|)^{-\beta}} \\
&\quad + \frac{1}{2\pi}\int_{-\infty}^{\infty} e^{ipx}\frac{K(p)(\hat{u}^{(l)}(p,t)-\hat{u}(p,t))\,dp}{(1+|p|)^\beta(1+|p|)^{-\beta}}\,.
\end{aligned}
\tag{2.28}
$$

Since $u^{(l)}(x,t)$ converges in the norm of $C^0([0,T];H^\beta(R_1))$ as $l\to\infty$, applying the Cauchy inequality and (2.17) we find that the integrals on the right-hand side of (2.28) tend to zero as $l\to\infty$, uniformly with respect to $x\in R_1$ and $t\in[0,T]$. We note that the time of existence T of the solution $u(x,t)$ is determined by the norm $\|u\|_{H^2(R_1)}$.

We now prove the uniqueness of the solution. Assuming the contrary, let there be two solutions $u^1(x,t)$ and $u^2(x,t)$. Then for the difference $w = u^1 - u^2$ we obtain

$$w_t + u^1 w_x + w u_x^2 + \mathbb{K}(w) = 0, \qquad w|_{t=0} = 0\,.$$

Multiplying by w and integrating over x from $-\infty$ to $+\infty$, we find

$$\dot{W} + 2\int_{-\infty}^{\infty} w^2(-\tfrac{1}{2}u_x^1 + u_x^2)\,dx + 2\int_{-\infty}^{\infty} w\mathbb{K}(w)\,dx = 0, \tag{2.29}$$

where $W = \int_{-\infty}^{\infty} w^2(x,t)\,dx$. Transforming the last term in (2.29) as in (2.23), we find

$$\dot{W} \le 2W(u_1^1 + u_1^2 + b), \qquad W(0) = 0\,.$$

Hence $W(t) \equiv 0$. The uniqueness is established. Theorem 2 is proved. □

2.3. Antidissipative operator. We prove now that in the case of an antidissipative operator $\mathbb{K}$ whose symbol $K(p)$ satisfies the condition

$$K^1(p) \ge -b(1+|p|)^\alpha \quad \text{for } p \in R_1, \tag{2.30}$$

where b and α are constants, $b > 0$, $0 < \alpha < 1$, there exists a solution of the Cauchy problem (1.1) on a finite interval of time $[0,T)$, if the initial perturbation $\overline{u}(x)$ is sufficiently small and belongs to some Gevrey class.

For $n \ge 0$ we denote

$$\overline{J}_n = \frac{1}{2\pi}\int_{-\infty}^{\infty} |\hat{\overline{u}}(p)|^2 |p|^{2n}\,dp, \quad A_n = (n+1)^{\sigma(2n-5)}, \qquad \sigma = 1/2\,.$$

THEOREM 3. *Suppose*

1) *Condition* (2.30) *is satisfied*;

2) *For the initial perturbation* $\overline{u}(x)$ *the following inequalities hold*:

$$\overline{J}_n \le A_n a^{-2n-1}, \qquad n \ge 0 \tag{2.31}$$

with a sufficiently large constant a,

$$a > c(T, b, \alpha)\,. \tag{2.32}$$

Then on the interval $[0,T)$ *there exists a solution* $u(x,t)$ *of problem* (1.1) *belonging to* $C^\infty([0,T]; H^\infty(R_1))$.

PROOF. Let us define the sequence of symbols:

$$K^{(l)}(p) = \begin{cases} K(p), & |p| \le l, \\ 0, & |p| > l,\ l = 1,2,\dots\,. \end{cases}$$

The operators $\mathbb{K}^{(l)}$ obtained are dissipative and, therefore, according to Theorem 2, there exist solutions $u^{(l)}(x,t)$ of problem (1.1) with the operators $\mathbb{K}^{(l)}$. Temporarily omitting the index l, for $t \in [0, T)$ we prove the following inequalities:

$$J_n < A_n e^{qt} a^{-2n}, \qquad n = 0, 2, \; q \equiv 10 + 2b \tag{2.33}$$

$$J_n < A_n a^{-2n} Z^{-\sigma(2n-5)}, \qquad n = 3, 4, \dots, \tag{2.34}$$

where

$$J_n(t) = \frac{1}{2\pi} \int_{-\infty}^{\infty} |\hat{u}(p,t)|^2 |p|^{2n}\, dp, \qquad Z(t) = 1 - \frac{t}{T}.$$

Owing to (2.31), the bounds (2.33) and (2.34) hold for $t = 0$. To prove (2.33) and (2.34), we apply inequalities (2.20)–(2.22) from the previous subsection. The regularity of the operators $\mathbb{K}^{(l)}$ implies

$$-I_n(t) = -\frac{1}{2\pi} \int_{-\infty}^{\infty} K^{(l),1}(p) |\hat{u}(p,t)|^2 |p|^{2n}\, dp \le \frac{\mathscr{N}}{a} J_n \tag{2.35}$$

for some natural number $\mathscr{N} > 3$. First, we prove (2.33) and (2.34) for $n \le \mathscr{N}$. Assuming the contrary, let $T_1 > 0$ be the first moment of time when one of these inequalities is violated. Then, in view of the continuity for $t \in [0, T]$, $n \le \mathscr{N}$, we have

$$\begin{aligned} &J_n(t) \le A_n e^{qt} a^{-2n}, \qquad n = 0, 2, \\ &J_n(t) \le A_n a^{-2n} Z(t)^{-\sigma(2n-5)}, \qquad n \ge 3. \end{aligned} \tag{2.36}$$

Taking into account (2.36) and (2.30) we estimate I_n:

$$\begin{aligned} -I_0 &\le b(J_0 + J_{\alpha/2}) \le b(J_0 + J_0^{1-\alpha/4} + J_2^{\alpha/4}) \\ &\le b e^{qt}(1 + a^{-\alpha}), \end{aligned} \tag{2.37}$$

$$\begin{aligned} -I_2 &\le b(J_2 + J_{2+\alpha/2}) \le b(J_2 + J_2^{1-\alpha/2} J_3^{\alpha/2}) \\ &\le b A_2 a^{-4} e^{qt}(1 + 4a^{-\alpha} Z(t)^{-1/2}), \end{aligned} \tag{2.38}$$

and for $3 \le n < \mathscr{N}$:

$$\begin{aligned} -I_n &\le b(J_n + J_{n+\alpha/2}) \le bJ_n + bJ_n^{1-\alpha/2} J_{n+1}^{\alpha/2} \\ &\le bJ_n + bA_n a^{-2n-\alpha} Z^{-\sigma(2n-5)-1}(n+2)\left(\frac{n+2}{n+1}\right)^{n-5/2} \\ &< bJ_n + be(n+2)Ana^{-2n-\alpha} Z^{-\sigma(2n-5)-1}. \end{aligned} \tag{2.39}$$

From (2.20) and (2.37), by (2.32) we obtain

$$\dot{J}_0 \le 2be^{qt}(1 + a^{-\alpha}) < qe^{qt}.$$

From this, using (2.31) we arrive at (2.33). By (2.32), it follows from (2.36) that $J_0 < a$, $J_2 < 4^{-2}a^{-3}$. Therefore, $J_1 \le \sqrt{J_0 J_2} < (4a)^{-1}$, so that for all $t \in [0, T_1]$

$$(4J_1 J_2)^{1/4} < \frac{1}{2a}. \tag{2.40}$$

Upon substitution of (2.40) and (2.38) in (2.21), by (2.36) we conclude that

$$\begin{aligned} \dot{J}_2 &\le \frac{5}{a} J_2 + 2bA_2 a^{-4} e^{qt}(1 + 4a^{-\alpha} Z^{-1/2}) \\ &\le A_2 a^{-4}(qe^{qt} + 8ba^{-\alpha} e^{qT} Z^{-1/2}). \end{aligned}$$

Whence, according to (2.31)

$$J_2(t) \le J_2(0) + A_2 a^{-4}(e^{qt} - 1 + 16bTa^{-\alpha}(1 - Z^{1/2})e^{qt}) < A_2 a^{-4} e^{qT},$$

since a is large. Thus, the bounds (2.33) hold on $[0, T_1]$. Substituting now (2.40) in (2.22), we obtain that for $3 \le n < \mathcal{N}$

$$\dot{J}_n < (2n+1)\frac{J_n}{a} + 2\mathcal{G}_n - 2I_n, \tag{2.41}$$

where

$$\begin{aligned}
\mathcal{G}_n &= \sum_{j=2}^{n-1} C_j^n (J_n J_{n+1-j})^{1/2} (4 J_j J_{j+1})^{1/4} \\
&\le \sqrt{J_n} \sum_{j=2}^{n-1} C_j^n a^{-n-1/2+j} A_{n+1-j}^{1/2} 2a^{-j} A_{j+1}^{1/2} Z^{-\sigma(n-j)+\beta-\sigma j+\beta} \\
&\le 2\sqrt{J_n} a^{-n-1/2} Z^{-\sigma(n-3)} \sum_{j=2}^{n-1} C_j^n A_{n+1-j}^{1/2} A_{j+1}^{1/2}, \qquad \beta = \frac{3\sigma}{2}.
\end{aligned} \tag{2.42}$$

For $3 \le n \le 4\beta/\chi$, where $\chi = \sigma - 1$ by (2.32) we can write

$$\sum_{j=2}^{n-1} C_j^n A_{n+1-j}^{1/2} A_{j+1}^{1/2} < n A_n^{1/2} \frac{a^{1/4}}{2}. \tag{2.43}$$

Otherwise, for $\chi n > 4\beta$ that same bound is obtained by using the Stirling formula and applying the explicit expressions for the numbers A_n. Substituting (2.43) in (2.42) we get

$$\mathcal{G}_n < n A_n a^{-2n-1/4} Z^{-\sigma(2n-5)}. \tag{2.44}$$

Using (2.32) and (2.39) we obtain from (2.41) that

$$\begin{aligned}
\dot{J}_n &< 2bJ_n + (4n+1) A_n a^{-2n-1/4} Z^{-\sigma(2n-5)} \\
&\quad + 2bA_n a^{-2n-\alpha} e(n+2) Z^{-\sigma(2n-5)-1} \\
&< 2bJ_n + A_n a^{-2n} Z^{-\sigma(2n-5)-1} T\sigma(2n-5) e^{-2bT}.
\end{aligned}$$

From this, integrating with respect to t and again applying (2.32), we obtain

$$\begin{aligned}
J_n(t) &\le J_n(0) e^{2bt} + \int_0^t e^{2b(t-\tau)} A_n a^{-2n} T\sigma(2n-5) e^{-2bT} Z^{-\sigma(2n-5)-1}(\tau)\, d\tau \\
&< A_n a^{-2n} \left(1 - \int_{Z(t)}^{Z(0)} \sigma(2n-5) Z^{-\sigma(2n-5)-1}\, dz\right) \\
&= A_n a^{-2n} Z^{-\sigma(2n-5)}(t).
\end{aligned} \tag{2.45}$$

The estimate (2.34) is thus established on $[0, T_1]$ for all n, $3 \le n < \mathcal{N}$.

By way of induction on $n \ge \mathcal{N}$ and assuming the contrary, we again have inequality (2.36) on some interval $[0, T_n]$. From (2.35) and (2.36) we find that

$$-I_n \le n J_n a^{-1} \le n A_n a^{-2n-1} Z^{-\sigma(2n-5)-1}.$$

Estimating the sum $\mathcal{G}_n$ in (2.41) as before with the aid of (2.44) we again obtain the estimate (2.45) on $[0, T_n]$. The contradiction obtained proves (2.34) on the entire interval $[0, T)$. Thus, the bounds (2.33) and (2.34) are established, and they, in turn, imply that the sequence $\{u^{(l)}(x,t)\}$ is bounded in the space $C^\infty([0,T]; H^\infty(R_1))$. Therefore, some subsequence converges, as $l \to \infty$, to the solution $u(x,t)$ of the Cauchy problem (1.1) with the antidissipative operator $\mathbb{K}$. Theorem 3 is proved. □

§3. The existence of a global in time solution

3.1. Strongly dissipative operator. We prove the existence of a classical solution $u(x,t)$ of the Cauchy problem (1.1) for all positive t, in the case when the operator $\mathbb{K}$ is of arbitrary finite order and is strictly dissipative, i.e., its conservative part $K^2(p) = \operatorname{Im} K(p)$ is arbitrary, and the following conditions hold for $K^1(p) = \operatorname{Re} K(p)$:

$$K^1(p) \geq 0 \quad \text{for } |p| < h; \qquad K^1(p) \geq \varepsilon > 0 \quad \text{for } |p| \geq h > 0\,. \tag{3.1}$$

THEOREM 4. *Suppose that*
1) *Condition* (3.1) *holds;*
2) *The initial perturbation* $\overline{u}(x) \in H^\infty(R_1)$ *is sufficiently small:*

$$\|\overline{u}\|_{H^2(R_1)} < c_1, \tag{3.2}$$

where $c_1 = c_1(\varepsilon, h)$.

Then the Cauchy problem (1.1) *has a unique solution* $u(x,t)$ *that belongs to* $C^\infty([0,\infty); H^\infty(R_1))$.

PROOF. In Theorem 2 we proved the existence and uniqueness of a solution for the problem (1.1) on a finite closed interval $[0, T]$, where $T > 0$ is determined by the norm $\|u\|_{H^2(R_1)}$. We show that the norm $\|u\|_{H^2(R_1)}$ does not increase with time. This means that we can extend a classical solution $u(x,t)$ of the problem (1.1) to any interval $[0, T]$ in a finite number of steps.

Let us first estimate $J_0(t) = \int_{-\infty}^{\infty} u^2(x,t)\,dx$. To this end we turn to inequality (2.20). Using (3.1) and the Plancherel Theorem we estimate the integral I_0:

$$I_0 = \frac{1}{2\pi}\int_{-\infty}^{\infty} |\hat{u}(p,t)|^2 K^1(p)\,dp \geq 0\,.$$

Therefore, from (2.20) we obtain $\dot{J}_0 \leq 0$, and thus for all $t \geq 0$

$$J_0(t) \leq J_0(0). \tag{3.3}$$

Let us estimate the function $J_2(t) = \frac{1}{2\pi}\int_{-\infty}^{\infty} |\hat{u}(p,t)|^2 p^4\,dp$ with the aid of inequality (2.21). This time we estimate the integral $I_2(t)$ as follows:

$$\begin{aligned} I_2(t) &= \frac{1}{2\pi}\int_{-\infty}^{\infty} K^1(p) p^4 |\hat{u}(p,t)|^2\,dp \\ &= \varepsilon J_2 + \frac{1}{2\pi}\int_{|p|\leq h} (K^1(p) - \varepsilon) p^4 |\hat{u}(p,t)|^2\,dp \\ &\quad + \frac{1}{2\pi}\int_{|p|\geq h} (K^1(p) - \varepsilon) p^4 |\hat{u}(p,t)|^2\,dp \\ &\geq \varepsilon J_2 - \varepsilon h^4 J_0\,. \end{aligned} \tag{3.4}$$

Substituting (3.4) in (2.21) and using (3.3) we get

$$\begin{aligned} \dot{J}_2 &\leq -2\varepsilon J_2 + 2\varepsilon h^4 J_0 + 5 J_2 (4 J_1 J_2)^{1/4} \\ &\leq -2\varepsilon J_2(t) + 2\varepsilon h^4 J_0(0) + 10 J_0^{1/8}(0) J_2^{11/8}(t)\,. \end{aligned} \tag{3.5}$$

We now fix the constant c_1 in (3.2) so that the following inequalities hold:

$$J_0(0) > \min\left\{\left(\frac{\varepsilon}{10}\right)^8, \frac{1}{2h^4}\right\}, \qquad J_2(0) < 1.$$

Then, by continuity, on some t-interval we have $J_2(t) < 1$ and from (3.5) we obtain

$$\dot{J}_2 \leq \varepsilon(2h^4 J_0(0) = J_2(t)),$$

whence

$$J_2(t) \leq \max(J_2(0), 2h^4 J_0(0)) < 1 .$$

Thus, $J_2(t)$ is always less than 1, the norm $\|u\|_{H^2(R_1)}$ does not increase with time, and the solution is global. Theorem 4 is proved. □

3.2. Strongly dissipative operator. It turns out that in the case of stronger dissipation in the Whitham equation, the condition (3.2) that the initial perturbation be small can be omitted. An operator $\mathbb{K}$ is called strongly dissipative if

$$K^1(p) \geq \varepsilon|p|^\alpha \quad \text{for } |p| \geq h > 0, \tag{3.6}$$

where $\varepsilon > 0$, $\alpha > 0$.

THEOREM 5. *Suppose that*
1) *Condition* (3.6) *with* $\alpha > \frac{3}{2}$ *holds;*
2) *The initial perturbation* $\overline{u}(x)$ *belongs to* $H^\infty(R_1)$.
Then there exists a unique solution $u(x,t)$ *of problem* (1.1) *which belongs to the class* $C^\infty([0,\infty); H^\infty(R_1))$.

PROOF. As in the previous theorem, we show that the norm $\|u\|_{H^2(R_1)}$ remains bounded for any finite t. We estimate the integral I_0 using condition (3.6) and the Plancherel Theorem

$$I_0(t) = \frac{1}{2\pi}\int_{-\infty}^{\infty} |\hat{u}(p,t)|^2 K^1(p)\, dp \geq \frac{1}{2\pi}\int_{|p|\leq h} |\hat{u}(p,t)|^2 K^1(p)\, dp$$
$$\geq -bJ_0(t), \quad \text{where } b = -\inf_{|p|\leq h}(0, K^1(p)) \geq 0 .$$

Then, from (2.20) we obtain $\dot{J} \leq 2bJ_0$, whence we arrive at the following estimate:

$$J_0(t) \leq J_0(0)e^{2bt}. \tag{3.7}$$

We apply the Fourier transform to the Whitham equation to get

$$\hat{u}(p,t) + \frac{ip}{2}\int_{-\infty}^{\infty} \hat{u}(p-q,t)\hat{u}(q,t)\, dq + K(p)\hat{u}(p,t) = 0,$$

whence integrating with respect to t, we find that

$$\hat{u}(p,t) = \hat{\overline{u}}(p)e^{-K(p)t} - \frac{ip}{2}\int_0^t e^{-K(p)(t-\tau)} V(p,\tau)\, d\tau, \tag{3.8}$$

where

$$V(p,\tau) = \int_{-\infty}^{\infty} \hat{u}(p-q,\tau)\hat{u}(q,\tau)\, dq .$$

Introducing the notation $w(p) = \varepsilon|p|^\alpha$, we obtain, by (3.6), $K^1(p) \geq -b - w(h) + w(p)$ for all $p \in R_1$. Multiplying (3.8) by p^2, we get

$$p^2|\hat{u}(p,t)| \leq p^2|\hat{\bar{u}}(p)|e^{bt} + \frac{|p|^3}{2}e^{bt+w(h)t}\int_0^t e^{-w(p)(t-\tau)}|V(p,\tau)|\,d\tau\,. \tag{3.9}$$

Put $\gamma = \min(\frac{1}{2}, (\alpha - \frac{3}{2})/(\frac{3}{2}\alpha + 2))$. Taking the inequality $|p|^{2-2\gamma} \leq 4(|p-q|^{2-2\gamma} + |q|^{2-2\gamma})$ into account and applying the Hölder inequality twice, we find that

$$\begin{aligned}|p|^{2-2\gamma}|V(p,\tau)| &\leq 8\int_{-\infty}^{\infty} |\hat{u}(p-q,\tau)|\,|\hat{u}(q,\tau)|\,|q|^{2-2\gamma}\,dq \\ &\leq 8\left(\int_{-\infty}^{\infty} |\hat{u}(p-q,\tau)|^{2/(1+\gamma)}|\hat{u}(q,\tau)|^{2\gamma/(1+\gamma)}\,dq\right)^{(1+\gamma)/2} \\ &\quad\cdot\left(\int_{-\infty}^{\infty} q^4|\hat{u}(q,\tau)|^2\,dq\right)^{(1-\gamma)/2} \\ &\leq 8J_0^{(1+\gamma)/2}(\tau)J_2^{(1-\gamma)/2}(\tau)\,.\end{aligned} \tag{3.10}$$

Substitution of (3.7) and (3.10) in (3.9) yields

$$p^2|\hat{u}(p,t)| \leq p^2|\hat{\bar{u}}(p)|e^{bt} + 4J^{(1+\gamma)/2}(0)e^{3bt+w(h)t}|p|^{1+2\gamma}\int_0^t e^{-w(p)(t-\tau)}J_2^{(1-\gamma)/2}(\tau)\,d\tau\,. \tag{3.11}$$

We show that

$$J_2(t) < 3J_2(0)e^{2Bt}, \tag{3.12}$$

where B is sufficiently large, so that

$$\gamma B \geq 3b + w(h), \quad B^\gamma > 16(1 + J_0^{(1+\gamma)/2}(0)J_2^{-\gamma/2}(0) + (3\gamma\alpha)^{-1}\varepsilon^{-2})\,. \tag{3.13}$$

The bound (3.12) is obviously valid for $t = 0$. Now, assuming the contrary, let $T_1 > 0$ be the first moment of time, when equality in (3.12) takes place. We substitute (3.12) in (3.11). By (3.13), for the second term on the right-hand side of (3.11) we obtain

$$\begin{aligned}&4J_0^{(1+\gamma)/2}(0)e^{3bt+w(h)t}|p|^{1+2\gamma}\int_0^t e^{-w(p)(t-\tau)}(J_2(\tau))^{(1-\gamma)/2}\,d\tau \\ &\quad\leq B^\gamma\sqrt{J_2(0)}e^{\gamma Bt}|p|^{1+2\gamma}\int_0^t e^{-w(p)(t-\tau)+(1-\gamma)B\tau}\,d\tau \\ &\quad\leq B^\gamma\sqrt{J_2(0)}e^{Bt}|p|^{1+2\gamma}(w(p) + B(1-\gamma))^{-1}.\end{aligned} \tag{3.14}$$

Taking squares, from (3.11) and (3.14) we obtain

$$p^4|\hat{u}(p,t)|^2 \leq 2p^4|\hat{\bar{u}}(p)|^2e^{2Bt} + 2B^{2\gamma}J_2(0)\frac{e^{2Bt}|p|^{2+4\gamma}}{(w(p) + B(1-\gamma))^2}\,.$$

Integrating with respect to p from $-\infty$ to $+\infty$, we find that

$$J_2(t) \leq J_2(0)e^{2Bt}\left(2 + 2B^{2\gamma}\int_{-\infty}^{\infty}\frac{|p|^{2+4\gamma}\,dp}{(w(p) + B(1-\gamma))^2}\right)\,. \tag{3.15}$$

The integral in (3.15) can be easily estimated. Introducing a new variable $\xi = pB^{-1/2}$, we obtain

$$
(3.16)\qquad 2B^{2\gamma}\int_{-\infty}^{\infty}\frac{|p|^{2+4\gamma}\,dp}{(w(p)+B(1-\gamma))^2}\le 4B^{-\gamma}\left(\int_0^1\frac{d\xi}{(1-\gamma)^2}+\frac{1}{\varepsilon^2}\int_1^{\infty}\xi^{2+4\gamma-2\alpha}\,d\xi\right)
$$
$$
\le 4B^{-\gamma}(4+\varepsilon^{-2}(3\gamma\alpha)^{-1})<1,
$$

by virtue of the definitions for B and γ and the evident inequality $2\alpha-3-4\gamma\ge 3\gamma\alpha$. It is obvious from (3.15) and (3.16) that equality in (3.12) is impossible, so that (3.12) is always true. Thus, the norm of the solution $\|u\|_{H^2(R_1)}$ remains bounded for all finite t and the classical solution exists for all $t\ge 0$. Theorem 5 is proved. □

§4. Smoothing of solutions

4.1. Smoothing for short time intervals. It was established in the previous section that in the case of a strongly dissipative operator $\mathbb{K}$ the Cauchy problem (1.1) has a global solution. Now we show that the solution smooths out in time, i.e., given nonsmooth initial perturbations $\bar{u}(x)\in H^{\rho}(R_1)$, $0\le\rho\le\frac{1}{2}$, at any instance of time $t>0$ the solution becomes infinitely smooth: $u(x,t)\in C^{\infty}([0,T];H^{\infty}(R_1))$.

THEOREM 6. *Suppose that*
1) *The initial perturbation* $\bar{u}(x)\in H^{\rho}(R_1)$, $0\le\rho\le\frac{1}{2}$;
2) *The condition of strong dissipation holds*:

$$
(4.1)\qquad K^1(p)\ge\varepsilon|p|^{\alpha}\quad \textit{for } |p|\ge h\ge 1,\ \varepsilon>0,\ \alpha>\max\{1,\tfrac{3}{2}-\rho\}\,.
$$

Then, for some $T>0$ *there exists a unique solution of the Cauchy problem* (1.1) *in* $C^{\infty}([0,T],H^{\infty}(R_1))\cap C^0([0,T];H^s(R_1))$.

PROOF. We fix a sequence of functions with compact support $\{\bar{u}^{(s)}(x)\}$, $\bar{u}^{(s)}(x)\in C_0^{\infty}(R_1)$, $s\ge 0$, converging in the norm of $H^s(R_1)$ to the function $\bar{u}(x)$ as $s\to\infty$, and such that

$$
(4.2)\qquad \|\bar{u}^{(s)}\|_{\rho}^2\le 2\|\bar{u}\|_{\rho}^2\,.
$$

Here we have introduced the norm

$$
\|\varphi\|_{\rho}^2\equiv\int_{-\infty}^{\infty}M^{2\rho}(p)|\hat{\varphi}(p)|^2\,dp,\qquad M(p)\equiv\max(1,|p|);
$$

$\hat{\varphi}(p)=\int_{-\infty}^{\infty}e^{-ipx}\varphi(x)\,dx$ is the Fourier transform.

We apply the method of successive approximations to problem (1.1). We set $w^0(x,t)$ equal to the inverse Fourier transform of the function $\hat{\bar{u}}^{(0)}(p)e^{-K(p)t}$, and define the function $w^{(s)}(x,t)$ to be the solution of the following linear problem:

$$
(4.3)\qquad w_t^{(s)}+w^{(s-1)}w_x^{(s-1)}+\mathbb{K}(w^{(s)})=0,\qquad w^{(s)}|_{t=0}=\bar{u}^{(s)}.
$$

Applying the Fourier transform to (4.3), we find that

$$
(4.4)\qquad \hat{w}_t^{(s)}+\frac{ip}{2}\int_{-\infty}^{\infty}\hat{w}^{(s-1)}(p-q,t)\hat{w}^{(s-1)}(q,t)\,dq+K(p)\hat{w}^{(s)}(p,t)=0,
$$

whence, integrating with respect to t, we get:

$$|\hat{w}^{(s)}(p,t)| \le |\hat{\bar{u}}^{(s)}(p)|e^{-K^1(p)t} + \frac{|p|}{2}\int_0^t e^{-K^1(p)(t-\tau)}\int_{-\infty}^{\infty}|\hat{w}^{(s-1)}(p-q,\tau)\hat{w}^{(s-1)}(q,\tau)|\,dq\,. \tag{4.5}$$

Introducing a new function $v^{(s)}(p,t) = M^\rho(p)|\hat{w}^{(s)}(p,t)|e^{w(p)t}$, $w(p) = \varepsilon_1 M(p)$, $\varepsilon_1 = \frac{\varepsilon}{2}$, we find from (4.5) that

$$\begin{aligned} v^{(s)}(p,t) \le\ & M^\rho(p)|\hat{\bar{u}}^{(s)}(p)|e^{(w(p)-K^1(p))t} \\ & + \int_0^t d\tau e^{-K^1(p)(t-\tau)}\frac{|p|}{2}\int_{-\infty}^{\infty} dq v^{(s-1)}(q,\tau)v^{(s-1)}(p-q,\tau) \\ & \cdot M^\rho(p)M^{-\rho}(p-q)M^{-\rho}(q)e^{(w(p)t-w(p-q)\tau-w(q)\tau)}\,. \end{aligned} \tag{4.6}$$

Carrying out induction on $s \ge 0$ we show that

$$\|v^{(s)}(\,\cdot\,,t)\|_0 \le a \equiv 4\chi\|\bar{u}\|_\rho \tag{4.7}$$

where $\chi = \exp\{\varepsilon_1 M^\alpha(h) + \varepsilon_1 M(h) + \sup_{|p|\le h}(0, -K^1(p))\}$. Since $w(p) - w(p-q) - w(q) \le 0$, in view of (4.1) we have

$$\begin{aligned} \exp\{-K^1(p)(t-\tau) + w(p)t - w(p-q)\tau - w(q)\tau\} &\le e^{(w(p)-K^1(p))(t-\tau)} \\ \le \chi e^{-\varepsilon_1 M^\alpha(p)(t-\tau)} &\le \chi(\varepsilon_1 M^\alpha(p)(t-\tau))^{\delta-1}, \end{aligned} \tag{4.8}$$

where $\gamma = (\alpha - \frac{3}{2} + \rho)/(\alpha+1) > 0$. Taking account of (4.8) and the inequality $M^\rho(p) \le M^\rho(p-q) + M^\rho(q)$, from (4.6) we obtain

$$\begin{aligned} \|v^{(s)}(\,\cdot\,,t)\|_0^2 \le\ & \chi\int_{-\infty}^{\infty} v^{(s)}(p,t)M^\rho(p)|\hat{\bar{u}}^{(s)}(p)|\,dp \\ & + \int_0^t d\tau\,\chi(\varepsilon_1(t-\tau))^{\gamma-}J, \end{aligned} \tag{4.9}$$

where

$$J = \iint_{-\infty}^{\infty} dp\,dq v^{(s)}(p,t)v^{(s-1)}(p-q,\tau)v^{(s-1)}(q,\tau)M^{\alpha\gamma+1-\alpha}(p)M^{-\rho}(q)\,.$$

Put $\lambda \equiv \alpha - \alpha\gamma - 1 + \rho = \gamma + \frac{1}{2}$. Making use of (4.7) and of the Cauchy inequality we estimate J:

$$\begin{aligned} J &\le \iint_{|p|\le|q|} + \iint_{|p|>|q|} \\ &\le \int_{-\infty}^{\infty} dp\,v^{(s)}(p,t)M^{-\lambda}(p)\int_{|p|\le|q|} dq\,v^{(s-1)}(p-q,\tau)v^{(s-1)}_{(q,\tau)} \\ &\quad + \int_{-\infty}^{\infty} dq v^{(s-1)}(q,\tau)M^{-\lambda}(q)\int_{|q|<|p|} dp\,v^{(s)}(p,t)v^{(s-1)}(p-q,\tau) \\ &\le 2a^2\|v^{(s)}(\,\cdot\,,t)\|_0\left(\int_{-\infty}^{\infty} M^{-2\lambda}(p)\,dp\right)^{1/2} \le \frac{4a^2}{\gamma}\|v^{(s)}(\,\cdot\,,t)\|_0\,. \end{aligned} \tag{4.10}$$

From (4.9) and (4.10), applying again the Cauchy inequality and (4.2) we thus obtain

$$\begin{aligned}\|v^{(s)}(\cdot,t)\|_0^2 &\le \|v^{(s)}(\cdot,t)\|_0\left(\chi\|\overline{u}^{(s)}\|_\rho + \int_0^t \frac{6a^2}{\gamma}\chi\varepsilon_1^{\gamma-1}(t-\tau)^{\gamma-1}\,d\tau\right)\\ &\le a\|v^{(s)}(\cdot,t)\|_0,\end{aligned}\tag{4.11}$$

if T is so small that

$$T^\gamma 12a\chi\varepsilon_1^{\gamma-1}\gamma^{-2} < 1.\tag{4.12}$$

Therefore, the bound (4.7) is established by induction.

We show now that

$$\left\|\frac{\partial^k w^{(s)}(\cdot,t)}{\partial t^k}\right\|_n \le c_{n,k}t^{-k\mu-n},\tag{4.13}$$

for all $n, s \ge 0$, $t \in (0,T]$, where $c_{n,k}$ are integers independent of s, and $\mu \ge 2+\alpha > 3$ is such that

$$|K(p)| \le bM^\mu(p), \qquad b>0,\ p\in R_1.\tag{4.14}$$

For $k=0$ the inequalities (4.13) follow from (4.7). We carry out induction on k. Differentiating (4.4) with respect to t, we find that

$$\begin{aligned}\frac{\partial^{k+1}\hat{w}^{(s)}}{\partial t^{k+1}} &= -\frac{ip}{2}\sum_{j=0}^k C_j^k\int_{-\infty}^\infty \frac{\partial^j\hat{w}^{(s-1)}(p-q,t)}{\partial t^j}\frac{\partial^{k-j}\hat{w}^{(s-1)}(q,t)}{\partial t^{k-j}}\,dq\\ &\quad - K(p)\frac{\partial^k\hat{w}^{(s)}(p,t)}{\partial t^k}.\end{aligned}\tag{4.15}$$

Squaring (4.15) and multiplying by $M^{2n}(p)$ we integrate with respect to p from $-\infty$ to $+\infty$. Using (4.13) for the preceding numbers k and (4.14), we conclude that

$$\begin{aligned}\left\|\frac{\partial^{k+1}w^{(s)}}{\partial t^{k+1}}\right\|_n^2 &\le \sum_{j=0}^k C_j^k\int_{-\infty}^\infty dp\\ &\quad\cdot M^{-2}(p)\left|\int_{-\infty}^\infty dq\frac{\partial^j\hat{w}^{(s-1)}}{\partial t^j}(p-q,t)\frac{\partial^{k-j}}{\partial t^{k-j}}\hat{w}^{(s-1)}(q,t)\right|^2\\ &\quad\cdot M^{2n+4}(p) + 2b^2\left\|\frac{\partial^k w^{(s)}}{\partial t^k}\right\|_{n+\mu}^2\\ &\le c_{n,k+1}\sum_{j=0}^k\left(\left\|\frac{\partial^j w^{(s-1)}}{\partial t^j}\right\|_0^2\left\|\frac{\partial^{k-j}w^{(s-1)}}{\partial t^{k-j}}\right\|_{2+n}^2\int_{-\infty}^\infty M^{-2}(p)\,dp + \left\|\frac{\partial^k w^{(s)}}{\partial t^k}\right\|_{n+\mu}^2\right)\\ &\le c_{n,k+1}(t^{-2-n-k\mu}+t^{-\mu-n-k\mu})^2.\end{aligned}$$

Thus, the bound (4.13) is established by induction. It follows from (4.13) that the sequence $\{w^{(s)}(x,t)\}$ is bounded in the space $C^\infty((0,T];H^\infty(R_1))$. Therefore, some subsequence converges to a limit $u(x,t)\in C^\infty((0,T];H^\infty(R_1))$, which satisfies the Whitham equation for $0<t<T$. Let us show that $u(x,t)\in C^0([0,T];H^\rho(R_1))$, i.e., that $u(x,t)$ is continuous with respect to t at $t=0$ in the norm of $H^\rho(R_1)$. It follows

from (4.5) that

$$|\hat{w}^{(s)}(p,t) - \hat{\bar{u}}^{(s)}(p)| \le |\hat{\bar{u}}^{(s)}(p)|\,|e^{-K(p)t} - 1| + \frac{|p|}{2}\int_0^t e^{-K^1(p)(t-\tau)} \int_{-\infty}^{\infty} |\hat{w}^{(s-1)}(p-q,\tau)\hat{w}^{(s-1)}(q,\tau)|\,dq\,d\tau\,.$$

Multiplying this inequality by $M^\rho(p)$, squaring, and integrating with respect to $p \in R_1$, we obtain

$$\|w^{(s)}(x,t) - \bar{u}^{(s)}(x)\|_\rho^2 \le 2\int_{-\infty}^{\infty} |\hat{\bar{u}}^{(s)}(p)|^2 M^{2\rho}(p)(e^{-K(p)t} - 1)^2\,dp + J_1, \tag{4.16}$$

where

$$\begin{aligned} J_1 &= \int_{-\infty}^{\infty} dp\,p^2\left(\int_0^t e^{-K^1(p)(t-\tau)} \int_{-\infty}^{\infty} |\hat{w}^{(s-1)}(p-q,\tau)\hat{w}^{(s-1)}(q,\tau)|\,dq\,d\tau\right)^2 M^{2\rho}(p) \\ &= \int_{-\infty}^{\infty} dp\,p^2 \\ &\quad\cdot\left(\int_0^t d\tau\,e^{-K^1(p)(t-\tau)} \int_{-\infty}^{\infty} dq\,v^{(s-1)}(q,\tau)v^{(s-1)}_{(p-q,\tau)}\,\frac{M^\rho(e)^{-w(p-q)\tau - w(q)\tau}}{M^\rho(p-q)M^\rho(q)}\right)^2 . \end{aligned}$$

Inequalities (4.8)–(4.12) show that

$$J_1 \le ct^\gamma,$$

where $c > 0$ is a constant independent of s. From this and (4.16), letting s tend to ∞ as $s \to \infty$ we obtain

$$\|u(x,t) - \bar{u}(x)\|_\rho^2 \le ct^\gamma + 2\int_{-\infty}^{\infty} dp|\hat{\bar{u}}(p)|^2 M^{2\rho}(p)|e^{-K(p)t} - 1|^2\,.$$

This, in turn, implies that $u \in C^0([0,T], H^\rho(R_1))$.

We prove the uniqueness of the solution. Assuming the contrary, let there be two solutions $u^1 \ne u^2$. For the difference $w = u^1 - u^2$, from (4.4) we obtain

$$\hat{w}_t(p,t) + K(p)\hat{w}(p,t) + \frac{ip}{2}\int_{-\infty}^{\infty} (\hat{u}^1(p-q,t) + \hat{u}^2(p-q,t))\hat{w}(q,t)\,dq = 0\,.$$

From this, after integration with respect to t we find that

$$|\hat{w}(p,t)| \le \frac{|p|}{2}\int_0^t d\tau e^{-K^1(p)(t-\tau)} \int_{-\infty}^{\infty} dq|\hat{u}^1(p-q,\tau) + \hat{u}^2(p-q,\tau)|\,|\hat{w}(q,\tau)|\,. \tag{4.17}$$

Multiply (4.17) by $|\hat{w}(p,t)|$ and integrate with respect to p:

$$\begin{aligned} \|w\|_0^2 \le \frac{1}{2}\int_{-\infty}^{\infty} dp \int_0^t d\tau\,e^{-K^1(p)(t-\tau)} \\ \cdot\int_{-\infty}^{\infty} dq|\hat{u}^1(p-q,\tau) + \hat{u}^2(p-q,\tau)|\,|\hat{w}(q,\tau)|\,|\hat{w}(p,t)|M(p)\,. \end{aligned}$$

By (4.8) we get

$$\begin{aligned} \|w\|_0^2 \le \int_0^t \chi(\varepsilon_1(t-\tau))^{\gamma-1} \int_{-\infty}^{\infty} dq|\hat{w}(q,\tau)|M^{1+\alpha\gamma-\alpha}(p) \\ \cdot\int_{-\infty}^{\infty} dp|\hat{w}(p,t)|\,|\hat{u}^1(p-q,\tau) + \hat{u}^2(p-q,\tau)|\,. \end{aligned} \tag{4.18}$$

Since $\|u^1 + u^2\|_\rho \leq 2a$, using the Cauchy inequality and (4.12), from (4.18) we derive

$$\|w\|_0^2 \leq \frac{6}{\gamma} a \|w\|_0^2 \int_0^t \chi(\varepsilon_1(t-\tau))^{\gamma-1}\, d\tau \leq 6a\chi\varepsilon_1^{\gamma-1}\gamma^{-1}t^\gamma \|w\|_0^2 < \|w\|_0^2 . \tag{4.19}$$

We have arrived at a contradiction. Theorem 6 is proved. □

4.2. The existence of smoothed global solution. Theorems 5 and 6 trivially imply

THEOREM 7. *Suppose that*
1) *Condition* (4.1) *is satisfied with* $\alpha > \frac{3}{2}$;
2) *The initial perturbation* $\overline{u}(x) \in L_2(R_1)$.

Then there exists a unique solution to the Cauchy problem (1.1) *belonging to* $C^\infty((0,\infty); H^\infty(R_1)) \cap C^0([0,\infty); L_2(R_1))$.

A corollary of Theorems 4 and 6 is the following

THEOREM 8. *Suppose that*
1) $K^1(p) \geq 0$ *for* $|p| \leq h : K^1(p) \geq \varepsilon|p|^\alpha$ *for* $|p| > h \geq 1$, $\alpha > 1$, $\varepsilon > 0$;
2) *The initial perturbation* $\overline{u}(x) \in H^\rho(R_1)$, *with* $\rho \in [0, \frac{1}{2}]$ *and* $\rho > \frac{3}{2} - \alpha$, *and*

$$\|\overline{u}\|_\rho \leq c_1, \tag{4.20}$$

where $c_1 > 0$ *is sufficiently small.*

Then there exists a unique solution to the Cauchy problem (1.1) *belonging to* $C^\infty((0,\infty); H^\infty(R_1)) \cap C^0([0,\infty); H^\rho(R_1))$.

PROOF. Theorem 6 guarantees the existence and uniqueness of solution on the interval $[0, T]$. In view of (4.12), the bounds (4.7) and (4.8) imply

$$\begin{aligned} \|w^{(s)}(\,\cdot\,, T)\|_2 &\leq \left\{ \int_{-\infty}^{\infty} |\hat{v}^{(s)}(p, T) e^{-w(p)T} M^{2-\rho}(p)|^2\, dp \right\}^{1/2} \\ &\leq 2a\varepsilon_1^{-2} T^{-2} \leq e\|\overline{u}\|_\rho^{1+2/\gamma} . \end{aligned} \tag{4.21}$$

The bounds (4.20) and (4.21) imply condition (3.2) of Theorem 4. Considering now the Cauchy problem (1.1) on the interval $[T,\infty)$ with initial perturbation $u(x,T) \in H^\infty(R_1)$ and applying Theorem 4 we obtain that $u(x,t) \in C^\infty([T,\infty); H^\infty(R_1))$. Theorem 8 is proved. □

§5. Breaking of waves for a conservative or dissipative operator of order less than 3/5

The purpose of this section is to prove the following theorem.

THEOREM 9. *Suppose that*
1) *The kernel* $k(x)$ *in the Whitham equation satisfies the following conditions*:

$$\begin{aligned} k(x) \in C^1(R_1\backslash 0) \cap L_1(R_1);\quad |k(x)| \leq c|x|^{-\alpha},\quad |k'(x)| \leq c|x|^{-1-\alpha}, \\ x \in [-a, a]\backslash 0; \end{aligned} \tag{5.1}$$

$$\int_{|x|\geq a} |k'(x)|\, dx \leq c, \tag{5.2}$$

where

$$\alpha = \tfrac{3}{5} - \gamma,\ \gamma \in (0, \tfrac{1}{10}],\ c > 0,\ a \in (0, 1];$$

2) *The integral operator* $\mathbb{K}(u)$ *in* (1.1) *is dissipative, i.e., for the Fourier transform* $C(p) = \int_{-\infty}^{\infty} e^{-ipx}k(x)\,dx$ *of the kernel for* $|p| \geq h > 0$ *satisfies the inequality*

$$K^1(p) \equiv \operatorname{Re}(ipC(p)) \geq 0 \quad \textit{for } |p| \geq h > 0\,. \tag{5.3}$$

3) *The initial perturbation* $\overline{u}(x)$ *belongs to* $H^\infty(R_1)$ *and has a sufficiently large steepness* $m_0 = |\min_{x\in R_1} \overline{u}'(x)|$:

$$m_0^2 > \frac{7c}{\gamma a}(u_1 + \sqrt{J(0)}) + \left(2\frac{b}{\gamma}\right)^2, \tag{5.4}$$

where

$$b = \sup_{|p|\leq h}(0, -K^1(p)), \qquad u_1 = \max_{x\in R_1} |\overline{u}'(x)|,$$

$$J(0) = \int_{-\infty}^{\infty} \left(\frac{d^3\overline{u}(x)}{dx^3}\right)^2 dx\,.$$

Then there exists a solution $u(x,t) \in C^\infty([0,T_0); H^\infty(R_1))$ *breaking at the moment of time* T_0. *The following two-sided inequalities hold for* T_0:

$$m_0^{-1}(1+\gamma)^{-1} \leq T_0 \leq m_0^{-1}(1-\gamma)^{-2}\,. \tag{5.5}$$

The kernel $k_g(x)$ indicated above, corresponding to the exact potential theory of water waves, satisfies conditions (5.1)–(5.3). It follows from Theorem 9, in particular, that the asymmetry condition for the initial perturbation discussed in **[10]** and **[160]** is redundant.

PROOF OF THEOREM 9. First, we note that in accordance with Theorem 2, a classical solution $u(x,t) \in C^\infty([0,T_0); H^\infty(R_1))$ exists on some interval $[0, T_0)$. We show that the bounds (5.5) hold for the time of breaking T_0. We introduce the following notation

$$u_{(j)}(x,t) = \frac{\partial^j u(x,t)}{\partial x^j}, \quad u_j(t) = \max_{x\in R_1} |u_{(j)}|, \qquad j = 0, 1, 2, \ldots,$$

$$m(t) = \min_{x\in R_1} u_{(1)}(x,t) = u_{(1)}(y_1(t), t), \qquad n(t) = \max_{x\in R_2} u_{(1)}(x,t) = u_{(1)}(y_2(t), t),$$

$$J(t) = \int_{-\infty}^{\infty} u_{(3)}^2(x,t)\,dx\,.$$

For the sake of simplicity, we assume that the curves $y_j(t) \in C^1([0,T_0))$. (We will show in §§6 and 7 below how one can avoid this assumption by making use of the characteristics.)

Take the x-derivative of (1.1) on the curves $x = y_j(t)$, $j = 1, 2$. Then, taking into account that $u_{(2)}(y_j(t), t) = 0$, we obtain

$$\dot{m} + m^2(1 + A_1(t)) = 0, \tag{5.6}$$

$$\dot{n} + n^2 + m^2 A_2(t) = 0, \tag{5.7}$$

where

$$A_j(t) = m^{-2}(t)\int_{-\infty}^{\infty} k(y_j - x)u_{(2)}(x,t)\,dx$$

(5.8)
$$= m^{-2}\left\{\int_{-\delta}^{\delta} F(x,\delta)u_{(3)}(y_j - x, t)\,dx + k(\delta)u_{(1)}(y_j - \delta, t) - k(-\delta)u_{(1)}(y_j + \delta, t) + \int_{|x|\geq\delta} k'(x)u_{(1)}(y_j - x, t)\,dx\right\}, \qquad j = 1,2;$$

(5.9)
$$P(x,\delta) = \begin{cases} -\int_x^{\delta} k(\xi)\,d\xi & \text{for } 0 \leq x < \delta, \\ \int_{-\delta}^{x} k(\xi)\,d\xi & \text{for } -\delta \leq x < 0\,. \end{cases}$$

The number $\delta \in (0, a]$ will be selected below. Denote

(5.10)
$$Z(t) = 1 - m_0\left(t + \int_0^t A_1(\tau)\,d\tau\right)$$

and integrate (5.6) (considering that $m_0 = -m(0)$):

(5.11)
$$m(t) = -m_0 Z^{-1}(t)\,.$$

Let us prove that if $|A_j(0)| < \gamma$, then $|A_j(t)| < \gamma$ for all $t \in [0, T_0]$, $j = 1, 2$. Assuming the contrary, let $T \in (0, T_0)$ be the first moment of time when $|A_j(T)| = \gamma$, so that by the continuity of the function $A_j(t)$ on the interval $[0, T]$ the following condition is satisfied:

(5.12)
$$|A_j(t)| \leq \gamma\,.$$

We establish a number of inequalities on the interval $[0, T]$. From (5.10) and (5.12) we have

(5.13)
$$-m_0(1+\gamma) \leq \dot{Z} \leq -m_0(1-\gamma), \qquad 0 < Z(t) \leq 1\,.$$

Applying (5.11)–(5.13), from (5.7) we obtain

$$\dot{n} \leq \gamma m_0^2 Z^{-2} \leq m_0 Z^{-2}\dot{Z},$$

integration of which yields

(5.14)
$$n(t) \leq u_1(0)Z^{-1}(t)\,.$$

Differentiating the Whitham equation three times and multiplying by $u_{(3)}(x,t)$, we integrate it with respect to x;

(5.15)
$$\dot{J} + 8\int_{-\infty}^{\infty} u_{(1)}u_{(3)}^2\,dx + 2\int_{-\infty}^{\infty} u_{(0)}u_{(3)}u_{(4)}\,dx + 6\int_{-\infty}^{\infty} u_{(2)}^2 u_{(3)}\,dx + 2\int_{-\infty}^{\infty} u_{(3)}\mathbb{K}(u_{(3)})\,dx = 0\,.$$

Integration by parts leads to two equalities

$$\int_{-\infty}^{\infty} u_{(2)}^2 u_{(3)}\,dx = 0, \qquad 2\int_{-\infty}^{\infty} u_{(0)}u_{(3)}u_{(4)}\,dx = -\int_{-\infty}^{\infty} u_{(1)}u_{(3)}^2\,dx\,.$$

From the Plancherel theorem, owing to the condition of dissipation (5.3), we find

$$\int_{-\infty}^{\infty} u_{(3)}(x,t)\mathbb{K}(u_{(3)})\,dx = \frac{1}{2\pi}\int_{-\infty}^{\infty} K^1(p)p^6|\hat{u}(p,t)|^2\,dp$$
$$\geq \frac{1}{2\pi}\int_{|p|\leq h} K^1(p)p^6|\hat{u}(p,t)|^2\,dp \geq -bJ\,.$$

Therefore, (5.15) gives

$$\dot{J} \leq -7\int_{-\infty}^{\infty} u_{(1)}u_{(3)}^2\,dx + 2bJ \leq (-7m(t)+2b)J\,. \tag{5.16}$$

By (5.4) we have $2b \leq m_0\gamma$. Therefore, substituting (5.11) and (5.13) in (5.16) we arrive at the inequality

$$\dot{J} \leq -\beta Z^{-1}\dot{Z}J,$$

where $\beta = (7+\gamma)/(1-\gamma)$. Integration of this inequality yields

$$J(t) \leq J(0)Z^{-\beta}(t)\,. \tag{5.17}$$

We estimate the functions $A_j(t)$, $j=1,2$, on the closed interval $[0,T]$. Substituting (5.11) in (5.8) and using the Cauchy inequality, we get

$$\begin{aligned}|A_j(t)| \leq m_0^{-2}Z^2\Bigg\{&\left(\int_{-\delta}^{\delta} F^2(x,\delta)\,dx\int_{y_j-\delta}^{y_j+\delta} u_{(3)}^2\,dx\right)^{1/2}\\ &+u_1(t)\left(|k(\delta)|+|k(-\delta)|+\int_{|x|\geq\delta}|k'(x)|\,dx\right)\Bigg\}.\end{aligned} \tag{5.18}$$

From (5.14) and (5.1), (5.2) we easily obtain

$$\int_{-\delta}^{\delta} F^2(x,\delta)\,dx \leq \int_{|x|\leq\delta}\left(\frac{c\delta^{1-\alpha}}{1-\alpha}\right)^2 dx \leq 7^2c^2\delta^{3-2\alpha}, \tag{5.19}$$

$$\int_{|x|\geq\delta}|k'(x)|\,dx \leq -\frac{2c}{\alpha}|x|^{-\alpha}|_{\delta}^{a} + \int_{|x|\geq a}|k'(x)|\,dx \leq 5c\delta^{-\alpha}. \tag{5.20}$$

Comparing (5.18)–(5.20) and (5.11), (5.14), (5.17) we conclude that

$$\begin{aligned}|A_j(t)| &\leq m_0^{-2}Z^2\{(7^2c^2\delta^{3-2\alpha}J(0)Z^{-\beta}(t))^{1/2} + 7c\delta^{-\alpha}u_1(0)Z^{-1}\}\\ &\leq 7cm_0^{-2}(\sqrt{J(0)}\delta^{3/2-\alpha}Z^{2-\beta/2}(t) + u_1(0)Z(t)\delta^{-\alpha}), \qquad j=1,2\,.\end{aligned} \tag{5.21}$$

Whence, setting $\delta = a^{1/2}Z^{1/2}(t)$ in view of the inequalities $\frac{3}{2\alpha}+1-\frac{\beta}{2} \geq 0$ and $0<a\leq 1$, by (5.4) we finally obtain that for $t\in[0,T]$

$$|A_j(t)| \leq 7cm_0^{-2}(u_1(0)+\sqrt{J(0)})a^{-1} < \gamma. \tag{5.22}$$

The contradiction obtained shows that $|A_j(t)| < \gamma$ for all $t \in (0,T_0)$. The assumption $|A_j(0)| < \gamma$, by (5.19)–(5.22) at $t=0$, is also true.

Turning to formulae (5.10), (5.11), and (5.22), we conclude that $m(t) \to -\infty$ as $t \to T_0$ and for the moment of time T_0 (5.5) is satisfied. Theorem 9 is proved. □

§6. Breaking of waves for arbitrary operators of order less than $2/3$

The question of breaking of solutions of (1.1) is of interest not only for conservative (i.e., $K^1(p) = 0$) or dissipative ($K^1(p) \geq 0$) operators $\mathbb{K}(u)$, but also for arbitrary operators. For an arbitrary integral operator in the Whitham equation, the class of initial perturbations should be narrowed. The following theorem holds:

THEOREM 10. *Suppose that*
1) *The kernel* $k(x)$ *in* (1.1) *satisfies* (5.1), (5.2) *with* $\alpha = \frac{2}{3} - 10\gamma$, $\gamma \in (0, \frac{1}{100}]$;
2) *The initial perturbation*(*) $\overline{u}(x) \in \mathcal{G}_b^\alpha$, $b \geq 1$

$$\mathcal{G}_b^\alpha = \left\{\overline{u}(x) \in H^\infty(R_1) : \left|\frac{d^{n+1}\overline{u}(x)}{dx^{n+1}}\right| \leq (bn)^{n/\alpha}, n \geq 1\right\};$$

3) *The value of the steepness* m_0 *is sufficiently large*:

$$m_0 \geq c_0 = c_0(a, b, c, \gamma, q), \qquad q = u_1 m_0^{-1}.$$

Then there exists a solution $u(x,t)$ *of the Cauchy problem for* (1.1) *belonging to* $C^\infty([0, T_0); H^\infty(R_1))$ *and breaking at the moment of time* T_0. *For* T_0 *the estimates* (5.5) *hold.*

Theorem 10 will be proved in §7. In the remaining part of this section we present lemmas necessary for this proof.

6.1.

LEMMA 1. *For all integers* $n \geq 3$ *the following inequality holds*:

$$I \equiv \sum_{j=2}^{n-1} C_j^n (j-1)^{\sigma(j-1)}(n-j)^{\sigma(n-j)} \leq \frac{e3^\sigma}{\sigma - 1} n(n-1)^{\sigma(n-1)},$$

where $\sigma > 1$, *and* $C_j^n = n!/j!(n-j)!$ *are the binomial coefficients.*

PROOF. The Stirling formula implies

$$C_j^n \leq n^n j^{-j}(n-j)^{j-n}.$$

Therefore, denoting $\chi = \sigma - 1$, we obtain

$$C_j^n (j-1)^{\sigma(j-1)}(n-j)^{\sigma(n-j)}(n-1)^{-\sigma(n-1)}$$
$$\leq \frac{n(\frac{n}{n-1})^{n-1}(j-1)^{\chi(j-1)}(n-j)^{\chi(n-j)}}{j(\frac{j}{j-1})^{j-1}(n-1)^{\chi(n-1)}} \leq \frac{en}{j}\left(\frac{j-1}{n-1}\right)^\chi.$$

Hence,

$$I(n-1)^{-\sigma(n-1)} \leq e\sum_{j=2}^{n-1} \frac{n}{j}\left(\frac{j-1}{n-1}\right)^\chi$$
$$\leq \frac{en}{(n-1)^\chi}\int_1^n y^{\chi-1}\,dy \leq \frac{en}{\chi}\left(\frac{n}{n-1}\right)^\chi \leq \frac{en3^\chi}{\chi}.$$

Lemma 1 is proved. □

LEMMA 2. *Suppose the kernel* $k(x)$ *of the operator* $\mathbb{K}$ *satisfies* (5.1) *and* (5.2).

(*) The $\mathcal{G}_b^\alpha$ are called Gevrey classes, see [**83**, p. 335; **40**].

Then for all $\delta \in (0, a]$ *and* $\varphi(x) \in H^\infty(R_1)$

$$|\mathbb{K}(\varphi)| \le c_1(\delta^{-\alpha}\varphi_0 + \delta^{1-\alpha}\varphi_1),$$

where $\varphi_n = \max_{x\in R_1} |\varphi_{(n)}|$, $\varphi_{(n)} = \partial^n\varphi/\partial x^n$.

PROOF. Integration by parts, with (5.1) and (5.2) taken into account, results in

$$\begin{aligned}
|\mathbb{K}(\varphi)| &= \left| \int_{-\infty}^{-\delta} + \int_{-\delta}^{\delta} + \int_{\delta}^{\infty} \right| \\
&= \left| -k(-\delta)\varphi(x+\delta) + k(\delta)\varphi(x-\delta) + \int_{-\delta}^{\delta} k(y)\varphi'(x-y)\,dy \right. \\
&\quad \left. + \int_{|y|\ge 0} k'(y)\varphi(x-y)\,dy \right| \\
&\le \varphi_1 \int_{-\delta}^{\delta} \frac{c\,dx}{|x|^\alpha} + 2c\varphi_0\left(\delta^{-\alpha} + \int_\delta^a \frac{dx}{x^{1+\alpha}}\right) + \varphi_0 \int_{|y|\ge a} |k'(y)|\,dy \\
&\le c_1(\delta^{-\alpha}\varphi_0 + \delta^{1-\alpha}\varphi_1), \qquad c_1 = 7c,
\end{aligned}$$

since $\alpha \in (\frac{1}{2}, \frac{2}{3})$. Lemma 2 is proved. □

Consider the following equation

$$\dot\psi + \psi^2 + \beta = 0, \qquad x \in R_1,\ t \in [0, T], \tag{6.1}$$

for the function $\psi(x,t)$. We shall assume that $\beta(x,t)$ and $\psi(x,t)$ belong to $\psi(x,t) \in C^1(R_1 \times [0,T])$, and that $\psi(x,t) \to 0$ as $|x| \to \infty$. We introduce the notation(*)

$$m(t) = \min_{x\in R_1} \psi(x,t) = -m_0 Z^{-1}(t), \qquad m_0 = -m(0), \tag{6.2}$$

$$Q_\lambda(t) = \{x \in R_1 : \psi(x,t) \le (1-\lambda)m(t)\}\,.$$

Here and below, λ is any number from $[\gamma, \frac{1}{3}]$, $\gamma \in (0, \frac{1}{100}]$.

LEMMA 3. *Suppose that*

1) *There exists a function* $\psi(x,t) \in C^1(R_1 \times [0,T])$ *satisfying* (6.1), *for which* $m(t) < 0$ *on* $[0,T]$;

2)

$$|\beta(x,t)| \le \gamma^2 m^2(t), \qquad t \in [0,T]\,. \tag{6.3}$$

Then for any t_1 *and* t_2, $0 \le t_1 \le t_2 \le T$, *we have the inclusion*

$$Q_\lambda(t_2) \subset Q_\lambda(t_1)\,. \tag{6.4}$$

PROOF. Fix λ. It follows from the first condition of the lemma that the functions $\psi(x,t)$ and $m(t)$ are uniformly continuous in the region $R_1 \times [0,T]$. Therefore, there exists $\delta > 0$ such that

$$\psi(x,t) \le \tfrac{1}{2}m(t), \tag{6.5}$$

for all $t \in [T_1 - \delta, T_1 + \delta] \cap [0,T]$ and all $x \in Q_\lambda(T_1)$, where T_1 is any value from the

(*) Let us clarify why we introduce the sets $Q_\lambda(t)$. In general, it is impossible to guarantee the smoothness of the curve $y_1(t)$ (see §5) on which $\min_{x\in R_1} u_x(x,t) = m(t)$ is achieved. Because of this, one needs to investigate all the characteristics in the immediate vicinity of the curve $y_1(t)$—they are certainly smooth.

interval $[0, T]$. Fix that δ. Then, it is clearly sufficient to prove the inclusion (6.4) on the closed interval $[T_1, T_1+\delta]$ for some $T_1 \in [0, T-\delta]$. We choose any $T_1 \in [0, T-\delta]$. Assuming the contrary, we suppose that there also exist t_1 and t_2, $T_1 \le t_1 < t_2 \le T_2+\delta$, such that $\xi_1 \in Q_\lambda(t_2)$, but $\xi_1 \notin Q_\lambda(t_1)$. That means that

$$\psi(\xi_1, t_1) > m(t_1)(1-\lambda)\,. \tag{6.6}$$

Denote by ξ_2 any point for which $\psi(\xi_2, t_1) = m(t_1)$. It follows from inequalities (6.3) and (6.5) that

$$|\beta(\xi_j, t)| \le \frac{\gamma}{2}\psi^2(\xi_j, t), \qquad j = 1, 2,\ t \in [T_1, T_1+\delta]\,. \tag{6.7}$$

Substitution of (6.7) in (6.1) yields

$$\dot\psi(\xi_2, t) < -\left(1-\frac{\gamma}{2}\right)\psi^2(\xi_2, t),$$
$$\dot\psi(\xi_1, t) \ge -\left(1+\frac{\gamma}{2}\right)\psi^2(\xi_1, t)\,.$$

Whence, integrating with respect to t from t_1 to t_2, we find that

$$\psi(\xi_2, t_2) \le \psi(\xi_2, t_1)(1+\psi(\xi_2, t_1)(1-\gamma/2)(t_2-t_1))^{-1}, \tag{6.8}$$

$$\psi(\xi_1, t_2) \ge \psi(\xi_1, t_1)(1+\psi(\xi_1, t_1)(1+\gamma/2)(t_2-t_1))^{-1}\,. \tag{6.9}$$

By the choice of the point ξ_2, from (6.8) we conclude that

$$m(t_2) \le m(t_1)(1+m(t_1)(1-\gamma/2)(t_2-t_1))^{-1}\,. \tag{6.10}$$

From (6.6), (6.9), (6.10), and the condition $m(t) < 0$ of the lemma, we obtain a chain of inequalities

$$\begin{aligned}\psi(\xi_1, t_2) &> m(t_1)(1-\lambda)(1+m(t_1)(1-\lambda)(1+\gamma/2)(t_2-t_1))^{-1}\\ &> m(t_1)(1-\lambda)(1+m(t_1)(1-\delta/2)(t_2-t_1))^{-1} \ge m(t_2)(1-\lambda)\,.\end{aligned}$$

This contradicts the assumption that $\xi_1 \in Q_\lambda(t_2)$. Lemma 3 is proved. □

LEMMA 4. *Under the conditions of Lemma* 3, *for any* $t \in [0, T]$, $x \in Q_\lambda(t)$, *the following representation holds*:

$$\psi(x, t) = m_0 Y^{-1}(x, t), \tag{6.11}$$

where

$$Y(x, t) = m_0\overline{\psi}^{-1}(x) + m_0\int_0^t (1+A)\,d\tau,$$
$$A(x, t) = \beta(x, t)\psi^{-2}(x, t), \qquad \overline{\psi}(x) = \psi(x, 0)\,.$$

Moreover, the following inequalities are true:

$$Y(x, t) < 0, \qquad |A(x, t)| \le \gamma, \tag{6.12}$$

$$m_0(1-\gamma) \le \dot Y(x, t) \le m_0(1+\gamma), \tag{6.13}$$

$$Z(t) \le |Y(x, t)| \le (1-\lambda)^{-1}Z(t)\,. \tag{6.14}$$

PROOF. Fix any $T_1 \in (0, T]$. The inclusion (6.4) implies that if $x \in Q_\lambda(T_1)$, then $x \in Q_\lambda(t)$ for all $0 \le t \le T_1$. Then from the definition of the set Q_λ and condition 1) of Lemma 3 we find that

$$\psi(x, t) < 0 \tag{6.15}$$

for all $t \in [0, T_1]$ and $x \in Q_\lambda(T_1)$. Denoting $A = \beta\psi^{-2}$, we rewrite (6.1) in the form

$$\dot{\psi} = -\psi^2(1 + A), \qquad x \in Q_\lambda(T_1),\ 0 \le t \le T_1 .$$

From this, after integration with respect to t, we obtain the representation (6.11). The first inequality in (6.12) follows from (6.15) and the boundedness of $\psi(x, t)$, while the second one follows from (6.3) and the definition of the set $Q_\lambda(t)$: $|A(x, t)| \le \gamma^2(1 - \lambda)^{-2} < \gamma$. The bounds (6.13) follow from (6.11) and (6.12), and (6.14) can easily be obtained from (6.2), (6.11), and the definition of Q_λ. Lemma 4 is proved. □

LEMMA 5. *Under the conditions of Lemma* 3, *for* $0 \le t_1 < t \le T$, $\kappa \ge 0$, $\kappa \ne 1$, $(t - t_1)$ *sufficiently small*

$$\int_{t_1}^{t} Z^{-\kappa}(\tau)\, d\tau \le m_0^{-1}(\kappa - 1)^{-1}(1 - \gamma)^{-1-\kappa}(Z^{1-\kappa}(t) - Z^{1-\kappa}(t_1)) . \tag{6.16}$$

PROOF. According to (6.4), if $x_1 \in Q_\lambda(t)$, then $x_1 \in Q_\lambda(\tau)$ for all $\tau \in [0, t]$. Choose $x_1 \in Q_j(t)$ such that $Z(t_1) = |Y(x_1, t_1)|$. Then, using (6.13) and (6.14), we find ($\lambda = \gamma$):

$$\begin{aligned}
\int_{t_1}^{t} Z^{-\kappa}(\tau)\, d\tau &\le (1 - \gamma)^{-\kappa} \int_{t_1}^{t} |Y(x_1, \tau)|^{-\kappa}\, d\tau \\
&\le (1 - \gamma)^{-\kappa-1} m_0^{-1} \int_{t_1}^{t} |Y(x_1, \tau)|^{-\kappa}\, \dot{Y}(x_1, \tau)\, d\tau \\
&\le -m_0^{-1}(1 - \gamma)^{-1-\kappa} \int_{|Y(x_1, t_1)|}^{|Y(x_1, t)|} \xi^{-\kappa}\, d\xi \\
&= m_0^{-1}(1 - \gamma)^{-1-\kappa}(1 - \kappa)^{-1}(|Y(x_1, t_1)|^{-\kappa+1} - |Y(x_1, t)|^{1-\kappa}) \\
&\le m_0^{-1}(1 - \gamma)^{-1-\kappa}(\kappa - 1)^{-1}(Z^{1-\kappa}(t) - Z^{1-\kappa}(t_1))
\end{aligned}$$

Lemma 5 is proved. □

6.2. In this subsection, we prove Theorem 10 under the additional assumption that the operator $\mathbb{K}(u)$ is regular, i.e., that its kernel $k(x) \in C^1(R_1) \cap W_1^1(R_1)$. Integrating by parts, we obtain

$$\mathbb{K}(\varphi) = \int_{-\infty}^{\infty} k'(x - s)\varphi(s)\, ds$$

for any function $\varphi \in H^\infty(R_1)$. If $\mathcal{N}$ is an integer exceeding $4 + \int_{-\infty}^{\infty} |k'(\alpha)|\, dx$, we have the following bound for $\mathbb{K}(\varphi)$:

$$|\mathbb{K}(\varphi)| \le \mathcal{N}\varphi_0 \equiv \mathcal{N} \sup_{x \in R_1} |\varphi(x)| .$$

LEMMA 6. *Suppose that the conditions of Theorem* 10 *are satisfied and, in addition, that the operator* $\mathbb{K}$ *is regular.*

Then Theorem 10 *is true, and the following inequalities hold for* $n \geq 0$ *and* $t \in [0, T_0)$:

$$0 < Z(t) \leq 1, \quad u_0(t) < V_0, \quad u_{n+1}(t) < V_{n+1} Z^{-1-\sigma n}(t), \tag{6.18}$$

where

$$\sigma = (\tfrac{2}{3} - 2\gamma)^{-1}, \qquad Z(t) = m_0 |m(t)|^{-1}, \quad m_0 = |m(0)|,$$

$$m(t) = \min_{x \in R_1} u_{(1)}(x, t), \quad u_{(n)}(x, t) = \frac{\partial^n u(x, t)}{\partial x^n}, \quad u_n(t) = \max_{x \in R_1} |u_{(n)}(x, t)|,$$

$$V_1 = 2u_1(0), \quad V_0 = V_1 + 2u_0(0),$$

$$V_{n+1} = m_0^{3/4} (nb)^{n/\alpha}, \qquad n \geq 1 .$$

PROOF. According to Theorem 1, there exists a unique solution $u(x, t) \in C^\infty([0, T_0); H^\infty(R_1))$. Define the function $y(\xi, t)$ to be the solution of the Cauchy problem for the ordinary differential equation

$$\dot{y} = u(y, t), \qquad y|_{t=0} = \xi \in R_1 .$$

As we know (see Assertion in §2), the function $y(\xi, t)$ realizes a diffeomorphism $y \in R_1 \leftrightarrow \xi \in R_1$. Introduce the notation $v_{(n)}(\xi, t) = u_{(n)}(y(\xi, t), t)$; then $v_n(t) = u_n(t)$.

At $t = 0$ (6.18) holds owing to the equality $Z(0) = 1$ and the conditions 2) and 3) of Theorem 10 ($m_0 \geq c_0 \geq 1$). We prove (6.18) for $n = 0, 1, \ldots, \mathcal{N}$ by contradiction. Let $T_1 \in [0, T_0)$ be the first moment of time when one of the inequalities (6.18) is violated for $n = 0, 1, \ldots, \mathcal{N}$. Then, in view of the continuity on $[0, T_1)$, $n \leq \mathcal{N}$,

$$u_0 \leq V_0, u_{n+1} \leq V_{n+1} Z^{-1-\sigma n} \tag{6.20}$$

and also $0 < Z(t) \leq 2$.

We differentiate the Whitham equation n times with respect to x and pass to the characteristics $x = y(\xi, t)$ to eliminate the term with the highest derivative $uu_{(n+1)}$. As a result, we have

$$\dot{v}_{(n)} + \sum_{j=1}^{n} C_j^n v_{(j)} v_{(n+1-j)} + \beta_n = 0, \tag{6.21}$$

where $\beta_n(\xi, t) = \int_{-\infty}^{\infty} k(y(\xi, t) - s) u_{(n+1)}(s, t)\, ds$; for $n = 0$ the sum in (6.21) is absent. Using (6.20) and the result of Lemma 2 with $\delta = \delta_n = aZ^\sigma 2^{-\sigma} (bn)^{-1/2}$ we estimate the function β_n ($\delta_n \in (0, a]$, since $0 < Z \leq 2$, $\sigma < \frac{1}{\alpha}$, $b \geq 1$). For $n \geq 2$ we have

(6.22)

$$\begin{aligned}
|\beta_n(\xi, t)| &\leq c_1 (\delta_n^{-\alpha} u_n + \delta_n^{1-\alpha} u_{n+1}) \\
&\leq c_1 (2a^{-\alpha} bn Z^{-\sigma\alpha} V_n Z^{-1-\sigma(n-1)} + 2a^{1-\alpha} (bn)^{1-1/\alpha} Z^{\sigma(1-\alpha)} V_{n+1} Z^{-1-\sigma n}) \\
&\leq a^{-\alpha} m_0^{3/4} 2c_1 bn ((n-1)b)^{(n-1)/\alpha} Z^{(1-n-\alpha)\sigma - 1} \left(1 + a \left(\frac{n}{n-1}\right)^{(n-1)/\alpha}\right) \\
&\leq n m_0 ((n-1)b)^{(n-1)/\alpha} Z^{(1-n-\alpha)\sigma - 1} .
\end{aligned}$$

Here and further, we constantly use condition 3) of Theorem 10, i.e., that the value of m_0 is sufficiently large, so that all constants independent of n are majorized by m_0. For

$n = 1$ we obtain

$$|\beta_1(\xi, t)| \le c_1(V_1 a^{-\alpha} 2bZ^{-1-\sigma\alpha} + 2a^{1-\alpha} b^{1-1/\alpha} V_2) \le m_0^{5/4} Z^{-1-\alpha\sigma} . \tag{6.23}$$

For $n = 0$ put $\delta = \delta_0 = \frac{a}{2}Z$ in Lemma 2; then

$$|\beta_0(\xi, t)| \le c_1(V_0 a^{-\alpha} 2^{\alpha} Z^{-\alpha} + a^{1-\alpha} 2^{\alpha-1} Z^{1-\alpha} V_1 Z^{-1-\sigma\alpha}) \le m_0^{1/2} V_0 Z^{-\alpha} . \tag{6.24}$$

It follows from (6.23) that all conditions of Lemma 3 are satisfied for the function $\psi(\xi, t) = v_{(1)}(\xi, t)$ on the interval $[0, T_1]$. Therefore, the results of Lemmas 3, 4, and 5 are true on $[0, T_1]$. According to (6.11) and (6.13) the function $m(t)$ decreases, so that $Z(t) \in (0, 1]$ everywhere on $[0, T_1]$. We now establish the second inequality in (6.18). Substitution of (6.24) in (6.21) results in $|\dot{v}_{(0)}| \le |\beta_0| \le m_0^{1/2} V_0 Z^{-\alpha}$. From this, as in (6.16), we find that

$$|V_{(0)}(\xi, t)| \le |v_{(0)}(\xi, 0)| + V_0 m_0^{1/2} \int_0^t Z^{-\alpha}(\tau)\, d\tau < \tfrac{1}{2} V_0 + \tfrac{1}{2} V_0 (Z^{1-\alpha}(0) - Z^{1-\alpha}(t)) \le V_0 .$$

The bound for u_0 is established. To estimate $u_1(t)$, we note that

$$u_{(1)}(x, t) \ge m(t) = -m_0 Z^{-1}(t) \ge V_1 Z^{-1}(t) .$$

On the other hand, (6.21) and (6.23) imply

$$\dot{v}_{(1)} \le |\beta_1| \le m_0^{5/4} Z^{-2}(t) .$$

Whence, as in (6.16), we obtain

$$v_{(1)}(\xi, t) \le v_{(1)}(\xi, 0) + \int_0^t m_0^{5/4} Z^{-2}(\tau)\, d\tau < V_1 + V_1(Z^{-1}(t) - Z^{-1}(0)) = V_1 Z^{-1}(t),$$

so that the bound (6.18) holds for u_1 as well. Let us verify (6.18) for $n = 1$ (i.e., for u_2). Define $\kappa \in R_1$ by the equality

$$|v_{(2)}(\kappa, T_1)| = v_2(T_1) \equiv \sup_{x \in R_1} |v_{(2)}(x, T_1)| .$$

First we analyze the case $\kappa \notin Q_{1/3}(T_1)$, i.e., $\psi(\kappa, T_1) > \frac{2}{3} m(T_1)$. By continuity

$$\psi(\kappa, t) > \tfrac{2}{3} m(t),$$

for all $t \in [T_2, T_1]$, where $T_2 \in [0, T_1)$ and is sufficiently close to T_1. We can also assume that $v_{(2)}(\kappa, t)$ maintains the same sign on $[T_2, T_1]$. For definiteness, let $v_{(2)}(\kappa, t) > 0$. Then, from (6.21) and (6.22) we find that

$$\dot{v}_{(2)}(\kappa, t) \le -3 v_{(1)} v_{(2)} + |\beta_2| \le (2 + \gamma) m_0 V_2 Z^{-2-\sigma}(t) .$$

Integrating this inequality with respect to t from T_2 to T_1 and using (6.16), we obtain

$$\begin{aligned} v_{(2)}(\kappa, T_1) &\le v_{(2)}(\kappa, T_2) + (2 + \gamma) V_2 (1 - \gamma)^{-3-\sigma} (1 + \sigma)^{-1} (Z^{-1-\sigma}(T_1) - Z^{-1-\sigma}(T_2)) \\ &< V_2 Z^{-1-\sigma}(T_2) + V_2 (Z^{-1-\sigma}(T_1) - Z^{-1-\sigma}(T_2)) < V_2 Z^{-1-\sigma}(T_1), \end{aligned}$$

since $2 + \gamma < (1 + \sigma)(1 - \gamma)^{3+\sigma}$.

Thus, the bound (6.18) with $n = 1$ holds up to time T_1 if $\kappa \notin Q_{1/3}(T_1)$. Let us verify (6.18) for the case $\kappa \in Q_{1/3}(T_1)$. By (6.4) we have $\kappa \in Q_{1/3}(t)$ for all $t \in [0, T_1]$. Integrating (6.21, $n = 0$) with respect to t and changing the variable of integration, we obtain

$$u(y(\xi, t), t) = \bar{u}(\xi) - \int_0^t \int_{-\infty}^{\infty} k(s) u_s(y(\xi, \tau) - s, \tau)\, ds\, d\tau .$$

Whence

$$v_{(2)} y_3^2 + v_{(1)} y_{\xi\xi} = \bar{u}''(\xi) - I_1(\xi, t), \tag{6.25}$$

$$v_{(3)} y_\xi^3 + 3v_{(2)} y_\xi y_{\xi\xi} + v_{(1)} y_{\xi\xi\xi} = \bar{u}'''(\xi) - I_2(\xi, t), \tag{6.26}$$

where

$$I_1 - \int_0^t (y_{\xi\xi}(\xi, \tau)\beta_1(\xi, \tau) + y_\xi^2(\xi, \tau)\beta_2(\xi, \tau))\, d\tau,$$

$$I_2 = \int_0^t (y_\xi^3 \beta_3 + 3\beta_2 y_\xi y_{\xi\xi} + \beta_1 y_{\xi\xi\xi})\, d\tau .$$

We prove the following inequality on $[0, T_1]$:

$$\tfrac{1}{2} Z^{1+2\gamma}(t) \le y_\xi(\kappa, t) \le 2Z^{1-\gamma}(t) . \tag{6.27}$$

For this, we differentiate (6.19) with respect to ξ at $\xi = \kappa$

$$\dot{y}_\xi = \psi(\kappa, t) y_\xi, \qquad y_\xi(\kappa, 0) = 1,$$

and take into account the representation (6.11) and the bounds (6.18) to get

$$(1 - \gamma)\dot{y} y^{-1} \le \dot{y}_\xi y_\xi^{-1} \le (1 + 2\gamma)\dot{y} y^{-1} .$$

From this

$$(Y(\kappa, t)/Y(\kappa, 0))^{1+2\gamma} \le y_\xi(\kappa, t) \le (Y(\kappa, t)/Y(\kappa, 0))^{1-\gamma} .$$

Now using (6.14), we finally arrive at (6.27).

We establish the following bounds on $[0, T_1]$:

$$|y_{\xi\xi}(\kappa, t)| < \frac{V_2}{\gamma m_0} Z^{2-\sigma-2\gamma}, \qquad |y_{\xi\xi\xi}(\kappa, t)| < \gamma m_0^{-2} V_2^2 . \tag{6.28}$$

Let there be, to the contrary, a first moment of time $T_2 > 0$ when there is equality in at least one of the bounds (6.28). For the function $y_{\xi\xi}$ from (6.25) we have

$$|y_{\xi\xi}| \le |v_{(1)}|^{-1}(|v_{(2)}| y_\xi^2 + |\bar{u}''(\kappa)| + |I_1(\kappa, t)|) . \tag{6.29}$$

Let us estimate the integral I_1 with the aid of (6.16), (6.21), (6.22), and (6.27):

$$\begin{aligned} |I_1| &\le \int_0^t (|y_{\xi\xi}|\, |\beta_1| + y_\xi^2 |\beta_2|)\, d\tau \\ &\le \int_0^t \left(\frac{V_2}{\gamma m_0} Z^{2-\sigma-2\gamma} m_0^{5/4} Z^{-1-\alpha\sigma} + 4Z^{2-2\gamma} 2m_0 b^{1/\alpha} Z^{-\sigma-1-\alpha\sigma} \right) d\tau \\ &\le \gamma^2 V_2 m_0 \int_0^t Z^{-\sigma+2\gamma}\, d\tau \le \gamma V_2 Z^{1-\sigma+2\gamma}, \end{aligned} \tag{6.30}$$

since $1 - 2\gamma - \alpha\sigma \ge 2\gamma$. Further, the condition $\kappa \in Q_{1/3}(t)$ implies $|m(t)| \ge |v_{(1)}(\kappa, t)| \ge$

$\frac{2}{3}|m(t)|$. Therefore, in view of (6.20), (6.30), and the condition $\overline{u}(x) \in \mathscr{G}_6^2$, from (6.29) we obtain

$$|y_{\xi\xi}| < \tfrac{3}{2} Z m_0^{-1}(V_2 Z^{-1-\sigma} 4Z^{2-2\gamma} + b^{1/\alpha} + \gamma V_2 Z^{1-\sigma+2\gamma}) \\ < V_2(\gamma m_0)^{-1} Z^{2-\sigma-2\gamma}.$$

Now we verify the bound (6.28) for $y_{\xi\xi\xi}$. Differentiating (6.10) three times with respect to ξ and using (6.26) we get

$$\dot{y}_{\xi\xi\xi} = \overline{u}'''(\xi) - I_2(\xi, t), \qquad y_{\xi\xi\xi}(\xi, 0) = 0,$$

whence

$$|y_{\xi\xi\xi}| \le \int_0^t (|\overline{u}'''(\xi)| + |I_2(\xi,\tau)|)\, d\tau. \tag{6.31}$$

We estimate the integral I_2 using (6.16), (6.21), (6.27), and the nonstrict inequality (6.28):

$$\begin{aligned} |I_2(\kappa, t)| \le \int_0^t & d\tau (8Z^{3-3\gamma} 3m_0 (2b)^{2/\alpha} Z^{-2\sigma-\alpha\sigma-1} \\ & + 6Z^{1-\gamma} V_2 (\gamma m_0)^{-1} Z^{2-\sigma-2\gamma} 2m_0 b^{1/\alpha} Z^{-1-\sigma-\sigma\alpha} + \gamma V_2^2 m_0^{-2} m_0^{5/4} Z^{-1-\alpha\sigma}) \\ \le \int_0^t & \gamma^4 V_2^2 Z^{\gamma-2}\, d\tau < \gamma^3 V_2^2 m_0^{-1} Z^{\gamma-1}, \end{aligned}$$

since $2 - 2\sigma - \alpha\sigma - 3\gamma \ge \gamma - 2$. Substitution of (6.32) in (6.31) yields

$$|y_{\xi\xi\xi}| \le \int_0^t (b^{2/\alpha} + \gamma^3 V_2^2 m_0^{-1}) Z^{\gamma-1}\, d\tau < \gamma V_2^2 m_0^{-2}.$$

Thus, the bounds (6.28) are proved on the whole interval $[0, T_1]$. From the necessary condition for extremum we have

$$v_{(3)}(\kappa, T_1) = u_{(3)}(y(\kappa, T_1), T_1) = 0;$$

therefore, putting together (6.20), (6.25), (6.26), (6.30), (6.32) we obtain

$$\begin{aligned} v_{(2)}^2 &\le v_{(1)}^2 |y_{\xi\xi\xi}| y_\xi^{-3} + |\overline{u}'''(\kappa)|\, |v_{(1)}| y_\xi^{-3} \\ &\quad + |v_{(1)}| y_\xi^{-3} |I_2| + |v_{(2)}| y_\xi^{-2} (|\overline{u}''(\kappa)| + |I_1|) \\ &\le 8Z^{-3-6\gamma} \gamma V_2^2 m_0^{-2} Z^{-2} m_0^2 + 8Z^{-3-6\gamma} (2b)^{2/\alpha} m_0 Z^{-1} \\ &\quad + m_0 8Z^{-4-6\gamma} \gamma^3 V_2^2 m_0^{-1} Z^{\gamma-1} + V_2 Z^{-1-\sigma} 4Z^{-2-4\gamma} (b^{1/2} + \gamma V_2 Z^{1-\sigma+2\gamma}) \\ &< V_2^2 Z^{-2-2\sigma}, \end{aligned}$$

since $-5 - 6\gamma > -2 - 2\sigma$. We conclude that the inequality (6.18) for $n = 1$ is strict up to and including the moment of time $t = T_1$.

We prove the bounds (6.18) for $n = 2, 3, \ldots, \mathscr{N} - 1$ by induction. Introduce $\overline{\kappa}$ by the equality $v_n(T_1) = |v_{(n)}(\overline{\kappa}, T_1)|$. By continuity, one can find a sufficiently close moment of time $T_2 \in (0, T_1)$ such that $v_{(n)}(\overline{\kappa}, t)$ maintains its sign on $[T_2, T_1]$. To be specific, let $v_{(n)}(\overline{\kappa}, t)$ be positive. Then from (6.21) for $\xi = \overline{\kappa}$ and $t \in [T_2, T_1]$ we obtain

$$\dot{v}_{(n)} \leq -(n+1)m(t)v_{(n)} + \sum_{j=2}^{n-1} C_j^n v_j v_{n+1-j} + |\beta_n| . \tag{6.33}$$

Applying the results of Lemma 1 and inequality (6.20), we estimate the sum on the right-hand side of (6.33) as follows:

(6.34)

$$\begin{aligned}
&\sum_{j=?}^{n-1} C_j^n v_j v_{n+1-j} \\
&\quad \leq \sum_{j=2}^{n-1} C_j^n (b(j-1))^{(j-1)/\alpha} Z^{-(j-1)\sigma-1} m_0^{3/4} (b(n-j))^{(n-j)/\alpha} m_0^{3/4} Z^{-(n-j)\sigma-1} \\
&\quad \leq b^{(n-1)/\alpha} Z^{(1-n)\sigma-2} \sum_{j=2}^{n-1} C_j^n (j-1)^{(j-1)/\alpha} (n-j)^{(n-j)/\alpha} m_0^{3/2} \\
&\quad < V_n n \gamma^3 m_0 Z^{(1-n)\sigma-2} .
\end{aligned}$$

Substitution of (6.34) and (6.21) in (6.33) yields

$$\dot{v}_{(n)} \leq m_0(n + \gamma^2 n + 1) V_n Z^{(1-n)\sigma-2} . \tag{6.35}$$

The bound (6.16) turns out to be coarse now, since it gives an increasing factor of the form $(1+\gamma)^n$. Let us argue differently. In view of the continuity of the functions $Z(t)$ and $Y(\xi_1, t)$, it is possible to choose a moment of time T_3 so close to T_1 that for all $t \in [T_3, T_1]$

$$Z(t) \leq |y(\xi_1, t)| \leq (1+\gamma)^{1/\mu} Z(t),$$

with $Z(T_3) = |Y(\xi_1, T_3)|$; $\mu = 2 - (1-n)\sigma$, $n \geq 3$. Hence, on the interval $[T_3, T_1]$, we approximate a generally nondifferentiable function $Z(t)$ by a definitely smooth function $Y(\xi_1, t)$ with a small error. Using the bounds from Lemma 4 we get

$$\begin{aligned}
\int_{T_3}^{T_1} Z^{-\mu}(\tau)\, d\tau &\leq \int_{T_3}^{T_1} (1+\gamma)|Y(\xi_1, t)|^{-\mu}\, dt \\
&\leq -\frac{1+\gamma}{(1-\gamma)m_0} \int_{T_3}^{T_1} |Y(\xi_1, t)|^{-\mu}\, \dot{Y}(\xi_1, t)\, dt \\
&\leq \frac{1+\gamma}{(1-\gamma)(1-\mu)m_0} (|Y(\xi_1, T_3)|^{1-\mu} - |Y\xi_1, T_1|^{1-\mu}) \\
&\leq \frac{1+\gamma}{(1-\gamma)(\mu-1)m_0} (Z^{1-\mu}(T_1) - Z^{1-\mu}(T_3)) .
\end{aligned}$$

By (6.36) and the bound (6.18), valid for $t = T_3$, we find from (6.35) that

$$v_{(n)}(\overline{\kappa}, T_1) \le v_{(n)}(\overline{\kappa}, T_3) + m_0(n + \gamma^2 n + 1) V_n \int_{T_3}^{T_1} Z^{(1-n)\sigma-2}(t)\,dt$$
$$\le V_n Z^{(1-n)\sigma-1}(T_3) + \frac{(1+\gamma)(n+\gamma n+1)m_0 V_n}{(1-\gamma)(1-\sigma+n\sigma)m_0} \cdot (Z^{(1-n)\sigma-1}(T_1) - Z^{(1-n)\sigma-1}(T_3))$$
$$< V_n Z^{(1-n)\sigma-1}(T_1),$$

since $(1+\gamma)(n+\gamma n+1) < (1-\gamma)(\sigma n+1-\sigma)$ for $n \ge 3$. Thus, (6.18) is established by induction for all $t \in [0, T_1]$ and $1 \le n < \mathcal{N}$. Now we prove (6.18) for $n \ge \mathcal{N}$. Arguing as above, we shall arrive at (6.33) from (6.21). This time we estimate the function $\beta_n(\xi, t)$ using (6.17):

$$|\beta_n(\xi, t)| \le \mathcal{N} v_n(t) \le n v_n(t),$$

since $n \ge \mathcal{N}$. Therefore, we again arrive at the inequality (6.35). All further arguments remain valid. Thus, all bounds (6.18) are established for $t \in [0, T_1]$. The contradiction obtained proves (6.18) on the whole interval $[0, T_0)$.

In order to obtain (5.5), we note that (6.18) with $n = 0$ and the equality $m(t) = -m_0 Z^{-1}(t)$ imply that the time of breaking T_0 is determined by the condition

$$Z(t) \to 0, \qquad t \to T_0\,.$$

From (6.11) and (6.12) we conclude that for any $x \in Q_j(t)$

$$m_0(\overline{\psi}^{-1}(x) + (1-\gamma)t) \le Y(x, t) \le m_0(\overline{\psi}^{-1}(x) + (1+\gamma)t),$$

where $\overline{\psi}(x) = \overline{u}'(x)$. From (6.4) we have $x \in Q_j(0)$ and, therefore, $-m_0 \le \overline{\psi}(x) \le -(1-\gamma)m_0$. Then

$$-(1-\gamma)^{-1} + m_0(1-\gamma)t \le Y(x, t) \le -1 + m_0(1+\gamma)t$$

and from (6.14) we obtain

$$1 - m_0(1+\gamma)t \le Z(t) \le (1 - m_0(1-\gamma)^2 t)(1-\gamma)^{-1},$$

whence (5.5) follows. Lemma 6 is proved. □

§7. Proof of Theorem 10

We define a sequence of regular kernels $k^{(l)}(x) \in C^1(R_1) \cap W_1^1(R_1)$ satisfying (5.1), (5.2) and such that $k^{(l)}(x) \to k(x)$ in the norm of $L_1(R_1)$ as $l \to \infty$.

For the Cauchy problem (1.1) with the regular operator $\mathbb{K}^{(l)}$ corresponding to the kernel $k^{(l)}(x)$, Lemma 6 guarantees the existence of solutions $u^{(l)}(x, t) \in C^\infty([0, T_0^{(l)}); H^\infty(R_1))$ for all l and provides l-uniform bounds for the solutions $u^{(l)}(x, t)$ and their breaking times $T_0^{(l)}$. The bounds (5.5) imply the possibility of extracting from the sequence of breaking times $\{T_0^{(l)}\}$ a subsequence of at least one of the following types: a monotone nondecreasing or a decreasing subsequence, converging to $T_0 = \liminf_{l\to\infty} T_0^{(l)}$. To avoid changing the notation, we assume that the whole sequence $\{T_0^{(l)}\}$ is of one of these types. We consider first the case of a monotone nondecreasing sequence $\{T_0^{(l)}\}$. Fix an arbitrary number $j > 0$. There is the number $L > 0$ such that, for all $i, l \geq L$,

$$\int_{-\infty}^{\infty} |k^{(l)}(x) - k^{(i)}(x)|\, dx \leq \frac{j}{2V_3}\left(\frac{m_0}{2j}\right)^{\sigma+\rho}, \tag{7.1}$$

where $\rho = u_1(0)/\gamma m_0$. Consider the sequence $\{u^{(l)}(x,t)\}$ from the index L on. Since $T_0^L \leq T_0$, for some $t_j \in (0, T_0)$ we have

$$m^{(L)}(t_j) = -2j\,. \tag{7.2}$$

The sequence $\{Z^{(l)}(t_j)\}$ is bounded from below uniformly with respect to $l \geq L$, for otherwise, in view of the bounds, from Lemma 4 we would have $t_j = T_0$, which is impossible. Therefore, the sequence $\{u^{(l)}(x,t)\}$, starting with the index L, has a common domain of definition $R_1 \times [0, t_j]$, on which, according to Lemma 6, this sequence is bounded in $C^\infty([0,t_j]; C^\infty(R_1))$ uniformly with respect to $l \geq L$. Thus, some subsequence of it converges to a limit $u(x,t) \in C^\infty(R_1 \times [0,t_j])$, which is a solution of the Cauchy problem (1.1) with the original operator $\mathbb{K}$. To prove that

$$m(t_j) < -j \tag{7.3}$$

we show that the sequence $\{u^{(l)}\}$ is fundamental ($\equiv$ Cauchy) on $[0, T_0^{(L)})$ for $l \geq L$. Let

$$p_{(n)} = u_{(n)}^{(l)} - u_{(n)}^{L}, \qquad p_n = \max_{x\in R_1} |p_{(n)}|,$$

$$\varepsilon = \int_{-\infty}^{\infty} |k^{(l)}(x) - k^{(L)}(x)|\, dx, \qquad U_{(n)} = \tfrac{1}{2}(u_{(n)}^{(l)} + u_{(n)}^{(L)}), \quad n \geq 0\,.$$

We define the function $\eta(\xi,t)$ as a solution of the Cauchy problem (6.19) with right-hand side $u_0(x,t)$. Then, for the functions $r_{(n)}(\xi,t) = p_{(n)}(\eta(\xi,t),t)$, $n \geq 0$, from the Whitham equation we find that

$$\dot r_{(n)} = -\sum_{i=0}^{n} C_j^n U_{(n+1-i)} r_{(i)} - I_n - J_n, \qquad r_{(n)}(\xi,0) = 0, \tag{7.4}$$

where

$$I_n = \frac{1}{2}\int_{-\infty}^{\infty} (k^{(l)}(s) + k^{(L)}(s)) p_{(n+1)}(\eta(\xi,t) - s, t)\, ds,$$

$$J_n = \int_{-\infty}^{\infty} (k^{(l)}(s) - k^{(L)}(s)) U_{(n+1)}(\eta(\xi,t) - s, t)\, ds\,.$$

We prove for all $t \in [0, T_0^{(L)})$, $n \geq 0$:

$$r_n < \varepsilon U_{n+2} Z^{-n\sigma-\rho}, \tag{7.5}$$

where $Z(t) = \min(Z^{(l)}(t), Z^{(L)}(t))$. Let $\mathscr{N} = \max(\mathscr{N}^{(l)}, \mathscr{N}^{(L)})$, where the constants $\mathscr{N}^{(l)}$ and $\mathscr{N}^{(L)}$ are taken from the bound (6.17) for the regular operators $\mathbb{K}^{(l)}$ and $\mathbb{K}^{(L)}$. We prove (7.5) by contradiction, first for $n = 0, 1, \ldots, \mathscr{N}$. Let T_1 be the first moment of time, when there is equality in one of the inequalities (7.5) with $n = 0, 1, \dots, \mathcal{N}$. Then, using (6.18), on the interval $[0, T_1]$ we obtain

$$|J_n| \le \varepsilon U_{n+1} \le \varepsilon V_{n+1} Z^{-n\sigma-\rho}(t)\,. \tag{7.6}$$

With the choice of $\delta = a(Z/b_n)^{1/\alpha}$ in Lemma 2, for $n = 0, 1, \dots, \mathcal{N} - 1$ we have

$$\begin{aligned} I_n &\le c_1(\delta^{-\alpha} r_n + \delta^{1-\alpha} r_{n+1}) \\ &\le c_1(a^{-\alpha} bnZ^{-1} \varepsilon V_{n+2} Z^{-n\sigma-\rho} + a^{1-\alpha}(bn)^{1-1/\alpha} Z^{1/\alpha-1}(t)\varepsilon V_{n+3} Z^{-\sigma-n\sigma-\rho}(t)) \\ &\le \gamma^2 m_0 \varepsilon n V_{n+2} Z^{-n\sigma-\rho-1}(t)\,. \end{aligned} \tag{7.7}$$

Using Lemma 1, (7.5), and (6.18), we estimate the sum in (7.4):

$$\begin{aligned} \left|\sum_{i=0}^{n-1} C_i^{n+1} U_{(n+1-i)} r_{(i)}\right| &\le \varepsilon \sum_{i=0}^{n-1} C_i^{n+1} V_{n+1-i} V_{i+2} Z^{-n\sigma-\rho-1} \\ &\le m_0^{3/2} \varepsilon b^{(1+n)/\alpha} Z^{-n\sigma-\rho-1} \sum_{i=0}^{n-1} C_i^{n+1} (n-i)^{(n-i)/\alpha} (i+1)^{(i+1)/\alpha} \\ &\le \gamma^2 m_0 n \varepsilon V_{n+2} Z^{-n\sigma-\rho-1}(t)\,. \end{aligned} \tag{7.8}$$

Substitution of (7.6) through (7.8) in (7.4) yields

$$|\dot r_{(n)} + (n+1) U_{(1)} r_{(n)}| \le \gamma \varepsilon m_0 n V_{n+2} Z^{-n\sigma-\rho-1}\,. \tag{7.9}$$

If $Z^{(l)}(T_1) = Z(T_1)$ (the case $Z(T_1) = Z^{(L)}(T_1)$ is completely analogous), we select an $x_1 \in Q_\gamma^{(l)}(T_1)$ such that $|y^{(l)}(x_1, T_1)| = Z(T_1)$. In view of the continuity of the functions $Y^{(l)}(x_1, t)$ and $Z(t)$, we can find $T_2 \in (0, T_1)$ such that for $t \in [T_2, T_1]$:

$$Z(t) \le |Y^{(l)}(x_1, t)| \le (1+\gamma)^{1/\nu} Z(t),$$

where $\nu = n\sigma + \rho + 1$. Moreover, one can also suppose, for instance, that the function $r_{(n)}(\xi, t)$ is nonnegative on $[T_2, T_1]$. Hence, (7.9) leads to

$$\dot r_{(n)} \le (n + \gamma n + 1) m_0 \varepsilon V_{n+2} Z^{-n\sigma-\rho-1},$$

whence, integrating with respect to t and using (6.33) for $\mu = \nu$, we get

$$\begin{aligned} r_{(n)}(\xi, T_1) &\le r_{(n)}(\xi, T_2) + \varepsilon V_{n+2} \frac{(1+\gamma)(n+\gamma n+1)}{(1-\gamma)(n\sigma+\rho)} (Z^{-n\sigma-\rho}(T_1) - Z^{-n\sigma-\rho}(T_2)) \\ &< \varepsilon V_{n+2} Z^{-n\sigma-\rho}(T_1)\,. \end{aligned}$$

Thus, (7.5) is established for $0 \le n < \mathcal{N}$ by induction. In the proof of (7.5) with $n \ge \mathcal{N}$, we use (6.17) to estimate the integral I_n:

$$|I_n| \le \mathcal{N} U_n \le n \varepsilon V_{n+2} Z^{-n\sigma-\rho}\,.$$

Therefore, we again obtain (7.9), from which (7.5) is derived for $n \ge \mathcal{N}$. The contradiction obtained proves (7.5) on $[0, t_j]$.

Take the bound (7.5) with $n = 1$ and let $l \to \infty$:

(7.10)
$$\sup_{x \in R_1} |u_x(x,t) - u_x^{(L)}(x,t)| < \varepsilon V_3 \widetilde{Z}^{-\sigma-\rho}(t), \\ \widetilde{Z} = \min(Z^{(\infty)}, Z^{(L)}), \qquad t \in [0, T_0^{(L)}) .$$

We prove (7.3) by contradiction. Let

$$Z^{(\infty)}(t_j) = m_0 |m^{(\infty)}(t_j)|^{-1} \geq \frac{m_0}{j} .$$

Then, from (7.2) we obtain

(7.11)
$$\widetilde{Z}(t_j) = Z^{(L)}(t_j) = \frac{m_0}{2j} .$$

Substituting (7.2) and (7.11) in (7.10), by (7.1) we find that

$$m(t_j) \leq m^{(L)}(t_j) + \varepsilon V_3 \left(\frac{m_0}{2j} \right)^{-\sigma-\rho} < -j$$

and we have (7.3).

Now letting $j \to \infty$, we find that $t_j \to T_0$, $\min_{x \in R_1} u_x(x,t) \to \infty$ as $t \to T_0$, i.e., the solution $u(x,t)$ breaks at the moment of time T_0.

In the case of a decreasing subsequence $\{T_0^{(l)}\}$, the existence of a solution $u(x,t)$ on $[0, T_0)$ is considered in the same way as above. We prove, by contradiction, that the solution breaks as $t \to T_0$. Let u_x be bounded, i.e.,

(7.12)
$$\max_{x \in R_1} |u_x(x, T_0)| < j$$

for some j. Since $u_x^{(l)} \rightrightarrows u_x$, and $T_0^{(l)} \to T_0$, for a sufficiently large L we have

(7.13)
$$u_1^{(L)}(T_0) < 2j \quad \text{and} \quad T_0^L - T_0 \leq \frac{1}{8j} .$$

Let us return to the estimates of the breaking time $T_0^{(L)}$, derived in Lemma 6. Fix an arbitrary $t \in (T_0, T_0^{(L)})$. Then, for any $x \in Q_\gamma^{(L)}(t)$, from Lemma 4 we get

$$|Y^{(L)}(x, T_0)| - m_0(1+\gamma)(t - T_0) \leq |Y^{(L)}(x,t)|,$$

whence

$$Z^{(L)}(t) \geq (1-\gamma)(Z^{(L)}(T_0) - m_0(1+\gamma)(t - T_0)) .$$

It follows from (7.13) that $Z^{(L)}(T_0) > m_0/2j$, and, thus, $Z^{(L)}(T_0^{(L)}) > (1-\gamma)m_0/4j > 0$. But this contradicts the fact that $T_0^{(L)}$ is the time of breaking. Thus, (7.12) is impossible for any j, and the solution $u(x,t)$ breaks at the moment of time T_0. Theorem 10 is proved.

CHAPTER 3

The Periodic Problem

§1. Introduction

In the case where the wave medium has a periodic structure, it is natural to consider the Whitham equation with the integral evaluated over the period. Our goal is to study the following Cauchy problem:

$$u_t + uu_x + \int_{-L}^{L} k(x-s)u_s(s,t)\,ds = 0, \qquad u|_{t=c} = \bar{u}(x) \tag{1.1}$$

where $k(x)$ is a $2L$-periodic real function and $\bar{u}(x)$ is a $2L$-periodic initial perturbation. It is possible to go over from the representation of the operator $\mathbb{K}(u)$ in the form of a convolution (1.1) to a more general definition of $\mathbb{K}(u)$ as a pseudodifferential operator

$$\mathbb{K}(u) = \frac{1}{2L}\sum_{p=-\infty}^{\infty} K_p \hat{u}_p e^{ipx\pi/L}, \tag{1.2}$$

where $u(x)$ is any function in $C_L^\infty(R_1) = \{u(x) \in C^\infty(R_1); u(x) = u(x+2L)\}$, and $\hat{u}_p$ are its Fourier coefficients. To simplify notation we assume below that $L = \pi$.

So far, little literature has been devoted to the periodic problem (1.1) (see **[10]** and **[82]**), where only very general observations concerning the properties of solutions of problem (1.1) can be found for the case of a smooth kernel $k(x)$. In this chapter we present the following results. For a conservative or dissipative operator $\mathbb{K}$ of order less than $3/5$ (the order of $\mathbb{K}$ is $\inf \alpha$, where α is such that $|pK_p| \le c(1+|p|)^\alpha$, $p \in z$, $\alpha \ge 0$, $c > 0$), we prove a theorem on the wave breaking in finite time provided the steepness of the initial perturbation is sufficiently large. For a dissipative operator $\mathbb{K}$ of any nonnegative order there exists a smooth solution to problem (1.1) for all times, provided the steepness of the initial perturbation is sufficiently small. If the operator $\mathbb{K}$ is strongly dissipative of order greater than $\frac{3}{2}$, then there is a global solution of problem (1.1) for any initial perturbation in L_2, and smoothing of the initial perturbation occurs. Finally, we consider the periodic problem for the Whitham equation with a weak interaction. These results have been published earlier (**[51, 54, 58, 72]**).

§2. Breaking of waves for a conservative or dissipative operator $\mathbb{K}$ of order $\alpha < 3/5$

This section is devoted to the proof of a theorem on breaking in finite time of waves, described by problem (1.1) with a conservative or dissipative operator $\mathbb{K}$ of order less than $\frac{3}{5}$, in the case where the initial perturbation $\bar{u}(x) \in C_\pi^\infty(R_1)$ is sufficiently steep. We introduce the notation:

$$\Omega = [-\pi, \pi], \quad m_0 = \left|\min_{x\in\Omega} \bar{u}'(x)\right|, \quad u_1 = \max_{x\in\Omega} |\bar{u}'(x)|,$$

$$K_p = \int_\Omega e^{-ipx} k(x)\,dx, \quad K_p^1 = \operatorname{Re} ipK_p, \quad K_p^2 = \operatorname{Im} ipK_p,$$

and for a smooth function $u(x,t)$

$$\begin{aligned}
&u_{(j)}(x,t) = \frac{\partial^j u(x,t)}{\partial x^j}^{x,t}, \qquad u_j(t) = \max_{x\in\Omega} |y_{(j)}(x,t)|,\\
(2.1)\qquad &J_j(t) = \int_\Omega u_{(j)}^2(x,t)\,dx, \qquad I_j(t) = \int_\Omega u_{(j)}(x,t)\mathbb{K}(u_{(j)})\,dx,\\
&\qquad j = 0, 1, 2, \dots.
\end{aligned}$$

THEOREM 1. *Let the following conditions hold*:

1) $k(x) \in C^1(\Omega\backslash 0)$, $|k(x)| \le C|x|^{-\alpha}$, $|k'(x)| \le c|x|^{-1-\alpha}$, $x \in \Omega\backslash 0$, $\alpha \le \frac{3}{5} - \gamma$, $\gamma \in (0, \frac{1}{10}]$, $c > 0$, $K_p^1 \ge 0$ *for* $|p| \ge p_0 > 0$, $p \in Z$;

2) $\bar{u}(x) \in C_\pi^\infty(R_1)$;

3) *the steepness* m_0 *of the initial perturbation is sufficiently large*

$$m_0^2 > \frac{b^2}{\gamma^2} + \frac{40c}{\gamma}(u_1 + J_3^{1/2}(0)), \qquad b = 2\max_{|p|\le p_0} |K_p^1|\,.$$

Then the solution $u(x,t)$ *of problem* (1.1) *belongs to the class* $C^\infty([0,T_0); C_\pi^\infty(R_1))$, *it is unique and breaks at time* $T_0 \le \frac{1}{m_0(1-\gamma)}$.

REMARK. Theorem 1 includes the case of symmetric kernels $k(x) = k(-x)$, important for physical applications, because the nonconservative part of the symbol K_p is equal to zero in this case: $K_p^1 = 0$, $p \in Z$.

We divide the proof of Theorem 1 into several steps.

2.1. Local (in time) existence of a solution of problem (1.1). We consider the Cauchy problem for the ordinary differential equation, $t \ge 0$,

$$\begin{aligned}
(2.2)\qquad & \begin{cases} \dot{y} = u(y,t),\\
y|_{t=0} = \xi \in R_1\,. \end{cases}\\
(2.3)\qquad &
\end{aligned}$$

LEMMA 1. *Let* $u(y,t) \in C^\infty([0,T]; C_\pi^\infty(R_1))$ *for some* $T > 0$. *Then,*

1) *There exists a unique solution* $y(\xi,t) \in C^\infty(R_1 \times [0,T])$ *of problem* (2.2), (2.3);

2) *For any* $x \in R_1$ *there exists a unique solution* $\xi(x,t)$ *of the equation* $y(\xi,t) = x$ *for each* $t \in [0,T]$, *so that the solution* $y(\xi,t)$ *of problem* (2.2), (2.3) *is a diffeomorphism* $R_1 \leftrightarrow R_1$ *for each* $t \in [0,T]$.

PROOF. Both assertions of the lemma follow immediately from the general theory of ordinary differential equations. □

We note that the function $y(\xi,t) + 2\pi k = y(\xi + 2\pi k, t)$, $k \in Z$, is also a solution of problem (2.2), (2.3).

In the next lemma we prove that the Cauchy problem (1.1) with a regular operator $\mathbb{K}$ has a solution $u(x,t) \in C^\infty([0,T_0); C_\pi^\infty(R_1))$, $T_0 > 0$, which can break down at time T_0 only as a result of overturning, i.e., $\max_{x\in\Omega} |u_{(1)}(x,t)| \to \infty$ as $t \to T_0 < \infty$.

An operator $\mathbb{K}$ is called regular if its kernel $k(x)$ has no singularity at $x = 0$, or, more precisely, if, in addition to condition 1) of Theorem 1, the regular kernel $k(x)$

satisfies the condition

$$|k(-0)| + |k(+0)| + \int_{-\pi}^{\pi} |k'(x)|\, dx < \mathcal{N}, \tag{2.4}$$

where $\mathcal{N}$ is some natural number.

LEMMA 2. *Let the kernel $k(x)$ be regular, and $\bar{u}(x) \in C_\pi^\infty(R_1)$. Then there is a $T_0 > 0$ such that there exists a solution $u(x,t)$ of the Cauchy problem* (1.1) *belonging to the class $C^\infty([0, T_0); C_\pi^\infty(R_1))$, and this solution $u(x,t)$ can break down only as a result of overturning.*

PROOF. The method of successive approximations will be used to prove the existence of a solution. Let $W^{(0)}(x,t) = \bar{u}(x)$; we define the functions $w^{(s+1)}(x,t)$ for $s \geq 0$ to be the solutions of the following problem:

$$w_t^{(s+1)} + w^{(s)} w_x^{(s+1)} + \mathbb{K}(w^{(s)}) = 0, \qquad w^{(s+1)}|_{t=0} = \bar{u}(x)\,. \tag{2.5}$$

In other words, they are solutions of a linear partial differential equation. Assume, by induction, that $w^{(s)}(x,t) \in C^\infty(R_+, C_\pi^\infty(R_1))$ has already been determined and, observe that due to the regularity of its kernel $\mathbb{K}(w^{(s)}) \in C^\infty(R_+, C_\pi^\infty(R_1))$; we conclude that the solution of problem (2.5), i.e., the function $w^{(s+1)}(x,t)$, also belongs to $C^\infty(R_+, C_\pi^\infty(R_1))$. Let us prove that for $T > 0$ the sequence $\{w^{(s)}(x,t)\}$ is uniformly bounded with respect to $s \geq 0$ in $C^\infty([0,T]; C_\pi^\infty(R_1))$. For this we differentiate (2.5) n times with respect to x and transfer the derivative to the kernel $k(x)$ in the integral term:

$$\begin{aligned} &\frac{\partial}{\partial t} w_{(n)}^{(s+1)} + \sum_{j=0}^{n} C_j^n w_{(j)}^{(s)} w_{(n+1-j)}^{(s+1)} \\ &\quad + k(x-y) w_{(n)}^{(s)}(y,t)\Big|_{y=x+0}^{y=x-0} + \int_{-\pi}^{\pi} k'(x-y) w_{(n)}^{(s)}(y,t)\, dy = 0\,. \end{aligned} \tag{2.6}$$

We define the function $y^{(s)}(\xi, t)$ to be the solution of problem (2.2), (2.3) with right-hand side equal to $w^{(s-1)}(x,t)$. According to Lemma 1, the function $y^{(s)}(\xi,t)$ establishes a diffeomorphism $R_1 \to R_1$ for each $t \in [0,T]$. Therefore, having defined the sequence of new functions $v_{(n)}^{(s)}(\xi,t) \equiv w_{(n)}^{(s)}(y^{(s)}(\xi,t),t)$, we have $v_n^{(s)}(t) = w_n^{(s)}(t)$ (see the notation 2.1). In view of (2.4), we find from (2.6) that

$$\begin{cases} |\dot{v}_{(0)}^{(s)}| \leq \mathcal{N} v_0^{(s-1)}, \\ |\dot{v}_{(n)}^{(s)}| \leq \sum_{j=1}^{n} C_j^n v_j^{(s-1)} v_{n+1-j}^{(s)} + \mathcal{N} v_n^{(s-1)}, \qquad n \geq 1, \\ v_{(n)}^{(s)}|_{t=0} = \bar{u}_{(n)}, \qquad n \geq 0, \end{cases} \tag{2.7}$$

since $\dot{v}_{(n)}^{(s+1)} = \dot{w}_{(n)}^{(s+1)} + w_{(n+1)}^{(s+1)} \dot{y}^{(s+1)} = \dot{w}_{(n)}^{(s+1)} + w_{(n+1)}^{(s+1)} w^{(s)}$.

We also analyze the system of equations for the functions $P_n(t)$ on the segment $[0,T]$

$$\begin{cases} \dot{P}_0 = \mathcal{N} P_0, \\ \dot{P}_n = 2\sum_{j=1}^{n} C_j^n P_j P_{n+1-j} + \mathcal{N} P_n, \qquad n \geq 1, \\ P_n(0) = \bar{u}_n, \qquad n \geq 0. \end{cases} \tag{2.8}$$

Here $T > 0$ determines the time of existence of a bounded solution of the system (2.8), i.e., $P_n(t) \in C^1([0,T])$ for all $n \geq 0$. In system (2.8) only the equation for $n = 1$ is nonlinear; all the others are linear. Integrating the equation with $n = 1$ we get

$$P_1(t) = \mathcal{N} P_1(0) e^{\mathcal{N} t}\{\mathcal{N} + 2P_1(0)(1 - e^{\mathcal{N} t})\}^{-1},$$

i.e., $P_1(t)$ remains bounded together with all of its derivatives with respect to t on $[0, t]$ if T is chosen from the condition $0 < T < \frac{1}{\mathcal{N}} \ln(1 + \mathcal{N}/2P_1(0))$. Integration of the remaining equations of (2.8) on $[0, T]$ results in the following equalities for some collection of numbers $\mathcal{D}_n > 0$:

(2.9) $$0 \leq P_n(t) \leq \mathcal{D}_n, \quad n \geq 0,\ t \in [0, T].$$

We now prove that the solution $P_n(t)$ of the system (2.8) majorizes the functions $v_{(n)}^{(s)}(x,t)$ on $[0, T]$ for all $n, s \geq 0$:

(2.10) $$v_n^{(s)}(t) \leq P_n(t).$$

Indeed, the inequalities (2.10) hold for $t = 0$, and also for all $n \geq 0$, $t \in [0, T]$, if $s = 0$. We carry out induction on $s \geq 0$. For $s > 0$ we prove (2.10) by contradiction. Let n_1 be the smallest index for which (2.10) is not satisfied. We denote by T_1 the last time $0 < T_1 \leq T$ when $v_{n_1}^{(s)}(T_1) \leq P_{n_1}(T_1)$. Since the functions $v_n^{(s)}$, P_n are continuous and nonnegative for all $n \leq n_1$, there exists a time T_2, $T_2 > T_1 > 0$ such that

(2.11) $$v_n^{(s)}(t) \leq 2P_n(t) \quad \text{for } n_1 \geq n,\ t \in [0, T_2].$$

Substituting (2.11) and (2.10), valid by the induction hypothesis for the index $(s-1)$, in (2.7), we obtain

$$|\dot{v}_{(n)}^{(s)}(\xi, t)| \leq \dot{P}_n(t), \quad n \leq n_1,\ \xi \in R_1,\ t \in (0, T_2],$$

$$|v_{(n)}^{(s)}||_{t=0} \leq P_n|_{t=0} = \bar{u}_n,$$

hence, (2.10) follows immediately for $t \in [0, T_2]$ for all $n \leq n_1$. The contradiction obtained proves (2.10) for all $n, s \geq 0$ and $t \in [0, T]$.

It follows from (2.9) and (2.10) that the $v_n^{(s)}$ are uniformly bounded with respect to $s \geq 0$ and $t \in [0, T]$ for each $n \geq 0$. It can be proved analogously that the expressions $\max_{t \in [0,T], x \in \Omega} |\partial^k w_{(n)}^{(s)}/\partial t^k|$ are uniformly bounded with respect to $s \geq 0$ for any fixed $n, k \geq 0$. Thus, the sequence $\{w^{(s)}(x,t)\}$ is uniformly bounded in $C^\infty([0,T], C_\pi^\infty(R_1))$ with respect to $s \geq 0$. Therefore, some subsequence of it converges to a function $u(x,t) \in C^\infty([0,T], C_\pi^\infty(R_1))$, which is evidently a solution of problem (1.1) in the case of a regular kernel $k(x)$. As was shown above, if $P_1(t)$ is bounded on $[0, T]$, then all the functions $P_n(t)$, $n \geq 0$, are bounded on $[0, T]$ as well. Therefore, a classical solution $u(x,t)$ of the Cauchy problem (1.1) with a regular operator $\mathbb{K}$ can be continued in time as long as the quantity $u_1(t)$ remains bounded, i.e., breakdown of a smooth solution can only occur by overturning. Lemma 2 is proved. □

According to condition 1) of Theorem 1, the kernel of the operator $\mathbb{K}$ is integrable on Ω; therefore the representations of $\mathbb{K}$ as the convolution (1.1) and as the pseudodifferential operator (1.2) are equivalent. Suppose $\mathbb{K}$ has an arbitrary conservative part $K_p^2 = \operatorname{Im} ipK_p$ of the symbol K_p, while the nonconservative part $K_p^1 = \operatorname{Re} ipK_p$ of the symbol satisfies the condition

(2.12) $$K_p^1 \geq 0, \qquad |p| \geq p_0 > 0, \qquad p \in Z;$$

then the following lemma is true.

LEMMA 3. *Suppose the symbol of the operator* $\mathbb{K}$ *satisfies condition* (2.12) *while the initial perturbation* $\bar{u}(x) \in C_\pi^\infty(R_1)$. *Then there is a time* $T > 0$, *depending only on the norm* $\|\bar{u}\|_{H^2(\Omega)}$, *such that there exists a unique solution of problem* (1.1) *of the class* $C^\infty([0, T]; C_\pi^\infty(R_1))$.

PROOF. The symbol K_p (i.e., the Fourier coefficients of the kernel $k(x)$, if $k(x)$ is integrable) may slowly tend to zero as $p \to \infty$. We define a sequence of symbols $\{K_p^{(l)}\}$, $l \geq 1$, as follows:

$$K_p^{(l)} = \begin{cases} K_p, & |p| \leq l, \\ 0, & |p| > l\,. \end{cases} \tag{2.13}$$

The operator $\mathbb{K}^{(l)}$ with symbol $K_p^{(l)}$ has order zero, and can be represented as the convolution

$$\mathbb{K}^{(l)}(\varphi) = \int_{-\pi}^{\pi} k^{(l)}(x - y)\varphi'(y)\,dy, \qquad \varphi \subset C_\pi^\infty(R_1),\ l \geq 1, \tag{2.14}$$

where the kernel $k^{(l)}(x) = \frac{1}{2\pi}\sum_{p=-\infty}^{\infty} e^{ipx}K_p^{(l)}$ is infinitely differentiable on Ω, and is therefore regular. Let $b = 2\max_{|p|\leq p_0}|K_p^1|$, then

$$|K_p^{(l)1}| \leq \frac{b}{2}, \quad |p| \leq p_0, \quad l \geq 1\,. \tag{2.15}$$

It follows from definition (2.13) and condition (2.12) that

$$K_p^{(l)1} \geq 0, \qquad |p| \geq p_0 > 0, \tag{2.16}$$

$$pK_p^{(l)}(1+|p|)^{-\alpha_1} \to pK_p(1+|p|)^{-\alpha_1} \quad \text{as } l \to \infty \tag{2.17}$$

in the norm of the space l_2: here $\alpha_1 = 2+\alpha$, and α is the order of the operator $\mathbb{K}$. From the representation of $\mathbb{K}^{(l)}$ as the convolution (2.14) with a regular kernel, it follows by Lemma 2 that there exists a solution $u^{(l)}(x,t) \in C^\infty([0, T^{(l)}]; C_\pi^\infty(R_1))$ of the Cauchy problem (1.1) with the regular operator $\mathbb{K}^{(l)}$. We now show that $T^{(l)} \geq T > 0$. We first establish some estimates for $u^{(l)}(x,t)$. We temporarily omit the index l. Multiplying the Whitham equation and its x-derivatives by $u_{(n)}(x,t)$, where $n \geq 0$ is any integer, and integrating with respect to x from $-\pi$ to $+\pi$, we obtain

$$\dot{J}_n + 2\sum_{j=0}^{n} C_j^n \int_\Omega u_{(n)}u_{(n+1-j)}u_{(j)}\,dx + 2I_n(t) = 0, \qquad n \geq 0, \tag{2.18}$$

where

$$J_n(t) = \int_\Omega u_{(n)}^2(x,t)\,dx, \qquad I_n(t) = \int_\Omega u_{(n)}(x,t)\mathbb{K}(u_{(n)})\,dx,$$

for any $n \geq 1$ the following inequality holds:

$$u_n(t) \leq 2T_n^{1/4}(t)(T_n^{1/4}(t) + T_{n+1}^{1/4}(t))\,. \tag{2.19}$$

Indeed, let a point $x_0 \in \Omega$ be such that $|u_{(n)}(x_0,t)| = u_n(t)$. We consider the following two cases:

1) there exists an $x_1 \in \Omega$ such that $|u_{(n)}(x_1,t)| = \frac{1}{2}u_n(t)$; then

$$\tfrac{3}{4}u_n^2(t) = |u_n^2(t) - u_{(n)}^2(x_1,t)| = \left|\int_{x_1}^{x_0} \frac{\partial}{\partial x}u_{(n)}^2(x,t)\,dx\right| \leq 2\sqrt{J_nJ_{n+1}},$$

whence $u_n(t) \leq (2J_nJ_{n+1})^{1/4}$,

2) no such x_1 exists then, by the continuity of $u_{(n)}(x,t)$, we have $|u_{(n)}(x,t)| > u_n(t)/2$, for all $x \in \Omega$, and thus $J_n(t) = \int_\Omega u_{(n)}^2\,dx > \frac{\pi}{2}U_N^2(t)$, whence $u_j(t) \le 2J_n^{1/2}$ and (2.19) is established. Applying (2.19), the evident identity

$$2\int_\Omega u_{(n)}u_{(n+1)}u_{(0)}\,dx = -\int_\Omega u_{(n)}^2 u_{(1)}\,dx, \qquad n \ge 0, \tag{2.20}$$

and the Cauchy inequality, we transform the equalities (2.18) to the form

$$\begin{aligned}
&\dot J_0 + 2I_0 = 0,\\
&\dot J_1 + 2I_1 \le 2J_1^{5/4}(J_1^{1/4} + J_2^{1/4}),\\
&\dot J_2 + 2I_2 \le 10J_1^{1/4}J_2(J_1^{1/4} + J_2^{1/4}),\\
&\dot J_N + 2I_N \le 2(2n+1)J_nJ_1^{1/4}(J_1^{1/4} + J_2^{1/4})\\
&\qquad + 4\sum_{j=2}^{n-1} C_j^n\sqrt{J_nJ_{n+j-1}}J_n^{1/4}(J_j^{1/4} + J_{j+1}^{1/4}), \qquad n \ge 3.
\end{aligned} \tag{2.21}$$

The operator $\mathbb{K}$ is real, i.e., $K_p = K_{-p}^*$; therefore, $(ipK_p)^* = i(-p)K_{-p}$. Using this relation, (2.15), (2.16), and the Parseval equality, we obtain a lower bound for the integral I_n:

$$\begin{aligned}
I_n(t) &= \int_\Omega u_{(n)}\mathbb{K}^{(l)}(u_{(n)})\,dx\\
&= \frac{1}{4\pi}\sum_{p=-\infty}^{\infty}(ipK_p^{(l)} + (ipK_p^{(l)})^*)((ip)^n\hat u_p(t))^*(ip)^n\hat u_p(t)\\
&= \frac{1}{2\pi}\sum_{p=-\infty}^{\infty}K_p^{(l)1}|p^n\hat u_p(t)|^2\\
&\le \sum_{|p|\le p_0}K_p^{(l)1}|p^n\hat u_p(t)|^2 \ge -\frac{b}{2}J_n(t), \qquad l \ge 1.
\end{aligned} \tag{2.22}$$

The relations (2.21), (2.22) yield

$$\begin{aligned}
&\dot J_0 \le bJ_0, \qquad \dot J_1 \le 2J_1^{5/4}(J_1^{1/4} + J_2^{1/4}) + bJ_1,\\
&\dot J_2 \le 10J_2J_1^{1/4}(J_1^{1/4} + J_2^{1/4}) + bJ_2,\\
&\dot J_n \le 2(2n+1)J_nJ_1(J_1^{1/4} + J_2^{1/4})\\
&\qquad + bJ_n + 4\sum_{j=2}^{n-1}C_j^n\sqrt{J_nJ_{n+1-j}}J_j^{1/4}(J_j^{1/4} + J_{j+1}^{1/4}), \qquad n \ge 3.
\end{aligned} \tag{2.23}$$

Adding the second and the third inequalities in (2.23) and introducing the notation $J(t) \equiv J_1(t) + J_2(t)$, we get the inequality $\dot J \le 20J^{3/2} + bJ$, the integration of which with respect to t gives

$$J(t) \le J(0)e^{bt}\left\{1 - \left(\frac{20}{b}\right)J^{1/2}(0)\left(e^{bt/2} - 1\right)\right\}^{-2}.$$

It follows that if we choose T from the condition

$$0 < T < \frac{2}{b}\ln\left(1+\left(\frac{b}{20}\right)J^{1/2}(0)\right), \tag{2.24}$$

then there exists a number $c_1 > 0$ such that $J(t) \leq c_1$, $t \in [0, T]$, for all $l \geq 1$ at once, since $J(0)$ is independent of l. Further, we integrate successively the first and fourth inequality in (2.23) with respect to t for $t \in [0, T]$. It suffices here to restrict our attention to the case $J_n \geq 1$. Then each of the inequalities considered can be majorized by the linear inequality $\dot{J}_n \leq b_n J_n$, where the $b_n > 0$ are constants independent of l. Hence, for all $n \geq 0$ and $t \in [0, T]$ we have

$$J_n(t) \leq c_n,$$

with constants $c_n > 0$ independent of l. One can prove completely analogously that for all $n, m \geq 0$ and $t \in [0, T]$

$$J_{n,m} \equiv \int_\Omega \left\{\frac{\partial m}{\partial t^m} u^{(l)}_{(n)}(x, t)\right\}^2 dx \leq c_{n,m} \tag{2.25}$$

where the $c_{n,m} > 0$ are independent of l.

By standard imbedding theorems, the estimates (2.25) imply that the sequence $\{u^{(l)}(x, t)\}$ has a subsequence converging to some function $u(x, t) \in C^\infty([0, T]; C^\infty_\pi(R_1))$ in the norm of $C^n([0, T]; C^m_\pi(R_1))$, where n and m are any natural numbers. To simplify the notation, we assume that the entire sequence $\{u^{(l)}(x, t)\}$ converges to $u(x, t)$. By the Parseval equality the sequence of Fourier coefficients $\hat{u}^{(l)}_p(t)$ converges in the norm of $C^m([0, T]; h^n)$ to the Fourier coefficients $\hat{u}_p(t)$ of the function $u(x, t)$ (h^n is the discrete analogue of the Sobolev space H^n); $n, m \geq 0$ are arbitrary. In order to prove that $u(x, t)$ is a solution of the Cauchy problem (1.1) with the operator $\mathbb{K}$, it suffices to establish that the sequence of functions $\mathbb{K}^{(l)}(u^{(l)})$ converges to the function $\mathbb{K}(u)$ as $l \to \infty$, uniformly with respect to $x \in \Omega$ and $t \in [0, T]$. We have

$$\begin{aligned} \mathbb{K}^{(l)}(u^{(l)}) &= \frac{1}{2\pi}\sum_{p=-\infty}^{\infty} e^{ipx} ip K^{(l)}_p \hat{u}^{(l)}_p(t) \\ &= \mathbb{K}(u) + \frac{1}{2\pi}\sum_{p=-\infty}^{\infty} e^{ipx} ip \frac{(K^{(l)}_p - K_p)}{(1+|p|)^{\alpha_1}} \hat{u}^{(l)}_p(t)(1+|p|)^{\alpha_1} \\ &\quad + \frac{1}{2\pi}\sum_{p=-\infty}^{\infty} e^{ipx} \frac{ipK_p}{(1+|p|)^{\alpha_1}}[\hat{u}^{(l)}_p(t) - \hat{u}_p(t)](1+|p|)^{\alpha_1}. \end{aligned} \tag{2.26}$$

Since $\hat{u}^{(l)}_p(t)$ converges to $\hat{u}_p(t)$ as $l \to \infty$ in the norm of $C^0([0, T], h^{\alpha_1})$, using the Cauchy inequality and (2.17) we conclude that the last two terms in (2.26) tend to zero as $l \to \infty$, uniformly with respect to $x \in R_1$ and $t \in [0, T]$. This completes the proof of the existence of a solution of problem (1.1) on $[0, T]$. From (2.24) it follows that T depends only on the norm $\|\bar{u}\|_{H^2(\Omega)}$ of the initial perturbation $\bar{u}(x)$. Let us prove the uniqueness. Assuming the contrary, suppose there exist two solutions $u^{(1)}$ and $u^{(2)}$. Then $w = u^{(1)} - u^{(2)}$ is a solution of the problem

$$w_t + u^{(1)}w_x + wu_x^{(2)} + \mathbb{K}(w) = 0, \qquad w|_{t=0} = 0\,.$$

After multiplying by $w(x,t)$, we integrate with respect to x over Ω:

$$\dot{J}_0 + 2\int_\Omega w^2(-\tfrac{1}{2}u_x^{(1)} + u_x^{(2)})\,dx + 2\int_\Omega w\mathbb{K}(w)\,dx = 0,$$

where $J_0(t) = \int_\Omega w^2(x,t)\,dx$. Transforming the last term according to (2.22), we obtain

$$\dot{J}_0 \leq (2u_1^{(2)} + u_1^{(1)} + b)J_0, \qquad J_0|_{t=0} = 0\,.$$

From this, in view of the estimates $u_1^{(1)}(t) \leq c_0$ and $u_1^{(2)}(t) \leq c_0$ for $t \in [0,T]$ (cf. (2.25) with $m = 0$), it follows that $J_0(t) \equiv 0$, $t \in [0,T]$. The uniqueness of the solution, and together with it Lemma 3, are proved. □

2.2.

PROOF OF THEOREM 1. It follows from Lemma 3 that there exists a solution $u(x,t)$ of the Cauchy problem up to the breaking time T_0 belonging to the class $C^\infty([0,T_0); C_\pi^\infty(R_1))$, where T_0 may be equal to $+\infty$. Thus, to prove Theorem 1 it is necessary to establish an upper bound for the breaking time T_0. Let

$$m(t) = \min_{x\in\Omega} u_{(1)}(x,t) = u_{(1)}(y_1(t),t),$$
$$n(t) = \max_{x\in\Omega} u_{(1)}(x,t) = u_{(1)}(y_2(t),t)\,.$$

We assume that the curves $y_k(t) \in C^1([0,T_0))$, $k = 1,2$. It is easy to avoid that assumption if we carry out the arguments based on the characteristics $y(\xi,t)$ of Lemma 1 (see also Chapter 2).

We differentiate the Whitham equation with respect to x on the curves $x = y_k(t)$, $k = 1,2$. Then, taking into account that $u_{(2)}(y_k(t),t) = 0$, $k = 1,2$, we obtain

$$\dot{m} + m^2(1 + A_1(t)) = 0, \qquad \dot{n} + n^2 + m^2A_2(t) = 0, \tag{2.27}$$

where

(2.28)
$$\begin{aligned} A_j(t) &= m^{-2}(t)\int_{-\pi}^{\pi} k(y_j(t) - x)u_{(2)}(x,t)\,dx \\ &= m^{-2}(t)\left\{\int_{-\delta}^{\delta} F(x,\delta)u_{(3)}(y_j - x,t)\,dx + k(\delta)u_{(1)}(y_j - \delta,t)\right. \\ &\qquad \left. - k(-\delta)u_{(1)}(y_j + \delta,t) + \int_{\delta\leq|x|\leq\pi} k'(x)u_{(1)}(y_j - x,t)\,dx\right\}, \\ &\qquad\qquad j = 1,2; \end{aligned}$$

$$F(x,\delta) = \begin{cases} -\int_x^\delta k(\xi)d\xi, & x \in (0,\delta], \\ \int_{-\delta}^x k(\xi)d\xi, & x \in [-\delta,0), \end{cases} \tag{2.29}$$

and $\delta \in (0, 1]$ will be chosen below. Setting

$$Z(t) = 1 - m_0 \left(t + \int_0^t A_1(\tau)\, dt \right), \qquad m_0 = -m(0), \tag{2.30}$$

we integrate the first equation in (2.27):

$$m(t) = -m_0 Z^{-1}(t). \tag{2.31}$$

Let us prove that if $|A_j(0)| < \gamma/4$, then $|A_j(t)| < \gamma/4$ for all $t \in [0, T_0]$, $j = 1, 2$, where T_0 is determined below. Assuming the contrary, let $t_1 \in (0, T_0)$ be the first moment of time for which $|A_j(t_1)| \geq \gamma/4$.

By the continuity of the functions $A_j(t)$ there exists a $t_2 \in (t_1, T_0)$ such that for all $t \in [0, t_2]$

$$|A_j(t)| < \frac{\gamma}{2}. \tag{2.32}$$

We establish some estimates on the intervals on $[0, t_2]$. From (2.30) and (2.32) we have

$$\dot{Z}(t) < -m_0 \left(1 - \frac{\gamma}{2}\right). \tag{2.33}$$

From the second equation in (2.27), with consideration of (2.31)–(2.33) we obtain the inequality $\dot{n} \leq -m_0 Z^{-2} \dot{Z}$, and integrating it, we find that

$$n(t) \leq u_1(0) Z^{-1}(t). \tag{2.34}$$

After differentiating the Whitham equation three times, we multiply it by $u_{(3)}(x, t)$ and integrate with respect to x from $-\pi$ to π. We obtain

$$\dot{J}_3 + 8 \int_\Omega u_{(1)} u_{(3)}^2\, dx + 2 \int_\Omega u_{(0)} u_{(3)} u_{(4)}\, dx + 2 \int_\Omega u_{(3)} \mathbb{K}(u_{(3)})\, dx = 0. \tag{2.35}$$

Using condition 1) of the theorem and the Parseval equality, we estimate the last integral in (2.35) by (2.22):

$$-2 \int_\Omega u_{(3)} \mathbb{K}(u_{(3)})\, dx \leq b J_3,$$

and with the help of (2.20) with $n = 3$ we convert (2.35) to the form

$$\dot{J}_3 \leq (-7m(t) + b) J_3(t). \tag{2.36}$$

By condition 3) of the theorem $b \leq |\gamma m_0|$, substituting also (2.31) and (2.33) in (2.36), we obtain $\dot{J}_3(t) \leq -7(1+\gamma) Z^{-1} \dot{Z} J_3$, integration of which yields

$$J_3(t) \leq J_3(0) Z(t)^{-7(1+\gamma)}. \tag{2.37}$$

We now estimate the functions $A_j(t)$, $j = 1, 2$, on the segment $[0, t_2]$. By (2.38) and (2.31) we have

$$|A_j(t)| \leq Z^2 m_0^{-2} \left\{ \left(\int_{-\delta}^{\delta} F^2\, dx \int_{y_j - \delta}^{y_j + \delta} u_{(3)}^2\, dx \right)^{1/2} + u_1 \left(|k(\delta)| + |k(-\delta)| + \int_{\delta \leq |x| \leq \pi} |k'(x)|\, dx \right) \right\}. \tag{2.38}$$

From condition 1) of the theorem we find that

$$\text{(2.39)}\qquad \int_{|x|\le\delta} F^2(x,\delta)\,dx \le 48c^2\delta^{3-2\alpha}, \qquad \int_{\delta\le|x|\le\pi} |k'(x)|\,dx \le 4c\delta^{-2}.$$

Combining (2.31), (2.34), and (2.37)–(2.39), we obtain

$$\begin{aligned}|A_j(t)| &\le Z^2 m_0^{-2}\{(49c^2\delta^{3-2\alpha}J_3(0)Z(t)^{-7(1+\gamma)})^{1/2} + u_1(0)Z^{-1}6c\delta^{-\alpha}\}\\ &\le 7cm_0^{-2}\{J_3^{1/2}(0)\delta^{3/2-\alpha}Z^{2-7/2(1+\gamma)} + \delta^{-\alpha}u_1(0)Z\}, \qquad j=1,2.\end{aligned}$$

From this, setting $\delta = Z^{1/2}(t)$ and noting that $0 < Z(t) \le 1$, for $t \in [0, t_2]$, we find that $(j = 1, 2)$

$$\text{(2.40)}\qquad |A_j(t)| \le 7cm_0^{-2}(u_1(0) + J_3^{1/2}(0)) < \gamma/4$$

by condition 3) of the theorem. This contradiction shows that $|A_j(t)| < \gamma/4$ for all $t \in [0, T_0)$. The assumption $|A_j(0)| < \gamma/4$ is also satisfied by (2.38)–(2.40) with $t = 0$. Turning now to (2.30), (2.31), and (2.40), we conclude that $m(t) \to -\infty$ as $t \to T_0 \le (m_0(1-\frac{\gamma}{4}))^{-1} < \infty$. Theorem 1 is proved. □

REMARK 2. Theorem 1 for the periodic problem gives a positive answer to a question of G. Whitham **[82]** concerning breaking of waves described by the Whitham equation with a kernel $k_g(x)$ corresponding to the potential theory of waves on water.

REMARK 3. For a somewhat more restricted class of initial perturbations Theorem 1 can be strengthened in the following way (see Chapter 2).

THEOREM 2. *Suppose the following conditions are satisfied*:

1) *The kernel* $k(x) \in C^1(\Omega\backslash 0)$, $|k(x)| \le c|x|^{-\alpha}$, $|k'(x)| \le C|x|^{-1-\alpha}$ *where* $x \in \Omega\backslash 0$, $c > 0$, $\alpha < \frac{2}{3}$;

2) $\bar{u}(x) \in M_b^\alpha$, *where*

$$M_b^\alpha = \left\{ v(x) \in C_\pi^\infty(R_1)\colon \max_{x\in\Omega} |v_{(n+1)}(x)| \le b(b_n)^{n/\alpha},\ n = 1, 2, \dots \right\}.$$

3) *The steepness* m_0 *of the initial perturbation is sufficiently large*:

$$m_0 \ge c_1 u_1^{6/7}, \qquad c_1 = c_1(c, b, \alpha) > 0.$$

Then up until the time of breaking T_0 *there exists a unique solution* $u(x,t)$ *of the Cauchy problem* (1.1) *belonging to the class* $C^\infty([0, T_0); C_\pi^\infty(R_1))$; *and for the time of breaking* T_0 *there is the following two-sided estimate*:

$$((1+\gamma)m_0)^{-1} \le T_0 \le ((1-\gamma)m_0)^{-1}, \qquad \gamma = \gamma(c, b, \alpha, c_1) > 0.$$

§3. On the existence of a global solution of the Cauchy problem

In this section we prove a theorem on the existence of a global solution (for all times $t \ge 0$) of the Cauchy problem (1.1) with a dissipative operator $\mathbb{K}$ of order $\alpha \ge 0$ in the case where the steepness of the initial perturbation is sufficiently small. Comparison of this result with Theorem 1 shows that the requirement that the steepness of the initial perturbation be large is not only sufficient but also necessary for the breaking of a wave in finite time. Since the order of the operator $\mathbb{K}$ is here any nonnegative number (in particular, the kernel of $\mathbb{K}$ need not be integrable on Ω), we shall view

$\mathbb{K}$ as a pseudodifferential operator defined by (1.2). Suppose the nonconservative component of the symbol $K_p^1 = \operatorname{Re} ipK_p$ satisfies the following dissipation conditions:

$$K_p^1 \geq 0, \quad p \in Z; \qquad K_p^1 \geq 40\varepsilon > 0, \quad |p| \geq p_0 \geq 1 \tag{3.1}$$

where ε and p_0 are any fixed numbers.

We introduce the notation

$$a = p_0^{-4}(80\varepsilon)^{-1}.$$

THEOREM 3. *Suppose the following conditions hold*:

1) *The symbol K_p of the operator $\mathbb{K}$ satisfies* (3.1);

2) $\bar{u}(x) \in C_\pi^\infty(R_1)$;

3) $\|\bar{u}\|^2_{H^2(\Omega)} \leq a\varepsilon^3$.

Then there exists a unique solution $u(x,t)$ of the Cauchy problem (1.1) *belonging to the class $C^\infty(R_+; C_\pi^\infty(R_1))$.*

PROOF. Lemma 3 of §2 guarantees the existence and uniqueness of the solution of the Cauchy problem (1.1) on the interval $[0, T]$, where T is determined by the norm $\|\bar{u}\|_{H^2(\Omega)}$ (see (2.24)). We show that the norm $\|u(x,t)\|_{H^2(\Omega)}$ of the solution does not increase with time. More precisely, we prove that if for some $t_0 \geq 0$

$$J_0(t_0) \leq a\varepsilon^3, \qquad J_n(t_0) \leq \varepsilon^2, \quad n = 1, 2, \tag{3.2}$$

then for all $t \in [t_0, t_0 + T]$, where T is defined in (2.24), the following inequalities hold:

$$J_0(t) \leq a\varepsilon^3, \qquad J_n(t) \leq \varepsilon^2, \qquad n = 1, 2\,. \tag{3.3}$$

We note that for $t_0 = 0$ the inequalities (3.2) follow from condition 3) of the theorem, since $a\varepsilon \leq 1$. Differentiating the Whitham equation n times with respect to x, multiplying by $u_{(n)}(x,t)$, and integrating with respect to x from $-\pi$ to π, we obtain

$$\dot{J}_n + 2\sum_{j-0}^{n} C_j^n \int_\Omega u_{(n)}u_{(n+1-j)}u_{(j)}\,dx + 2I_n = 0, \qquad n = 0, 1, \ldots, \tag{3.4}$$

where J_n and I_n are defined in (2.1).

Applying (2.19) and (2.20), we transform (3.4) to the following system of inequalities ($n = 0, 1, 2$):

$$\begin{gathered} \dot{J}_0 = -2I_0, \qquad \dot{J}_1 \leq 2J_1^{5/4}(J_1^{1/4} + J_2^{1/4}) - 2I_1, \\ \dot{J}_2 \leq 10J_2J_1^{1/4}(J_1^{1/4} + J_2^{1/4}) - 2I_2\,. \end{gathered} \tag{3.5}$$

Transforming the integrals I_n as in (2.22) and using (2.1) we obtain

$$\begin{aligned} &I_0(t) \geq 0, \\ &I_n(t) = \int_\Omega u_{(n)}\mathbb{K}(u_{(n)})\,dx = \frac{1}{2\pi}\sum_{p=-\infty}^{\infty} K_p^1|p^n\hat{u}_p(t)|^2 \\ &\quad = 40\varepsilon J_n + \frac{1}{2\pi}\sum_{|p|>p_0}(K_p^1 - 40\varepsilon)|p^n\hat{u}_p(t)|^2 + \frac{1}{2\pi}\sum_{|p|\leq p_0}(K_p^1 - 40\varepsilon)|p^n\hat{u}_p(t)|^2 \\ &\quad \geq 40\varepsilon J_n - 40\varepsilon\frac{p_0^4}{2\pi}\sum_{|p|\leq p_0}|\hat{u}_p(t)|^2 \geq 40\varepsilon J_n - \frac{1}{2a}J_0, \qquad n = 1, 2\,. \end{aligned} \tag{3.6}$$

Substitution of (3.6) in (3.5) yields

$$\dot{J}_0 \le 0, \qquad \dot{J}_1 \le -80\varepsilon J_1 + 2J_1^{5/4}(J_1^{1/4} + J_2^{1/4}) + \frac{1}{a}J_0, \tag{3.7}$$
$$\dot{J}_2 \le -80\varepsilon J_2 + 10J_2J_1^{1/4}(J_1^{1/4} + J_2^{1/4}) + \frac{1}{a}J_0\,.$$

Integrating the first inequality in (3.7) and accounting for (3.2), we obtain

$$J_0(t) \le J_0(t_0) \le \dot{a}\varepsilon^3, \qquad t \in [t_0, t_0 + T]. \tag{3.8}$$

The remaining inequalities in (3.3) are proved by contradiction.

Suppose $T_1 \in (t_0, t_0 + T]$ is the first time when any of the inequalities (3.3) with $n = 1, 2$ is violated. By (3.2) and the continuity of $J_n(t)$ we evidently have

$$J_n(t) \le 2\varepsilon^2, \qquad t \in [t_0, T_1],\ n = 1, 2\,. \tag{3.9}$$

Substituting (3.8) and (3.9) in the right-hand sides of the second and third inequalities in (3.7), we obtain

$$\dot{J}_n(t) \le -\varepsilon J_n(t) + \varepsilon^3, \qquad t \in [t_0, T_1],\ n = 1, 2\,.$$

Integrating it with respect to t and employing (3.2), we find that

$$J_n(t) \le \varepsilon^2 + (J_n(t_0) - \varepsilon^2)e^{-\varepsilon t} \le \varepsilon^2, \qquad t \in [t_0, T_1],\ n = 1, 2,$$

which contradicts the assumption. The inequalities (3.3) have thus been proved. They show that a classical solution $u(x, t)$ of the Cauchy problem (1.1) after a finite number of steps can be extended to any time-interval $[0, T_0]$, since at each step the time T of existence of the solution $u(x, t)$, guaranteed by Lemma 3 of §2, depends only on the norm $\|u\|_{H^2(\Omega)}$ at the initial time t_0 of the step, and, according to (3.3), it can be chosen to be constant for all steps. Theorem 3 is proved. □

REMARK. Condition 3) of the theorem indicates that the initial perturbation $\bar{u}(x)$ and its steepness (the minimum of the first derivative) are sufficiently small.

§4. Smoothing of solutions of the Cauchy problem

In this section we consider strongly dissipative operators $\mathbb{K}$ for which the conservative part K_p^2 of the symbol K_p is arbitrary, while the nonconservative part K_p^1 satisfies the conditions

$$K_p^1 \ge 0, \quad p \in Z; \qquad K_p^1 \ge \varepsilon|p|^\alpha, \quad |p| \ge p_0 > 0,\ \alpha, \varepsilon > 0\,. \tag{4.1}$$

We prove that a solution of the Cauchy problem (1.1) with a strongly dissipative operator $\mathbb{K}$ satisfying (4.1) with $\alpha > \frac{3}{2}$ belongs to the class $C^\infty((0,\infty); C_\pi^\infty(R_1))$, if the initial perturbation $\bar{u}(x)$ is 2π-periodic and belongs to the class $L_2^{\mathrm{loc}}(R_1)$. Thus, a nonsmooth initial condition at any time $t > 0$ becomes and remains infinitely smooth for all subsequent times; thus, breaking of waves is impossible in the class of strongly dissipative operators $\mathbb{K}$ of order greater than $\frac{3}{2}$.

4.1. Global existence of a solution of the Cauchy problem for smooth initial conditions. The theorem proved below guarantees the global existence of a classical solution of the Cauchy problem (1.1) with a strongly dissipative operator and with an arbitrary initial condition $\bar{u}(x) \in C_\pi^\infty(R_1)$, which, in contrast to Theorem 3 of §3, is not small in general.

THEOREM 4. *Suppose $\mathbb{K}$ satisfies condition (4.1) with $\alpha > \frac{3}{2}$, and $\bar{u}(x) \in C_\pi^\infty(R_1)$. Then there exists a unique solution of the Cauchy problem (1.1) belonging to the class $C^\infty([0,\infty); C_\pi^\infty(R_1))$.*

PROOF. Since by Lemma 3 of §2 there exists a unique solution $u(x,t)$ of the Cauchy problem (1.1) belonging to $C^\infty([0,T]; C_\pi^\infty(R_1))$, where $T > 0$ depends only on the norm $\|\bar{u}\|_{H^2(\Omega)}$, it is sufficient to prove that the norm $\|u(x,t)\|_{H^2(\Omega)}$ of the solution is bounded for all finite t. We show that for any $t \in [0,T]$, where T is the time of existence of a classical solution of problem (1.1), the following estimate is true:

$$J_n(t) \le 4J_n(0)e^{2ct}, \qquad n = 0, 1, 2, \tag{4.2}$$

where $c > 0$ is a constant. Inequality (4.2) for $n = 0$ is readily obtained by multiplying the Whitham equation by $u(x,t)$ and integrating with respect to x. We get

$$\dot{J}_0 + 2\int_\Omega u^2 u_x\,dx + 2\int_\Omega u\mathbb{K}(u)\,dx = 0\,. \tag{4.3}$$

Since $u(x,t) \in C_\pi^\infty(R_1)$, the first integral in (4.3) is equal to zero; we transform the second integral as in (2.22) using (4.1):

$$\int_\Omega u\mathbb{K}(u)\,dx = \frac{1}{2\pi}\sum_{p=-\infty}^{\infty} K_p^1|\hat{u}_p(t)|^2 \ge 0\,.$$

Therefore, from (4.3) we obtain $\dot{J}_0 \le 0$, whence

$$J_0(t) \le J_0(0), \qquad t \in [0,T]\,. \tag{4.4}$$

Thus, (4.2) with $n = 0$ is proved. We next substitute in the Whitham equation in the Fourier series expansion of $u(x,t)$: $u(x,t) = \frac{1}{2\pi}\sum_{p=-\infty}^{\infty} e^{ipx}\hat{u}_p(t)$, and equate the Fourier coefficients corresponding to the harmonic e^{ipx} to zero. We obtain

$$\dot{\hat{u}}_p(t) + \frac{ip}{2}\sum_{q=-\infty}^{\infty}\hat{u}_{p-q}(t)\hat{u}_q(t) + ipK_p\hat{u}_p(t) = 0, \qquad p \in Z\,. \tag{4.5}$$

Adjoining to (4.5) the initial condition $\hat{u}_p(t)|_{t=0} = \hat{\bar{u}}_p$, integrating it with respect to t, and multiplying the result by $(ip)^n$, $n = 1, 2$, we find that for any $p \in Z$

$$(ip)^n\hat{u}_p(t) = (ip)^n\hat{\bar{u}}_p e^{-ipK_pt} - \frac{(ip)^{n+1}}{2}\int_0^t e^{-ipK_p(t-\tau)}\sum_{q=-\infty}^{\infty}\hat{u}_{p-q}(\tau)\hat{u}_q(\tau)\,d\tau\,. \tag{4.6}$$

We transform the term with the integral in (4.6) as follows: we move one factor p inside the sum write it in the form $p = (p-q) + q$ and apply the Cauchy inequality, and subsequently the Parseval equality; as a result, in place of (4.6), we obtain the inequality

$$(4.7)\qquad |p^n\hat{u}_p(t)| \le |p^n\hat{\bar{u}}_p| + 2|p|\int_0^t e^{-K_p^1(t-\tau)}\sqrt{J_0(\tau)J_n(\tau)}\,d\tau \qquad p\in Z, n=1,2\,.$$

We introduce the notation

$$(4.8)\qquad \begin{gathered} A^2 = 40J_0(0), \qquad \beta = \tfrac{1}{2}(\alpha - \tfrac{3}{2}) > 0, \\ \mathcal{N} = \max\{p_0+4, (2^{\alpha+3}A\varepsilon^{-1}\beta^{-1/2})^{1/\beta}\}, \qquad c = 2A\mathcal{N}^{3/2}. \end{gathered}$$

Let us prove (4.2) with $n = 1, 2$ and the constant c defined in (4.8). Excluding the trivial case, we assume that $\bar{u}(x) \not\equiv \text{const}$. We now prove (4.2) by contradiction. Suppose for some $t_1 \in [0, T]$ the inequalities (4.2) are violated. In view of the continuity of J_1 and J_2 with respect to t and the fact that $J_n(0) \ne 0$, $n = 1, 2$, for all $t \in [0, t_1]$, we have

$$(4.9)\qquad J_n(t) \le 6J_n(0)e^{2ct}, \qquad n = 1,2\,.$$

Substituting (4.4) and (4.9) in (4.7) and taking the l_2-norm of both sides, we obtain

$$J_n^{1/2}(t) \le J_n^{1/2}(0) + 2\sqrt{6}\left\| |p|\int_0^t e^{-K_p^1(t-\tau)}J_0^{1/2}(0)J_n^{1/2}(0)e^{c\tau}\,d\tau\right\|_{l_2}.$$

Using the notation (4.8), from this we obtain

$$(4.10)\qquad J_n(t) \le J_n(0)\left\{1 + A\left\|\int_0^t |p|e^{-K_p^1(t-\tau)+c\tau}\,d\tau\right\|_{l_2}\right\}^2, \qquad n = 1,2\,.$$

We estimate the integral in (4.10) taking into account the properties (4.1)

$$(4.11)\qquad |p|\int_0^t e^{-K_p^1(t-\tau)+c\tau}\,d\tau \le \begin{cases} \frac{|p|e^{ct}}{c+K_p^1}, & \text{if } |p|\ge\mathcal{N}, \\ \mathcal{N}e^{ct}c^{-1}, & \text{if } |p|\le\mathcal{N}. \end{cases}$$

Because of condition (4.1) and the choice of the numbers $\mathcal{N}$ and β (see (4.8)), for $|p| \ge \mathcal{N}$ we find that

$$(4.12)\qquad 0 < \frac{|p|}{c+K_p^1} \le \frac{1}{\varepsilon|p|^\beta|p|^{1/2+\beta}} \le \frac{2^\alpha}{\varepsilon\mathcal{N}^\beta(1+|p|)^{\beta+1/2}} \le \frac{\beta^{1/2}}{8A(1+|p|)^{\beta+1/2}}.$$

By (4.11) and (4.12), for $n = 1, 2$ we obtain

$$(4.13)\qquad \begin{aligned} \sum_{p=-\infty}^{\infty}\left(|p|\int_0^t e^{-K_p^1(t-\tau)+c\tau}\,d\tau\right)^2 &= \sum_{|p|\le\mathcal{N}} + \sum_{|p|>\mathcal{N}} \\ &\le \left(\frac{2\mathcal{N}^3}{C^2} + \frac{\beta}{8^2A^2\beta}\right)e^{2ct} \le A^{-2}e^{2ct}. \end{aligned}$$

Substituting (4.13) in (4.10), we see that inequalities (4.2) are satisfied for all $t \in [0, t_1]$. This contradiction proves the validity of inequalities (4.2) on the entire inverval $[0, T]$.

The bound (4.2), in turn, makes it possible to assert that

$$\|u(x,t)\|^2_{H^2(\Omega)} \le 4\|\bar{u}(x)\|^2_{H^2(\Omega)}e^{2ct}, \qquad t\in[0,T]\,.$$

Therefore, according to Lemma 3 of §2, the solution of the Cauchy problem (1.1) can be extended in time to the entire interval $[0, \infty)$. Theorem 4 is proved. □

4.2. Smoothing.

LEMMA. *Suppose the operator* $\mathbb{K}$ *satisfies condition* (4.1) *with* $\alpha > \frac{3}{2}$ *and* $\bar{u}(x) \in L_2^{\rm loc}(R_1)$ *is periodic with period* 2π. *Then there is a* $t_0 > 0$ *such that there exists a unique solution of problem* (1.1)

$$u(x,t) \in C^\infty((0,t_0]; C_\pi^\infty(R_1)) \cap C^0([0,t_0], L_2^{\rm loc}(R_1)).$$

PROOF. Choose a sequence $\{\bar{u}^{(s)}(x)\}$, $s = 0,1,2,\ldots$, of functions such that

$$\bar{u}^{(s)}(x) \in C_\pi^\infty(R_1), \quad \|\bar{u}^{(s)} - \bar{u}\|_{L_2} \to 0, \quad s \to \infty, \quad \|\bar{u}^{(s)}\|_{L_2} \le \sqrt{2}\|\bar{u}\|_{L_2}. \tag{4.14}$$

Applying the method of successive approximations to problem (1.1), we set

$$w^{(0)}(x,t) = F^{-1}(\hat{w}_p^{(0)}(t)) = F^{-1}(\hat{\bar{u}}_p^{(0)} e^{-ipK_pt}),$$

where F^{-1} is a representation of the function in the form of its Fourier series, while for $s \ge 1$ we define the function $w^{(s)}(x,t)$ to be the solutions of the linear Cauchy problem

$$w_t^{(s)} + w^{(s-1)}w_x^{(s-1)} + \mathbb{K}(w^{(s)}) = 0, \qquad w^{(s)}|_{t=0} = \bar{u}^{(s)}(x). \tag{4.15}$$

It follows from the general theory of linear differential equations that there is a solution $w^{(s)}(x,t) \in C^\infty([0,\infty); C_\pi^\infty(R_1))$, $s \ge 1$, of problem (4.15).

As in (4.5), we represent the solution $w^{(s)}(x,t)$ in the form of a Fourier series; from (4.15) we then obtain

$$\begin{gathered} \dot{\hat{w}}_p^{(s)}(t) + \frac{ip}{2}\sum_{q=-\infty}^{\infty} \hat{w}_{p-q}^{(s-1)}\hat{w}_q^{(s-1)} + ipK_p\hat{w}_p^{(s)}(t) = 0, \\ \hat{w}_p^{(s)}(t)|_{t=0} = \hat{\bar{u}}_p^{(s)}. \end{gathered} \tag{4.16}$$

Integration of (4.16) with respect to t gives

$$\hat{w}_p^{(s)}(t) = \hat{\bar{u}}_p^{(s)} e^{-ipK_pt} - \frac{ip}{2}\int_0^t e^{-ipK_p(t-\tau)} \sum_{q=-\infty}^{\infty} \hat{w}_{p-q}^{(s-1)}(\tau)\hat{w}_q^{(s-1)}(\tau)\,d\tau,$$

whence

$$|\hat{w}_p^{(s)}(t)| \le |\hat{\bar{u}}_p^{(s)}|e^{-K_p^1t} + \frac{|p|}{2}\int_0^t e^{-K_p^1(t-\tau)} \sum_{q=-\infty}^{\infty} |\hat{w}_{p-q}^{(s-1)}(\tau)\hat{w}_q^{(s-1)}(\tau)|\,d\tau. \tag{4.17}$$

We now prove that for some t_0, $0 < t_0 \le 1$, independent of s, and for all $s \ge 0$, $t \in [0,t_0]$, $p \in Z$ the following inequalities hold:

$$|\hat{w}_p^{(s)}(t)| \le \varphi_p^{(s)} e^{-\varepsilon|p|t}, \tag{4.18}$$

where

$$\begin{gathered} \varphi_p^{(s)} = \kappa|\hat{\bar{u}}_p^{(s)}| + (1+|p|)^{-\beta-1/2}, \qquad \beta = \tfrac{1}{2}(\alpha - \tfrac{3}{2}) > 0, \\ \varphi_p = \kappa|\hat{\bar{u}}_p| + (1+|p|)^{-\beta+1/2} \\ \kappa = 2^{\varepsilon p_0 + b/2}, \qquad b = 2\max_{|p|\le p_0} K_p^1. \end{gathered}$$

For $s=0$ inequality (4.18) is satisfied by the definition of $\hat{w}^{(0)}$. Assume, by induction, that (4.18) holds for $s \le s_0$; let us estimate $\hat{w}_p^{(s)}(t)$ for $s = s_0+1$ on the basis of (4.17):

$$
\begin{aligned}
\left|\sum_{q=-\infty}^{\infty} \hat{w}_{p-q}^{(s-1)}(\tau)\hat{w}_q^{(s-1)}(\tau)\right| &\le \sum_{q=-\infty}^{\infty} \varphi_q^{(s-1)} e^{-\varepsilon|p-q|\tau-\varepsilon|q|\tau} \\
&\le e^{-e|p|\tau} \sum_{q=-\infty}^{\infty} \varphi_{p-q}^{(s-1)}\varphi_q^{(s-1)} \le e^{-\varepsilon|p|\tau}\|\varphi^{(s-1)}\|_{l_2}^2 \\
&\le 4\|\varphi\|_{l_2}^2 e^{-\varepsilon|p|\tau},
\end{aligned}
\tag{4.19}
$$

where we have applied (4.14) in the last inequality. Substituting this in (4.17), we obtain

$$
|\hat{w}_p^{(s)}(t)| \le \tfrac{1}{2}\varphi_p^{(s)} e^{-\varepsilon|p|t} + 2|p|\,\|\varphi\|_{l_2}^2 \int_0^t e^{-K_p^1(t-\tau)-\varepsilon|p|\tau}\,d\tau, \tag{4.20}
$$

since for $0 \le t \le t_0 \le 1$ we have

$$
|\hat{\bar{u}}_p^{(s)}|e^{-K_p^1 t} \le \tfrac{1}{2}\kappa|\hat{\bar{u}}_p^{(s)}|e^{-\varepsilon|p|t} \le \tfrac{1}{2}\varphi_p^{(s)} e^{-\varepsilon|p|t}.
$$

In view of (4.1), for the integral in (4.20) we have the estimate

$$
|p|\int_0^t e^{-K_p^1(t-\tau)-\varepsilon|p|\tau}\,d\tau \le \begin{cases} \frac{|p|e^{-\varepsilon|p|t}}{K_p^1-\varepsilon|p|} & \text{for } |p| \ge \mathcal{N}, \\ \mathcal{N}t & \text{for } |p| \le \mathcal{N}. \end{cases} \tag{4.21}
$$

Choose $\mathcal{N}$ sufficiently large: $\mathcal{N} = \max\{p_0+4, (2^{3+\alpha}\|\varphi\|_{l_2}^2\varepsilon^{-1})^{1/\beta}\}$, so that for $p \ge \mathcal{N}$ we have

$$
\begin{aligned}
0 < \frac{|p|}{K_p^1 - \varepsilon|p|} &\le \frac{2}{\varepsilon\mathcal{N}^\beta|p|^{\beta+1/2}} \le \frac{2^\alpha}{\varepsilon\mathcal{N}^\beta(1+|p|)^{\beta+1/2}} \\
&\le \frac{1}{8\|\varphi\|_{l_2}^2(1+|p|)^{\beta+1/2}},
\end{aligned}
$$

and then fix a sufficiently small t_0: $t_0 = \min\{(2^{3+\alpha}\mathcal{N}^{1+\alpha}e^{\varepsilon\mathcal{N}}\|\varphi\|_{l_2}^2)^{-1}, 1\}$, so that for $t \in [0, t_0]$ and $|p| \le \mathcal{N}$ we have

$$
\mathcal{N}t \le \mathcal{N}te^{\varepsilon\mathcal{N}t_0-\varepsilon|p|t} \le \frac{e^{-\varepsilon|p|t}}{\|\varphi\|_{l_2}^2 2^{3+\alpha}\mathcal{N}^\alpha} \le \frac{e^{-\varepsilon|p|t}}{8\|\varphi\|_{l_2}^2(1+|p|)^{\beta+1/2}}. \tag{4.22}
$$

From (4.21), with the inequality $\varphi_p^{(s)} \ge (1+|p|)^{-\beta-1/2}$ taken into account, we find that

$$
|p|\int_0^t e^{-K_p^1(t-\tau)-\varepsilon|p|\tau}\,d\tau \le \varphi_p^{(s)}\frac{e^{-\varepsilon|p|t}}{8\|\varphi\|_{l_2}^2}. \tag{4.23}
$$

Inequality (4.18) with $s = s_0+1$ now follows immediately from (4.20) and (4.23); i.e.,

(4.18) is satisfied for all $s \geq 0$. We now prove that for all $n, s \geq 0$, all $p \in Z$, and all $t \in (0, t_0]$,

$$\left|\frac{\partial^n \hat{w}_p^{(s)}}{\partial t^n}(t)\right| \leq c_n \varphi_p^{(s)} e^{-\varepsilon|p|t}(1+|p|)^{\alpha_1 n} t^{-\alpha_1 n^2}, \tag{4.24}$$

where the c_n are positive numbers independent of s and $\alpha_1 - 4 = \alpha$ is the order of the symbol K_p, i.e.,

$$|pK_p| \leq \gamma |p|^{\alpha_1 - 4}, \quad \gamma \geq 0, \quad |p| \geq p_0 . \tag{4.25}$$

By assumption $\alpha_1 - 4 > \frac{3}{2}$. For $n = 0$ and any $s \geq 0$ the estimate (4.24) with $c_0 = 1$ follows from (4.18). We carry out induction on n: suppose (4.24) is satisfied for $n \leq n_0$. Then for $n = n_0 + 1$ and for any $s \geq 0$ we differentiate (4.16) $(n-1)$ times with respect to t:

$$\frac{\partial^n \hat{w}_p^{(s)}}{\partial t^n} = -ipK_p \frac{\partial^{n-1} \hat{w}_p^{(s)}}{\partial t^{n-1}} - \frac{ip}{2} \sum_{q=-\infty}^{\infty} \sum_{j=0}^{n-1} C_j^{n-1} \frac{\partial^j \hat{w}_{p-q}^{(s-1)}}{\partial t^j} \frac{\partial^{n-1-j} \hat{w}_q^{(s-1)}}{\partial t^{n-1-j}} . \tag{4.26}$$

From this, using (4.24) for $n \leq n_0$ and $s \geq 0$, for $t \in (0, t_0]$ we obtain

$$\begin{aligned} \left|\frac{\partial^n \hat{w}_p^{(s)}}{\partial t^n}\right| &\leq |pK_p| c_n \varphi_p^{(s)} e^{-\varepsilon|p|t}(1+|p|)^{\alpha_1(n-1)} t^{-\alpha_1(n-1)^2} \\ &\quad + c_n |p| \sum_{j=0}^{n-1} \sum_{q=-\infty}^{\infty} t^{-\alpha_1 j^2 - \alpha_1(n-j-1)^2} e^{-\varepsilon|p-q|t - \varepsilon|q|t} \\ &\quad \cdot (1+|p-q|)^{\alpha_1 j}(1+|q|)^{\alpha_1(n-1-j)} \varphi_{p-q}^{(s-1)} \varphi_q^{(s-1)} . \end{aligned} \tag{4.27}$$

Here and below we agree to include in the constant c_n all numbers independent of p, s, and t. To be specific, we suppose $p > 0$, and estimate the sum in (4.27):

(4.28)
$$\begin{aligned} &\sum_{q=-\infty}^{\infty} e^{-\varepsilon|p-q|t - \varepsilon|q|t}(1+|p-q|)^{\alpha_1 j}(1+|q|)^{\alpha_1(n-1-j)} \varphi_{p-q}^{(s-1)} \varphi_q^{(s-1)} \\ &\quad = \sum_{q=0}^{p} + \sum_{q=p+1}^{\infty} + \sum_{q=-\infty}^{-1} \\ &\quad \leq e^{-\varepsilon|p|t}(1+|p|)^{\alpha_1(n-1)} \sum_{q=0}^{p} \varphi_{p-q}^{(s-1)} \varphi_q^{(s-1)} + e^{-\varepsilon|p|t} \sum_{q=-\infty}^{-1} e^{-2\varepsilon|q|t} \varphi_{p-q}^{(s-1)} \varphi_q^{(s-1)} \\ &\quad \cdot \{(1+|p-q|)^{\alpha_1 j}(1+|q|)^{\alpha_1(n-1-j)} + (1+|p-q|)^{\alpha_1(n-1-j)}(1+|q|)^{\alpha_1 j}\} \\ &\quad \leq c_n e^{-\varepsilon|p|t} \|\varphi\|_{l_2}^2 (1+|p|)^{\alpha_1(n-1)} t^{-\alpha_1(n-1)}, \end{aligned}$$

by (4.14) and the obvious estimate

$$\begin{aligned} &e^{-2\varepsilon|q|t}(1+|p-q|)^{\alpha_1(n-1-j)}(1+|q|)^{\alpha_1 j} \\ &\quad \leq e^{-2\varepsilon|q|t}(1+|p|)^{\alpha_1(n-1)}(1+|q|)^{\alpha_1(n-1)} \leq c_n(1+|p|)^{\alpha_1(n-1)} t^{-\alpha_1(n-1)}, \end{aligned} \tag{4.29}$$

valid for $n \geq 1$, $0 \leq j \leq n-1$, $p, q \in Z$, $t > 0$. Substituting (4.28) in (4.27) and noting (4.25), we obtain

$$\left|\frac{\partial^n \hat{w}_p^{(s)}}{\partial t^n}\right| \leq c_n \varphi_p^{(s)} e^{-\varepsilon|p|t}(1+|p|)^{\alpha_1 n} t^{-\alpha_1 n^2} + \sum_{j=0}^{n-1} c_n |p|(1+|p|)^{\alpha_1(n-1)} e^{-\varepsilon|p|t} t^{-\alpha_1 j^2 - \alpha_1(n-j-1)^2 - \alpha_1(n-1)} \leq c_n \varphi_p^{(s)} e^{-\varepsilon|p|t}(1+|p|)^{\alpha_1 n} t^{-\alpha_1 n^2}, \tag{4.30}$$

since $\varphi_p^{(s)}(1+|p|)^{\alpha_1} \geq |p|$. Thus, (4.24) is proved by induction.

The bounds (4.18) and (4.24) show that from the sequence $\{w^{(s)}(x,t)\}$, $s \geq 0$, consisting of the Fourier transforms $w^{(s)}(x,t) = F^{-1}(\hat{w}_p^{(s)}(t))$, it is possible to extract a subsequence converging to a limit $u(x,t) \in C^\infty((0,t_0]; C_\pi^\infty(R_1))$ in the norm of the spaces $C^m((0,t_0]; h^n(\Omega))$ for any n and m. It can be proved in the same way as in §2 that $u(x,t)$ satisfies the Whitham equation for $t > 0$. In order to show that the initial condition is satisfied, we have to prove that some subsequence of $\{w^{(s)}(x,t)\}$ converges as $s \to \infty$ in the norm of $C^0([0,t_0], L_2(\Omega))$ as well. From the sequence $\{\bar{u}^{(s)}(x)\}$ we extract a subsequence such that $\|\bar{u}^{(s)} - \bar{u}^{(s-1)}\|_{L_2(\Omega)} \leq 2^{-s}\|\varphi\|_{l_2}$ for all $s \geq 1$ (we retain the same notation for this subsequence). We claim that for $s \geq 1$

$$\max_{t\in[0,t_0]} \|\hat{w}^{(s)} - \hat{w}^{(s-1)}\|_{l_2} \leq 2^{3-s}\|\varphi\|_{l_2}. \tag{4.31}$$

For $s = 1$ this is a consequence of (4.18). Let by induction (4.31) be proved for $s \leq s_0$; then for $s = s_0 + 1$ from (4.16) integrated with respect to t we find that

$$|\hat{w}_p^{(s)} - \hat{w}_p^{(s-1)}| \leq |\hat{\bar{u}}_p^{(s)} - \hat{\bar{u}}_p^{(s-1)}|e^{-K_p^1 t} + \frac{|p|}{2}\int_0^t e^{-K_p^1(t-\tau)} \sum_{q=-\infty}^{\infty} |\hat{w}_{p-q}^{(s-1)}(\tau) - \hat{w}_{p-q}^{(s-2)}(\tau)||\hat{w}_q^{(s-1)}(\tau) + \hat{w}_q^{(s-2)}(\tau)|\,d\tau. \tag{4.32}$$

We estimate the sum in (4.32) by means of the Cauchy inequality, (4.18), and (4.31):

$$\sum_{q=-\infty}^{\infty} \leq 2^{5-s}\|\varphi\|_{l_2}^2. \tag{4.33}$$

Substituting (4.33) in (4.32) and noting (4.23) with $\xi = 0$, we obtain

$$|\hat{w}_p^{(s)}(t) - \hat{w}_p^{(s-1)}(t)| \leq \hat{\bar{u}}_p^{(s)} - \hat{\bar{u}}_p^{(s-1)}| + 2^{5-s}\|\varphi\|_{l_2}^2|p| \int_0^t e^{-K_p^1(t-\tau)}\,d\tau \leq |\hat{\bar{u}}_p^{(s)} - \hat{\bar{u}}_p^{(s-1)}| + 2^{2-s}\varphi_p. \tag{4.34}$$

Summing (4.34) over all $p \in Z$, we find that

$$\|\hat{w}^{(s)} - \hat{w}^{(s-1)}\|_{l_2} \leq \|\hat{\bar{u}}^{(s)} - \hat{\bar{u}}^{(s-1)}\|_{l_2} + 2^{2-s}\|\varphi\|_{l_2} \leq 2^{3-s}\|\varphi\|_{l_2}$$

and (4.31) is proved by induction. From (4.31) it follows immediately that $u(x,t) \in C^0([0,t_0]; L_2(\Omega))$. To complete the proof of the lemma, the uniqueness of the solution of problem (1.1) remains to be proved. If there are two solutions $u^{(1)}(x,t)$ and $u^{(2)}(x,t)$, we write them as Fourier series, substitute them in problem (1.1), subtract, and then integrate with respect to t. We obtain

$$\begin{aligned} &|\hat{u}_p^{(1)}(t) - \hat{u}_p^{(2)}(t)| \\ &\leq \frac{|p|}{2}\int_0^t e^{-K_p^1(t-\tau)} \sum_{q=-\infty}^{\infty} |\hat{u}_{p-q}^{(1)}(\tau) + \hat{u}_{p-q}^{(2)}(\tau)|\,|\hat{u}_q^{(1)}(\tau) - \hat{u}_q^{(2)}(\tau)|\,d\tau\,. \end{aligned} \tag{4.35}$$

In view of (4.14) and (4.18) we have

$$\|\hat{u}^{(1)} + \hat{u}^{(2)}\|_{l_2} \leq 4\|\varphi\|_{l_2},$$

which, together with (4.35), gives

$$\begin{aligned} &\max_{t\in[0,t_0]} \|\hat{u}^{(1)} - \hat{u}^{(2)}\|_{l_2} \\ &\leq \max_{t\in[0,t_0]} \|\hat{u}^{(1)} - \hat{u}^{(2)}\|_{l_2} 2\|\varphi\|_{l_2} \|\,|p|\int_0^t e^{-K_p^1(t-\tau)}\,d\tau\|_{l_2}\,. \end{aligned} \tag{4.36}$$

As in (4.23), we obtain the bound

$$\|\,|p|\int_0^t e^{-K_p^1(t-\tau)}\,d\tau\|_{l_2} \leq \left\|\frac{1}{8}\frac{\varphi_p^{(s)}}{\|\varphi\|_{l_2}^2}\right\| \leq \tfrac{1}{4}\|\varphi\|_{l_2}^{-1}. \tag{4.37}$$

Substitution of (4.37) in (4.36) yields

$$\max_{t\in[0,t_0]} \|\hat{u}^{(1)} - \hat{u}^{(2)}\|_{l_2} \leq \tfrac{1}{2} \max_{t\in[0,t_0]} \|\hat{u}^{(1)} - \hat{u}^{(2)}\|_{l_2},$$

whence it follows that $\hat{u}_p^{(1)}(t) \equiv \hat{u}_p^{(2)}(t)$ and $u^{(1)}(x,t) \equiv u^{(2)}(x,t)$. The lemma is completely proved. □

The next result is a simple corollary of the lemma and Theorem 4.

THEOREM 5. *Suppose the operator* $\mathbb{K}$ *satisfies condition* (4.1) *with* $\alpha > \frac{3}{2}$, *and the initial perturbation* $\bar{u}(x) \in L_2^{\mathrm{loc}}(R_1)$ *is* 2π*-periodic. Then there exists a unique solution* $u(x,t)$ *of the Cauchy problem* (1.1) *belonging to the class* $C^\infty((0,\infty); C_\pi^\infty(R_1)) \cap C^0([0,\infty; L_2(\Omega))$.

PROOF. Applying the lemma, we find that for some $t_0 > 0$ problem (1.1) has a unique solution $u(x,t) \in C^\infty((0,t_0]; C_\pi^\infty(R_1)) \cap C^0([0,t_0]; L_2(\Omega))$. Considering the Cauchy problem (1.1) for $t \geq t_0$ with initial condition $\bar{u}(x) = u(x,t_0) \in C_\pi^\infty(R_1)$, $t = t_0$, and applying Theorem 4, we arrive at the required assertion. The theorem is proved. □

EXAMPLE. The symbol K_p corresponding to the Korteweg-de Vries-Burgers $u_t + uu_x + u_{xxx} - u_{xx} = 0$ is $K_p = -p^2 - ip$, so that the dissipation condition (4.1) holds and Theorems 4 and 5 hold.

REMARK. Consider the Cauchy problem for the Whitham equation on the entire line

$$u_t + uu_x + \int_{-\infty}^{\infty} k(x-s)u_s(s,t)\,ds = 0, \qquad u|_{t=0} = \bar{u}(x), \tag{4.38}$$

with a periodic initial function $\bar{u}(x) \in C_\pi^\infty(R_1)$. Theorems 1 through 5 remain in force for 2π-periodic solutions of problem (4.38) if the operator $\mathbb{K}$ is subject to the same conditions as in these theorems, but the periodicity requirement on the kernel is replaced by the requirement that the kernel decay at infinity together with its derivatives.

§5. The periodic problem with a weak interaction

To account for weakness of a nonlocal interaction it is natural to consider the Whitham equation with a small parameter μ in front of the integral term. Thus, we consider the problem

$$u_t + uu_x + \mu \int_{-\pi}^{\pi} k(x-s)u_s(s,t)\,ds = 0, \qquad u|_{t=0} = \bar{u}(x)\,. \tag{5.1}$$

The objective of this section is to study the asymptotics of a solution of problem (5.1) as $\mu \to 0$. The following theorem is true.

THEOREM 6. *Suppose the following conditions are satisfied*:

1) $k(x) \in C^1(\Omega\backslash 0)$, $|k(x)| \leq c|x|^{-\alpha}$, $|k'(x)| \leq c|x|^{-1-\alpha}$, $x \in \Omega\backslash 0$, $\alpha \leq \frac{3}{5} - \gamma$, $\gamma \in (0, \frac{1}{10}]$, $c > 0$, $K_p^1 \geq 0$ *for* $|p| \geq p_0 > 0$, $p \in Z$;

2) $\bar{u}(x) \in C_\pi^\infty(R_1)$;

3) *The steepness* m_0 *of the initial perturbation is sufficiently large*:

$$m_0^2 > \frac{b^2}{\gamma^2} + \frac{40c}{\gamma}(u_1 + J_3^{1/2}(0))\,.$$

Then the following assertions are true:

1) *For any* $\mu \in [0,1]$ *there exists a unique* 2π*-periodic solution* $u_\mu(x,t)$ *of problem* (5.1), $u_\mu(x,t) \in C^\infty([0,T_\mu); C_\pi^\infty(R_1))$, *where* T_μ *is the time of breaking.*

2) $u_\mu(x,t)$ *has the asymptotic expansion* $u_\mu(x,t) = \sum_{k=0}^{\infty} \mu^k u_k(x,t)$, *where the remainder term* δ_n *satisfies*

$$\max_{t\in[0,T_0-\beta]} \left\| \frac{\partial^k \delta_n}{\partial t^k} \right\|_{(l)} \leq c_0 \mu^{n+1};$$

$\beta > 0$ *is arbitrary*, $k, l \geq 0$ *are arbitrary integers, and* $\|\cdot\|_{(l)}$ *is the norm of the space* $C^l([0,2\pi])$, $c_0 = c_0(l,k,n,\beta)$.

Before proving Theorem 6 we present two simple lemmas.

LEMMA 1. *Let* $u(x,t)$ *be a solution of the problem*

$$\begin{gathered} u_t + a(x,t)u + b(x,t)u_x = f(x,t), \qquad u(x,0) = 0,\ x \in \Omega, t \in [0,T], \\ a(x,t), b(x,t), f(x,t) \in C^\infty([0,T]; C_\pi^\infty(R_1))\,. \end{gathered} \tag{5.2}$$

Then $u(x,t) \in C^\infty([0,T]; C_\pi^\infty(R_1))$ *and* $\|u\|_{(k)} \leq c\|f\|_{(k)}$, *where*

$$\|u\|_{(k)} = \sum_{l=0}^{k} \max_{x\in\Omega} \left| \frac{\partial^l u(x,t)}{\partial x^l} \right|; \ k \geq 0$$

is an arbitrary integer.

PROOF. The Cauchy problem (5.2) is linear, and the existence of a solution follows from the general theory of linear partial differential equations. We show that $\|u\|_{(k)} \leq c\|f\|_{(k)}$. Denote by $x(s,t)$ the solution of the problem

$$\dot{x} = b(x,t), \qquad x|_{t=0} = s\,. \tag{5.3}$$

According to Lemma 1 of §2, problem (5.3) has a unique solution $x(s,t)$, which establishes a diffeomorphism $R_1 \to R_1$ for each $t \in [0, T]$. Put $v(s,t) = u(x(s,t),t)$. The function $v(s,t)$ is a solution of the problem

$$v_t + av = F(s,t), \qquad v(s,0) = 0, \tag{5.4}$$

where $F(s,t) = f(x(s,t),t)$. Solving (5.4) we obtain

$$v(s,t) = \exp\left(-\int_0^t a(x(s,t),\tau)\,d\tau\right)\int_0^t F(s,\tau)\exp\left(\int_0^\tau a(x(s,t'),t')\,dt'\right)\,d\tau,$$

whence we find that $\|v\|_{(0)} \le c\|f\|_{(0)}$. Therefore, since $x(s,t)$ establishes a diffeomorphism $R_1 \to R_1$, it follows that also

$$\|u\|_{(0)} \le c\|f\|_{(0)}. \tag{5.5}$$

To estimate $\|u\|_{(1)}$ we differentiate (5.2) with respect to x. We obtain

$$\dot{u}_x + a_x u + a u_x + b_x u_x + b u_{xx} = f_x,\ u_x(x,0) = 0,$$

or, setting $u_x(x,t) \equiv w(x,t)$, we have the problem

$$w_t + a_1 w + b w_x = f_1, \qquad w(x,0) = 0 \tag{5.6}$$

for $w(x,t)$, where $a_1 = a + b_x$, $f_1 = f_x - a_x u$. Applying to problem (5.6) the same arguments as to problem (5.2), we obtain the bound $\|w\|_{(0)} \le c\|f_1\|_{(0)}$, or, considering the form of w and f_1 and the bound (5.5), $\|u\|_{(1)} \le c\|f\|_{(1)}$. Differentiating problem (5.2) further, we establish the desired estimate

$$\|u\|_{(k)} \le c\|f\|_{(k)}, \qquad k \ge 0 \text{ an arbitrary integer}. \ \square \tag{5.7}$$

LEMMA 2. *Let $u(x,t)$ be a solution of the problem*

$$u_t + u u_x = 0, \quad u(x,0) = \bar{u}(x), \quad x \in \Omega,\ \bar{u}(x) \in C_\pi^\infty(R_1). \tag{5.8}$$

Then $u(x,t) \in C^1([0,T_0]; C_\pi^\infty(R_1))$ for some $T_0 > 0$, and

$$\|u\|_{(k)} \le c_k, \qquad k \ge 0 \text{ an arbitrary integer}, \tag{5.9}$$

for $t \in [0, T_0 - \beta]$, $\beta > 0$, where the constant c_k depends on k and the norm $\|\bar{u}\|_{(k)}$.

PROOF. The existence of a solution of problem (5.8) is well known (see, for example **[74]**). We prove (5.9). Denote by $x(s,t)$ the solution of problem

$$\dot{x} = u(x,t), \qquad x|_{t=0} = s. \tag{5.10}$$

According to Lemma 1 of §2, problem (5.10) has a unique solution $x(s,t)$ which establishes a diffeomorphism $R_1 \to R_1$ for each $t \in [0, T_0)$. The function $v(s,t) =$

$u(x(s,t),t)$ is a solution of the problem $\dot{v}=0,\ v(s,0)=\bar{u}(s)$. Thus, $v(s,t)=\bar{u}(s)$ and hence

$$\|u\|_{(0)} \le \|\bar{u}\|_{(0)}\,. \tag{5.11}$$

Next, differentiating (5.8) with respect to x, we obtain $\dot{u}_x + u_x^2 + uu_{xx} = 0,\ u_x(x,0) = \bar{u}_x(x)$, or setting $u_x(x,t) \equiv w(x,t)$ we have the problem

$$\dot{w} + uw_x + w^2 = 0, \qquad w(x,0) = \bar{u}_x(x) \tag{5.12}$$

for $w(x,t)$. We use the same change of variable x: $v_1(s,t) = w(x(s,t),t)$. In view of (5.12) for v_1 we obtain the problem

$$\dot{v}_1 + v_1^2 = 0, \qquad v_1(s,0) = \bar{u}_x(s)\,.$$

From this we find that $v_1(s,t) = \bar{u}_x(s)/1 + t\bar{u}_x(s)$. From the explicit form of the solution it follows that $t\in[0,\,T_0-\beta],\,\beta>0$, or $\|v_1\|_{(0)} \le c\|\bar{u}\|_{(1)}$,

$$\|u\|_{(1)} \le c\|\bar{u}\|_{(1)}, \tag{5.13}$$

for $t\in[0,\,T_0-\beta]$, $\beta>0$, where $T_0 = (|\min_{s\in\Omega}\bar{u}_x(s)|)^{-1}$. The equations for the leading derivatives with respect to x have the same form. We differentiate problem (5.8) n times with respect to x and obtain for $n\ge 2$

$$\dot{u}_n + \sum_{k=0}^{n} C_k^n u_{(k+1)}u_{(n-k)} = 0, \qquad u_{(n)}(x,0) = \bar{u}_{(n)}(x),$$

or

$$\dot{u}_{(n)} + uu_{(n+1)} + au_{(n)} = f, \qquad u_{(n)}(x,0) = \bar{u}_{(n)}(x), \tag{5.14}$$

where $f = -\sum_{k=1}^{n-2} C_k^n u_{(k+1)}u_{(n-k)}$, $a = (n+1)u_x$. Making in problem (5.14) the same change as above, we obtain the following problem for $v_{(n)}(s,t) = u_{(n)}(x(s,t),t)$:

$$\dot{v}_{(n)} + av_{(n)} = f, \qquad v_{(n)}(s,0) = \bar{u}_{(n)}(s)\,.$$

Integration of it yields

$$v_{(n)}(s,t) = \exp\left(-\int_0^t a(x(s,\tau),\tau)\,d\tau\right)\left(\bar{u}_{(n)}(s) + \int_0^t f(x(s,\tau),\tau)e^{\int_0^\tau a dt'}\,d\tau\right).$$

From this we conclude that

$$\|u\|_{(n)} \le c_n, \qquad n\ge 2\,. \tag{5.15}$$

Comparing (5.11), (5.13), and (5.15), we see that (5.9) holds, and Lemma 2 is proved. □

Proof of Theorem 6. The first assertion of the theorem follows immediately from the proof of Theorem 1 given in §2, where the time of breaking T_μ of the solution $u_\mu(x,t)$ satisfies $T_\mu \le 1/m_0(1-\mu\gamma)$ (the lower bound $T_\mu \ge 1/m_0(1+\mu\delta)$ for the time of breaking can be proved in an entirely similar way). We now prove assertion 2). We define successively the functions $u_j(x,t)$ in the asymptotic expansion of the solution $u_\mu(x,t)$ as the solutions of the problems

$$
\begin{gathered}
\dot u_0 + u_0 u_0' = 0, \qquad u_0|_{t=0} = \bar u(x), \\
\dot u_1 + u_0 u_1' + u_1 u_0' + \int_{-\pi}^{\pi} k(x-s) u_0'(s,t)\,ds = 0, \qquad u_1|_{t=0} = 0, \\
\dot u_j + \sum_{l+r=j} u_l u_r' + \int_{-\pi}^{\pi} k(x-s) u_{j-1}'(s,t)\,ds = 0, \qquad u_j|_{t=0} = 0, \quad j \ge 2
\end{gathered}
\tag{5.16}
$$

(the dot denotes differentiation with respect to t and the prime differentiation with respect to x). All the problems (5.16) with $j \ge 1$ are linear. It is obvious that there exist solutions of all problems (5.16) with $j \ge 0$ and $u_j(x,t) \in C^\infty([0,T_0); C_\pi^\infty(R_1))$, where $T_0 = (|\min_{x\in\Omega} \bar u'(x)|)^{-1}$ is the time of breaking of problem (5.16) with $j=0$ or of the degenerate problem (5.1) with $\mu = 0$. We set $U_n(x,t) = \sum_{j=0}^n \mu^j u_j(x,t)$ and $\delta_n(x,t) = u_\mu(x,t) - U_n(x,t)$. In view of the definition of the functions $u_j(x,t)$ as solutions of problems (5.16), $U_n(x,t)$ and $\delta_n(x,t)$ are solutions of the problems

$$
\begin{aligned}
&\dot U_n + U_n U_n' + \mu \int_{-\pi}^{\pi} k(x-s) U_n'(s,t)\,ds \\
&\quad = \mu^{n+1} P(u_1,\dots,u_n') + \mu^{n+1} \int_{-\pi}^{\pi} k(x-s) u_n'(s,t)\,ds, \qquad U_n|_{t=0} = \bar u(x)\,.
\end{aligned}
\tag{5.17}
$$

$$
\begin{aligned}
&\dot\delta_n + u'\delta_n + U\delta_n' = -\mu \int_{-\pi}^{\pi} k(x-s)\delta_n'(s,t)\,ds \\
&\quad - \mu^{n+1} F(u_1,\dots,u_n') - \mu^{n+1} \int_{-\pi}^{\pi} k(x-s) u_n'(s,t)\,ds, \qquad \delta_n|_{t=0} = 0,
\end{aligned}
\tag{5.18}
$$

where

$$
P(u_1,\dots,u_n') = \sum_{\substack{n+1\le l+r\le 2n,\\ 0\le l,r\le n}} \mu^{l+r-(n+1)} u_l u_r'\,.
$$

Successively integrating (5.16), we conclude that all $u_j \in C^\infty([0,T_0-\beta]; C_\pi^\infty(R_1))$, where the integral terms in (5.16) belong to the class $C^\infty([0,T_0-\beta], C_\pi^\infty(R_1))$. Therefore, they are bounded together with all their derivatives with respect to x and t, since

$$
\int_{-\pi}^{\pi} k(x-s) u_{j-1}'(s,t)\,ds = \int_{-\pi}^{\pi} k(s) u_{j-1}'(x-s,t)\,ds,
$$

while u_{j-1}, as a solution of problem (5.16) for the preceding index, belongs to $C^\infty([0,T_0-\beta]; C_\pi^\infty(R_1))$ for any $\beta > 0$, and the kernel $k(s)$ is absolutely integrable on $[-\pi,\pi]$. It follows from here, in view of Lemmas 1 and 2, that

$$
\|u_j\|_{(l)} \le c \quad \text{for all } j \text{ and } l \ge 0, \tag{5.19}
$$

where the constant c depends on j, l, $\|\bar u\|_{(l)}$ and $\int_{-\pi}^{\pi} |k(s)\,ds$. From (5.19) it follows that for any integer $l \ge 0$

$$
\|U_n\|_{(l)} \le c. \tag{5.20}
$$

Therefore, we have the estimate $\|\delta_n\|_{(l)} \le \|u_\mu\|_{(l)} + \|U_n\|_{(l)} \le c$ for any $l \ge 0$, for

$\delta_n(x,t) = u_\mu(x,t) - U_n(x,t)$ as well. Returning now to problem (5.18) and using Lemma 1, we obtain

$$\|\delta_n\|_{(l)} \le c\|\Phi\|_{(l)} \le c\mu, \tag{5.21}$$

where Φ denotes the entire right-hand side of (5.18). Using (5.21) with $l = 1$ to estimate the first term on the right in (5.18), we find that $\|\Phi\|_{(l)} \le c\mu^2$. Substitution of this in (5.21) yields $\|\delta_n\|_{(l)} \le c\mu^2$, etc. Finally, we obtain $\|\delta_n\|_{(l)} \le c\mu^{n+1}$ for all $t \in [0, T_0 - \beta]$.

The bound

$$\max_{t\in[0,T_0-\beta]} \left\| \frac{\partial^k \delta_n}{\partial t^k} \right\|_{(l)} \le c\mu^{n+1}, \qquad k = 1, 2, \ldots,$$

can be proved in exactly the same way. Theorem 6 is proved. □

CHAPTER 4

The System of Equations of Surface Waves

Our goal in this chapter is to study the system of equations describing surface waves on water

$$\begin{cases} \eta_t + (\nabla, \eta\nabla\varphi) + \mathbb{K}_1(\eta) + \mathbb{K}_2(\varphi) = 0, \\ \varphi_t + \frac{1}{2}(\nabla\varphi)^2 + \mathbb{K}_3(\eta) + \mathbb{K}_4(\varphi) = 0; \end{cases}$$

where $\eta(x, t)$ is the free surface of water, $\varphi(x, t)$ is the velocity potential, $x \in R_2$, $t \geq 0$, and the $\mathbb{K}_j$ are linear pseudodifferential operators defined by

$$\mathbb{K}_j(\psi) = \frac{1}{4\pi^2} \iint_{-\infty}^{\infty} e^{i(p,x)} K_j(p)\hat{\psi}(p)\, dp, \qquad j = 1, \dots, 4,$$

where $\hat{\psi}(p) = \iint_{-\infty}^{\infty} e^{-i(p,x)}\psi(x)\, dx$ is the Fourier transform of the function $\psi(x)$. The symbols $K_j(p)$ of the operators $\mathbb{K}_j$ represent the dispersion relation for the system of surface waves, and for different choices of these relations the operators $\mathbb{K}_j$ correspond to different processes of wave propagation: conservative, dissipative, or processes related to an energy pumping. In view of this, the system of surface wave equations is very general.

The following system of equations is more general:

$$\begin{cases} \eta_t + (\nabla, \eta u) + \mathbb{K}_1(\eta) + \mathbb{K}_2(u) = 0, \\ u_t + (u, \nabla)u + \mathbb{K}_3(\eta) + \mathbb{K}_4(u) = 0, \end{cases}$$

where η is the free surface and u is the velocity vector. If we differentiate the first system with respect to x_1 and x_2 and put $u = \nabla\varphi$, we obtain the second system of equations.

The content of this chapter is as follows. In §1 the conservation laws are presented, in §2 the Cauchy problem for the system of equations of surface waves with regular operators $\mathbb{K}$ is considered, in §3 the Cauchy problem is studied for these systems in the case of a singular dissipative or conservative operator $\mathbb{K}$. Section 4 is devoted to the problem of breaking of waves. In §5 a theorem on the existence of a global solution of the Cauchy problem for the system of equations of surface waves is proved. In §§6 and 7 the problem of smoothing initial perturbations is considered in the case of a strongly dissipative operator $\mathbb{K}$. In §8 a theorem on asymptotic behavior of solutions for the system of equations for surface waves with respect to a parameter of a weak nonlocal interaction is formulated.

The results of this chapter have been published earlier (see **[55, 60, 85, 86]**).

§1. Conservation laws

We derive the basic conservation laws for solutions of the systems of equations for surface waves, assuming that the functions η and φ decay sufficiently fast as $|x| \to \infty$.

1) The law of conservation of mass $I_1 \equiv \iint_{-\infty}^{\infty} \eta(x,t)\,dx = \text{const}$ is valid if

$$\iint_{-\infty}^{\infty} \mathbb{K}_j(\psi)\,dx = 0, \quad \psi(x) \in H^\infty(R_2), \quad j = 1, 2,$$

as follows immediately from the first equation of the system

$$\eta_t + (\nabla, \eta\nabla\varphi) + \mathbb{K}_1(\eta) + \mathbb{K}_2(\varphi) = 0\,.$$

2) The law of conservation of momentum $I_2 \equiv \iint_{-\infty}^{\infty} \eta\nabla\varphi\,dx = \text{const}$ holds if the following condition is satisfied (the adjoint is taken in $L_2(R_2)$):

$$\mathbb{K}_1 = -\mathbb{K}_4^*, \quad \mathbb{K}_2 = \mathbb{K}_2^*, \quad \mathbb{K}_3 = \mathbb{K}_3^*\,. \tag{1.1}$$

In fact, we show that, for example, the first component $I_2^{(1)}$ of the momentum is conserved

$$\begin{aligned}
\frac{d}{dt}I_2^{(1)} &= \iint_{-\infty}^{\infty} (\eta_t\varphi_{x_1} + \eta\varphi_{x_1t})\,dx \\
&= -\iint_{-\infty}^{\infty} \Big\{\varphi_{x_1}[(\nabla, \eta\nabla\varphi) + \mathbb{K}_1(\eta) + \mathbb{K}_2(\varphi)] \\
&\qquad\qquad + \eta\frac{\partial}{\partial x_1}\left[\frac{(\nabla\varphi)^2}{2} + \mathbb{K}_3(\eta) + \mathbb{K}_4(\varphi)\right]\Big\}\,dx \\
&= -\iint_{-\infty}^{\infty} \{\varphi_{x_1}(\mathbb{K}_1(\eta) + \mathbb{K}_2(\varphi)) + \eta(\mathbb{K}_3(\eta_{x_1}) + \mathbb{K}_4(\varphi_{x_1}))\}\,dx,
\end{aligned} \tag{1.2}$$

since

$$\begin{aligned}
&\iint_{-\infty}^{\infty} \{\varphi_{x_1}(\eta_{x_1}\varphi_{x_1} + \eta\varphi_{x_1x_1} + \eta_{x_2}\varphi_{x_2} + \eta\varphi_{x_2x_2}) + \eta(\varphi_{x_1}\varphi_{x_1x_1} + \varphi_{x_2}\varphi_{x_1x_2})\}\,dx \\
&\qquad = \iint_{-\infty}^{\infty} \left\{\frac{\partial}{\partial x_1}(\varphi_{x_1}^2\eta) + \frac{\partial}{\partial x_2}(\varphi_{x_1}\varphi_{x_2}\eta)\right\}\,dx = 0\,.
\end{aligned}$$

In view of (1.1), the expression (1.2) is equal to zero. We observe that the systems of the shallow water equations, Boussinesq and Dobrokhotov (see Introduction), satisfy (1.1).

3) The law of conservation of energy also holds if (1.1) is satisfied:

$$I_3 \equiv \iint_{-\infty}^{\infty} \{\eta(\nabla\varphi)^2 - 2\varphi\mathbb{K}_1(\eta) - \varphi\mathbb{K}_2(\varphi) + \eta\mathbb{K}_3(\eta)\}\,dx = \text{const}\,.$$

The proof is analogous to that of 2). The law of conservation of energy I_3 makes it possible to write the system of the shallow water equations in Hamiltonian form with the Hamiltonian $\mathscr{H} = -I_3/2$:

$$\eta_t = -\frac{\delta\mathscr{H}}{\delta\varphi}, \qquad \varphi_t = \frac{\delta\mathscr{H}}{\delta\eta}\,.$$

4) Motion of the center of mass. If the symbols $K_1(p)$ and $K_2(p)$ have a zero of second order at $p = 0$, then under conditions (1.1) the following integral is conserved:

$$I_4 \equiv \iint_{-\infty}^{\infty} (x\eta(x,t) - t\eta(x,t)\nabla\varphi(x,t))\,dx = \text{const}\ .$$

Indeed,

$$\dot{I}_4 = \iint_{-\infty}^{\infty} (x\eta_t - \eta\nabla\varphi)\,dx - t\dot{I}_2 = \iint_{-\infty}^{\infty} (x\eta_t - \eta\nabla\varphi)\,dx,$$

since I_2 is conserved in time. Therefore, for instance, upon integration by parts for the first component $I_4^{(1)}$, we have

$$\dot{I}_4 = \iint_{-\infty}^{\infty} \left\{ x_1 \left[\frac{\partial}{\partial x_1}(\eta\varphi_{x_1}) + \frac{\partial}{\partial x_2}(\eta\varphi_{x_2}) \right] - \eta\varphi_{x_1} - x_1\mathbb{K}_1(\eta) - x_1\mathbb{K}_2(\varphi) \right\} dx$$

$$= - \iint_{-\infty}^{\infty} x_1(\mathbb{K}_1(\eta) + \mathbb{K}_2(\varphi))\,dx = i\frac{\partial}{\partial p_1}[K_1(p)\hat{\eta}(p) + K_2(p)\hat{\varphi}(p)]|_{p=0} = 0\,.$$

The constancy of the integral I_4 implies the equation for the radius vector of the center of mass

$$R \equiv \iint_{-\infty}^{\infty} x\eta(x,t)\,dx = I_4 + tI_2, \qquad \frac{dR}{dt} = I_2 = \text{const},$$

i.e., the velocity of motion of the center of mass is constant.

5) The conservation law for the following equation

$$\varphi_t + \tfrac{1}{2}(\nabla\varphi)^2 + \mathbb{K}_4(\varphi) = 0$$

$$I_5 = \iint_{-\infty}^{\infty} \varphi_{x_1}\varphi_{x_2}\,dx = \text{const}$$

holds if $\mathbb{K}_4 = -\mathbb{K}_4^*$. Indeed,

$$\dot{I}_5 = - \iint_{-\infty}^{\infty} \left\{ \frac{1}{2}\varphi_{x_2}\frac{\partial}{\partial x_1}(\nabla\varphi)^2 + \varphi_{x_2}\mathbb{K}_4(\varphi_{x_1}) + \frac{1}{2}\varphi_{x_1}\frac{\partial}{\partial x_2}(\nabla\varphi)^2 + \varphi_{x_1}\mathbb{K}_4(\varphi_{x_2}) \right\} dx$$

$$= \iint_{-\infty}^{\infty} \varphi_{x_1x_2}(\nabla\varphi)^2\,dx = \frac{1}{3}\iint_{-\infty}^{\infty} \left[\frac{\partial}{\partial x_2}(\varphi_{x_1}^3) + \frac{\partial}{\partial x_1}(\varphi_{x_2}^3) \right] dx = 0\,.$$

§2. The Cauchy problem for the system of equations of surface waves with a regular operator

We consider the Cauchy problem for the system of equations of surface waves

$$\begin{cases} \eta_t + (\nabla, \eta\nabla\varphi) + \mathbb{K}_1(\eta) + \mathbb{K}_2(\varphi) = 0, \\ \varphi_t + \frac{1}{2}(\nabla\varphi)^2 + \mathbb{K}_3(\eta) + \mathbb{K}_4(\varphi) = 0, \\ \eta|_{t=0} = \bar{\eta}(x), \qquad \varphi|_{t=0} = \bar{\varphi}(x), \end{cases} \tag{2.1}$$

where $\eta(x,t)$ is the height of the free surface of the water measured from the zeroth unperturbed level, $\varphi(x,t)$ is the velocity potential, and $u = \nabla\varphi$.

Together with (2.1) we consider the system of equations for η and u obtained when the operator ∇ is applied to the second equation in (2.1)

$$\begin{cases} \eta_t + (\nabla, \eta u) + \mathbb{K}_1(\eta) + \mathbb{K}_2(u) = 0, \\ u_t + (u, \nabla)u + \mathbb{K}_3(\eta) + \mathbb{K}_4(u) = 0, \\ \eta|_{t=0} = \bar{\eta}(x), \qquad u|_{t=0} = \bar{u}(x), \end{cases} \tag{2.2}$$

where $u(x, t)$ is a two-dimensional vector, and the previous notation is preserved for the operators $\mathbb{K}_j$ (here the operators $\mathbb{K}_2$, $\mathbb{K}_3$, and $\mathbb{K}_4$ are to be considered matrix operators).

In this section we show that the Cauchy problem (2.2) with a regular operator $\mathbb{K}$ has a classical solution $(\eta, u) \in C^\infty([0, T]; (H^\infty(R_2))^3)$, which exists up to a time $T > 0$ and can destruct only as a result of breaking.

We first consider an auxiliary system of equations:

$$\dot{y} = u(y, t), \qquad t > 0, \tag{2.3}$$

$$y|_{t=0} = \xi \in R_2 . \tag{2.4}$$

LEMMA 1. *Let* $u(y, t) \in C^\infty([0, T]; (H^\infty(R_2))^2)$ *for some* $T > 0$. *Then*

1) *There exists a unique solution* $y(\xi, t) \in C^\infty(R_2 \times [0, T]))$ *of problem* (2.3), (2.4).

2) *For any* $x \in R_2$, *there exists a unique solution* $\xi(x, t)$ *of the equation* $y(\xi, t) = x$ *for each* $t \in [0, T]$, *so that the solution* $y(\xi, t)$ *of problem* (2.3), (2.4) *is a diffeomorphism* $R_2 \to R_2$ *for each* $t \in [0, T]$.

PROOF. Both assertions follow directly from the general theory of systems of ordinary differential equations, since by the condition of the lemma the right-hand side $u(y, t)$ of (2.3) is bounded together with all of its derivatives in the region $R_2 \times [0, T]$. □

We introduce the notion of a regular operator.

DEFINITION. An operator $\mathbb{K}(\psi)$ is called regular if in the x-representation it has the form of a convolution

$$\mathbb{K}(\psi) = \iint_{-\infty}^{\infty} k(x - y)\psi(y)\, dy, \tag{2.5}$$

for any $\psi \in H^\infty(R_2)$, where its kernel $k(x) \in L_1(R_2)$, i.e., for some $\mathcal{N} > 0$

$$\iint_{-\infty}^{\infty} |k(x)|\, dx \le \mathcal{N} . \tag{2.6}$$

THEOREM 1. *Suppose the following conditions are satisfied*:

1) *The operators*

$$\mathbb{K} \equiv \begin{pmatrix} \mathbb{K}_1 & \mathbb{K}_2 \\ \mathbb{K}_3 & \mathbb{K}_4 \end{pmatrix}$$

and $\nabla\mathbb{K}_3$ *are regular*;

2) $(\bar{\eta}(x), \bar{u}(x)) \equiv \bar{v}(x) \in (H^\infty(R_2))^3$.

Then the Cauchy problem (2.2) *has a solution in the class* $C^\infty([0, T_0); (H^\infty(R_2))^3)$, $T_0 > 0$. *This solution can destruct only as a result of breaking.*

PROOF. To prove the existence of a solution of the Cauchy problem (2.2) we apply the method of successive approximations. We set $v^{(0)}(x, t) = \bar{v}(x)$ and define the vector-valued function $v^{(s+1)}(x,t) \equiv (\eta^{(s+1)}, u^{(s+1)})$ for $s \geq 0$ as the solution of the following problem:

$$
\left\{
\begin{array}{l}
\eta_t^{(s+1)} + (\nabla, \eta^{(s+1)} u^{(s)}) + \mathbb{K}_1(\eta^{(s)}) + \mathbb{K}_2(u^{(s)}) = 0, \\
u_t^{(s+1)} + (u^{(s)}, \nabla) u^{(s+1)} + \mathbb{K}_3(\eta^{(s)}) + \mathbb{K}_4(u^{(s)}) = 0, \\
v^{(s+1)}(x,0) = \bar{v}(x) .
\end{array}
\right.
\tag{2.7}
$$

We shall prove that for all $s \geq 0$ problem (2.7) has a solution $v^{(s+1)}(x,t) \in C^\infty(R_+; (H^\infty(R_2))^3)$. Omitting temporarily the index $(s+1)$, we rewrite (2.7) in the form

$$
\left\{
\begin{array}{l}
\eta_t + (\nabla, \eta g) + f = 0, \\
u_t + (g, \nabla) u + h = 0, \qquad v(x,0) = \bar{v}(x)
\end{array}
\right.
\tag{2.8}
$$

where $g = u^{(s)}$, $f = \mathbb{K}_1(\eta^{(s)}) + \mathbb{K}_2(u^{(s)})$, and $h = \mathbb{K}_3(\eta^{(s)}) + \mathbb{K}_4(u^{(s)})$. Suppose $v^{(s)} \in C^\infty(R_+; (H^\infty(R_2))^3)$ has already been determined by induction; then, since the operator $\mathbb{K}$ is regular, we have $g, h \in C^\infty(R_+; (H^\infty(R_2))^2)$, $f \in C^\infty(R_+; H^\infty(R_2))$, $\bar{v} \in (H^\infty(R_2))^3$. Therefore, from the general theory of linear partial differential equations it follows that there exists a solution of problem (2.8), a vector-valued function $v(x,t) \in C^\infty(R_+; (C^\infty(R_2))^3)$. Moreover, the linear system (2.8) splits into three independent equations, and thus, by integrating each of them along the characteristics and using Lemma 1, it is easy to show that $v(x,t) \in C^\infty(R_+; (H^\infty(R_2))^3)$, since the coefficients and the right-hand side of (2.8) belong to $H^\infty(R_2)$ for each $t \geq 0$. In an entirely similar manner it can be proved that any derivative $\partial^k v(x,t)/\partial t^k \in (H^\infty(R_2))^3$. Thus, we have proved that for all $s \geq 0$ problem (2.7) has a solution $v^{(s+1)}(x,t) \in C^\infty(R_+; (H^\infty(R_2))^3)$. We now prove that for some $T > 0$ the sequence $\{v^{(s)}(x,t)\}$, $s = 0, 1, \dots$, is bounded in $C^\infty([0,T]; (C^\infty(R_2))^3)$ uniformly with respect to $s \geq 0$. We differentiate (2.7) n times with respect to x_1 and m times with respect to x_2 and obtain

$$
\begin{array}{l}
\dfrac{\partial}{\partial t} \eta_{n,m}^{(s+1)} + \displaystyle\sum_{i=0}^{n} \sum_{j=0}^{m} C_i^n C_j^m (\nabla, \eta_{i,j}^{(s+1)} u_{n-i,m-j}^{(s)}) + \mathbb{K}_1(\eta_{n,m}^{(s)}) + \mathbb{K}_2(u_{n,m}^{(s)}) = 0, \\
\dfrac{\partial}{\partial t} u_{n,m}^{(s+1)} + \displaystyle\sum_{i=0}^{n} \sum_{j=0}^{m} C_i^n C_j^m (u_{n-i,m-j}^{(s)}, \nabla) u_{i,j}^{(s+1)} + \mathbb{K}_3(\eta_{n,m}^{(s)}) + \mathbb{K}_4(u_{n,m}^{(s)}) = 0;
\end{array}
\tag{2.9}
$$

here C_i^n, C_j^m are the binomial coefficients. We denote

$$
\psi_{n,m}(x,t) \equiv \frac{\partial^{n+m} \psi(x,t)}{\partial x_1^n \partial x_2^m}, \qquad \psi_{(n,m)}(t) \equiv \max_{x \in R_2} |\psi_{n,m}(x,t)| .
\tag{2.10}
$$

We define the vector-valued function $x^s(y,t)$ as the solution of problem (2.3), (2.4) with the right-hand side equal to $u^{(s-1)}(x,t) \in C^\infty(R_+; (H^\infty(R_2))^2)$. By Lemma 1, $x^s(y,t)$ realizes a diffeomorphism $R_2 \to R_2$ for each $t \geq 0$. Therefore, if we introduce the new functions $\tilde{\eta}_{n,m}^{(s)}(y,t) \equiv \eta_{n,m}^{(s)}(x^{(s)}(y,t),t)$, $\tilde{u}_{n,m}^{(s)}(y,t) \equiv u_{n,m}^{(s)}(x^s(y,t),t)$, and $\tilde{v}_{n,m}^{(s)}(y,t) \equiv (\tilde{\eta}_{n,m}^{(s)}, \tilde{u}_{n,m}^{(s)})$, then $\tilde{v}_{(n,m)}^{(s)}(t) = v_{(n,m)}^{(s)}(t)$. Introduction of the functions $x^{(s)}(y,t)$ removes the leading terms in the sums in (2.9). We also denote

$$
\sum_{c \leq n+m \leq k} \tilde{\eta}_{(n,m)}^{(s)}(t) = \eta_{(k)}^{(s)}(t), \quad \sum_{c \leq n+m \leq k} \tilde{u}_{(n,m)}^{(s)}(t) = u_{(k)}^{(s)}(t) .
\tag{2.11}
$$

Using the regularity of the operator $\mathbb{K}$, we estimate the integral terms in (2.9) as follows ($n+m=k$):

$$\begin{aligned} |\mathbb{K}_1(\eta_{n,m}^{(s)})| &\le \mathscr{N}\eta_{(k)}^{(s)}(t), \qquad |\mathbb{K}_2(u_{n,m}^{(s)})| \le \mathscr{N}u_{(k)}^{(s)}(t), \\ |\mathbb{K}_3(\eta_{n,m}^{(s)})| &\le \mathscr{N}\eta_{(k-1)}^{(s)}(t), \qquad |\mathbb{K}_4(u_{n,m}^{(s)})| \le \mathscr{N}u_{(k)}^{(s)}(t) \end{aligned} \tag{2.12}$$

(in the operator $\mathbb{K}_3$ one derivative of the function $\eta_{n,m}^{(s)}$ is transferred into the kernel). From (2.9), (2.11), and (2.12) we obtain ($n+m=k$)

$$\begin{aligned} |\dot{\tilde{\eta}}_{n,m}^{(s+1)}| &\le c_k\left(\sum_{0\le j\le k}\eta_{(j)}^{(s+1)}u_{(k+1-j)}^{(s)}+\eta_{(k)}^{(s)}+u_{(k)}^{(s)}\right), \\ |\dot{\tilde{u}}_{n,m}^{(s+1)}| &\le c_k\left(\sum_{0\le j\le k-1}u_{(j+1)}^{(s+1)}u_{(k-j)}^{(s)}+\eta_{(k-1)}^{(s)}+u_{(k)}^{(s)}\right), \end{aligned} \tag{2.13}$$

where c_k denotes all constants not depending on s, and the modulus sign denotes the sum of the moduli of the vector components. Integrating (2.13) with respect to t we obtain

(2.14)

$$|\tilde{\eta}_{n,m}^{(s+1)}(x,t)| \le |\tilde{\eta}_{n,m}^{(s+1)}(x,c)| + c_k\int_0^t\left[\sum_{0\le j\le k}\eta_{(j)}^{(s+1)}u_{(k+1-j)}^{(s)}+\eta_{(k)}^{(s)}+u_{(k)}^{(s)}\right]dt,$$

$$|\tilde{u}_{n,m}^{(s+1)}(x,t)| \le |\tilde{u}_{n,m}^{(s+1)}(x,0)| + c_k\int_0^t\left[\sum_{1\le j\le k}u_{(j)}^{(s+1)}u_{(k+1-j)}^{(s)}+\eta_{(k-1)}^{(s)}+u_{(k)}^{(s)}\right]dt\,.$$

We take the maximum over $x\in R_2$ of both sides of (2.14), sum the first inequality (2.14) over all n and m, $0\le n+m\le k-1$, and the second over all n and m, $0\le n+m\le k$ ($k\ge 1$ is fixed), and, adding the results, we arrive at the inequality

(2.15)

$$\begin{aligned} w_{(k)}^{(s+1)}(t) &\le w_{(k)}^{(s+1)}(0) + c_k\int_0^t\left\{\sum_{1\le j\le k}[\eta_{(j-1)}^{(s+1)}u_{(k+1-j)}^{(s)}+u_{(j)}^{(s+1)}u_{(k+1-j)}^{(s)}]+w_{(k)}^{(s)}\right\}dt \\ &\le w_{(k)}^{(s+1)}(0) + c_k\int_0^t\left\{\sum_{1\le j\le k}w_{(j)}^{(s+1)}w_{(k+1-j)}^{(s)}+w_{(k)}^{(s)}\right\}dt, \end{aligned}$$

where the notation $w_{(j)}^{(s)}=\eta_{(j-1)}^{(s)}+u_{(j)}^{(s)}$ has been introduced, and the same for the index $(s+1)$, $j=0,1,\dots,k$.

Consider the following system of equations on $[0, T]$:

$$\begin{cases} \dot{p}_k = 2c_k\left(\displaystyle\sum_{1\le j\le k}p_jp_{k+1-j}+p_k\right), \qquad k\ge 0, \\ p_k(0)=2w_{(k)}(0)\,. \end{cases} \tag{2.16}$$

Here $T>0$ determines the time of existence of a bounded solution of (2.16), i.e., $p_k(t)\in C^1([0,T])$, $k=0,1,\dots$. Observe that in (2.16) only the equation with $k=1$ is nonlinear, while all the others are linear. Integration of (2.16) with $k = 1$ gives

$$p_1(t) = p_1(0)e^{2c_1 t}\{1 + p_1(0)(1 - e^{2c_1 t})\}^{-1},$$

whence it follows that the function $p_1(t)$ remains bounded together with its derivatives on the entire interval $[0, T]$, provided T is chosen from the condition

$$0 < T < \frac{1}{2c_1} \ln\left(1 + \frac{1}{p_1(0)}\right). \tag{2.17}$$

Integrating the remaining (linear) equations in (2.16) over $[0, T]$, we obtain

$$p_k(t) \leq \mathscr{D}_k, \qquad k = 0, 1, \dots, \ t \in [0, T], \tag{2.18}$$

where $\mathscr{D}_k \geq 0$ are some fixed numbers. We shall prove that system (2.16) majorizes the system of inequalities (2.15) on $[0, T]$, i.e., that

$$w^{(s)}_{(k)}(t) < p_k(t), \qquad t \in [0, T],\ k, s = 0, 1, 2, \dots . \tag{2.19}$$

Indeed, the inequalities (2.19) hold for $t = 0$, and also for all $k \geq 0$ and $t \in [0, T]$ if $s = 0$. We apply induction on s. We prove (2.19) for $s > 0$ by contradiction. Suppose $k_1 \geq 0$ is the least index for which (2.19) is not satisfied. Let $T_1 < T$ be the first time (2.19) is violated for the index k_1. In view of the continuity of the functions $w^{(s+1)}_{(k)}(t)$ and $p_k(t)$, there exists some T_2, $T_1 < T_2 < T$, such that for all $t \in [0, T_2]$ and $k \leq k_1$

$$w^{(s+1)}_{(k)}(t) \leq 2p_k(t). \tag{2.20}$$

Substituting (2.20) and (2.19) with index s, valid by the induction hypothesis, in (2.15), we find that for $k \leq k_1$ and $t \in [0, T_2]$

$$w^{(s+1)}_{(k)}(t) \leq w^{(s+1)}_{(k)}(0) + 2c_k \int_0^t \left[\sum_{1 \leq j \leq k} p_j(\tau)p_{k+1-j}(\tau) + p_k(\tau)\right] d\tau \leq p_k(t), \tag{2.21}$$

where the last inequality is satisfied by (2.16). It follows from (2.21) that for all $k \leq k_1$ the inequality (2.19) with index $s + 1$ is satisfied also for $t = T_1$, which contradicts our assumption. Thus, (2.19) has been proved on $[0, T]$ for all $k, s \geq 0$.

The estimates (2.18) and (2.19) together with the definition of the $w^{(s)}_{(k)}(t)$ imply that the functions $v^{(s)}_{(n,m)}(t)$ are bounded for each $n, m \geq 0$ uniformly with respect to $s \geq 0$ and $t \in [0, T]$. It can be proved by similar arguments that the expressions

$$\max_{0 \leq t \leq T, x \in R_2} \left|\frac{\partial k}{\partial t^k} v^{(s)}_{n,m}(x, t)\right|,$$

are bounded for all $n, m, k \geq 0$ uniformly with respect to $s \geq 0$. Thus, it has been proved that the sequence of vector-valued functions $\{v^{(s)}(x, t)\}$, $s = 0, 1, \dots$, is bounded in $C^\infty([0, T]; (C^\infty(R_2))^3)$ uniformly with respect to $s \geq 0$.

We now prove that the sequence $\{v^{(s)}(x, t)\}$, $s \geq 0$, is bounded also in the class $C^\infty([0, T]; (H^\infty(R_2))^3)$ uniformly with respect to $s \geq 0$. Multiplying the equations in

(2.9) by $\eta_{n,m}^{(s+1)}$ and $u_{n,m}^{(s+1)}$, respectively, and integrating over all $x \in R_2$, we obtain

$$\tag{2.22} \begin{aligned} &\tfrac{1}{2}\dot{H}_{n,m}^{(s+1)} + \sum_{i=0}^{n}\sum_{j=0}^{m} C_i^n C_j^m \iint_{-\infty}^{\infty} \eta_{n,m}^{(s+1)}(\nabla, \eta_{i,j}^{(s+1)} u_{n-i,m-j}^{(s)})\,dx \\ &\qquad + \iint_{-\infty}^{\infty} \eta_{n,m}^{(s+1)}[\mathbb{K}_1(\eta_{n,m}^{(s)}) + \mathbb{K}_2(u_{n,m}^{(s)})]\,dx = 0, \\ &\tfrac{1}{2}\dot{U}_{n,m}^{(s+1)} + \sum_{i=0}^{n}\sum_{j=0}^{n} C_i^n C_j^m \iint_{-\infty}^{\infty} (u_{n,m}^{(s+1)}, (u_{n-i,m-j}^{(s)}, \nabla) u_{i,j}^{(s+1)})\,dx \\ &\qquad + \iint_{-\infty}^{\infty} (u_{n,m}^{(s+1)}, [\mathbb{K}_3(\eta_{n,m}^{(s)}) + \mathbb{K}_4(u_{n,m}^{(s)})])\,dx = 0\,. \end{aligned}$$

Here we have used the notation

$$\tag{2.23} \iint_{-\infty}^{\infty} (\eta_{n,m}^{(s+1)})^2\,dx = H_{n,m}^{(s+1)}, \qquad \iint_{-\infty}^{\infty} (u_{n,m}^{(s+1)})^2\,dx = U_{n,m}^{(s+1)}\,.$$

With the aid of the Cauchy inequality, (2.18), (2.19), and the identity ($g \in (H^1(R_2))^2$)

$$\tag{2.24} \iint_{-\infty}^{\infty} \psi(g, \nabla\psi)\,dx = -\tfrac{1}{2}\iint_{-\infty}^{\infty} \psi^2 \operatorname{div} g\,dx, \qquad \psi \in H^1(R_2),$$

we transform (2.22) to the form ($n + m = k$)

$$\tag{2.25} \begin{aligned} \dot{H}_{n,m}^{(s+1)} &\le c_k \Bigg\{ \sum_{0\le j\le k-1} [\mathscr{D}_{k-j}(H_k^{(s+1)} H_{j+1}^{(s+1)})^{1/2} + \mathscr{D}_{k-j+1}(H_k^{(s+1)} H_j^{(s+1)})^{1/2}] \\ &\qquad + \mathscr{D}_1 H_k^{(s+1)} \Bigg\} \\ &\quad + \mathscr{N}((H_k^{(s+1)} H_k^{(s)})^{1/2} + (H_k^{(s+1)} U_k^{(s)})^{1/2}), \\ \dot{U}_{n,m}^{(s+1)} &\le c_k \Bigg\{ \sum_{0\le j\le k-1} \mathscr{D}_{k-j}(U_k^{(s+1)} U_{j+1}^{(s+1)})^{1/2} \Bigg\} \\ &\quad + \mathscr{N}((U_k^{(s+1)} H_{k-1}^{(s)})^{1/2} + (U_k^{(s+1)} U_k^{(s)})^{1/2}); \end{aligned}$$

here, in estimating the integral terms we have used the condition (2.6) of regularity of the kernels and introduced the notation $\sum_{0\le n+m\le k} H_{n,m}^{(s)} = H_k^{(s)}$, $\sum_{0\le n+m\le k} U_{n,m}^{(s)} = U_k^{(s)}$, and the same for the index $s+1$; c_k does not depend on s. Summing (2.25) over all n and m such that $0 \le n+m \le k$, ($k \ge 0$ is fixed) and introducing the notation $J_k^{(s)} = H_k^{(s)} + U_k^{(s)}$, from (2.25) we find that

$$\tag{2.26} \dot{J}_k^{(s+1)} \le c_k\{J_k^{(s+1)}(J_k^{(s+1)} + J_k^{(s)})\}^{1/2}, \qquad k = 0, 1, 2, \dots\,.$$

As in (2.15) and (2.16), from the system of inequalities (2.26) we go over to the majorant system

$$\tag{2.27} \dot{R}_k = 4c_k R_k, \quad R_k(0) = J_k^{(0)}(0), \qquad k = 0, 1, 2, \dots\,.$$

Integrating (2.27), we get

$$\tag{2.28} J_k^{(s)}(t) \le R_k(0) = J_k^{(0)}(0)e^{4c_k t} \le B_k, \qquad t \in [0, T],$$

for all $k \geq 0$ and $s \geq 0$, where the constants B_k are independent of s. Similar assertions are valid for the vector-valued functions $(\partial^k/\partial t^k)v_{n,m}^{(s)}(x,t)$. Thus, the sequence of vector-valued functions $\{v^{(s)}(x,t)\}$ is bounded in $C^\infty([0,T];(H^\infty(R_2))^3)$ uniformly with respect to $s \geq 0$. Therefore, some subsequence of $\{v^{(s)}(x,t)\}$ converges to a vector-valued function $v(x,t) \in C^\infty([0,T];(H^\infty(R_2))^3)$, which is obviously a solution of the Cauchy problem (2.2). In proving the existence of a solution we have seen that if the function $p_1(t)$ is bounded on $[0,T]$, then the remaining functions $p_k(t)$, $k = 0, 1, \dots$, are also bounded, and hence, in view of the inequality (2.19), the same goes for the subsequence $\{v^{(s)}(x,t)\}$ and the solution $v(x,t)$ together with all their derivatives. A classical solution of problem (2.2) with a regular operator $\mathbb{K}$ can be continued in time until the first derivatives with respect to x of $v(x,t)$ become infinite, i.e., a classical solution can destruct only by breaking. Theorem 1 is completely proved. □

It should be noted that, in view of the connection between the systems (2.1) and (2.2) indicated above, Theorem 1 holds also for the Cauchy problem (2.1).

§3. The Cauchy problem for the system of equations of surface waves with a dissipative or conservative operator

In this section we prove the existence of a classical solution of the Cauchy problem (2.2) with a singular dissipative or conservative operator $\mathbb{K}$, i.e., such that

$$K_j(p) + K_j^*(p) \geq 0, \qquad |p| \geq p_0 > 0, \; j = 1, 4\,. \tag{3.1}$$

The symbol $K_4(p)$ is a 2×2 matrix, and $K_4^*(p)$ is the Hermitian conjugate matrix. Here and below, we assume that the operator $\mathbb{K}$ has finite order, i.e., there exists $\alpha_j < \infty$ for which

$$|K_j(p)| \leq c_j(1+|p|)^{\alpha_j}, \qquad p \in R_2,\; c_j > 0,\; j = 1, \dots, 4\,. \tag{3.2}$$

The infimum $\bar{\alpha}_j$ of such numbers α_j is called the order of the operator $\mathbb{K}_j$.

THEOREM 2. *Suppose the operator* $\mathbb{K}$ *satisfies the dissipative condition* (3.1) *and* $|K_2(p)| \leq c(1+|p|)$, $|K_3(p)(1+|p|)| \leq c$, $p \in R_2$, $c > 0$, $\bar{v}(x) = (\bar{\eta}(x), \bar{u}(x)) \in (H^\infty(R_2))^3$.

Then, for some $T > 0$ *problem* (2.2) *has a unique solution* $v(x,t) = (\eta(x,t), u(x,t))$ *in the class* $C^\infty([0,T];(H^\infty(R_2))^3)$.

PROOF. We first consider problem (2.2) with a regularized operator $\mathbb{K}$. We define the sequence of symbols $K_j^{(l)}(p)$, $j = 1, \dots, 4$, as follows:

$$K_j^{(l)}(p) = (1+|p|)^{\alpha_j} \iint_{|q| \leq l} \rho_l(p-q) \frac{K_j(q)}{(1+|q|)^{\alpha_j}}\, dq, \qquad l \geq 1; \tag{3.3}$$

here $\alpha_j = 3 + \bar{\alpha}_j$, $\bar{\alpha}_j$ is the order of the symbol $K_j(p)$, and $\rho_l(x) = l^2 \rho(lx)$, $l \geq 1$, $\rho(x) \in C_0^\infty(R_2)$ is such that

$$\rho(x) \geq 0, \quad \rho(x) \equiv 0 \quad \text{for } |x| \geq 1, \iint_{-\infty}^{\infty} \rho(x)\, dx = 1\,. \tag{3.4}$$

We note that the operators $\mathbb{K}_j^{(l)}$ with symbols $K_j^{(l)}(p)$ are regular, since in the x-representation they have the form

$$\mathbb{K}_j^{(l)}(t) + \iint_{-\infty}^{\infty} k_j^{(l)}(x-y)\psi(y)\, dy, \qquad \psi \in H^\infty(R_2),$$

where the kernels $k_j^{(l)}(x) = F^{-1}(K_j^{(l)}(p))$ belong to the Schwartz space S and F^{-1} is the inverse Fourier transform. From the definition (3.3) it follows that there exists a number $b > 0$, not depending on l, such that

$$|K_j^{(l)}(p)| \le b \quad \text{for } |p| \le p_0 + 2,\ l \ge 1,\ j = 1, \dots, 4. \tag{3.5}$$

It follows directly from (3.3), (3.4), and (3.1) that

$$\begin{gathered} K_j^{(l)}(p) + K_j^{(l)^*}(p) \ge 0 \quad \text{for } |p| \ge p_0 + 2,\ l \ge 1,\ j = 1, 4, \\ |K_2^{(l)}(p)| \le 2c(1 + |p|), \qquad |K_3^{(l)}(p)(1 + |p|)| \le 2c; \end{gathered} \tag{3.6}$$

$$\|(1 + |p|)^{-\alpha_j} |K_j^{(l)}(p) - K_j(p)|\,\|_{L_2(R_2)} \to 0, \quad l \to \infty, \quad j = 1, \dots, 4. \tag{3.7}$$

As usual, by the modulus of a vector or a matrix we understand the sum of the moduli of the components. In view of the regularity of the operators $\mathbb{K}_j^{(l)}$ Theorem 1 holds, i.e., there exists a solution $v^{(l)}(x, t) \in C^\infty([0, T^{(l)}]; (H^\infty(R_2))^3)$ of the Cauchy problem (2.2), $l \ge 1$. Omitting temporarily the index l, we differentiate (2.2) n times with respect to x_1 and m times with respect to x_2, take the inner product of the first equation with $\eta_{n,m}$ and of the second equation with $u_{n,m}$, and integrate over all $x \in R_2$; we obtain

$$\begin{gathered} \tfrac{1}{2}\dot{H}_{n,m} + \sum_{i=0}^{n}\sum_{j=0}^{m} C_i^n C_j^m \iint_{-\infty}^{\infty} \eta_{n,m}(\nabla, \eta_{i,j} u_{n-i,m-j})\, dx + I_{n,m}^{(1)} = 0, \\ \tfrac{1}{2}\dot{U}_{n,m} + \sum_{i=0}^{n}\sum_{j=0}^{m} C_i^n C_j^m \iint_{-\infty}^{\infty} (u_{n,m}(u_{n-i,m-j}, \nabla) u_{i,j})\, dx + I_{n,m}^{(2)} = 0, \end{gathered} \tag{3.8}$$

where

$$\begin{gathered} H_{n,m} = \iint_{-\infty}^{\infty} \eta_{n,m}^2\, dx, \qquad U_{n,m} = \iint_{-\infty}^{\infty} (u_{n,m}, u_{n,m})\, dx, \\ I_{n,m}^{(1)} = \iint_{-\infty}^{\infty} \eta_{n,m}[\mathbb{K}_1(\eta_{n,m}) + \mathbb{K}_2(u_{n,m})]\, dx, \\ I_{n,m}^{(2)} = \iint_{-\infty}^{\infty} (u_{n,m}, [\mathbb{K}_3(\eta_{n,m}) + \mathbb{K}_4(u_{n,m})])\, dx. \end{gathered} \tag{3.9}$$

We estimate the integrals $I_{n,m}^{(j)}$ using the Plancherel equality, the properties (3.5) and (3.6) of the operators $\mathbb{K}_j$, $j = 1, \dots, 4$, and the Cauchy inequality(*):

$$\begin{aligned} I_{n,m}^{(1)} &= \tfrac{1}{2} \iint_{-\infty}^{\infty} \hat{\eta}_{n,m}(p, t)[K_1^*(p) + K_1(p)]\hat{\eta}_{n,m}^*(p, t)\, dp \\ &\quad + \iint_{-\infty}^{\infty} \hat{\eta}_{n,m}(p, t)(K_2^*(p), \hat{u}_{n,m}^*(p, t))\, dp \\ &\ge -b \iint_{|p| \le p_0 + 2} |\hat{\eta}_{n,m}|^2\, dp - \iint_{-\infty}^{\infty} |\hat{\eta}_{n,m}(p, t)|\, |\hat{u}_{n,m}(p, t)|\, |K_2^*(p)|\, dp \\ &\ge -bH_{n,m}(t) - 2c\{H_{n,m}(U_{n+1,m} + U_{n,m+1})\}^{1/2} \\ &\ge -bH_k - 2cH_k^{1/2} U_{k+1}^{1/2}, \qquad k = n + m \ge 0; \end{aligned} \tag{3.10}$$

(*) To shorten the formulas everywhere below we will omit the constant $(4\pi)^{-2}$ in the inverse Fourier transform.

$$
\begin{aligned}
I^{(2)}_{n,m} &= \iint_{-\infty}^{\infty} (\hat{u}_{n,m}(p,t), K_3^*(p))\hat{\eta}^*_{n,m}(p,t)\,dp \\
&\quad + \tfrac{1}{2}\iint_{-\infty}^{\infty} (\hat{u}_{n,m}(p,t), [K_4^*(p)+K_4(p)]\hat{u}^*_{n,m}(p,t))\,dp \\
&\geq -8b\iint_{|p|\leq p_0+2} |\hat{u}_{n,m}(p,t)|^2\,dp \\
&\quad - \iint_{-\infty}^{\infty} |\hat{u}_{n,m}(p,t)|\,|K_3^*(p)(1+|p|)|\frac{|\hat{\eta}^*_{n,m}(p,t)|\,dp}{1+|p|} \\
&\geq -16bU_k - 2cU_k^{1/2}H_{k-1}^{1/2}, \qquad k=n+m\geq 1\,.
\end{aligned}
\tag{3.11}
$$

Here we have introduced the notation

$$
H_k = \sum_{0\leq i+j=k} H_{i,j}, \qquad U_k = \sum_{0\leq i+j\leq k} U_{i,j}\,. \tag{3.12}
$$

We estimate the integrals under the summation sign in (3.8). We set

$$
J_k = H_{k-1}+U_k, \qquad S_k = \sum_{0\leq i+j\leq k-1}\eta_{(i,j)} + \sum_{0\leq i+j\leq k} u_{(i,j)}\,. \tag{3.13}
$$

For $0\leq i+j<n+m\equiv k$ we have

$$
\begin{aligned}
&\left|\iint_{-\infty}^{\infty} \eta_{n,m}(\nabla, \eta_{i,j}u_{n-i,m-j})\right| dx \\
&\leq \iint_{-\infty}^{\infty} |\eta_{n,m}|[|\nabla\eta_{i,j}|\,|u_{n-i,m-j}| + |\eta_{i,j}|\,|\nabla u_{n-i,m-j}|]\,dx \\
&\leq J_{k+1}S_{\bar{k}}, \qquad \bar{k} = \left[\frac{k+2}{2}\right];
\end{aligned}
\tag{3.14}
$$

here, in each term in the square brackets the factor with lowest order derivatives, always less than $\bar{k}=[\frac{k+2}{2}]$ (the integral part of the number), can be estimated in modulus and the maximum modulus taken out from under the integral sign, which gives the factor $S_{\bar{k}}$ in (3.14); the other factor J_{k+1} is obtained as a result of applying the Cauchy inequality to the remaining integral. For $i+j=n+m=k$, integrating by parts we find that

$$
\left|\iint_{-\infty}^{\infty} \eta_{n,m}(\nabla, \eta_{n,m}u)\,dx\right| = \left|\tfrac{1}{2}\iint_{-\infty}^{\infty} \eta^2_{n,m}\operatorname{div} u\,dx\right| \leq J_{k+1}S_1 \leq J_{k+1}S_{\bar{k}}\,. \tag{3.15}
$$

In exactly the same manner we obtain, $0\leq i+j\leq k$,

$$
\left|\iint_{-\infty}^{\infty} u_{n,m}(u_{n-i,m-j}, \nabla)u_{i,j}\,dx\right| \leq J_kS_{\bar{k}}\,. \tag{3.16}
$$

Substituting (3.10), (3.11), and (3.14)–(3.16) in (3.8), we have

$$
\begin{aligned}
\dot{H}_{n,m} &\leq c_k(J_{k+1}S_{\bar{k}} + H_k + H_k^{1/2}U_{k+1}^{1/2}) \leq c_kJ_{k+1}(1+S_{\bar{k}}), \\
\dot{U}_{n,m} &\leq c_k(J_kS_{\bar{k}} + U_k + H_{k-1}^{1/2}U_k^{1/2}) \leq c_kJ_k(1+S_{\bar{k}});
\end{aligned}
\tag{3.17}
$$

here and below c_k are constants depending only on k. Summing the first inequality in

(3.17) over all n and m such that $0 \leq n+m \leq k-1$, and the second over all n and m such that $0 \leq n+m \leq k$, $k \geq 1$ a fixed number, and adding the results, we obtain

$$\dot{J}_k \leq c_k J_k (1 + S_{\bar{k}}) \,. \tag{3.18}$$

Integration by parts and the Cauchy inequality yield the following inequality for any function $\psi \in H^\infty(R_2)$:

$$\begin{aligned} \psi_{i,j}^2(x_1,x_2) &= \left| \iint_{-\infty}^{x_1 x_2} \frac{\partial^2}{\partial y_1 \partial y_2} \psi_{i,j}^2(y_1,y_2)\, dy \right| \\ &\leq 2 \left\{ \iint_{-\infty}^{\infty} \psi_{i+1,j}^2 \, dx \iint_{-\infty}^{\infty} \psi_{i,j+1}^2 \, dx \right\}^{1/2} \\ &\quad + 2 \left\{ \iint_{-\infty}^{\infty} \psi_{i,j}^2 \, dx \iint_{-\infty}^{\infty} \psi_{i+1,j+1}^2 \, dx \right\}^{1/2} . \end{aligned} \tag{3.19}$$

Using (3.13) and (3.19), we find that $S_{\bar{k}} \leq c_k J_{\bar{k}+2}^{1/2}$. Substitution of this in (3.18) implies

$$\dot{J}_k \leq c_k J_k (1 + J_{\bar{k}+2}^{1/2}) \,. \tag{3.20}$$

Noting that $\bar{k}+2 < k$ for $k > 6$, we write (3.20) in the form

$$\dot{J}_6 \leq c_6 J_6 (1 + J_6^{1/2}), \tag{3.21}$$

$$\dot{J}_k \leq c_k J_k (1 + J_{k-1}^{1/2}), \qquad k > 6; \tag{3.22}$$

the inequalities for J_k with $k > 6$ are linear. Integrating (3.21) and (3.22), we conclude that there exists a $T > 0$ such that for all $t \in [0, T]$

$$J_k(t) \leq c_k, \qquad k = 0, 1, 2, \dots \,. \tag{3.23}$$

The number T is determined from the inequality (3.21) for J_6. It can be proved similarly that the quantities

$$\iint_{-\infty}^{\infty} \left| \frac{\partial^r}{\partial t^r} v_{n,m}^{(l)}(x,t) \right|^2 dx$$

are bounded for $t \in [0, T]$ uniformly with respect to l.

Based on (3.23) we conclude that from the sequence of vector-valued functions $\{v^{(l)}(x,t)\}$ it is possible to extract a subsequence converging to a vector-valued function $v(x,t) \in C^\infty([0,T]; (H^\infty(R_2))^3)$ in the norm of the space $C^m([0,T]; (H^n(R_2))^3)$; n and m are any nonnegative integers. We now prove that $v(x,t)$ is a solution of the Cauchy problem (2.2). For this, it suffices to show that as $l \to \infty$ the sequence of vector-valued functions $\mathbb{K}^{(l)}(v^{(l)}) \to \mathbb{K}(v)$ uniformly with respect to $x \in R_2$ and $t \in [0, T]$. We have

(3.24)

$$\begin{aligned} \mathbb{K}^{(l)}(v^{(l)}) &= \iint_{-\infty}^{\infty} e^{i(p,x)} K^{(l)}(p) \hat{v}^{(l)}(p,t)\, dp \\ &= \iint_{-\infty}^{\infty} e^{i(p,x)} \frac{(K^{(l)}(p) - K(p)) \hat{v}^{(l)}(p,t)\, dp}{(1+|p|)^{\alpha_5}(1+|p|)^{-\alpha_5}} + \mathbb{K}(v) \\ &\quad + \iint_{-\infty}^{\infty} e^{i(p,x)} \frac{K(p)(\hat{v}^{(l)}(p,t) - \hat{v}(p,t))}{(1+|p|)^{\alpha_5}(1+|p|)^{-\alpha_5}}\, dp, \qquad \alpha_5 = \max_{1 \leq j \leq 4} \bar{\alpha}_j + 3, \end{aligned}$$

where $\bar{\alpha}_j$ is the order of the operator $\mathbb{K}_j$, $j = 1, \dots, 4$. Since $v^{(l)}(x,t)$ converges to $v(x,t)$ as $l \to \infty$ in the norm of $C^0([0,T]; (H^{\alpha_5}(R_2))^3)$, using property (3.7) and the Cauchy inequality, we conclude that the first two terms on the right in (3.24) tend to zero as $l \to \infty$ uniformly with respect to $x \in R_2$ and $t \in [0,T]$.

It remains to prove the uniqueness of the solution of (2.2). Assuming the contrary, suppose there exists two solutions $v^{(1)}(x,t)$ and $v^2(x,t)$; their difference $v = (\eta, u)$ is then a solution of the problem

$$(3.25)\qquad \begin{cases} \eta_t + (\nabla, \eta u^{(1)}) + (\nabla, \eta^{(2)} u) + \mathbb{K}_1(\eta) + \mathbb{K}_2(u) = 0, \\ u_t + (u, \nabla) u^{(1)} + (u^{(2)}, \nabla) u + \mathbb{K}_3(\eta) + \mathbb{K}_4(u) = 0, \\ \eta|_{t=0} = 0, \qquad u|_{t=0} = 0\,. \end{cases}$$

We differentiate the second equation in (3.25) with respect to x:

$$(3.26)\qquad \begin{aligned} &\frac{\partial^2 u}{\partial t \partial x_i} + \left(\frac{\partial u}{\partial x_i}, \nabla\right) u^{(1)} + (u, \nabla)\frac{\partial u^{(1)}}{\partial x_i} + \left(\frac{\partial u^{(2)}}{\partial x_i}, \nabla\right) u \\ &\qquad + (u^{(2)}, \nabla)\frac{\partial u}{\partial x_i} + \mathbb{K}_3\left(\frac{\partial \eta}{\partial x_i}\right) + \mathbb{K}_4\left(\frac{\partial u}{\partial x_i}\right) = 0\,. \end{aligned}$$

Taking the inner product of (3.25) with v and of (3.26) with $\partial u/\partial x_i$, and integrating over all x, we add the results and obtain

$$(3.27)\qquad \tfrac{1}{2}\dot{H}_0 + \iint_{-\infty}^{\infty} [\eta(\nabla, \eta u^{(1)}) + \eta(\nabla, \eta^{(2)} u)]\, dx + I_{0,0}^{(1)} = 0;$$

(3.28)

$$\begin{aligned} &\tfrac{1}{2}\dot{U}_1 + \iint_{-\infty}^{\infty} u[(u,\nabla)u^{(1)} + (u^{(2)}, \nabla)u]\, dx + I_{0,0}^{(2)} \\ &\quad + \sum_{i+j=1} \Big\{ I_{i,j}^{(2)} + \iint_{-\infty}^{\infty} u_{i,j}[(u_{i,j}, \nabla)u^{(1)} + (u,\nabla)u_{i,j}^{(1)} + (u_{i,j}^{(2)}, \nabla)u \\ &\qquad\qquad\qquad + (u^{(2)}, \nabla)u_{i,j}]\, dx \Big\} = 0. \end{aligned}$$

Integrating by parts the first integral term in (3.27) and the fourth term under the summation sign in (3.28), we find, using (3.10), (3.11), and the Cauchy inequality, that

$$\dot{H}_0 \le c_1(H_0 + H_0^{1/2} U_1^{1/2}), \qquad \dot{U}_1 \le c_1(U_1 + H_0^{1/2} U_1^{1/2})\,.$$

Whence, for $J_1 = H_0 + U_1$, we find that $\dot{J}_1 \le c J_1$ and $J_1(0) = 0$, i.e. $J_1(t) \equiv 0$ for $t \in [0,T]$. The uniqueness of the solution, and with it Theorem 2, is proved. □

From Theorem 2 we easily obtain a similar result for the Cauchy problem (2.1).

THEOREM 3. *Suppose the operator* $\mathbb{K}$ *satisfies the dissipative condition* (3.1), *and that* $|K_2(p)| \le c(1+|p|)^2$, $|K_3(p)(1+|p|)^2| \le c$, *where* $p \in R_2$ *and* $c > 0$. *Let* $\bar{w}(x) = (\bar{\eta}(x), \bar{\varphi}(x)) \in (H^\infty(R_2))^2$ *be the initial perturbation. Then, for some* $T > 0$ *there exists a unique solution* $w(x,t) = (\eta(x,t), \varphi(x,t))$ *of the Cauchy problem* (2.1) *in the class* $C^\infty([0,T]; (H^\infty(R_2))^2)$.

§4. Breaking of waves

In this section we prove that the system of equations of surface waves (2.1) describes the effect of breaking of waves. Two-sided bounds for the time of breaking T are found.

4.1. The theorem on breaking of waves. We consider the Cauchy problem for the system of equations of surface waves

$$
\left\{
\begin{array}{l}
\eta_t + (\nabla, \eta\nabla\varphi) + \mathbb{K}_1(\eta) + \mathbb{K}_2(\nabla\varphi) = 0, \\
\varphi_t = \frac{1}{2}(\nabla\varphi)^2 + \mathbb{K}_3(\eta) + \mathbb{K}_4(\nabla\varphi) = 0, \\
\eta|_{t=0} = \bar{\eta}(x), \qquad \varphi|_{t=0} = \bar{\varphi}(x)\,.
\end{array}
\right.
\tag{4.1}
$$

We introduce the notation

$$
m_1(t) = \min_{x\in R_2} \varphi_{0,2}(x,t), \quad m_3(t) = \min_{x\in R_2} \varphi_{20}(x,t), \quad m(t) = \min\{m_1(t), m_3(t)\},
$$

$$
n(t) = \max_{x\in R_2}\{\varphi_{0,2}(x,t), \varphi_{2,0}(x,t)\}, \qquad r(t) = \max_{x\in R_2} |\varphi_{1,1}(x,t)| \equiv \varphi_{(1,1)}(t),
$$

$$
\begin{gathered}
\eta_{(0)}(t) + \max_{x\in R_2} |\eta(x,t)|, \qquad H_0(t) = \iint_{-\infty}^{\infty} \eta^2(x,t)\,dx, \\
P(t) = 1 + \iint_{-\infty}^{\infty} \left\{ \sum_{i+j=4} \varphi_{i,j}^2 + \varphi_{1,3}^2 + \varphi_{3,1}^2 \right\} dx, \qquad m_0 = |m_1(0)|, \\
\Phi_1(t) = 1 + \iint_{-\infty}^{\infty} (\nabla\varphi, \nabla\varphi)\,dx, \qquad H(t) = \eta_{(0)}(t) + H_0(t) + \Phi_1(t)\,.
\end{gathered}
\tag{4.2}
$$

All these quantities can be expressed at the initial time ($t = 0$) in terms of the initial conditions $\bar{\eta}(x)$ and $\bar{\varphi}(x)$.

THEOREM 4. *Suppose the following conditions are satisfied*:

1) *The operator* $\mathbb{K}$ *is such that*

$$
\operatorname{Re} \widetilde{K}_4(p) \geq 0 \quad \textit{for } |p| \geq p_0 > 0, \quad \textit{where } \widetilde{K}_4(p) \equiv i\sum_{j=1}^{2} p_j K_4^{(j)}(p),
\tag{4.3}
$$

$K_1(x), K_2^{(j)}(x), K_3(x)$, *and* $K_4^{(j)}(x)$, *are the kernels of* $\mathbb{K}$ *in the x-representation, and*

$$
\begin{gathered}
\iint_{-\infty}^{\infty} |K_1(x)|\,dx \leq c, \qquad \iint_{-\infty}^{\infty} |K_2^{(j)}(x)|^q\,dx \leq c, \quad q = 1, 2, \\
\iint_{-\infty}^{\infty} |K_{3;n,m}(x)|\,dx \leq c, \qquad 0 \leq n + m \leq 4, \\
K_4^{(j)}(x) \in C^1(R_2 \backslash 0); \\
|K_4^{(j)}(x)| \leq c|x|^{-1-\alpha}, \qquad |\nabla K_4^{(j)}(x)| \leq c|x|^{-2-\alpha}, \\
x \neq 0,\ c \geq 1,\ 0 < \alpha < \tfrac{2}{5} - 10\gamma,\ \gamma \in (0, \tfrac{1}{100})\,.
\end{gathered}
\tag{4.4}
$$

2) *The initial perturbations* $(\bar{\varphi}(x), \bar{\eta}(x)) \in (H^\infty(R_2))^2$.

3) *The inequalities*

$$
m_3(0) > -\tfrac{1}{3}m_0, \qquad r(0) \leq \gamma^3 m_0,
\tag{4.5}
$$

hold, and m_0 *is sufficiently large*:

$$
m_0 \geq \frac{10^6 c}{\alpha\gamma^6}\left(1 + \frac{n(0)}{m_0} + \frac{\sqrt{P(0)}}{m_0}c_3 + c_2\frac{H(0)}{m_0}\right),
\tag{4.6}
$$

where

$$b = \max_{|p|\le p_0} |\widetilde{K}_4(p)|, \qquad c_1 = 2b + 8c,$$

$$c_2 = 2\exp\frac{2c_1 + 2n(0)}{m_0}, \qquad c_3 = \exp\frac{n(0) + 2b + 28c\sqrt{c_2 H(0)}}{m_0}.$$

Then there is a unique solution of the Cauchy problem (4.1) *belonging to the class* $C^\infty([0, T_0); (H^\infty(R_2))^2)$. *This solution breaks*(*) *at time* T_0, *and for* T_0 *the following two-sided bound holds*:

$$(4.7) \qquad (m_0(1+\gamma)^2)^{-1} \le T_0 \le (m_0(1-\gamma)^2)^{-1}.$$

PROOF. The existence and uniqueness of solution of problem (4.1) are proved in Theorem 3(**). It remains to prove (4.7). We differentiate the second equation of system (4.1) n times with respect to x_1 and m times with respect to x_2:

$$(4.8) \qquad \begin{aligned} \dot{\varphi}_{n,m} + \tfrac{1}{2}\sum_{i=0}^{n}\sum_{j=0}^{m} C_i^n C_j^m (\varphi_{n-i,m+1-j}\varphi_{i,j+1} + \varphi_{n+1-i,m-j}\varphi_{i+1,j}) \\ + \mathbb{K}_3(\eta_{n,m}) + \mathbb{K}_4(\nabla\varphi_{n,m}) = 0, \end{aligned}$$

whence, for $n + m = 2$ we have

$$(4.9) \qquad \begin{aligned} &\dot{\varphi}_{0,2} + \varphi_{0,2}^2 + \varphi_{1,1}^2 + \varphi_{0,1}\varphi_{0,3} + \varphi_{1,0}\varphi_{1,2} + \mathbb{K}_3(\eta_{0,2}) + \mathbb{K}_4(\nabla\varphi_{0,2}) = 0, \\ &\dot{\varphi}_{1,1} + \varphi_{1,1}(\varphi_{0,2} + \varphi_{2,0}) + \varphi_{1,2}\varphi_{0,1} + \varphi_{2,1}\varphi_{1,0} + \mathbb{K}_3(\eta_{1,1}) + \mathbb{K}_4(\nabla\varphi_{1,1}) = 0, \\ &\dot{\varphi}_{2,0} + \varphi_{2,0}^2 + \varphi_{1,1}^2 + \varphi_{1,0}\varphi_{3,0} + \varphi_{0,1}\varphi_{2,1} + \mathbb{K}_3(\eta_{2,0}) + \mathbb{K}_4(\nabla\varphi_{2,0}) = 0. \end{aligned}$$

We define the vector-valued function $y = y(\xi, t)$ as a solution of the problem

$$(4.10) \qquad \begin{cases} \dot{y} = u(y, t), \\ y|_{t=0} = \xi \in R_2, \qquad u(x, t) = \nabla\varphi \in C^\infty([0, T_0); (H^\infty(R_2))^2). \end{cases}$$

The system (4.10) was considered in §2. According to Lemma 1 of §2, the function $y(\xi, t)$ establishes a diffeomorphism $R_2 \to R_2$ for each $t \in [0, T_0)$. Therefore, introducing new functions by means of the equalities $\tilde{\varphi}_{n,m}(x, t) = \varphi_{n,m}(y(x, t), t)$, we obtain $\tilde{\varphi}_{n,m} \in C^\infty([0, T_0); (H^\infty(R_2))$, $\tilde{\varphi}_{(n,m)}(t) = \varphi_{(n,m)}(t)$, where $n, m \ge 0$ are arbitrary. From (4.9) and (4.10) we find

$$(4.11) \qquad \begin{aligned} \dot{\tilde{\varphi}}_{0,2} + \tilde{\varphi}_{0,2}^2 + \tilde{\varphi}_{1,1}^2 + A_1 = 0, \\ \dot{\tilde{\varphi}}_{1,1} + \tilde{\varphi}_{1,1}(\tilde{\varphi}_{0,2} + \tilde{\varphi}_{2,0}) + A_2 = 0, \\ \dot{\tilde{\varphi}}_{2,0} + \tilde{\varphi}_{2,0}^2 + \tilde{\varphi}_{1,1}^2 + A_3 = 0, \end{aligned}$$

(*) Breaking of waves in terms of the velocity potential φ means that

$$\max_{\substack{x\in R_2,\\ n+m=2}} \left|\frac{\partial^2\varphi}{\partial x_1^n \partial x_2^m}\right| \to \infty$$

as $t \to T_0$; $T_0 < \infty$ is called the time of breaking.

(**) The dissipativity condition for the operator $\mathbb{K}_1$ in Theorem 3 can easily be replaced by the condition $\iint_{-\infty}^{\infty} |K_1(x)|\,dx \le c$.

where

$$\begin{aligned} A_1 &= \mathbb{K}_3(\eta_{0,2}) + \mathbb{K}_4(\nabla\varphi_{0,2})|_{x=y(\xi,t)}, \\ A_2 &= \mathbb{K}_3(\eta_{1,1}) + \mathbb{K}_4(\nabla\varphi_{1,1})|_{x=y(\xi,t)}, \\ A_3 &= \mathbb{K}_3(\eta_{2,0}) + \mathbb{K}_4(\nabla\varphi_{2,0})|_{x=y(\xi,t)} . \end{aligned} \tag{4.12}$$

Multiplying (4.8) by $\varphi_{n,m}(x,t)$ and integrating with respect to $x \in R_2$ we obtain

$$\begin{aligned} \dot{\Phi}_{n,m} + \sum_{i=0}^{n}\sum_{j=0}^{m} C_i^n C_j^m \iint_{-\infty}^{\infty} \varphi_{n,m}(\varphi_{n-i,m+1-j}\varphi_{i,j+1} + \varphi_{n+1-i,m-j}\varphi_{i+1,j})\,dx \\ + 2I_{n,m} = 0, \end{aligned} \tag{4.13}$$

where

$$\Phi_{n,m}(t) = \iint_{-\infty}^{\infty} \varphi_{n,m}^2(x,t)\,dx, \qquad I_{n,m} = \iint_{-\infty}^{\infty} \varphi_{nm}(\mathbb{K}_3(\eta_{nm}) + \mathbb{K}_4(\nabla\varphi_{nm}))\,dx.$$

Denote

$$\Phi_k = 1 + \sum_{n+m-k} \Phi_{n,m}, \quad I_k = \sum_{n+m=k} I_{n,m}, \quad R(t) = 2(I_4 + I_{3,1} + I_{1,3}) . \tag{4.14}$$

Integration by parts leads to the following two identities

$$\begin{aligned} 2\iint_{-\infty}^{\infty} \varphi_{n,m}\varphi_{n+1,m}\varphi_{1,0}\,dx &= -\iint_{-\infty}^{\infty} \varphi_{n,m}^2\varphi_{2,0}\,dx, \\ 2\iint_{-\infty}^{\infty} \varphi_{n,m}\varphi_{n,m+1}\varphi_{0,1}\,dx &= -\iint_{-\infty}^{\infty} \varphi_{n,m}^2\varphi_{0,2}\,dx . \end{aligned} \tag{4.15}$$

Summing up the equalities (4.13) for $n+m=1$, we obtain

$$\dot{\Phi}_1 + 2\iint_{-\infty}^{\infty} (2\varphi_{1,0}\varphi_{1,1}\varphi_{0,1} + \varphi_{1,0}^2\varphi_{2,0} + \varphi_{0,1}^2\varphi_{0,2})\,dx + 2I_1 = 0,$$

and from this, by (4.15), we find

$$\dot{\Phi}_1 = \iint_{-\infty}^{\infty} (\varphi_{0,1}^2\varphi_{2,0} + \varphi_{1,0}^2\varphi_{0,2})\,dx - 2I_1 . \tag{4.16}$$

We write out the equalities (4.13) for $n+m=4$ and use the identities

$$2\iint_{-\infty}^{\infty} \varphi_{1,3}\varphi_{1,2}\varphi_{0,3}\,dx = -\iint_{-\infty}^{\infty} \varphi_{1,2}^2\varphi_{0,4}\,dx,$$

$$\iint_{-\infty}^{\infty} \varphi_{1,2}\varphi_{3,1}\varphi_{2,1}\,dx = 0,$$

$$2\iint_{-\infty}^{\infty} \varphi_{1,2}\varphi_{2,2}\varphi_{3,0}\,dx = \iint_{-\infty}^{\infty} \varphi_{1,1}(\varphi_{4,0}\varphi_{1,3} - \varphi_{2,2}\varphi_{3,1})\,dx,$$

and also the identities obtained from them by permutation of the indices; we find that

$$\dot{\Phi}_{0,4} + \iint_{-\infty}^{\infty} (7\varphi_{0,2}\varphi_{0,4}^2 + 8\varphi_{1,1}\varphi_{1,3}\varphi_{0,4} + 6\varphi_{0,4}\varphi_{1,2}^2 - \varphi_{0,4}^2\varphi_{2,0})\,dx + 2I_{0,4} = 0,$$

$$\dot{\Phi}_{1,3} + \iint_{-\infty}^{\infty} [\varphi_{1,3}^2(5\varphi_{0,2} + \varphi_{2,0}) + 2\varphi_{1,1}\varphi_{1,3}(3\varphi_{2,2} + \varphi_{0,4}) - 3\varphi_{1,2}^2\varphi_{0,4}]\,dx + 2I_{1,3} = 0,$$

$$\dot{\Phi}_{2,2} + \iint_{-\infty}^{\infty} \{3\varphi_{2,2}^2(\varphi_{0,2} + \varphi_{2,0}) + \varphi_{1,1}[3\varphi_{2,2}(\varphi_{3,1} + \varphi_{1,3}) + \varphi_{4,0}\varphi_{1,3} + \varphi_{3,1}\varphi_{0,4}]\}\,dx + 2I_{2,2} = 0,$$

$$\dot{\Phi}_{3,1} + \iint_{-\infty}^{\infty} \{\varphi_{3,1}^2(5\varphi_{2,0} + \varphi_{0,2}) + 2\varphi_{1,1}\varphi_{3,1}(3\varphi_{2,2} + \varphi_{4,0}) - 3\varphi_{2,1}^2\varphi_{4,0}\}\,dx + 2I_{3,1} = 0,$$

$$\dot{\Phi}_{4,0} + \iint_{-\infty}^{\infty} \{7\varphi_{2,0}\varphi_{4,0}^2 + 8\varphi_{1,1}\varphi_{3,1}\varphi_{4,0} + 6\varphi_{4,0}\varphi_{2,1}^2 - \varphi_{4,0}^2\varphi_{0,2}\}\,dx + 2I_{4,0} = 0\,.$$

Summing up all these equalities and taking the definition of the function $P(t)$ into account, we find that

$$\begin{aligned}\dot{P} + \iint_{-\infty}^{\infty} &\{7\varphi_{0,2}\varphi_{0,4}^2 + 7\varphi_{2,0}\varphi_{4,0}^2 + 8\varphi_{1,1}(\varphi_{1,3}\varphi_{0,4} + \varphi_{3,1}\varphi_{4,0}) \\ &+ 2\varphi_{1,3}^2(5\varphi_{0,2} + \varphi_{2,0}) + 2\varphi_{3,1}^2(5\varphi_{2,0} + \varphi_{0,2}) \\ &- \varphi_{0,4}^2\varphi_{2,0} - \varphi_{4,0}^2\varphi_{0,2} + 4\varphi_{1,1} \\ &\cdot [\varphi_{1,3}(3\varphi_{2,2} + \varphi_{0,4}) + \varphi_{3,1}(3\varphi_{2,2} + \varphi_{4,0})] + 3\varphi_{2,2}^2(\varphi_{0,2} + \varphi_{2,0}) \\ &+ 3\varphi_{1,1}\varphi_{2,2}(\varphi_{3,1} + \varphi_{1,3}) + \varphi_{1,1}(\varphi_{4,0}\varphi_{1,3} + \varphi_{3,1}\varphi_{0,4})\}\,dx + R = 0\,.\end{aligned}$$

Hence, using the Cauchy inequality, we obtain

$$\dot{p} \le (-7m + n + 40r)P - R\,. \tag{4.17}$$

Applying the Plancherel theorem and inequality 3), we have

$$\begin{aligned}\iint_{-\infty}^{\infty} \varphi_{n,m}\mathbb{K}_4(\nabla\varphi_{n,m})\,dx &= \iint_{-\infty}^{\infty} |\hat{\varphi}_{n,m}(p,t)|^2 \operatorname{Re}\widetilde{K}_4(p)\,dp \\ &\ge -\iint_{|p|\le p_0} |\hat{\varphi}_{n,m}(p,t)|^2 |\operatorname{Re}\widetilde{K}_4(p)|\,dp \ge -b\Phi_{n,m}\,.\end{aligned} \tag{4.18}$$

We now estimate the integral ($n + m \le 4$):

$$\begin{aligned}\left|\iint_{-\infty}^{\infty} \varphi_{n,m}\mathbb{K}_3(\eta_{n,m})\,dx\right| &= \left|\iint_{-\infty}^{\infty} \varphi_{n,m}\,dx \iint_{-\infty}^{\infty} K_{3;n,m}(x-y)\eta(y,t)\,dy\right| \\ &\le c\Phi_{n,m}^{1/2}H_0^{1/2}\,.\end{aligned} \tag{4.19}$$

Summation of (4.18) and (4.19) gives $I_{n,m} \ge -b\Phi_{n,m} - c\Phi_{n,m}^{1/2}H_0^{1/2}$, whence, using the definition of I_1 and R, we find that

$$I_1 \ge -b\Phi_1 - 2c\Phi_1^{1/2}H_0^{1/2}, \qquad R \ge -2bP - 14cP^{1/2}H_0^{1/2}\,. \tag{4.20}$$

Substitution of (4.20) in (4.16) and (4.17) yields

$$\begin{aligned} \dot{\Phi}_1 &\leq \varphi_{(2)}\Phi_1 + 2b\Phi_1 + 4c\Phi_1^{1/2}H_0^{1/2}, \\ \dot{P} &\leq (-7m + n + 40r)P + 2bP + 14cP^{1/2}H_0^{1/2}, \end{aligned} \tag{4.21}$$

where we have set $\varphi_{(2)} = \varphi_{(2,0)} + \varphi_{(0,2)} + \varphi_{(1,1)}$. We shall establish an estimate for $\eta_{(0)}(t)$. We again turn to system (4.10). Introducing the new function $\tilde{\eta}(x,t) = \eta(y(x,t),t)$, from the first equation of (4.1) we find

$$\dot{\tilde{\eta}} + \tilde{\eta}(\tilde{\varphi}_{2,0} + \tilde{\varphi}_{0,2}) + \mathbb{K}_1(\eta) + \mathbb{K}_2(\nabla\varphi) = 0\,. \tag{4.22}$$

Using conditions (4.4), we estimate the integral terms in (4.22)

$$|\mathbb{K}_1(\eta)| \leq c\eta_{(0)}, \qquad |\mathbb{K}_2(\nabla\varphi)| = \left|\iint_{-\infty}^{\infty} K_2(x-y)\nabla\varphi(y,t)\,dy\right| \leq 2c\Phi_1^{1/2}\,. \tag{4.23}$$

Substituting (4.23) in (4.22) we get

$$|\dot{\tilde{\eta}}_{(0)}| \leq \varphi_{(2)}\eta_{(0)} + c\eta_{(0)} + 2c\Phi_1^{1/2},$$

and integrating it with respect to t, we obtain

$$\eta_{(0)}(t) \leq \eta_{(0)}(0) + \int_0^t (\varphi_{(2)}\eta_{(0)} + c\eta_{(0)} + 2c\Phi_1^{1/2})\,dt\,. \tag{4.24}$$

We multiply the first equation in (4.1) by η and integrate with respect to $x \in R_2$:

$$\tfrac{1}{2}\dot{H}_0 + \iint_{-\infty}^{\infty} \eta(\nabla, \eta\nabla\varphi)\,dx + \iint_{-\infty}^{\infty} \eta(\mathbb{K}_1(\eta) + \mathbb{K}_2(\nabla\varphi))\,dx = 0\,.$$

Integrating the second term by parts and using (4.4), we find

$$\dot{H}_0 + \iint_{-\infty}^{\infty} \eta^2(\varphi_{2,0} + \varphi_{0,2})\,dx \leq 2cH_0 + 2cH_0^{1/2}\Phi_1^{1/2},$$

or

$$\dot{H}_0 \leq \varphi_{(2)}H_0 + 2c(H_0 + \sqrt{H_0\Phi_1})\,. \tag{4.25}$$

Summing up (4.25) and the first inequality in (4.21), and denoting $\widetilde{H} = H_0 + \Phi_1$, we obtain the inequality $\dot{\widetilde{H}} \leq \varphi_{(2)}\widetilde{H} + (2b + 6c)\widetilde{H}$; integrating the latter with respect to t, we have

$$\widetilde{H}(t) \leq \widetilde{H}(0) + \int_0^t (\varphi_{(2)} + 2b + 6c)\widetilde{H}\,dt\,. \tag{4.26}$$

Denote $H(t) = \widetilde{H}(t) + \eta_{(0)}(t)$. From (4.24) and (4.26) we obtain

$$H(t) \leq H(0) + \int_0^t (\varphi_{(2)} + c_1)H\,dt, \qquad c_1 + 2b + 8c\,.$$

From this, by Gronwall's inequality,

$$H(t) \leq H(0)\exp\left(\int_0^t (\varphi_{(2)}(t) + c_1)\,dt\right)\,. \tag{4.27}$$

We now prove that for all $t \in [0, T_0)$

$$\sup_{\xi \in R_2} |A_j(\xi, t)| \le \tfrac{1}{2}\gamma^6 m_0^2 \widetilde{Z}^{\gamma-2}(t), \qquad j = 1, 2, 3, \tag{4.28}$$

$$r(t) \le \gamma^2 |m_1(t)|, \qquad m_1(t) < 0, \tag{4.29}$$

$$0 < \widetilde{Z}(t) \le 1, \tag{4.30}$$

where $\widetilde{Z}(t) \equiv m_0 |m_1(t)|^{-1}$, so that $\widetilde{Z}(0) = 1$. We note first of all that the inequalities (4.28) and (4.29) hold for $t = 0$. The inequalities (4.18) follow from conditions (4.4) and (4.6) and Lemma 3 for $\delta = 1$ (see subsection 4.2 below)

$$|A_j| \le \frac{10^4 c}{\alpha} \{\bar{\varphi}_{(2)} + \overline{\Phi}_4^{1/2} + \bar{\eta}_{(0)}\} \le \tfrac{1}{2}\gamma^6 m_0^2, \qquad j = 1, 2, 3,$$

and the inequalities (4.29) for $t = 0$ follow from (4.5) and (4.6). For $t > 0$ we prove (4.28), (4.29), and (4.30) by contradiction. Suppose $T_1 > 0$ is the first time when at least one of these inequalities is violated. In view of the continuity of all functions in these inequalities, there exists $T > T_1$ such that for all $t \in [0, T]$

$$\begin{gathered} \sup_{x \in R_2} |A_j(x, t)| \le \gamma^6 m_0^2 \widetilde{Z}^{\gamma-2}(t), \qquad r(t) \le 2\gamma^2 |m_1(t)|, \\ m_1(t) < 0, \qquad 0 < \widetilde{Z}(t) \le 2\,. \end{gathered} \tag{4.31}$$

In view of this, on $[0, T]$ there is also the following estimate for the function $\beta(x, t) \equiv \tilde{\varphi}_{1,1}^2 + A_1$:

$$\sup_{x \in R_2} |\beta(x, t)| \le r^2 + \sup_{x \in R_2} |A_1| \le \frac{\gamma}{100} m_1^2(t)\,. \tag{4.32}$$

Thus, on $[0, T]$ the function $\psi_1(t) \equiv \tilde{\varphi}_{0,2}$ satisfies all conditions of Lemma 1 (see subsection 4.2 below). Indeed, the first equation of (4.11) and (4.31), (4.32) are satisfied. System (4.11) for the functions $\psi_1 = \tilde{\varphi}_{0,2}$, $\psi_2 = \tilde{\varphi}_{1,1}$, $\psi_3 = \tilde{\varphi}_{2,0}$, inequality (4.27) and the second inequality in (4.21) are considered in Lemma 2 (see below), whose conditions, by (4.31), (4.32) and (4.5), are satisfied. It follows from Lemmas 1 and 2 that (4.29) and (4.30) hold for all $t \in [0, T]$. To estimate the integrals A_j, we apply Lemma 3, whose conditions are satisfied by condition 1) on the operator $\mathbb{K}$. Therefore, by Lemmas 2 and 3 and condition (4.6) on m_0, if we choose $\delta = \widetilde{Z}^{\frac{1-\gamma}{\alpha}}$ and note that $\Phi_4 \le P$ and $\eta_{(0)} \le H$, we obtain the following estimate for A_j:

$$\begin{aligned} |A_j| &\le \frac{10^4 c}{\alpha} \{\varphi_{(2)} \delta^{-\alpha} + \delta^{1-\alpha} \Phi_4^{1/2} + H\} \\ &\le \frac{10^4 c}{\alpha} \Bigg\{ 2\left(\frac{m_0}{\widetilde{Z}} + n(0)\right) \widetilde{Z}^{\gamma-1} + c_3 \sqrt{2P(0)} \widetilde{Z}^{(1-\alpha)(1-\gamma)/\alpha - 7/2 - 35\gamma/2} \\ &\qquad\qquad + c_2 H(0) \widetilde{Z}^{-2+\gamma} \Bigg\} \\ &\le \frac{10^4 c}{\alpha} \widetilde{Z}^{\gamma-2} \{2m_0 + 2n(0) + 2\sqrt{P(0)} c_3 + c_2 H(0)\} \le \tfrac{1}{2}\gamma^6 m_0^2 \widetilde{Z}^{\gamma-2}, \end{aligned}$$

since $\frac{1-\gamma}{\alpha} + \gamma - 1 - (7 + 35\gamma)/2 > \gamma - 2$, because $\alpha < \frac{2}{5} - 10\gamma$, $\gamma \in (0, \frac{1}{100})$.

Thus, inequalities (4.28) are also satisfied on the entire interval $[0, T]$. This contradiction proves the estimates (4.28)–(4.30) on the entire interval $[0, T_0)$. We note that Lemmas 1 and 2 are true also up to the time of breaking T_0. Therefore, from inequalities (4.39) of Lemma 1 (see below) it follows that the breaking of the wave (the solution of the Cauchy problem (4.1)) is characterized by the condition $\widetilde{Z}(t) \to 0$ as $t \to T_0$, i.e., $Z(x,t) \to 0$ as $t \to T_0$ for any $x \in Q_\gamma(t)$. By the estimate (4.37) of Lemma 1 we now find that $-m_0(1+\gamma) \le \dot{Z}(x,t) \le -m_0(1-\gamma)^2$ or, integrating these inequalities, $1 - m_0(1+\gamma)t \le Z(x,t) \le 1 - m_0(1-\gamma)^2 t$, whence the two-sided bound (4.7) for the time of breaking T_0 readily follows. Theorem 4 is proved. □

4.2. Technical lemmas. We consider the Cauchy problem for the ordinary differential equation

$$\dot{\psi}_1 + \psi_1^2 + \beta(x,t) = 0, \qquad x \in R_2,\ t > 0,\ \psi_1|_{t=0} = \bar{\psi}_1(x). \tag{4.33}$$

Suppose the solution of problem (4.33) is a function $\psi_1(x,t)$ belonging to $C^\infty([0,T]; H^\infty(R_2))$. We introduce the notation

$$\begin{gathered} m_1(t) = \min_{x \in R_2} \psi_1(x,t), \qquad m_0 = -m_1(0), \\ \widetilde{Z}^{-1}(t) = -\frac{m_1(t)}{m_0}, \qquad m_0 \ne 0, \\ Q_\lambda(t) = \{x \in R_2 : \psi_1(x,t) \le (1-\lambda)m_1(t)\}, \qquad \lambda \in [\gamma, \tfrac{1}{2}],\ \gamma \in \left(0, \frac{1}{100}\right). \end{gathered} \tag{4.34}$$

LEMMA 1. *Suppose the following conditions are satisfied*:
1) $\psi_1(x,t) \in C^\infty([0,T]; H^\infty(R_2))$ *is a solution of problem* (4.33);
2) $m_1(t) < 0$ *on* $[0,T]$;
3) $\sup_{x \in R_2} |\beta(x,t)| \le \frac{\gamma}{100} m_1^2(t)$; $t \in [0,T]$.
Then the following assertions are true:
1) *For any* t_1, t_2, $0 \le t_1 \le t_2 \le T$, $Q_\lambda(t_2) \subset Q_\lambda(t_1)$;
2) *For any* $t \in [0,T]$ *and any* $x \in Q_\lambda(t)$

$$\begin{gathered} \psi_1(x,t) = \bar{\psi}_1(x) Z^{-1}(x,t), \quad \textit{where } Z(x,t) = 1 + \bar{\psi}_1(x) \int_0^t (1 + A(x,\tau))\,dr, \\ A = \beta(x,t)\psi_1^{-2}(x,t). \end{gathered} \tag{4.35}$$

3) *For any* $t \in [0,T]$ *and any* $x \in Q_\lambda(t)$

$$0 < Z(x,t) \le 1, \quad |A(x,t)| \le \frac{\gamma}{10}, \quad 0 < \widetilde{Z}(t) \le 1, \tag{4.36}$$

$$\bar{\psi}_1(x)\left(1 + \frac{\gamma}{10}\right) \le \dot{Z}(x,t) \le \bar{\psi}_1(x)\left(1 - \frac{\gamma}{10}\right), \tag{4.37}$$

$$(1-\lambda)\widetilde{Z}(t) \le Z(x,t) \le (1-\lambda)^{-1}\widetilde{Z}(t). \tag{4.38}$$

From (4.37) *and* (4.38) *it follows that*

$$1 \le \frac{\dot{Z}(x,t)}{-m_0(1-\gamma)^2}, \qquad \widetilde{Z}^{-1}(t) \le (1-\gamma)^{-1} Z^{-1}(x,t). \tag{4.39}$$

PROOF. Fix $\lambda \in [\gamma, \frac{1}{2}]$. We prove 1). From the first condition of the lemma it follows that $\psi_1(x,t)$ and $m_1(t)$ are uniformly continuous with respect to t uniformly in

$x \in R_2$. Therefore, there exists a $\delta > 0$ such that for any $t \in [T_1 - \delta, T_1 + \delta] \cap [0, T]$ for all $x \in Q_\lambda(T_1)$

$$\psi_1(x, t) \leq \tfrac{1}{3} m_1(t), \tag{4.40}$$

since $\lambda \leq \frac{1}{2}$, and T_1 is any point of $[0, T]$. We fix such a $\delta > 0$. It is clear that 1) will be proved if for every $T_1 \in [0, T - \delta]$ we show that $Q_\lambda(t_2) \subset Q_\lambda(t_1)$ for $t_1, t_2 \in [T_1, T_1 + \delta]$. We therefore choose any $T_1 \in [0, T - \delta]$ and suppose, arguing by contradiction, that there exist t_2 and $t_1 \in [T_1, T_1 + \delta]$, $t_2 > t_1$, such that

$$\xi_1 \in Q_\lambda(t_2), \quad \text{but } \xi_1 \notin Q_\lambda(t_1)\,. \tag{4.41}$$

Denote by ξ_2 any point for which $\psi_1(\xi_2, t_1) = m_1(t_1)$. From (4.40) and condition 3) of the lemma it follows that

$$|\beta(\xi_k, t)| \leq \frac{\gamma}{10} \psi_1^2(\xi_k, t), \qquad k - 1, 2,\ t \in [T_1, T_1 + \delta]\,. \tag{4.42}$$

Substituting (4.42) in (4.33), we find

$$\dot\psi_1(\xi_2, t) \leq -\left(1 - \frac{\gamma}{10}\right) \psi_1^2(\xi_2, t), \qquad \dot\psi_1(\xi_1, t) \geq -\left(1 + \frac{\gamma}{10}\right) \psi_1^2(\xi_1, t)\,. \tag{4.43}$$

Integration of (4.43) with respect to t, $t \in [t_1, T_1 + \delta]$, yields

$$\psi_1(\xi_2, t) \leq \psi_1(\xi_2, t_1) \left[1 + \psi_1(\xi_2, t_1) \left(1 - \frac{\gamma}{10}\right)(t - t_1)\right]^{-1}, \tag{4.44}$$

$$\psi_1(\xi_1, t) \geq \psi_1(\xi_1, t_1) \left[1 + \psi_1(\xi_1, t_1) \left(1 + \frac{\gamma}{10}\right)(t - t_1)\right]^{-1}. \tag{4.45}$$

Using the definition of the point ξ_2 and the function $m(t)$, from (4.44) we have

$$m_1(t) \leq m_1(t_1) \left[1 + m_1(t_1) \left(1 - \frac{\gamma}{10}\right)(t - t_1)\right]^{-1} \tag{4.46}$$

By (4.41) we have $\psi_1(\xi_1, t_1) > m_1(t_1)(1 - \lambda)$; hence, from (4.45), (4.46), and condition 2) for $t \in [t_1, T_1 + \delta]$ we obtain the chain of inequalities

$$\begin{aligned}\psi_1(\xi_1, t) &> m_1(t_1)(1 - \lambda) \left[1 + m_1(t_1)(1 - \lambda) \left(1 + \frac{\gamma}{10}\right)(t - t_1)\right]^{-1} \\ &\geq m_1(t_1)(1 - \lambda) \left[1 + m_1(t_1) \left(1 - \frac{\gamma}{10}\right)(t - t_1)\right]^{-1} \geq m_1(t)(1 - \lambda),\end{aligned}$$

since $\lambda \geq \gamma$. From this it follows that $\psi_1(\xi_1, t_2) > m_1(t_2)(1 - \lambda)$ and the assumption (4.41) is impossible. Assertion 1) is proved.

We prove 2). We fix any $T_1 \in (0, T]$. From assertion 1) it follows that if $x \in Q_\lambda(T_1)$, then $x \in Q_\lambda(T)$ for all $t \in [0, T_1]$. Therefore, the definition (4.34) and condition 2) imply that

$$\psi_1(x, t) < 0 \quad \text{for all } x \in Q_\lambda(T_1) \text{ and } t \in [0, T_1]\,. \tag{4.47}$$

According to the definition of the function $A(x, t)$, (4.33) can be written in the form $\dot\psi_1 = -\psi_1^2(1 + A)$, $t \in [0, T_1]$, $x \in Q_\lambda(T_1)$. Hence, after integration with respect to t from 0 to T_1, in view of the fact that $T_1 \in (0, T]$ is arbitrary, we obtain the representation (4.35). Thus, assertion 2) is proved.

We prove 3). The first inequality in (4.36) follows from assertion 1) and (4.47) in view of the boundness of $\psi_1(x, t)$. Using the definition (4.34) and condition 3) we can prove the second inequality in (4.36). Indeed, since $t \in [0, T]$ and $x \in Q_\lambda(t)$, it follows that

$$|A(x,t)| \le \frac{\gamma}{100} m_1^2((1-\lambda)m_1)^2 \le \frac{\gamma}{10}.$$

In order to obtain (4.37), we differentiate the second equality in (4.35) with respect to t and consider (4.36) and (4.47). Finally, to prove (4.38) we choose any t and x, $t \in [0, T]$, $x \in Q_\lambda(t)$. According to assertion 1), $x \in Q_\lambda(0)$, and hence, using (4.34) and (4.35), we find

$$-m_0\widetilde{Z}^{-1}(t) \le \bar{\psi}_1 Z^{-1}(x,t) \le -(1-\lambda)m_0\widetilde{Z}^{-1}(t); \qquad -m_0 \le \bar{\psi}_1(x) \le -(1-\lambda)m_0.$$

Hence $-m_0 Z(x,t) \le \bar{\psi}_1(x)\widetilde{Z}(t) \le -(1-\lambda)m_0\widetilde{Z}(t)$; $-m_0\widetilde{Z}(t) \le \bar{\psi}_1(x)\widetilde{Z}(t) \le -(1-\lambda)m_0 Z(x,t)$. Comparing the last inequalities, we arrive at (4.38). The third inequality in (4.36) is valid since the function $m_1(t)$ is decreasing, as follows from the decrease of $\psi_1(x,t)$, $x \in Q_\lambda(t)$. Lemma 1 is proved. □

On $[0, T]$ we consider the system of differential equations and inequalities for the functions $\psi_j(x,t)$, $P(t)$, $H(t)$, $t \ge 0$, $x \in R_2$:

$$\text{(4.48)} \qquad \begin{cases} \dot\psi_1 + \psi_1^2 + \psi_2^2 + A_1 = 0, \\ \dot\psi_2 + \psi_2(\psi_1 + \psi_3) + A_2 = 0, \\ \dot\psi_3 + \psi_3^2 + \psi_2^2 + A_3 = 0, \\ \dot P \le (-7m + 40r + n + 2b)P + 14c\sqrt{PH}, \\ H(t) \le H(0)\exp\left\{\int_0^t (\varphi_2(t') + c_1)\,dt'\right\}. \end{cases}$$

In this formula $A_j(x,t)$ are given functions, $m_j(t) = \min_{x\in R_2}\psi_j(x,t)$, $n_j(t) = \max_{x\in R_2}\psi_j(x,t)$, $j = 1,2,3$, $m(t) = \min\{m_1(t), m_s(t)\}$, $n(t) = \max\{n_1(t), n_3(t)\}$, $r(t) = \max\{n_2(t), |m_2(t)|\}$, $\varphi_2(t) = |m(t)| + n(t) + r(t)$, and b, c, c_1 are some positive numbers.

LEMMA 2. *Suppose the following conditions are satisfied*:

1) *There exist functions* $\psi_j(x,t) \in C^\infty([0,T]; H^\infty(R_2))$, $P(t) \in C^1([0,T])$, $P(t) \ge 1$, $H(t) \in C^0([0,T])$, *satisfying the system* (4.48);

2) *For the function* $\psi_1(x,t)$ *the conditions of Lemma* 1 *are satisfied on* $[0,T]$.

3) *The following inequalities are satisfied* ($\gamma \in (0, \frac{1}{100})$):

$$\text{(4.49)} \qquad m_3(0) > -\tfrac{1}{3}m_0,$$

$$\text{(4.50)} \qquad r(0) \le \gamma^3 m_0,$$

$$\text{(4.51)} \qquad \sup_{x\in R_2, j=1,2,3} |A_j(x,t)| \le \gamma^6 m_0^2 \widetilde{Z}^{\gamma-2}(t), \qquad t \in [0,T].$$

Then for all $t \in [0, T]$

$$\text{(4.52)} \qquad m_3(t) > \tfrac{1}{3}m_1(t),$$

$$\text{(4.53)} \qquad r(t) < \gamma^2|m_1(t)|,$$

$$\text{(4.54)} \qquad n(t) \le n(0) + \gamma^2|m_1(t)|,$$

$$\text{(4.55)} \qquad P(t) \le 2c_3^2 p(0)\widetilde{Z}(t)^{-7-35\gamma},$$

$$\text{(4.56)} \qquad H(t) \le c_2 H(0)\widetilde{Z}(t)^{\gamma-2},$$

where

$$c_2 = 2\exp\left(\frac{2(c_1+n(0))}{m_0}\right), \qquad c_3 = \exp\left(\frac{n(0)+2b+28c\sqrt{c_2 H(0)}}{m_0}\right).$$

PROOF. We prove the first two estimates by contradiction. For $t=0$ the inequalities (4.52) and (4.53) are satisfied by (4.49) and (4.50). In view of the continuity of the functions contained in these inequalities, there exists a first time $T_1 > 0$ at which at least one of inequalities (4.52) and (4.53) is violated. Thus, for all $t \in [0, T_1]$

$$m_3(t) \geq \tfrac{1}{3} m_1(t), \qquad r(t) \leq \gamma^2 |m_1(t)|, \tag{4.57}$$

where for $t = T_1$ equality holds in at least one of the inequalities (4.57). We first prove (4.52). Since $m_1(t) < 0$, it suffices to consider a time $T_2 < T_1$ and points $x \in R_2$ for which

$$\tfrac{1}{3} m_1(t) \leq \psi_3(x,t) < 0, \qquad t \in [T_2, T_1]. \tag{4.58}$$

We fix $\xi \in Q_\gamma(T_1)$ so that $(1-\gamma)m_1(T_1) = \psi_1(\xi, T_1)$. Considering (4.58) and substituting (4.57) and (4.51) in the third equation of (4.48), we find that for $t \in [T_2, T_1]$

$$\begin{aligned} \dot\psi_3 &\geq -\left(\tfrac{1}{9}+\gamma^4+\gamma^6\right)\frac{m_0^2}{\widetilde{Z}^2(t)} \geq -\frac{\frac{1}{3}(1-\gamma)^5 m_0^2}{(1-\gamma)^2 Z^2(\xi,t)} \\ &\geq \tfrac{1}{3}(1-\gamma) m_0 Z^{-2}(\xi,t)\dot Z(\xi,t) \geq -\tfrac{1}{3}\bar\psi_1(\xi) Z^{-2}(\xi,t)\dot Z(\xi,t), \end{aligned} \tag{4.59}$$

where we have used the estimate (4.38) and (4.39) of Lemma 1. Integrating (4.59) with respect to t, $t \in [T_2, T_1]$, and considering (4.52), valid for $t = T_2$, and (4.39), we obtain

$$\begin{aligned} \psi_3(x, T_1) &\geq \psi_3(x, T_2) - \tfrac{1}{3}\bar\psi_1(\xi)(Z^{-1}(\xi, T_2) - Z^{-1}(\xi, T_1)) \\ &> \tfrac{1}{3} m_1(T_2) - \tfrac{1}{3}\bar\psi_1(\xi) Z^{-1}(\xi, T_2) + \tfrac{1}{3}\bar\psi_1(\xi) Z^{-1}(\xi, T_1) \\ &\geq \tfrac{1}{3} m_1(T_2) + \tfrac{1}{3}(1-\gamma) m_1(T_1) - \tfrac{1}{3}(1-\gamma) m_1(T_2) \geq \tfrac{1}{3} m_1(T_1), \end{aligned}$$

since $\xi \in Q_\gamma(T_2)$ has been chosen so that $(1-\gamma)m_1(T_1) = \bar\psi_1(\xi)Z^{-1}(\xi, T_1)$ and $|m_1(t)| \geq |m_1(T_2)|$. Thus, (4.52) is satisfied on the whole interval $[0, T_1]$. Hence, there remains only one possibility: $r(t) = \gamma^2 |m_1(t)|$ for $t = T_1$. If we denote by $x_0 \in R_2$ a point at which $\max_{x \in R_2} |\psi_2(x, T_1)|$ is attained, then by hypothesis we have

$$|\psi_2(x_0, T_1)| = \gamma^2 |m_1(T_1)|. \tag{4.60}$$

We prove (4.53) first for the case when $x_0 \in Q_{1/2}(T_1)$. In view of conditions (4.50), (4.57), and (4.60) and the continuity of the functions $m_1(t)$ and $\psi_2(x_0, t)$ in t, we can choose $T_3 \geq 0$ so that on $[T_3, T_1]$

$$\gamma^2 |m_1(t)| \geq |\psi_2(x_0, t)| \geq \gamma^3 |m_1(t)| > 0, \tag{4.61}$$

and for $t = T_3$, by (4.50),

$$|\psi_2(x_0, T_3)| = \gamma^3 |m_1(T_3)|. \tag{4.62}$$

From assertion 1) of Lemma 1 it follows that $x_0 \in Q_{1/2}(t)$ for $t \in [T_3, T_1]$. We multiply the first and the third equations in (4.48), taken at the point (x_0, t), by $\psi_2(x_0, t)$ and subtract one from the other

$$\psi_2(\dot\psi_1 - \dot\psi_3) = -\psi_2(\psi_1 - \psi_3)(\psi_1 + \psi_3) - \psi_2(A_1 - A_3)\,. \tag{4.63}$$

We now multiply the second equation in (4.48) by $\psi_1 - \psi_3$ and subtract from (4.63); we get

$$\psi_2(\dot\psi_1 - \dot\psi_3) - \dot\psi_2(\psi_1 - \psi_3) = -\psi_2(A_1 - A_3) + A_2(\psi_1 - \psi_3)\,. \tag{4.64}$$

Since $x_0 \in Q_{1/2}(T_1)$, by (4.52) we have

$$\psi_1(x_0, t) - \psi_3(x_0, t) \le \tfrac{1}{6} m_1(t) < 0 \tag{4.65}$$

for $t \in [T_3, T_1]$. Therefore, dividing (4.64) by the product $\psi_2(\psi_1 - \psi_3)(x_0, t)$, which is not zero by (4.61) and (4.65), we find that

$$\frac{\dot\psi_1 - \dot\psi_3}{\psi_1 - \psi_3} = \frac{\dot\psi_2}{\psi_2} + \frac{A_2}{\psi_2} - \frac{A_1 - A_3}{\psi_1 - \psi_3}\,.$$

Integrating the last equality with respect to $t \in [T_3, T_1]$, we obtain

$$\begin{aligned} |\psi_2(x_0, t)| =& |\psi_2(x_0, T_3)| \left| \frac{\psi_1(x_0, t) - \psi_3(x_0, t)}{\psi_1(x_0, T_3) - \psi_3(x_0, T_3)} \right| \\ &\cdot \exp\left(\int_{T_3}^{t} \left(\frac{A_1 - A_3}{\psi_1 - \psi_3} - \frac{A_2}{\psi_2} \right) d\tau \right)\,. \end{aligned} \tag{4.66}$$

We estimate the expressions in (4.66). In view of (4.51), (4.61), and (4.65) we have

$$\left| \frac{A_2}{\psi_2} \right| \le \gamma^3 m_0 \widetilde{Z}^{\gamma - 1}, \qquad \left| \frac{A_1 - A_3}{\psi_1 - \psi_3} \right| \le \gamma^3 m_0 \widetilde{Z}^{\gamma - 1}\,.$$

From this and from (4.37)–(4.39) it follows that for any $x \in Q_\gamma(T_1)$

$$\begin{aligned} &\left| \int_{T_3}^{t} \left(\frac{A_1 - A_3}{\psi_1 - \psi_3} - \frac{A_2}{\psi_2} \right) d\tau \right| \le 2\gamma^3 \int_{T_3}^{t} m_0 \widetilde{Z}^{\gamma - 1}(\tau)\, d\tau \\ &\le \gamma^2 \left| \int_{T_3}^{t} Z^{\gamma - 1}(x, \tau) \dot Z(x, \tau)\, d\tau \right| \le \gamma |Z^\gamma(x, t) - Z^\gamma(x, T_3)| < \gamma, \end{aligned} \tag{4.67}$$

since $0 < Z(x, t) \le 1$. Substitution of (4.62) and (4.67) in (4.66) implies

$$|\psi_2(x_0, t)| \le \gamma^3 |m_1(T_3)| e^\gamma \left| \frac{\psi_1(x_0, t) - \psi_3(x_0, t)}{\psi_1(x_0, T_3) - \psi_3(x_0, T_3)} \right|\,. \tag{4.68}$$

Inequality (4.52) gives a lower bound for $\psi_3(x_0, t)$, while an upper bound can easily be obtained from the third equation in (4.48) in view of (4.38), (4.39), and (4.51):

$$\dot\psi_3 \le |A_3| \le -\gamma^6 m_0 (1 - \lambda)^{-4} Z^{-2}(x, t) \dot Z(x, t), \qquad x \in Q_\lambda(T_1),$$

whence, after integration with respect to t, we find

$$\begin{aligned} \psi_3(x_0, t) &\le \psi_3(x_0, T_3) + \frac{\gamma^6}{(1 - \gamma)^4} m_0 (Z^{-1}(x, t) - Z^{-1}(x, T_3)) \\ &\le \psi_3(x_0, T_3) + \gamma |m_1(t)|\,. \end{aligned} \tag{4.69}$$

In view of the choice of $x_0 \in Q_{1/2}(T_1)$ (and hence $x_0 \in Q_{1/2}(T_3)$, since $T_3 < T_1$), we have $\psi_1(x_0, T_3) \le \frac{1}{2}m_1(T_3) < 0$. Moreover, it is obvious that $\psi_1(x_0, t) \ge m_1(t)$, and hence, substituting (4.69) in (4.68), we obtain

$$
\begin{aligned}
|\psi_2(x_0, t)| &\le \gamma^3 e^{\gamma} \frac{|m_1(t)|\,|1 + \gamma - \psi_3(x_0, T_3)m_1^{-1}(t)|}{|\psi_1(x_0, T_3) - \psi_3(x_0, T_3)|\,|m_1(T_3)|^{-1}} \\
&\le \gamma^3 e^{\gamma}|m_1(t)| \left| \frac{1 + \gamma - \psi_3(x_0, T_3)m_1^{-1}(t)}{\frac{1}{2} - \psi_3(x_0, T_3)m_1^{-1}(T_3)} \right| .
\end{aligned} \tag{4.70}
$$

If $\psi_3(x_0, T_3) \ge 0$, then it is possible to replace $-m_1^{-1}(t)$ by the larger value $-m_1^{-1}(T_3)$ in the numerator in (4.70), and (4.70) assumes the following form:

$$
|\psi_2(x_0, t)| \le \gamma^3 e^{\gamma}|m_1(t)| \frac{1 + \gamma + d}{\frac{1}{2} + d} \le \gamma^3 e^{\gamma} 2(1 + \gamma)|m_1(t)| < \gamma^2|m_1(t)|,
$$

where $d = \psi_3(x_0, T_3)/|m_1(T_3)| \ge 0$. But if $\psi_3(x_0, T_3) \le 0$, then the term ψ_3/m_1 in the numerator in (4.70) can be omitted, and (4.65) can be used to estimate the denominator; in this case

$$
|\psi_2(x_0, t)| \le 6\gamma^3 e^{\gamma}(1 + \gamma)|m_1(t)| < \gamma^2|m_1(t)| .
$$

Thus, if $x_0 \in Q_{1/2}(T_1)$, then (4.53) holds. We consider the remaining case where $x_0 \in R_2 \backslash Q_{1/2}(T_1)$. By the continuity of the functions $\psi_1(x_0, t)$ and $m_1(t)$, appearing in the definition (4.34) of the set $Q_\lambda(t)$, the set $\{Q_{1/2}(t) \times [0, T_1]\}$ is closed in R_3, and hence there exists $\delta > 0$ such that the point x_0 does not belong to the set $Q_{1/2}(t)$ for any $t \in [T_4, T_1]$, where $T_4 = T_1 - \delta$. In other words,

$$
\psi_1(x_0, t) \ge \tfrac{1}{2}m_1(t), \qquad t \in [T_4, T_1] . \tag{4.71}
$$

We now fix such a $\delta > 0$ and consider the second equation in (4.48) for $x = x_0$ and $t \in [T_4, T_1]$. We choose $\xi \in Q_\gamma(T_1)$ so that $(1 - \gamma)m_1(T_1) = \psi_1(\xi, T_1)$, and we represent $\psi_2(x_0, t)$ in the form

$$
\psi_2(x_0, t) = \chi(t) Z^{\gamma - 1}(\xi, t) . \tag{4.72}
$$

Substituting (4.72) in (4.48) we get $\dot{\chi} = -B\chi - A_2 Z^{1-\gamma}(\xi, t)$, where $B(t) = \psi_1(x_0, t) + \psi_3(x_0, t) - (1 - \gamma)\psi_1(\xi, t)(1 + A(\xi, t))$, and $A(\xi, t)$ is defined in (4.35). Integrating the resulting equation with respect to t, we find that

$$
\begin{aligned}
\chi(T_1) =& \chi(T_4) \exp\left(-\int_{T_4}^{T_1} B(\tau)\, d\tau \right) \\
&- \int_{T_4}^{T_1} \exp\left(-\int_t^{T_1} B(\tau)\, d\tau \right) A_2(x_0, t) Z^{1-\gamma}(\xi, t)\, dt .
\end{aligned} \tag{4.73}
$$

Using (4.52), (4.71), and the definition of the set $Q_\gamma(t)$, we estimate the function $B(t)$:

$$
B(t) \ge \tfrac{1}{2}m_1(t) + \tfrac{1}{3}m_1(t) - (1 - \gamma)^2\psi_1(\xi, t) \ge (\tfrac{5}{6} - (1 - \gamma)^3)m_1(t) > 0,
$$

since $m_1(t) < 0$ and $\gamma < \frac{1}{100}$. From (4.73) we now find that

(4.74)
$$\begin{aligned}
|\chi(T_1)| &\leq |\chi(T_4)| + \int_{T_4}^{T_1} \gamma^3 \bar{\psi}_1(\xi) Z^{-\gamma-1}(\xi,t)\dot{Z}(\xi,t)\,dt \\
&< \gamma^2 |m_1(T_4)| Z^{1-\gamma}(\xi,T_4) + \gamma^2\bar{\psi}_1(\xi)(Z^{-\gamma}(\xi,T_4) - Z^{-\gamma}(\xi,T_1)) \\
&\leq \gamma^2(1-\gamma)^{-1}|\bar{\psi}_1(\xi)|Z^{-\gamma}(\xi,T_4) + \gamma^2\bar{\psi}_1(\xi)Z^{-\gamma}(\xi,T_4) - \gamma^2\bar{\psi}_1(\xi)Z^{-\gamma}(\xi,T_1) \\
&\leq \frac{\gamma^2}{1-\gamma}|\psi_1(\xi,T_1)|Z^{1-\gamma}(\xi,T_1) \equiv \gamma^2|m_1(T_1)|Z^{1-\gamma}(\xi,T_1),
\end{aligned}$$

since $Z(\xi,t)$ decreases with time; moreover, here we have used the representation (4.72) and the fact that (4.53) is valid for $t = T_4$. Substitution of (4.74) in (4.72) leads to (4.53) for $t = T_1$ also in the case where $x_0 \notin Q_{1/2}(T_1)$. This contradiction proves the validity of (4.52) and (4.53) on the entire segment $[0, T]$.

Inequalities (4.54)–(4.56) can be proved directly on the basis of (4.51)–(4.53). Indeed, fix any $\xi \in Q_\gamma(t)$; then for all $t \in [0,T]$, according to Lemma 1, $\xi \in Q_\gamma(t)$, and from (4.48) we obtain

$$\dot{\psi}_1(x,t) \leq |A_1| \leq \gamma^6 m_1^2 \leq \frac{-\gamma^4 m_0 \dot{Z}(\xi,t)}{Z^2(\xi,t)}, \qquad \dot{\psi}_3(x,t) \leq -\gamma^4 m_0 \frac{\dot{Z}(\xi,t)}{Z^2(\xi,t)}, \tag{4.75}$$

for $x \in R_2$, $t \in [0,T]$; here we have used (4.39). Integrating (4.75) with respect to t we find

$$\psi_j(x,t) \leq \psi_j(x,0) + \gamma^4 m_0 (Z^{-1}(\xi,t) - 1), \qquad j = 1,3,\ t \in [0,T],$$

whence we obtain $n(t) = \max_{x \in R_2}(\psi_1,\psi_3) \leq n(0) + \gamma^2|m_1(t)|$, and (4.54) is established. We prove (4.56). Observe that (4.52) implies $|m(t)| = |m_1(t)| = m_0|\widetilde{Z}(t)|$; therefore, from (4.53) and (4.54) we find an estimate for $\varphi_2(t)$:

$$\varphi_2(t) \leq \frac{m_0}{\widetilde{Z}(t)} + n(0) + 2\gamma^2\frac{m_0}{\widetilde{Z}(t)} \leq \frac{(1+\gamma)m_0}{\widetilde{Z}(t)} + n(0)\,.$$

Hence, integration with respect to t yields

$$\begin{aligned}
\int_0^t (\varphi_2(t) + c_1)\,dt &\leq \int_0^t \{(1+\gamma)m_0\widetilde{Z}^{-1}(t) + c_1 + n(0)\}\,dt \\
&\leq \int_0^t \left\{\frac{(1+\gamma)m_0}{(1-\gamma)Z(\xi,t)} + c_4\right\}dt \leq \int_0^t \left\{\frac{(1+\gamma)m_0}{(1-\gamma)Z(\xi,t)} + c_4\right\}\frac{\dot{Z}(\xi,t)\,dt}{-(1-\gamma)^2 m_0} \\
&= -\frac{(1+\gamma)}{(1-\gamma)^3}\ln Z(\xi,t) + c_4\frac{(1-Z(\xi,t))}{(1-\gamma)^2 m_0} \\
&\leq -\frac{1+\gamma}{(1-\gamma)^3}\ln Z(\xi,t) + \frac{c_4}{(1-\gamma)^2 m_0},
\end{aligned}$$

where $c_4 = c_1 + n(0)$ and in estimating the integral we have used (4.38) and (4.39). This makes it possible to estimate the function $H(t)$:

$$\begin{aligned}
H(t) &\leq H(0)\exp\left\{-\frac{1+\gamma}{(1-\gamma)^3}\ln Z(\xi,t) + \frac{c_4}{(1-\gamma)^2 m_0}\right\} \leq c_5 H(0) Z^{-(1+\gamma)/(1-\gamma)^3} \\
&\leq c_5 H(0) Z^{-2+\gamma}(\xi,t) \leq c_2 H(0)\widetilde{Z}^{-2+\gamma}(t),
\end{aligned}$$

since $(1+\gamma)/(1-\gamma)^3 < 2-\gamma$. Here $c_2 = 2\exp\{2(c_1 + n(0))/m_0\} = 2c_5$. Thus, (4.56) is established. We finally prove (4.55). Substituting (4.53), (4.54), and (4.56) in the penultimate inequality of the system (4.48), we obtain

$$\dot{P}(t) \leq -(7+41\gamma^2)m_1(t)P(t) + (n(0)+2b)P(t) + 14c\sqrt{c_2 H(0)\widetilde{Z}^{\gamma-2}P(t)},$$

since $P(t) \geq 1$. Integrating over t, $t \in [0, T]$, again using (4.39), we find

$$\begin{aligned} P(t) \leq P(0)\exp\Bigg\{ &-\int_0^t (7+41\gamma^2)(1-\gamma)^{-3}Z^{-1}(\xi,\tau)\dot{Z}(\xi,\tau)\,d\tau \\ &- 2\int_0^t \frac{n(0)+2b}{m_0}\dot{Z}(\xi,\tau)\,d\tau \\ &- 28c\frac{\sqrt{c_2H(0)}}{m_0}\int_0^t Z^{\gamma/2-1}(\xi,\tau)\dot{Z}(\xi,\tau)\,d\tau\Bigg\} \\ \leq P(0)\exp\Bigg\{&\ln Z^{-7-35\gamma}(\xi,t) + \frac{2}{m_0}(n(0)+2b) + \frac{56c\sqrt{c_2H(0)}}{\gamma m_0}\Bigg\} \\ \leq 2c_3^2 P(0)&\widetilde{Z}^{-7(1+5\gamma)}(t)\,. \end{aligned}$$

The estimate (4.55) is now established, and Lemma 2 is proved. □

Suppose the functions $\varphi(x)$ and $\eta(x)$ belong to $H^\infty(R_2)$. We introduce the notation:

(4.76)

$$\begin{aligned} A_{n,m}(x) = &\iint_{-\infty}^{\infty} K_4^{(1)}(x-y)\varphi_{n,m+1}(y)\,dy + \iint_{-\infty}^{\infty} K_4^{(2)}(x-y)\varphi_{n+1,m}(y)\,dy \\ &+ \iint_{-\infty}^{\infty} K_3(x-y)\eta_{n,m}(y)\,dy, \qquad n+m=2; \end{aligned}$$

$$\varphi_{(2)} = \max_{x\in R_2, n+m=2}|\varphi_{n,m}(x)|, \quad \eta_{(0)} = \max_{x\in R_2}|\eta(x)|, \quad \Phi_4 = \iint_{-\infty}^{\infty}\sum_{n+m=4}\varphi_{n,m}^2(x)\,dx\,.$$

LEMMA 3. *Suppose the kernels* $K_3(x)$ *and* $K_4^{(j)}(x)$, $j=1,2$, *in the integrals* (4.76) *satisfy the conditions*

$$\begin{gathered} K_4^{(j)}(x) \in C^1(R_2\backslash 0), \quad |K_4^{(j)}(x)| \leq c|x|^{-1-\alpha}, \\ |\nabla K_4^{(j)}(x)| \leq c|x|^{-2-\alpha}, x \neq 0,\ 0<\alpha<\tfrac{2}{5}, \\ \iint_{-\infty}^{\infty} |K_{3;n,m}(x)|\,dx \leq c, \qquad n+m=2,\ c\geq 0\,. \end{gathered}$$

Then for any $\delta > 0$

$$(4.77) \qquad |A_{n,m}(x)| \leq \frac{10^4 c}{\alpha}\{\varphi_{(2)}\delta^{-\alpha} + \varphi_4^{1/2}\delta^{1-\alpha} + \eta_{(0)}\}, \qquad x \in R_2\,.$$

PROOF. We begin with an estimate of the integrals in (4.76) containing the kernels $K_4^{(j)}$. They are all estimated in a similar way, and we consider only one of them, for example, the first. We introduce a compactly supported function $\rho(x) \in C_0^\infty(R_2)$ with the properties

$$\rho(x) + \rho(r) = \begin{cases} 1 & \text{for } r \leq \frac{1}{2}, \\ 0 & \text{for } r \geq 1 \end{cases}, \qquad 0 \leq \rho(r) \leq 1,$$

$$\left| \frac{d^k \rho(r)}{dr^k} \right| \leq 10^k, \qquad k = 1, 2,\ r = |x| .$$

Then

$$\begin{aligned} I &\equiv \iint_{-\infty}^{\infty} K_4^{(1)}(x-y)\varphi_{n,m+1}(y)\,dy \\ &= \iint_{-\infty}^{\infty} K_4^{(1)}(y)\varphi_{n,m+1}(x-y)\rho\left(\frac{y}{\delta}\right)\,dy \\ &\quad + \iint_{-\infty}^{\infty} K_4^{(1)}(y)\varphi_{n,m+1}(x-y)\left(1-\rho\left(\frac{y}{\delta}\right)\right)\,dy \\ &\equiv I_1 + I_2, \end{aligned} \tag{4.78}$$

where $\delta > 0$ is any fixed number. We estimate the integral I_2 by integrating by parts, $K_4^{(1)} \equiv K$:

$$\begin{aligned} I_2 &= \iint_{|y|\geq\delta/2} K(y)\varphi_{n,m+1}(x-y)\left(1-\rho\left(\frac{y}{\delta}\right)\right)\,dy \\ &= \iint_{|y|\geq\delta/2} \varphi_{n,m}(x-y)\left\{K_{y_2}^1(y)\left(1-\rho\left(\frac{y}{\delta}\right)\right) - K(y)\frac{\partial\rho}{\partial y_2}^{(y/\delta)}\right\}\,dy\,. \end{aligned}$$

From this, passing to polar coordinates and considering the properties of the kernel $K(x)$ and the function $\rho(x)$, we have

$$|I_2| \leq \varphi_{(2)} \int_0^{2\pi} d\theta \left\{ \int_{\delta/2}^{\infty} cr^{-2-\alpha} r\,dr + \int_{\delta/2}^{\delta} 10c\delta^{-1} r^{-1-\alpha} r\,dr \right\} \leq \frac{200c}{\alpha}\varphi_{(2)}\delta^{-\alpha}\,. \tag{4.79}$$

We set

$$F(y) = \frac{1}{r}\int_0^r rK(r\cos\theta, r\sin\theta)\,dr, \tag{4.80}$$

so that $rK(r\cos\theta, r\sin\theta) = \frac{\partial}{\partial r}(rF(r\cos\theta, r\sin\theta))$. We write the integral I_1 in the polar coordinate system and integrate by parts with respect to r:

$$\begin{aligned} I_1(x) &= \iint_{-\infty}^{\infty} K(y)\varphi_{n,m+1}(x-y)\rho\left(\frac{y}{\delta}\right)\,dy \\ &= \int_0^{2\pi}\int_0^{\delta} rK(r\cos\theta, r\sin\theta)\rho\left(\frac{r}{\delta}\right)\cdot\varphi_{n,m+1}(x_1 - r\cos\theta, x_2 - r\sin\theta)\,dr\,d\theta \\ &= \int_0^{2\pi}\int_0^{\delta} rF[\varphi_{n,m+2}\sin\theta + \varphi_{n+1,m+1}\cos\theta]\cdot\rho\left(\frac{r}{\delta}\right)\,dr\,d\theta \\ &\quad - \int_0^{2\pi}\int_0^{\delta} rF\varphi_{n,m+1}\frac{1}{\delta}\rho'\left(\frac{r}{\delta}\right)\,dr\,d\theta \\ &\equiv I_3 + I_4\,. \end{aligned} \tag{4.81}$$

We estimate I_3 by means of the Cauchy inequality:

(4.82)
$$|I_3| \le \int_0^{2\pi}\int_0^{\delta} |F|(|\varphi_{n,m+2}| + |\varphi_{n+1,m+1}|) r\,dr\,d\theta \le 2\Phi_4^{1/2}\left(\iint_{|y|\le 0} |F(y)|^2\,dy\right)^{1/2}.$$

Using the definition (4.80) and the properties of the kernel $K(x)$, we obtain

$$|F(y)| \le \frac{1}{r}\int_0^r r|K|\,dr \le \frac{c}{r}\int_0^r r^{-\alpha}\,dr \le 2cr^{-\alpha}, \tag{4.83}$$

so that

$$\iint_{|y|\le\delta} |F(y)|^2\,dy \le 4c^2 2\pi \int_0^{\delta} r^{1-2\alpha}\,dr \le 40c^2\delta^{2-2\alpha}. \tag{4.84}$$

Comparing (4.82) and (4.84), we find that

$$|I_3| \le 100c\delta^{1-\alpha}\Phi_4^{1/2}. \tag{4.85}$$

We estimate I_4 integrating by parts:

$$\begin{aligned} I_4 &= -\int_0^{2\pi}\int_0^{\delta} rF(r\cos\theta, r\sin\theta)\varphi_{n,m+1}\frac{1}{\delta}\rho'\left(\frac{r}{\delta}\right) dr\,d\theta \\ &= -\iint_{|y|\le\delta} \frac{1}{\delta}F(y)\varphi_{n,m+1}(x-y)\rho'\left(\frac{y}{\delta}\right) dy \\ &= -\frac{1}{\delta}\iint_{|y|\le\delta} \varphi_{n,m}(x-y)\left\{F'_{y_2}\rho'\left(\frac{y}{\delta}\right) + P(y)\frac{\partial\rho'^{\,(y/\delta)}}{\partial y_2}\right\} dy. \end{aligned}$$

Therefore,

$$|I_4| \le \frac{1}{\delta}\varphi_{(2)}\iint_{|y|\le\delta}\left\{10|F'_{y_2}(y)| + |F(y)|\left|\frac{\partial\rho'^{\,(y/\delta)}}{\partial y_2}\right|\right\} dy. \tag{4.86}$$

We estimate $|F'_{y_2}(y)|$ using the definition (4.80) and the properties of the kernel K:

$$\begin{aligned} |F'_{y_2}| \equiv |F_{0,1}| &= \left|F'_r\sin\theta + \frac{1}{r}P'_\theta\cos\theta\right| \\ &= \left|\left(-\frac{1}{r^2}\int_0^r rK\,dr + K\right)\sin\theta + \frac{1}{r^2}\int_0^r r\frac{\partial K}{\partial\theta}(r\cos\theta, r\sin\theta)\,dr\cos\theta\right| \\ &\le 7cr^{-1-\alpha}. \end{aligned} \tag{4.87}$$

Next,

$$\left|\frac{\partial\rho'^{\,(y/\delta)}}{\partial y_2}\right| = \left|\frac{1}{\delta}\rho''\left(\frac{y}{\delta}\right)\sin\theta\right| \le 100\delta^{-1}. \tag{4.88}$$

Comparing (4.83) and (4.86)–(4.88), we find that

$$|I_4| \le \varphi_{(2)}\frac{1}{\delta}\int_0^{2\pi}\int_0^{\delta}(70cr^{-\alpha} + \frac{100}{\delta}2cr^{1-\alpha})\,dr\,d\theta \le 2000c\varphi_{(2)}\delta^{-\alpha}. \tag{4.89}$$

Combining (4.79), (4.85), and (4.89), we obtain

$$\left|\iint_{-\infty}^{\infty} K_4^{(1)}(x-y)\varphi_{n,m+1}(y)\,dy\right| \le \frac{2200c}{\alpha}\varphi_{(2)}\delta^{-\alpha} + 100c\delta^{1-\alpha}\Phi_4^{1/2}. \tag{4.90}$$

We estimate the remaining integral in (4.76):

$$\left|\iint_{-\infty}^{\infty} K_3(x-y)\eta_{n,m}(y)\,dy\right| \le \iint_{-\infty}^{\infty} |K_{3;n,m}(x-y)|\,|\eta(y)|\,dy \le c\eta_{(0)}. \tag{4.91}$$

From (4.90) and (4.91) we obtain (4.77), and Lemma 3 is proved. □

4.3. The second theorem on breaking of waves. The theorem on breaking of waves, proved above, does not include the system of the Dobrokhotov **[13]** and Broer **[100]** equations, which is of interest for applications. In this subsection we consider the system of equations for surface waves, one-dimensional in space

$$\begin{cases} \eta_t + (\eta u)_x + \mathbb{K}_1(\eta) + \mathbb{K}_2(u) = 0, \\ u_t + uu_x + \eta_x + \mathbb{K}_3(\eta) + \mathbb{K}_4(u) = 0, \end{cases} \tag{4.92}$$

where $\eta(x,t)$ is the disturbed level of water surface, while the level of water at rest is equal to $h = \text{const} > 0$. The linear operators $\mathbb{K}_j$, $j = 1, 3$, are defined by the formulas ($\varphi(x) \in C_0^\infty(R_1)$, arbitrary):

$$\mathbb{K}_j(\varphi) = \int_{-\infty}^{\infty} k_j(x-y)\varphi(y)\,dy, \qquad \mathbb{K}_{j+1}(\varphi) = \int_{-\infty}^{\infty} k_{j+1}(x-y)\varphi'(y)\,dy,$$

where the kernels $k_j(x), k_{j+1}(x) \in L_1(R_1)$.

In the particular case $k_1 = k_3 = k_4 = 0$, and the kernel $k_2(x)$ has the form

$$k_2(x) = \frac{1}{2\pi}\int_{-\infty}^{\infty} e^{ipx}\frac{\tanh p}{ip}\,dp,$$

(4.92) represents the system of equations of Dobrokhotov and Broer. We note that the kernel $k_2(x)$ decays exponentially at infinity, is symmetric, and has a logarithmic singularity at the origin. Besides, it follows from the Plancherel theorem that the operator $\mathbb{K}_2$ is bounded in the space $L_2(R_1)$.

The method used to prove the breaking of solutions for system (4.92) is different from the method developed before and is based on introducing Riemann variables. We show that for a sufficiently large steepness of the initial perturbation a solution of the Cauchy problem for (4.92) breaks within a limited period of time if the kernels $k_1(x)$ and $k_3(x)$ belong to $L_1(R_1)$, while $k_2(x)$ and $k_4(x)$ have a singularity of order not higher than $\alpha < \frac{1}{3.5}$ at the origin, and, in addition, the operator $\mathbb{K}_2$ is bounded in the space $L_2(R_1)$, while $\mathbb{K}_4$ is conservative or dissipative, i.e., for any function $\varphi(x) \in C_0^\infty(R_1)$ and some $\nu > 0$

$$\|\mathbb{K}_2(\varphi)\|_{L_2(R_1)} \le \nu\|\varphi\|_{L_2(R_1)}, \qquad \int_{-\infty}^{\infty} \varphi\mathbb{K}_4(\varphi)\,dx \ge 0. \tag{4.93}$$

Let us, then consider the Cauchy problem for the system (4.92) with the initial perturbation $u(x,0) = \bar{u}(x)$, $\eta(x,0) = \bar{\eta}(x)$. Using the estimates established below, by methods of §§2 and 3 one can prove the existence, until the time of breaking T_0, of a solution $(\eta - h, u)$ to the Cauchy problem for (4.92) belonging to the space $C^\infty([0,T_0);(H^\infty(R_1))^2)$, if the initial perturbation $(\bar{\eta} - h, \bar{u}) \in (H^\infty(R_1))^2$. In what follows, we assume that the last condition holds. We introduce the following notation for the initial condition:

$$\bar r_{1,2}(x) = \bar u(x) \pm 2\sqrt{\bar\eta(x)}, \qquad \bar\eta(x) > 0,$$
$$m_0 = -\inf_{x\in R_1} \bar r_1'(x) > 0, \qquad \bar g_j = \sup_{x\in R_1} |\bar r_j'(x)|, \quad j = 1, 2,$$
$$\overline{J} = \int_{-\infty}^{\infty} \sum_{j=1}^{2} (\bar r_j''(x))^2\, dx, \qquad q = \frac{\bar g_1}{m_0} \geq 1,$$
$$b = 1 + \sup_{x\in R_1}(|\bar r_1|, |\bar r_2|), \qquad a = \inf_{x\in R_1} \bar r_1(x) - \sup_{x\in R_1} \bar r_2(x),$$
$$\overline{V} = m_0 \int_{-\infty}^{\infty} ((\bar r_1(x) - 2\sqrt{h})^2 + (\bar r_2(x) + 2\sqrt{h})^2)\, dx\,.$$

THEOREM 5. *Assume that*
1) *The kernels* $k_j(x)$ *satisfy the conditions* ($j = 1, 3$):

$$k_j(x) \in L_1(R_1), \qquad k_{j+1}(x) \in C^1(R_1\backslash 0) \cap L_1(R_1),$$
$$|k_{j+1}(x)| \leq \nu|x|^{-\alpha}, \quad |k'_{j+1}(x)| \leq \nu|x|^{-1-\alpha}, \qquad x \neq 0,\ \alpha \in (0, \tfrac{1}{3.5})\,.$$

2) *The operators* $\mathbb{K}_2$ *and* $\mathbb{K}_4$ *satisfy the requirements* (4.93).
3) *The initial condition is such that* $a > 0$, $\bar g_2 \leq \gamma m_0 q^{-1}$, *where* γ *is a small constant*:

$$0 < \gamma < \tfrac{1}{120} \min(a, \tfrac{1}{3.5} - \alpha)\,.$$

4) *The quantity* m_0 *is sufficiently large*: $m_0 \geq c + \overline{J}^{2/7}$, *where* $c = c(\nu, b, a, \gamma, q, \overline{V}) > 0$ *is some constant.*

Then a solution of the Cauchy problem for (4.92) *breaks at a moment of time* T_0,

$$T_0 \leq m_0^{-1}(\tfrac{3}{4} - \gamma)^{-1}\,. \tag{4.94}$$

REMARK. Condition 4) of Theorem 5 can be satisfied as follows. Suppose that $\bar\eta(x)$ and $\bar u(x)$ are any initial perturbations for which the requirements 3) of the theorem are met. Then for the new functions $\check{\bar\eta}(x) = \bar\eta(\lambda x)$ and $\check{\bar u}(x) = \bar u(\lambda x)$, $\lambda \geq 1$, the quantities $\check a$, $\check b$, $\check q$, $\check{\overline{V}}$ remain the same as before for any $\lambda \geq 1$, so that the magnitude of c in condition 4) of the theorem is unchanged. After the change of the variable of integration $\lambda x = y$, for the integral $\check{\overline{J}}$ we obtain $\check{\overline{J}} = \lambda^3 \overline{J}$, so that condition 4) takes the form $\check m_0 = \lambda m_0 \geq c + \lambda^{6/7}\overline{J}^{2/7}$, and is satisfied if the number λ is chosen sufficiently large.

PROOF. We rewrite (4.92) with the notation $v_1 = \eta$, $v_2 = u$ in the following form:

$$\vec v_t + A\vec v_x + \vec{\mathscr{G}} = 0, \tag{4.95}$$

where the vectors $\vec v$ and $\vec g$ are

$$\vec v = \begin{pmatrix} v_1 \\ v_2 \end{pmatrix}, \qquad \vec{\mathscr{G}} = \begin{pmatrix} \mathscr{G}_1 \\ \mathscr{G}_2 \end{pmatrix} = \begin{pmatrix} \mathbb{K}_1(v_1) + \mathbb{K}_2(v_2) \\ \mathbb{K}_3(v_1) + \mathbb{K}_4(v_2) \end{pmatrix}$$

and the matrix A has the form

$$A = \begin{pmatrix} v_2 & v_1 \\ 1 & v_2 \end{pmatrix}.$$

We find the eigenvalues of the matrix A: $\lambda_{1,2} = v_2 \pm \sqrt{v_1}$ and the left eigenvectors corresponding to these eigenvalues: $\vec{e}_1 = (1, \sqrt{v_1})$, $\vec{e}_2 = (1, -\sqrt{v_1})$. Multiplying (4.95) by $\vec{e}_1$ and $\vec{e}_2$ from the left, we obtain the following system of equations:

$$
\left\{
\begin{aligned}
&\eta_t + \lambda_1\eta_x + \sqrt{\eta}(u_t + \lambda_1 u_x) + \mathcal{G}_1 + \sqrt{\eta}\mathcal{G}_2 = 0, \\
&\eta_t + \lambda_2\eta_x - \sqrt{\eta}(u_t + \lambda_2 u_x) + \mathcal{G}_1 - \sqrt{\eta}\mathcal{G}_2 = 0 .
\end{aligned}
\right. \tag{4.96}
$$

In order to convert (4.96) to the Riemann variables r_j, we choose integrating factors μ_j such that $(j = 1, 2)$:

$$
\mu_j(\vec{e}_j, d\vec{v}) = dr_j = \sum_{i=1}^{2} \frac{\partial r_j}{\partial v_i}\, dv_i .
$$

It is not difficult to find such quantities:

$$
\mu_{1,2} = \pm v_1^{-1/2}, \qquad r_{1,2} = v_2 \pm 2\sqrt{v_1} = u \pm 2\sqrt{\eta} .
$$

Multiplying the first equation in (4.96) by μ_1 and the second by μ_2, and introducing the new functions r_1 and r_2, we arrive at the following system:

$$
\left\{
\begin{aligned}
&r_{1t} + \lambda_1 r_{1x} + \mathcal{G}_2 + 4(r_1 - r_2)^{-1}\mathcal{G}_1 = 0, \\
&r_{2t} + \lambda_2 r_{2x} + \mathcal{G}_2 - 4(r_1 - r_2)^{-1}\mathcal{G}_1 = 0,
\end{aligned}
\right. \tag{4.97}
$$

where

$$
\begin{aligned}
\lambda_1 &= \tfrac{1}{4}(3r_1 + r_2), \qquad \lambda_2 = \tfrac{1}{4}(3r_2 + r_1), \\
u &= \tfrac{1}{2}(r_1 + r_2), \qquad \sqrt{\eta} = \tfrac{1}{4}(r_1 - r_2), \\
\mathcal{G}_1 &= \tfrac{1}{16}\mathbb{K}_1((r_1 - r_2)^2) + \tfrac{1}{2}\mathbb{K}_2(r_1 + r_2), \\
\mathcal{G}_2 &= \tfrac{1}{16}\mathbb{K}_3((r_1 - r_2)^2) + \tfrac{1}{2}\mathbb{K}_4(r_1 + r_2) .
\end{aligned}
$$

We note that if $\mathcal{G}_1 = \mathcal{G}_2 = 0$, then owing to (4.97) the Riemann variables r_1 and r_2 are independent of time on the characteristics y_j defined by the equations $d_j^y/dt = \lambda_j$. In this case, the functions r_1 and r_2 are called the Riemann invariants (see **[74, 106]**).

Differentiating (4.97) with respect to x, for the function $g_j \equiv \partial r_j(x,t)/\partial x$ we obtain

$$
\left\{
\begin{aligned}
&g_{1t} + \lambda_1 g_{1x} + \tfrac{3}{4}g_1^2 + \tfrac{1}{4}g_1 g_2 + \mathcal{G}_{2x} + 4((r_1 - r_2)^{-1}\mathcal{G}_1)_x = 0, \\
&g_{2t} + \lambda_2 g_{2x} + \tfrac{3}{4}g_2^2 + \tfrac{1}{4}g_1 g_2 + \mathcal{G}_{2x} - 4((r_1 - r_2)^{-1}\mathcal{G}_1)_x = 0 .
\end{aligned}
\right. \tag{4.98}
$$

We use the notation:

$$
\begin{aligned}
m(t) &= \min_{x \in R_1} g_1 \equiv g_1(X_1(t), t), \qquad m_0 = -m(0) > 0, \\
n(t) &= \max_{x \in R_1} g_1 \equiv g_1(X_2(t), t), \\
A_j(t) &= m^{-2}(t)\{\tfrac{1}{4}g_1 g_2 + \mathcal{G}_{2x} + 4((r_1 - r_2)^{-1}\mathcal{G}_1)_x\}|_{x = X_j(t)}, \qquad j = 1, 2 .
\end{aligned}
$$

Then, from the first equation of (4.98) on the curves $x = X_1(t)$ and $x = X_2(t)$ we find

$$
\dot{m} + m^2(\tfrac{3}{4} + A_1) = 0, \qquad \dot{n} + \tfrac{3}{4}n^2 + m^2 A_2 = 0 . \tag{4.99}
$$

Integration of the first equation in (4.99) with respect to t yield

$$
m(t) = m_0 Z^{-1}(t), \quad \text{where } Z(t) = 1 - m_0 \int_0^t (\tfrac{3}{4} + A_1(\tau))\, d\tau .
$$

We introduce the following notation

$$B(t) = \max_{x\in R_1}(|\mathcal{G}_2| + 4|\mathcal{G}_1|(r_1 - r_2)^{-1}),$$

$$P(t) = \max_{x\in R_1}\{|\mathcal{G}_{2x}| + 4|((r_1 - r_2)^{-1}\mathcal{G}_1)_x| + \tfrac{3}{4}\tilde{g}_2^2\}m^{-2}(t),$$

$$J(t) = \int_{-\infty}^{\infty}\sum_{i=1}^{2} f_i^2(x,t)\,dx, \quad f_j(x,t) = \frac{\partial g_j(x,t)}{\partial x},$$

$$\tilde{g}_j(t) = \max_{x\in R_1}|g_j|, \qquad j = 1, 2.$$

We prove that for all $t \in [0, T_0)$, $j = 1, 2$,

$$|A_j(t)| \le \frac{\gamma}{2}, \quad F(t) < \frac{\gamma}{4q}, \quad B(t) < \gamma m_0 Z^{-2/3}(t), \quad J(t) < m_0^{7/2}Z^{-5.5}(t). \tag{4.100}$$

Assume not. Let $T_1 > 0$ be the first moment of time when at least one of the inequalities (4.100) is violated. (At $t = 0$ the inequalities (4.100) hold due to requirements 3) and 4) of the theorem.) Then, in view of continuity with respect to t, the nonstrict inequalities (4.100) hold on the interval $[0, T_1]$. Therefore, using the definition of the function $Z(t)$, we obtain the following bound on $[0, T_1]$:

$$\dot{Z} \le -m_0(\tfrac{3}{4} - \gamma)\exp. \tag{4.101}$$

Also, integrating the second equation in (4.99) with respect to t, by (4.101) we find

$$\begin{aligned} n(t) &\le n(0) + \int_0^t \gamma m^2\,d\tau \le n(0) - 2\gamma m_0\int_0^t Z^{-2}\dot{Z}\,d\tau \\ &\le n(0) + 2\gamma m_0(Z^{-1} - 1) \le 2\gamma|m(t)| + q m_0 - 2\gamma m_0, \end{aligned} \tag{4.102}$$

so that on $[0, T_1]$ we have

$$\tilde{g}_1(t) \le q|m(t)|. \tag{4.103}$$

To estimate $\tilde{g}_2(t) = \max_{x\in R_1}|g_2(x,t)| \equiv g_2(X_3(t), t)$, we analyze the second equation of the system (4.98) on the curve $x = X_3(t)$. By the inequality (4.101) and the nonstrict inequalities (4.100), we obtain on the interval $[0, T_1]$:

$$\dot{\tilde{g}}_2 \le \frac{\gamma}{4q}|m|^2 - \frac{1}{4}m\tilde{g}_2,$$

whence, due to condition 3) of the theorem, integrating with respect to t on $[0, T_1]$, we find

$$\tilde{g}_2(t) \le \tilde{g}_2(0) + \frac{\gamma m_0}{q}(Z^{-1} - 1) \le \frac{\gamma|m(t)|}{q}. \tag{4.104}$$

For the functions $w_j(t) = \min_{x\in R_1} r_j(x,t) = r_j(Y_j(t), t)$ and $W_j(t) = \max_{x\in R_1} r_j(x,t)$

$= r_j(\widetilde{Y}_j(t), t)$, from (4.97) on the curves $x = Y_j(t)$ and $x = \widetilde{Y}_j(t)$ we obtain, by the nonstrict inequalities (4.100), on $[0, T_1]$, $j = 1, 2$,

$$\dot{w}_j \geq -\gamma m_0 Z^{-2/3} \quad \text{and} \quad \dot{W}_j \leq \gamma m_0 Z^{-2/3},$$

whence, using (4.101) and integrating with respect to t, we find

$$W_j(t) \leq W_j(0) + 2\gamma \int_0^t Z^{-2/3} \dot{Z}\, dt \leq W_j(0) + 6\gamma,$$

and, similarly,

$$w_j(t) \geq w_j(0) - 6\gamma\,.$$

It follows from these inequalities that

$$\tilde{r}_j = \max_{x \in R_1} |r_j| \leq b, \tag{4.105}$$

and also

$$0 < a - 12\gamma \leq w_1 - W_2 \leq r_1 - r_2 \leq W_1 - w_2 \leq 2b\,. \tag{4.106}$$

We now estimate the quantity

$$V(t) = m_0 \int_{-\infty}^{\infty} \{(r_1 - 2\sqrt{h})^2 + (r_2 + 2\sqrt{h})^2\}\, dx\,.$$

Integrating by parts, from (4.97) we obtain

$$\dot{V} - \frac{m_0}{4} \int_{-\infty}^{\infty} \{(3g_1 + g_2)(r_1 - 2\sqrt{h})^2 + (3g_2 + g_1)(r_2 + 2\sqrt{h})^2\}\, dx$$
$$+ 2m_0 \int_{-\infty}^{\infty} \{(r_1 + r_2)\mathcal{G}_2 + 4\mathcal{G}_1(r_1 - r_2 - 4\sqrt{h})(r_1 - r_2)^{-1}\}\, dx = 0\,.$$

Using (4.102)–(4.106), condition 2) of the theorem, and the Cauchy inequality, we find that

$$\dot{V} \leq (cm_0 + 3\gamma|m|)\, V\,. \tag{4.107}$$

Here and below we denote by c various constants that depend only on a, b, q, γ, v, and $\overline{V}$. Integration of (4.107) with respect to t, by (4.101), yields

$$V(t) \leq \frac{c}{Z} e^{cm_0 T_0} \leq cZ^{-1}(t)\,. \tag{4.108}$$

From this and the nonstrict inequality (4.100) we have

$$\int_{-\infty}^{\infty} (g_1^2 + g_2^2)\, dx \leq \left(\frac{V}{m_0} J\right)^{1/2} \leq cm_0^{5/4} Z^{-3}(t)\,. \tag{4.109}$$

Differentiating (4.98) with respect to x we obtain

$$\begin{cases} f_{1t} + \lambda_1 f_{1x} + \frac{1}{4}(9g_1 + 2g_2) f_1 + \frac{1}{4} g_1 f_2 + \mathcal{G}_{2xx} + 4(\mathcal{G}_1 (r_1 - r_2)^{-1})_{xx} = 0, \\ f_{2t} + \lambda_2 f_{2x} + \frac{1}{4}(9g_2 + 2g_1) f_2 + \frac{1}{4} g_2 f_1 + \mathcal{G}_{2xx} - 4(\mathcal{G}_1 (r_1 - r_2)^{-1})_{xx} = 0, \end{cases} \tag{4.110}$$

where $f_j(x, t) = \partial^2 r_j(x, t)/\partial x^2$. Multiply the first equation in (4.110) by f_1 and the second equation by f_2, add them up and integrate with respect to x from $-\infty$ and $+\infty$, to get

$$\begin{aligned} &\dot{J} + \tfrac{1}{2}\int_{-\infty}^{\infty} \{3(3g_1 f_1^2 + 3g_2 f_2^2 + r_1 f_1 f_{1x} + r_2 f_2 f_{2x}) \\ &\qquad + (f_1 f_2(g_1+g_2) + 2f_1^2 g_2 + 2g_1 f_2^2 + r_2 f_1 f_{1x} + r_1 f_2 f_{2x})\}\, dx \\ &+ I_1 + I_2 = 0, \end{aligned} \tag{4.111}$$

where

$$\begin{aligned} I_1 &= \tfrac{1}{4}\int_{-\infty}^{\infty} (f_1+f_2)\mathbb{K}_3((g_1-g_2)^2 + (r_1-r_2)(f_1-f_2))\, dx \\ &\quad + \int_{-\infty}^{\infty} (f_1+f_2)\mathbb{K}_4(f_1+f_2)\, dx, \\ I_2 &= 8\int_{-\infty}^{\infty} (f_1-f_2)\{\mathcal{G}_{1xx}(r_1-r_2)^{-1} - 2\mathcal{G}_{1x}(g_1-g_2)(r_1-r_2)^{-2} \\ &\qquad + 2\mathcal{G}_1(g_1-g_2)^2(r_1-r_2)^{-3} - g_1(f_1-f_2)(r_1-r_2)^{-2}\}\, dx\,. \end{aligned}$$

By (4.100)–(4.109) and conditions 1) and 2) of the theorem we obtain the lower bound

$$I_1 \geq -cm_0^4 Z^{-5.5}\,. \tag{4.112}$$

Similarly, accounting for

$$\mathcal{G}_{1x} + \tfrac{1}{8}\mathbb{K}_1((r_1-r_2)(g_1-g_2)) + \tfrac{1}{2}\mathbb{K}_2(g_1+g_2),$$

and

$$\mathcal{G}_{1xx} = \tfrac{1}{8}\mathbb{K}_1((g_1-g_2)^2 + (r_1-r_2)(f_1-f_2)) + \tfrac{1}{2}\mathbb{K}_2(f_1+f_2),$$

we obtain the following bound for I_2:

$$|I_2| \leq cm_0^4 Z^{-5.5}\,. \tag{4.113}$$

Substituting (4.112) and (4.113) in (4.111), by (4.103), (4.104) and integrating by parts we get

$$\dot{J} \leq -m(4+5\gamma)J + cm_0^4 Z^{-5.5},$$

whence, integrating with respect to t, using (4.101) and the nonstrict inequality (4.100), we obtain

$$J(t) \leq J(0) - \int_0^t 5.5 m_0^{7/2} Z^{-6.5}\dot{Z}(\tau)\, d\tau < m_0^{7/2} Z^{-5.5}(t)\,.$$

Thus, the strict inequality (4.100) for the function $J(t)$ holds on the entire interval $[0, T_1]$. (Here and below, we constantly use condition 4) of the theorem.)

We now verify the estimates for $B(t)$ and $F(t)$ on the interval $[0, T_1]$. Accounting for condition 1) of the theorem and integration by parts, with account taken of condition

1) in the theorem, implies that for any function $\varphi(x) \in C_0^\infty(R_1)$, ($\delta > 0$ will be chosen later), $j = 2, 4$:

$$
\begin{aligned}
|\mathbb{K}_j(\varphi)| \le & \int_{-\delta}^{\delta} |k_j(y)|\,|\varphi'(x-y)|\,dy \\
& + \{|k_j(\delta)| + |k_j(-\delta)| + \int_{|y|\ge\delta} |k_j'(y)|\,dy\} \max_{x\in R_1} |\varphi(x)| \\
\le & \; c\left(\delta^{1-\alpha} \max_{x\in R_1} |\varphi'(x)| + \delta^{-\alpha} \max_{x\in R_1} |\varphi(x)|\right).
\end{aligned}
\tag{4.114}
$$

Now, taking $\delta = (Z/m_0)^{2/3\alpha}$ and substituting (4.103)–(4.105) in (4.114), we find, $j = 2, 4$,

$$
|\mathbb{K}_j(r_1 + r_2)| \le c m_0^{2/3} Z^{-2/3}. \tag{4.115}
$$

Therefore, by (4.105) and (4.106) we get $B(t) < \gamma m_0 Z^{-2/3}(t)$ up to the moment of time $t = T_1$. It remains to estimate $F(t)$ and $A_j(t)$. To this end, we use condition 1) of the theorem to estimate $\mathbb{K}_j(\varphi)$ for any $\varphi(x) \in C_0^\infty(R_1)$ as follows ($\delta > 0$, $j = 2, 4$):

$$
\begin{aligned}
|\mathbb{K}_j(\varphi)| \le & \left\{\int_{-\delta}^{\delta} k_j^2(y)\,dy\right\}^{1/2} \left\{\int_{-\infty}^{\infty} \varphi'^2(y)\,dy\right\}^{1/2} + c\delta^{-\alpha} \max_{x\in R_1} |\varphi(x)| \\
\le & \; c\delta^{-\alpha}\left\{\delta^{1/2}\left(\int_{-\infty}^{\infty} \varphi'^2(y)\,dy\right)^{1/2} + \max_{x\in R_1} |\varphi(x)|\right\}.
\end{aligned}
$$

Choose here $\delta = Z^{1/\alpha}(t)$; then, by (4.100), (4.103), and (4.104), we obtain, $j = 2, 4$:

$$
|\mathbb{K}_j(g_1 + g_2)| \le c m_0^{7/4} Z^{-2}(t).
$$

Therefore, accounting for

$$
\mathscr{G}_{2x} = \tfrac{1}{8}\mathbb{K}_3((r_1 - r_2)(g_1 - g_2)) + \tfrac{1}{2}\mathbb{K}_4(g_1 + g_2),
$$

and using the inequalities (4.103)–(4.106), (4.115) we find that for all $t \in [0, T_1]$

$$
F(t) < \frac{\gamma^2}{q^2} + c m_0^{-1/4} < \frac{\gamma}{49}.
$$

From this and the bounds (4.103), (4.104) we have also

$$
|A_j(t)| < \frac{\gamma}{2}, \qquad j = 1, 2,
$$

on the entire interval $[0, T_1]$. Hence, we have arrived at a contradiction, because all the estimates (4.100) turned out to be true also at $t = T_1$. Therefore, the inequalities (4.100) take place until the time of breaking, T_0.

By the definition of $Z(t)$ and inequalities (4.103) and (4.104) we obtain the following characteristic of the time of breaking T_0 : $Z(t) \to 0$ as $t \to T_0$. The estimate for the time of breaking (4.94) follows from this and (4.100). Theorem 5 is proved. □

§5. Existence of a global solution of the Cauchy problem

In this section we prove that under stronger dissipative conditions than in Theorem 2 a classical solution of the Cauchy problem (2.2) for the systems of surface waves exists globally (i.e., for all $t \geq 0$), if the initial perturbation is small in some sense.

THEOREM 6. *Assume that the following conditions are satisfied:*
1) *The operator* $\mathbb{K}$ *satisfies the dissipation condition*

$$
\begin{gathered}
K_j(p) + K_j^*(p) \geq \varepsilon > 0 \quad \text{for } p \in R_2,\ j = 1, 4, \\
|K_2(p)| \leq c(1 + |p|), \quad |K_3(p)(1 + |p|)| \leq c, \qquad p \in R_2,\ c \geq 0;
\end{gathered} \tag{5.1}
$$

and

2) $\bar{v}(x) = (\bar{\eta}(x), \bar{u}(x)) \in (H^\infty(R_2))^3$;
3) $J_6(0) \leq ((\varepsilon - \beta)/\sqrt{2}\beta)^2$, $\varepsilon \geq \beta > 0$; β *is a certain constant, indicated below.*

Then a solution $v(x, t)$ *of the Cauchy problem* (2.2) *exists, is unique, and belongs to* $C^\infty([0, \infty); (H^\infty(R_2))^3)$, *i.e., it exists globally (the notation is that from* §3).

PROOF. Theorem 2 guarantees the existence of a unique classical solution for the Cauchy problem (2.2) on some time interval $[0, T]$, where $T > 0$ depends only on $J_6(t)$. We refine the arguments in the proof of Theorem 2 and show that the quantity $J_6(t)$ does not increase with time. We turn to the system (3.8) and estimate the terms $I_{n,m}^{(j)}$, $j = 1, 2$, using (5.1):

$$
\begin{aligned}
I_{n,m}^{(1)} &= \iint_{-\infty}^{\infty} |\hat{\eta}_{n,m}(p, t)|^2 \operatorname{Re} K_1(p)\, dp + \iint_{-\infty}^{\infty} \hat{\eta}_{n,m} K_2^*(p) \hat{u}_{n,m}^*\, dp \\
&\geq \frac{\varepsilon}{2} H_{n,m} - 2c H_{n,m}^{1/2} (U_{n+1,m} + U_{n,m+1})^{1/2},
\end{aligned} \tag{5.2}
$$

$$
\begin{aligned}
I_{n,m}^{(2)} &= \iint_{-\infty}^{\infty} (\hat{u}_{n,m}(p, t), K_3^*(p)) \hat{\eta}_{n,m}^*(p, t)\, dp \\
&\quad + \tfrac{1}{2} \iint_{-\infty}^{\infty} (\hat{u}_{n,m}(p, t), [K_4^*(p) + K_4(p)] \hat{u}_{n,m}^*(p, t))\, dp \\
&\geq \frac{\varepsilon}{2} U_{n,m} - 2c U_k^{1/2} H_{k-1}^{1/2}, \qquad k = n + m \geq 1\,.
\end{aligned} \tag{5.3}
$$

Estimating the sums in (3.8) exactly as in §3, and using (5.2) and (5.3) rather than (3.9) and (3.20), we obtain, instead of (3.21) and (3.22), the system of inequalities

$$
\dot{J}_6 \leq c_6 J_6 (1 + J_6^{1/2}) - \frac{\varepsilon}{2} J_6, \tag{5.4}
$$

$$
\dot{J}_k \leq c_k J_k (1 + J_{k-1}^{1/2}) - \frac{\varepsilon}{2} J_k, \qquad k > 6\,. \tag{5.5}
$$

The constants c_k, $k \geq 6$, depend only on k and can be written out explicitly, if desired. We show that for $t \in [0, T]$

$$
J_6(t) \leq \frac{(\varepsilon - \beta)^2}{2\beta^2}, \qquad \beta \equiv 2c_6\,. \tag{5.6}
$$

Indeed, by the hypothesis of the theorem, (5.6) holds for $t = 0$. Assuming the contrary, let $T_1 > 0$ be the first moment of time when (5.6) is violated. By the continuity of $J_6(t)$ on the interval $[0, T_1]$ the following inequality holds:

$$J_6(t) \leq \frac{(\varepsilon - \beta)^2}{\beta^2}. \tag{5.7}$$

Substituting (5.7) in (5.4), we obtain for $t \in [0, T_1]$:

$$\dot{J}_6(t) \leq \frac{1}{2}\left(\beta - \varepsilon + \beta\frac{\varepsilon - \beta}{\beta}\right) J_6 = 0,$$

whence it follows that $J_6(t) \leq J_6(0)$, and thus $J_6(t)$ satisfies (5.6) for all $t \in [0, T_1]$, and, in particular, for $t = T_1$. The contradiction obtained proves that (5.6) is satisfied on the entire interval $[0, T]$ considered in Theorem 2. Integrating the inequalities (5.5) successively, we obtain $J_k(t) \leq \bar{c}_k$ for all $t \in [0, T]$, $k \geq 6$, where $\bar{c}_k > 0$ are constants. Therefore, applying Theorem 2, after a finite number of steps, we can extend the classical solution $v(x, t)$ of the Cauchy problem (2.2) to any time interval. Theorem 6 is proved. □

§6. Smoothing of the initial perturbations

In this section we prove that a solution of the Cauchy problem (2.1) or (2.2) with a nonsmooth initial perturbation is infinitely differentiable with respect to t and x at any time $t > 0$, if the operator $\mathbb{K}$ is strongly dissipative. We consider the Cauchy problem (2.2) for the system of equations of surface waves. We write this system in the vectorial form ($v = (\eta, u)$, $v_1 = \eta$, $v_2 = u^1$, $v_3 = u^2$);

$$v_t + A(v) + B(v) + \mathbb{K}(v) = 0, \qquad v|_{t=0} = \bar{v}(x), \tag{6.1}$$

where

$$\begin{gathered}\mathbb{K}(v) = \begin{pmatrix} \mathbb{K}_1 & \mathbb{K}_2 \\ \mathbb{K}_3 & \mathbb{K}_4 \end{pmatrix}\begin{pmatrix} \eta \\ u \end{pmatrix} = \begin{pmatrix} \mathbb{K}_1 & \mathbb{K}_2^1 & \mathbb{K}_2^2 \\ \mathbb{K}_3^1 & \mathbb{K}_4^{11} & \mathbb{K}_4^{12} \\ \mathbb{K}_3^2 & \mathbb{K}_4^{21} & \mathbb{K}_4^{22} \end{pmatrix}\begin{pmatrix} v_1 \\ v_2 \\ v_3 \end{pmatrix}, \\ A(v) = \sum_{k=1}^{2}\sum_{i,j=1}^{3} a_{ij}^k \frac{\partial}{\partial x_k}(v_i v_j), \qquad B(v) = \sum_{k=1}^{2}\sum_{i,j=1}^{3} b_{ij}^k \frac{\partial v_i}{\partial x_k} v_j,\end{gathered} \tag{6.2}$$

and a_{ij}^k and b_{ij}^k are sets of constant three-dimensional vectors:

$$\begin{gathered} a_{12}^1 = a_{13}^2 = \begin{pmatrix} 1 \\ 0 \\ 0 \end{pmatrix}, \quad a_{22}^1 = \begin{pmatrix} 0 \\ \frac{1}{2} \\ 0 \end{pmatrix}, \quad a_{33}^2 = \begin{pmatrix} 0 \\ 0 \\ \frac{1}{2} \end{pmatrix}, \\ b_{32}^1 = \begin{pmatrix} 0 \\ 0 \\ 1 \end{pmatrix}, \quad b_{23}^2 = \begin{pmatrix} 0 \\ 1 \\ 0 \end{pmatrix}, \end{gathered} \tag{6.3}$$

while the remaining a_{ij}^k and b_{ij}^k are equal to zero, $k = 1, 2$, $i, j = 1, 2, 3$.

We also consider the Cauchy problem for the system of linear equations

$$\psi_t(p, t) + K(p)\psi(p, t) = 0, \qquad \psi(p, t)|_{t=0} = \bar{\psi}(p), \tag{6.4}$$

where $K(p)$ is the symbol of the operator $\mathbb{K}$, and $p \in R_2$ is a parameter. We denote by $R(p, t) \equiv e^{-K(p)t}$ the fundamental matrix of the Cauchy problem (6.4), by $\|a\|_0 = \{a_1^2 + a_2^2 + a_3^2\}^{1/2}$ $a = (a_1, a_2, a_3)$, the norm of any vector $a = (a_1, a_2, a_3)$, and by $\|A\|_0 = \sup_{\|a\|_0=1} \|Aa\|_0$ the norm of the linear operator $A\colon R_3 \to R_3$. Suppose the operator $\mathbb{K}$ satisfies the following dissipation condition:

$$\|R(p,t)\|_0 \leq \tilde{c}e^{-\kappa(p)t}, \tag{6.5}$$

where

$$\tilde{c} \geq 1, \qquad \kappa(p) = \begin{cases} \varepsilon|p|^\alpha & \text{for } |p| \geq p_0 \geq 1,\ \alpha > 2,\ \varepsilon > 0, \\ -c_0 & \text{for } |p| \leq p_0,\ c_0 \geq 0\,. \end{cases} \tag{6.6}$$

We assume, as above, that the operator $\mathbb{K}$ is of finite order, i.e.

$$\|K(p)\|_0 \leq c_1(1+|p|)^{\alpha_1}, \qquad \alpha \leq \alpha_1 < \infty,\ p \in R_2\,. \tag{6.7}$$

THEOREM 7. *Suppose the operator $\mathbb{K}$ satisfies conditions* (6.5), (6.7), *and the initial perturbation $\bar{v}(x) \in (H^{1/2}(R_2))^3$. Then there is a $T_0 > 0$ such that there exists a unique solution of the Cauchy problem* (6.1) *in the class*

$$C^\infty((0,T_0];(H^\infty(R_2))^3) \cap C^0([0,T_0];(H^{1/2}(R_2))^3)\,.$$

REMARK. The following requirement is a sufficient condition for (6.5):

$$\lambda_0(p) \leq -\tfrac{3}{2}\kappa(p), \tag{6.8}$$

where $\lambda_0(p) = \max_{1\leq j\leq 3} \operatorname{Re}\lambda_j(p)$, and $\lambda_j(p)$ are the characteristic roots of the matrix $-K(p)$. Indeed, it is well known that the fundamental matrix $R(p,t)$ exists for all $t \geq 0$ and that it satisfies the following inequality:

$$\|R(p,t)\|_0 \leq ce^{\lambda_0(p)t}\sum_{l=0}^{2} t^l\|K(p)\|_0^l\,. \tag{6.9}$$

Since (6.7) holds for the matrix $K(p)$, analyzing separately the two cases $|p| \leq p_0$ and $|p| \geq p_0 \geq 1$, we readily obtain the bound

$$\|K(p)\|_0^l t^l \leq c_2 e^{\kappa(p)t/2}, \qquad c_2 > 0,\ t \geq 0,\ l = 0,1,2\,. \tag{6.10}$$

Comparing (6.8)–(6.10), we arrive at (6.5).

PROOF OF THEOREM 6. Take a sequence of functions $\{\bar{v}^{(s)}(x)\}$, $s = 0,1,2,\dots$, such that $\bar{v}^{(s)}(x) \in (C_0^\infty(R_2))^3$, and

$$\begin{gathered} \|\bar{v}^{(s)}(x) - \bar{v}(x)\|_1 \equiv \|\bar{v}^{(s)}(x) - \bar{v}(x)\|_{(H^{1/2}(R_2))^3} \to 0 \quad \text{as } s \to \infty, \\ \|\bar{v}^{(s)}\|_1 \leq 2\|\bar{v}\|_1\,. \end{gathered} \tag{6.11}$$

Applying the method of successive approximations to the Cauchy problem (6.1), we set $v^{(0)}(x,t) = F^{-1}(\hat{v}^{(0)}(p,t)) = F^{-1}(R(p,t)\hat{\bar{v}}^{(0)}(p))$, and for $s \geq 1$ we define the vector-valued function $v^{(s)}(x,t)$ as a solution of the following problem:

$$v_t^{(s)} + A(v^{(s-1)}) + B(v^{(s-1)}) + \mathbb{K}(v^{(s)}) = 0, \qquad v^{(s)}|_{t=0} = \bar{v}^{(s)}(x)\,. \tag{6.12}$$

Problem (6.12) is linear in the unknown vector-valued function $v^{(s)}(x,t)$, and the initial condition $\bar{v}^{(s)}(x) \in (H^\infty(R_2))^3$. Therefore, there exists a solution of the Cauchy problem (6.12) in $C^\infty([0,\infty);(H^\infty(R_2))^3)$ for every $s \geq 1$. Indeed, the vector-valued function

$$v^{(s)}(x,t) = \iint_{-\infty}^{\infty} e^{i(p,x)} \left\{ R(p,t)\hat{\overset{\frown}{v}}{}^{(s)}(p) - \int_0^t R(p,t-\tau)\hat{f}(p,\tau)\,d\tau \right\} dp$$

is a solution of (6.12) which, in view of (6.5), belongs to $C^\infty([0,\infty);(H^\infty(R_2))^3)$; here

$$f(x,t) = (B(v^{(s-1)}) + A(v^{(s-1)})) \in C^\infty([0,\infty);(H^\infty(R_2))^3)$$

by induction. We establish certain estimates for $\hat{v}^{(s)}(p,t)$. We apply the Fourier transform to problem (6.12):

$$\tag{6.13}\begin{aligned}
\hat{v}_t^{(s)}(p,t) + \sum_{k=1}^{2}\sum_{i,j=1}^{3} \Big\{ p_k a_{ij}^k \iint_{-\infty}^{\infty} \hat{v}_i^{(s-1)}(p-y,t)\hat{v}_j^{(s-1)}(y,t)\,dy& \\
+ b_{ij}^k \iint_{-\infty}^{\infty} (p_k - y_k)\hat{v}_i^{(s-1)}(p-y,t)\hat{v}_j^{(s-1)}(y,t)\,dy \Big\}& \\
+ K(p)\hat{v}^{(s)}(p,t) = 0,& \\
\hat{v}^{(s)}|_{t=0} = \hat{\overset{\frown}{v}}{}^{(s)}(p)\,.&
\end{aligned}$$

Integrating this equality with respect to t, we obtain ($s \geq 1, p \in R_2$):

$$\tag{6.14}\begin{aligned}
\hat{v}^{(s)}(p,t) = R(p,t)\hat{\overset{\frown}{v}}{}^{(s)}(p) + \int_0^t R(p,t-\tau)\,d\tau& \\
\cdot \sum_{k=1}^{2}\sum_{i,j=1}^{3} \Big\{ p_k a_{ij}^k \iint_{-\infty}^{\infty} \hat{v}_i^{(s-1)}(p-y,\tau)\hat{v}_j^{(s-1)}(y,\tau)\,dy& \\
+ b_{ij}^k \iint_{-\infty}^{\infty} (p_k - y_k)\hat{v}_i^{(s-1)}(p-y,\tau)\hat{v}_j^{(s-1)}(y,\tau)\,dy \Big\},&
\end{aligned}$$

whence, by (6.5) we find

$$\tag{6.15}\begin{aligned}
\|\hat{v}^{(s)}(p,t)\|_0 \leq \|R\|_0 \|\hat{\overset{\frown}{v}}{}^{(s)}\|_0 + b\int_0^t \|R(p,t-\tau)\|_0\,d\tau& \\
\cdot \iint_{-\infty}^{\infty} (|p| + |p-y|)\|\hat{v}^{(s-1)}(p-y,\tau)\|_0 \|\hat{v}^{(s-1)}(y,\tau)\|_0\,dy& \\
\leq \tilde{c}e^{-\kappa(p)t}\|\hat{\overset{\frown}{v}}{}^{(s)}(p)\|_0 + \tilde{b}\int_0^t e^{-\kappa(p)(t-\tau)}\,d\tau& \\
\cdot \iint_{-\infty}^{\infty} (|p| + |p-y|)\|\hat{v}^{(s-1)}(p-y,\tau)\|_0 \|\hat{v}^{(s-1)}(y,\tau)\|_0\,dy,&
\end{aligned}$$

where $b = 36\max_{i,j,k}(\|a_{ij}^k\|_0 + \|b_{ij}^k\|_0)$, $\tilde{b} = b\tilde{c}$.

We show that for some T_0, $0 < T_0 \le 1$, independent of s, for all $t \in [0, T_0]$ and all $s \ge 0$, $p \in R_2$, the following inequality holds:

$$\|\hat{v}^{(s)}(p,t)\|_0 \le \psi^{(s)}(p)e^{-\varepsilon|p|t}, \tag{6.16}$$

where

$$\psi^{(s)}(p) = \chi\|\hat{\bar{v}}^{(s)}(p)\|_0 + (1+|p|)^{-\beta-3/2}, \quad \beta = \frac{\alpha-2}{2} > 0, \quad \chi = 2\tilde{c}e^{\varepsilon p_0 + c_0}.$$

For $s = 0$ inequality (6.16) is satisfied by the definition of the vector-valued function $\hat{v}^0(p,t)$. Assuming by induction that (6.16) is true for $s < s_1$, we consider (6.15) for $s = s_1 + 1$. Using (6.11), we estimate the integral in (6.15) as follows ($|p| + |p-y| \le |y| + 2|p-y|$, and in the integral with $|y|$ we make the change $y = p - z$):

(6.17)

$$\begin{aligned}
&\iint_{-\infty}^{\infty} (|p| + |p-y|)\|\hat{v}^{(s-1)}(p-y,\tau)\|_0\|\hat{v}^{(s-1)}(y,\tau)\|_0\,dy \\
&\le 3\iint_{-\infty}^{\infty} |p-y|\,\|\hat{v}^{(s-1)}(p-y,\tau)\|_0\|\hat{v}^{(s-1)}(y,\tau)\|_0\,dy \\
&\le 3\sqrt{1+|p|}\iint_{-\infty}^{\infty} \sqrt{1+|y|}\sqrt{1+|p-y|}\,\|\hat{v}^{(s-1)}(p-y;\tau)\|_0\|\hat{v}^{(s-1)}(y;\tau)\|_0\,dy \\
&\le 3\sqrt{1+|p|}\iint_{-\infty}^{\infty} \sqrt{1+|y|}\sqrt{1+|p-y|} \\
&\qquad\qquad \cdot\psi^{(s-1)}(p-y)\psi^{(s-1)}(y)e^{-\varepsilon|p-y|\tau-\varepsilon|y|\tau}\,dy \\
&\le 3\sqrt{1+|p|}e^{-\varepsilon|p|\tau}\iint_{-\infty}^{\infty} \sqrt{1+|y|} \\
&\qquad\qquad \cdot[\chi\|\hat{\bar{v}}^{(s-1)}(y)\|_0 + (1+|y|)^{-\beta-3/2}]\sqrt{1+|p-y|} \\
&\qquad\qquad \cdot[\chi\|\hat{\bar{v}}^{(s-1)}(p-y)\|_0 + (1+|p-y|)^{-\beta-3/2}]\,dy \\
&\le 6\sqrt{1+|p|}e^{-\varepsilon|p|\tau}\iint_{-\infty}^{\infty} (1+|y|)[\chi^2\|\hat{\bar{v}}^{(s-1)}(y)\|_0^2 + (1+|y|)^{-2\beta-3}]\,dy \\
&\quad 24e^{-\varepsilon|p|\tau}\sqrt{1+|p|}\,\|\psi\|_1^2,
\end{aligned}$$

where $\psi(p) = \chi\|\hat{\bar{v}}(p)\|_0 + (1+|p|)^{-\beta-3/2}$. It follows from (6.15) and (6.17) that

$$\|\hat{v}^{(s)}(p,t)\|_0 \le \tfrac{1}{2}\psi^{(s)}(p)e^{-\varepsilon|p|t} + 24\tilde{b}\|\psi\|_1^2\sqrt{1+|p|}\int_0^t e^{-\kappa(p)(t-\tau)-\varepsilon|p|\tau}\,d\tau. \tag{6.18}$$

Here we have taken into account that

$$\tilde{c}\|\hat{\bar{v}}^{(s)}(p)\|_0 e^{-\kappa(p)t} \le \frac{\psi^{(s)}(p)}{2}e^{-\varepsilon|p|t}.$$

Considering that $0 \le t \le 1$, for the integral in (6.18) we get the following bound:

$$\int_0^t e^{-\kappa(p)(t-\tau)-\varepsilon|p|\tau}\,d\tau \le \begin{cases} \dfrac{e^{-\varepsilon|p|t} - e^{-\kappa(p)t}}{\kappa(p) - \varepsilon|p|}, & |p| \ge \mathcal{N}, \\ te^{c_0}, & |p| \le \mathcal{N}. \end{cases} \tag{6.19}$$

Choosing now a sufficiently large $\mathcal{N}$, for instance

$$\mathcal{N} = \max\{4 + p_0, (2^{7+\alpha}24\tilde{b}\|\psi\|_1^2\varepsilon^{-1})^{1/\beta}\},$$

so that for $|p| \geq \mathcal{N}$

$$0 < \frac{1}{\kappa(p) - \varepsilon|p|} \leq \frac{2 \cdot 2^{\alpha}}{\varepsilon \mathcal{N}^{\beta}(1 + |p|)^{\beta+2}} \leq \frac{(1 + |p|)^{-2-\beta}}{24 \cdot 2^6 \tilde{b} \|\psi\|_1^2},$$

and then fixing a sufficiently small $T_0 : T_0 = \min\{1, (2^{7+\alpha} \mathcal{N}^{\alpha+1} e^{c_0 + \varepsilon \mathcal{N}} 24 \tilde{b} \|\psi\|_1^2)^{-1}\}$, so that for $t \in [0, T_0]$ and for $|p| \leq \mathcal{N}$

$$te^{c_0} \leq Te^{c_0 + \varepsilon \mathcal{N} T_0 - \varepsilon|p|t} \leq \frac{e^{-\varepsilon|p|t}(1 + |p|)^{-2-\beta}}{2^6 24 \tilde{b} \|\psi\|_1^2},$$

from (6.19) we obtain

$$\int_0^t e^{-\kappa(p)(t-\tau) - \varepsilon|p|\tau} \, dr \leq \frac{e^{-\varepsilon|p|t}(1 + |p|)^{-\beta-2}}{24 \cdot 2^6 \tilde{b} \|\psi\|_1^2} \tag{6.20}$$

and the bound (6.16) follows directly from (6.18) and (6.20). Inequality (6.16) has thus been proved by induction for all $s \geq 0$.

We now show that for all integral $n, s \geq 0$

$$\left\| \frac{\partial^n \hat{v}^{(s)}(p, t)}{\partial t^n} \right\|_0 \leq c_n \psi^{(s)}(p) e^{-\varepsilon|p|t} (1 + |p|)^{\alpha_2 n} t^{-\alpha_2 n^2}, \qquad t \in (0, T_0], \ p \in R_2, \tag{6.21}$$

with some $c_n \geq 0$ independent of s, $\alpha_2 = 1 + \alpha_1$. For $n = 0$ and for all $s \geq 0$ the inequality (6.21) with $c_0 = 1$ follows from (6.16). Applying induction on n, suppose that (6.21) is true for $n \leq n_1$. Differentiating (6.13) n_1 times with respect to t, we obtain, for $n = n_1 + 1$ and for any $s \geq 0$,

$$\begin{aligned} \frac{\partial^n \hat{v}^{(s)}(p, t)}{\partial t^n} &+ \sum_{k=1}^{n} \sum_{i,j=1}^{3} \iint_{-\infty}^{\infty} \{p_k a_{ij}^k + (p_k - y_k) b_{ij}^k\} \\ &\cdot \sum_{l=0}^{n-1} c_{n-1}^l \frac{\partial^l \hat{v}_i^{(s-1)}(p - y, t)}{\partial t^l} \frac{\partial^{n-1-l} \hat{v}_j^{(s-1)}(y, t)}{\partial t^{n-1-l}} \, dy + K(p) \frac{\partial^{n-1} \hat{v}^{(s)}(p, t)}{\partial t^{n-1}} = 0, \end{aligned}$$

whence, as in (6.17), we find

$$\begin{aligned} \left\| \frac{\partial^n \hat{v}^{(s)}(p, t)}{\partial t^n} \right\|_0 &\leq c_n (1 + |p|)^{\alpha_2 n} \psi^{(s)}(p) t^{-\alpha_2 n^2} e^{-\varepsilon|p|t} \\ &+ c_n \sqrt{1 + |p|} \sum_{j=0}^{n-1} \iint_{-\infty}^{\infty} (1 + |p - y|)^{\alpha_2 j + 1/2} (1 + |y|)^{\alpha_2 (n-1-j) + 1/2} \psi^{(s)}(p - y) \\ &\cdot \psi^{(s-1)}(y) e^{-\varepsilon|p-y|t - \varepsilon|y|t} t^{-\alpha_2 (n-1-j)^2 - \alpha_2 j^2} \, dy; \end{aligned} \tag{6.22}$$

here and below, all constants independent of s, p, y, t are denoted by c_n. The integral in (6.22) is estimated as follows:

(6.23)
$$\begin{aligned}
&\iint_{-\infty}^{\infty} (1+|p-y|)^{\alpha_2 j+1/2}(1+|y|)^{\alpha_2(n-1-j)+1/2}\psi^{(s-1)}(p-y)\psi^{(s-1)}(y)e^{-\varepsilon|p-y|t} \\
&\qquad \cdot e^{-\varepsilon|y|t}\,dy \\
&\le \int_0^{2\pi} d\theta \Big\{ e^{-\varepsilon|p|t}(1+|p|)^{\alpha_2(n-1)} \\
&\qquad \cdot \int_0^{2|p|} \psi^{(s-1)}(p-y)\psi^{(s-1)}(y)(1+|y|)^{1/2}(1+|p-y|)^{1/2}|y|\,d|y| \\
&\qquad + e^{-\varepsilon|p|t}\int_0^{\infty} e^{-\varepsilon|y|t}(1+|p-y|)^{\alpha_2(n-1-j)+1/2}(1+|y|)^{\alpha_2 j+1/2} \\
&\qquad\qquad \cdot \psi^{(s-1)}(p-y)\psi^{(s-1)}(y)|y|\,d|y| \Big\} \\
&\le c_n e^{-\varepsilon|p|t}\|\psi\|_1^2 t^{-\alpha_2(n-1)}(1+|p|)^{\alpha_2(n-1)},
\end{aligned}$$

by (6.11) and the evident inequality

$$\begin{aligned}
&e^{-\varepsilon|y|t}(1+|p-y|)^{\alpha_2(q-j)}(1+|y|)^{\alpha_2 j} \\
&\qquad \le e^{-\varepsilon|y|t}((1+|p|)(1+|y|))^{\alpha_2 q} \le c_q t^{-\alpha_2 q}(1+|p|)^{\alpha_2 q},
\end{aligned}$$

which is valid for $n-1 \equiv q \ge 0$ and for all $0 \le j \le q$, $0 < t \le 1$, and $p, y \in R_2$. Substituting (6.23) in (6.22) we get

$$\begin{aligned}
\left\| \frac{\partial^n \hat{v}^{(s)}(p,t)}{\partial t^n} \right\|_0 &\le c_n(1+|p|)^{\alpha_2 n}\psi^{(s)}(p)t^{-\alpha_2 n^2}e^{-\varepsilon|p|t} \\
&\qquad + c_n \sum_{j=0}^{n-1} e^{-\varepsilon|p|t}(1+|p|)^{\alpha_2(n-1)+1/2}t^{-\alpha_2 j^2-\alpha_2(n-1-j)^2-\alpha_2(n-1)} \\
&\le c_n(1+|p|)^{\alpha_2 n}t^{-\alpha_2 n^2}e^{-\varepsilon|p|t}\psi^{(3)}(p),
\end{aligned}$$

since $(1+|p|)^{-\alpha_2+1/2} \le \psi^{(s)}(p)$, as $\beta + 2 \le \alpha \le \alpha_2 - \frac{1}{2}$.

Thus, the bound (6.21) is proved by induction for all s and n. By (6.16) and (6.21), from the sequence $\{v^{(s)}(x,t)\}$, $s = 0, 1, \dots$, consisting of the Fourier inverse transforms $v^{(s)}(x,t) = F^{-1}(\hat{v}^{(s)}(p,t))$, we can extract a subsequence converging to a limit $v(x,t) \in C^\infty((0,T_0]; (H^\infty(R_2))^3)$ in the norm of the space $C^m((0,T_0]; (H^n(R_2))^3)$ with any n and m. Evidently, $v(x,t)$ satisfies system (6.1) for all $t > 0$. To prove that the initial condition is satisfied, we show that some subsequence of $\{v^{(s)}(x,t)\}$ converges also in the norm of $C^0([0,T_0]; (H^{1/2}(R_2))^3)$. From the sequence $\{\bar{v}^{(s)}(x)\}$ of the initial vector-valued functions we choose a subsequence such that

$$\|\bar{v}^{(s)}(x) - \bar{v}^{(s-1)}(x)\|_1 \le (\tilde{c}e^{c_0})^{-1}2^{-s}\|\psi\|_1, \qquad s = 1, 2, \dots,$$

(we retain the previous notation for this subsequence). We show that for all $s \ge 1$ the following inequality holds:

$$\max_{t\in[0,T_0]} \|v^{(s)}(x,t) - v^{(s-1)}(x,t)\|_1 \le 2^{5-s}\|\psi\|_1\,. \tag{6.24}$$

For $s = 1$ this follows from (6.16), since

$$\max_{t\in[0,T_0]} \|v^{(1)}(x,t) - v^{(0)}(x,t)\|_1 \le \max_{t\in[0,T_0]} (\|\psi^{(1)}\|_1 + \|\psi^{(0)}\|_1) \le 4\|\psi\|_1,$$

(this is obtained by squaring (6.16), multiplying it by $(1+|p|)$, and integrating with respect to p). Let (6.24) be established by induction for all $s \le s_1$; then for $s = s_1 + 1$ we find from (6.14) that

$$\begin{aligned} &\|\hat{v}^{(s)}(p,t) - \hat{v}^{(s-1)}(p,t)\|_0 \\ &\quad \le \tilde{c}\|\hat{\bar{v}}^{(s)}(p) - \hat{\bar{v}}^{(s-1)}(p)\|_0 e^{-\kappa(p)t} + 24\tilde{b}4 \int_0^t e^{-\kappa(p)(t-\tau)}\,d\tau \\ &\quad \cdot \sqrt{1+|p|} \iint_{-\infty}^{\infty} (1+|p-y|)^{1/2}(1+|y|)^{1/2} \\ &\quad \cdot \|\hat{v}^{(s-1)}(p-y,\tau) - \hat{v}^{(s-2)}(p-y,\tau)\|_0 \\ &\quad \cdot \|\hat{v}^{(s-1)}(y,\tau) + \hat{v}^{(s-2)}(y,\tau)\|_0\,dy\,. \end{aligned} \tag{6.25}$$

We estimate the inner integral in (6.25) using (6.16), (6.24), and the Cauchy inequality:

$$\begin{aligned} &\iint_{-\infty}^{\infty} (1+|p-y|)^{1/2}(1+|y|)^{1/2}\|\hat{v}^{(s-1)}(p-y,\tau) - \hat{v}^{(s-2)}(p-y,\tau)\|_0 \\ &\qquad \cdot \|\hat{v}^{(s-1)}(y,\tau) + \hat{v}^{(s-2)}(y,\tau)\|_0\,dy \le 2^{8-s}\|\psi\|_1^2\,. \end{aligned} \tag{6.26}$$

Substituting (6.26) in (6.25), for $0 \le t \le T_0$ we obtain

$$\begin{aligned} &\|\hat{v}^{(s)}(p,t) - \hat{v}^{(s-1)}(p,t)\|_0 \\ &\quad \le 2^{8-s}\|\psi\|_1^2 96\tilde{b}\sqrt{1+|p|} \int_0^t e^{-\kappa(p)(t-\tau)}\,d\tau + \tilde{c}e^{c_0}\|\hat{\bar{v}}^{(s)}(p) - \hat{\bar{v}}^{(s-1)}(p)\|_0 \\ &\quad \le 2^{4-3}\psi(p) + \tilde{c}e^{c_0}\|\hat{\bar{v}}^{(s)}(p) - \hat{\bar{v}}^{(s-1)}(p)\|_0, \end{aligned} \tag{6.27}$$

in view of (6.20) with $\varepsilon = 0$. Multiplying (6.27) by $(1+|p|)^{1/2}$ and integrating with respect to p, we obtain

$$\begin{aligned} &\|v^{(s)}(x,t) - v^{(s-1)}(x,t)\|_1 \\ &\quad \le \tilde{c}e^{c_0}\|\bar{v}^{(s)}(x) - \bar{v}^{(s-1)}(x)\|_1 + 2^{4-s}\|\psi\|_1 \le 2^{5-s}\|\psi\|_1\,. \end{aligned}$$

Thus, (6.24) is established for all s by induction. It follows from (6.24) that $v(x,t) \in C^0([0,T_0];(H^{1/2}(R_2))^3)$. Thus, the initial condition is satisfied. Theorem 7 is proved. $\square$

Let us strengthen the dissipativity condition (6.5) for the operator $\mathbb{K}$ by replacing $\kappa(p)$ in (6.6) with

$$\kappa(p) = \varepsilon(1+|p|)^{\alpha}, \qquad p \in R_2,\ \alpha > 2,\ \varepsilon > 0\,. \tag{6.28}$$

In this case, a solution of the Cauchy problem (6.1) with a nonsmooth initial perturbation exists globally. Indeed, the following theorem applies.

THEOREM 8. *Suppose the operator $\mathbb{K}$ satisfies (6.5) and (6.7) with $\kappa(p)$ from (6.28), while $\bar{v}(x) \in (H^{1/2}(R_2))^3$ and, in addition, the norm $\|\bar{v}\|_1$ is sufficiently small: $\|\bar{v}\|_1 \le \varepsilon\beta(2^{\alpha+9}\pi\chi 2r\tilde{b})^{-1}$.*

Then the Cauchy problem (6.1) *in the class* $C^\infty((0,\infty);(H^\infty(R_2))^3)\cap C^0([0,\infty);(H^{1/2}(R_2))^3)$ *has a unique solution.*

PROOF. We show that (6.16) holds for all $t>0$, if for $\psi^{(s)}(p)$ we take the function $(\sqrt{\beta}\equiv\bar{\beta})$

$$\psi^{(s)}(p)=\chi\|\hat{\bar{v}}^{(s)}(p)\|_0+\frac{\bar{\beta}\chi}{2\pi}\|\bar{v}_1\|(1+|p|)^{-\beta-3/2}. \tag{6.29}$$

Indeed, (6.17) and (6.18) still hold. For the integral in (6.18) we have the bound

$$\begin{aligned}\int_0^t e^{-\kappa(p)(t-\tau)-\varepsilon|p|\tau}\,d\tau&=\frac{e^{-\varepsilon|p|t}-e^{-\varepsilon(1+|p|)^\alpha t}}{\varepsilon(1+|p|)^\alpha-\varepsilon|p|}\\ &\le\frac{e^{-\varepsilon|p|t}2^\alpha}{\varepsilon(1+|p|)^\alpha}\le\frac{e^{-\varepsilon|p|t}(1+|p|)^{-\beta-2}\bar{\beta}\chi\|\bar{v}\|_2}{2^6\cdot 24\tilde{b}\|\psi\|_{\tilde{1}}^2\cdot 2\pi},\end{aligned} \tag{6.30}$$

since the function

$$\psi(p)\equiv\chi\|\hat{\bar{v}}(p)\|_0+\frac{\bar{\beta}\chi}{2\pi}\|\bar{v}\|_2(1+|p|)^{-\beta-3/2}$$

has the bound

$$\|\psi\|_{\tilde{1}}^2\le 2^2\chi^2\|\bar{v}\|_{\tilde{1}}^2\le\frac{\varepsilon\bar{\beta}\chi\|\bar{v}\|_1}{2^{\alpha+6}24\tilde{b}2\pi}$$

by the condition of the theorem. The bound (6.30) is analogous to (6.20), and by the same token the bound (6.16) with $\psi^{(s)}(p)$ from (6.29) is established; (6.16) in turn implies that

$$\|v\|_{\tilde{1}}^2\le\frac{\varepsilon\bar{\beta}\chi\|\bar{v}\|_1}{2\pi}(2^{\alpha+6}24\tilde{b})^{-1}$$

for all $t\ge 0$. Using this estimate and applying Theorem 7 at each step in time, we can extend the solution of the Cauchy problem (6.1) to an arbitrarily large time interval. Theorem 8 is proved. □

§7. Smoothing of initial perturbations from L_2

We go back to problem (2.2) introducing the velocity potential φ, $u=\nabla\varphi$. Then the system of equations (2.2) takes the form (6.1) with $B(v)\equiv 0$, and the constant three-dimensional vectors a_{ij}^k have the following form:

$$a_{12}^1=a_{13}^2=\begin{pmatrix}1\\0\\0\end{pmatrix},\quad a_{22}^1=a_{33}^1=\begin{pmatrix}0\\\frac12\\0\end{pmatrix},\quad a_{22}^2=a_{33}^2=\begin{pmatrix}0\\0\\\frac12\end{pmatrix},$$

while the remaining a_{ij}^k are equal to zero.

We consider the Cauchy problem for the system of linear equations (6.4) with the dissipation condition (6.5), (6.6) for the operator $\mathbb{K}$. Then for (6.1) with $B(\gamma)\equiv 0$ the following theorem holds.

THEOREM 9. *Suppose the operator* $\mathbb{K}$ *satisfies the dissipation condition* (6.5), (6.6), *and the initial perturbation* $\bar{v}(x) \in (L_2(R_2))^3$. *Then, for some* $T_0 > 0$ *the Cauchy problem* (6.1) *has a unique solution in the class*

$$C^\infty((0, T_0]; (H^\infty(R_2))^3) \cap C^0([0, T_0]; (L_2(R_2))^3).$$

PROOF. Take a sequence of functions $\{\bar{v}^{(s)}(x)\}$, $s = 0, 1, 2, \ldots$, such that $\bar{v}^{(s)}(x) \in (C_0^\infty(R_2))^3$,

$$\tag{7.1} \begin{gathered} \|\bar{v}^{(s)}(x) - \bar{v}(x)\|_{(L_2(R_2))^3} \equiv \|\bar{v}^{(s)}(x) - \bar{v}(x)\|_1 \to 0 \quad \text{as } s \to \infty, \\ \text{and} \quad \|\bar{v}^{(s)}(x)\|_1 \leq 2\|\bar{v}\|_1 . \end{gathered}$$

Applying the method of successive approximations to the Cauchy problem (6.1), we set $v^{(0)}(x, t) = F^{-1}(\hat{v}^{(0)}(p, t)) = F^{-1}(R(p, t)\hat{\bar{v}}^{(0)}(p))$, and for $s \geq 1$ we define the vector-valued function $v^{(s)}(x, t)$ as a solution of the linear problem (6.12) with $B \equiv 0$. As in the proof of Theorem 7, we find that there is a solution of (6.12) in the class $C^\infty([0, \infty); (H^\infty(R_2))^3)$ for each $s \geq 1$.

Let us establish bounds for $\hat{v}^{(s)}(p, t)$. Applying the Fourier transform to (6.12), we obtain (6.13) with $B \equiv 0$. Integrating the latter with respect to t, we arrive at (6.14) with $b_{ij}^k \equiv 0$. In view of (6.5), from (6.14) we find that

$$\tag{7.2} \begin{aligned} \|\hat{v}^{(s)}(p, t)\|_0 \leq{} & \|R\|_0\|\hat{\bar{v}}^{(s)}\|_0 + b \int_0^t \|R(p, t - \tau)\|_0 \, d\tau \\ & \cdot \iint_{-\infty}^{\infty} |p|\|\hat{v}^{(s-1)}(p - y, \tau)\|_0\|\hat{v}^{(s-1)}(y, \tau)\|_0 \, dy \\ \leq{} & \tilde{c}e^{-\kappa(p)t}\|\hat{\bar{v}}^{(s)}(p)\|_0 + b_1 \int_0^t e^{-\kappa(p)(t-\tau)} \, d\tau \\ & \cdot \iint_{-\infty}^{\infty} |p| \, \|\hat{v}^{(s-1)}(p - y, \tau)\|_0\|\hat{v}^{(s-1)}(y,\tau)\|_0 \, dy, \qquad \tilde{c}, b_1 = \text{const} . \end{aligned}$$

We show that for some T_0, $0 < T_0 \leq 1$, independent of s, for all $t \in [0, T_0]$ and all $s \geq 0$, $p \in R_2$, the following inequality holds:

$$\tag{7.3} \|\hat{v}^{(s)}(p, t)\|_0 \leq \psi^{(s)}(p)e^{-\varepsilon|p|t},$$

where

$$\psi^{(s)}(p) = \chi\|\hat{\bar{v}}^{(s)}(p)\|_0 + (1 + |p|)^{-\beta-1}, \quad \beta = \frac{\alpha - 2}{2} > 0, \quad \chi = 2\tilde{c}e^{\varepsilon p_0 + c_0} .$$

The inequality (7.3) holds for $s = 0$ by the definition of the vector-valued function $\hat{v}^{(0)}(p, t)$. Suppose that (7.3) holds for $s \leq s_1$ by induction; then consider (7.2) for $s = s_1 + 1$. In view of (7.1), we can estimate the integral in (7.2) as follows:

$$\tag{7.4} \begin{aligned} & |p| \iint_{-\infty}^{\infty} \|\hat{v}^{(s-1)}(p - y, \tau)\|_0\|\hat{v}^{(s-1)}(y, \tau)\|_0 \, dy \\ & \quad \leq 2|p|e^{-\varepsilon|p|\tau} \iint_{-\infty}^{\infty} [\chi\|\hat{\bar{v}}^{(s-1)}(y)\|_0 + (1 + |y|))^{-\beta-1}]^2 \, dy \\ & \quad \leq 8|p|e^{-\varepsilon|p|\tau}\|\psi\|_1^2, \end{aligned}$$

where $\psi(p) = \chi\|\hat{\bar{v}}(p)\|_0 + (1+|p|)^{-\beta-1}$. From (7.4) and (7.2) we find that

$$\|\hat{v}^{(s)}(p,t)\|_0 \le \frac{\psi^{(s)}(p)}{2}e^{-\varepsilon|p|t} + 8b_1|p|\,\|\psi\|_1^2\int_0^t e^{-\kappa(p)(t-\tau)-\varepsilon|p|\tau}\,d\tau\,. \tag{7.5}$$

Assuming $0\le t\le 1$, for the integral in (7.5) we obtain the bound

$$\int_0^t e^{-\kappa(p)(t-\tau)-\varepsilon|p|\tau}\,d\tau \le \begin{cases} \dfrac{e^{-\varepsilon|p|t}-e^{-\kappa(p)t}}{\kappa(p)-\varepsilon|p|}, & |p|\ge\mathcal{N}, \\ te^{c_0}, & |p|\le\mathcal{N}. \end{cases} \tag{7.6}$$

Choosing now a sufficiently large $\mathcal{N}$: $\mathcal{N} = \max\{4+p_0, (2^{7+\alpha}24b_1\|\psi\|_1^2\varepsilon^{-1})^{1/\beta}\}$, so that for $|p|\ge\mathcal{N}$

$$0 < \frac{1}{\kappa(p)-\varepsilon|p|} \le \frac{2^{\alpha+1}}{\varepsilon\mathcal{N}^\beta(1+|p|)^{\beta+2}} \le \frac{(1+|p|)^{-2-\beta}}{242^6b_1\|\psi\|_1^2},$$

and then fixing a sufficiently small T_0 : $T_0 = \min\{1, (2^{7+\alpha}\mathcal{N}^{\alpha+1}e^{c_0+\varepsilon\mathcal{N}}24b_1\|\psi\|_1^2)^{-1}\}$, so that for $t\in[0,T_0]$ and for $|p|\le\mathcal{N}$

$$te^{c_0} \le T_0e^{c_0+\varepsilon\mathcal{N}T_0-\varepsilon|p|t} \le e^{-\varepsilon|p|t}\frac{(1+|p|)^{-2-\beta}}{2^6\cdot 24b_1\|\psi\|_1^2},$$

from (7.6) we obtain

$$\int_0^t e^{-\kappa(p)(t-\tau)-\varepsilon|p|\tau}\,d\tau \le e^{-\varepsilon|p|t}\frac{(1+|p|)^{-\beta-2}}{24\cdot 2^6b_1\|\psi\|_1^2}\,. \tag{7.7}$$

Substituting (7.7) in (7.5) we arrive at (7.3). Inequality (7.3) has thus been proved by induction for all $s\ge 0$.

We now prove that, for all integers $n, s\ge 0$,

$$\left\|\frac{\partial^n\hat{v}^{(s)}(p,t)}{\partial t^n}\right\|_0 \le c_n\psi^{(s)}(p)e^{-\varepsilon|p|t}(1+|p|)^{\alpha_2 n}t^{-\alpha_2 n^2}, \qquad t\in(0,T_0),\ p\in R_2, \tag{7.8}$$

with some $c_n\ge 0$, not depending on s, $\alpha_2 = 1+\alpha_1$.

For $n=0$ and for all $s\ge 0$ the inequality (7.8) with $c_0=1$ follows from (7.3). Suppose by induction, that (7.8) has been established for $n\le n_1$. Then for $n=n_1+1$ and for any $s\ge 0$ we differentiate (6.3) n_1 times with respect to t and obtain

$$\frac{\partial^n\hat{v}^{(s)}(p,t)}{\partial t^n} + \sum_{k=1}^2\sum_{i,j=1}^3\iint_{-\infty}^\infty p_k a_{ij}^k\sum_{l=0}^{n-1}C_l^{n-1}\frac{\partial^l\hat{v}_i^{(s-1)}(p-y,t)}{\partial t^l}\frac{\partial^{n-1-l}\hat{v}_j^{(s-1)}(y,t)}{\partial t^{n-1-l}}\,dy + K(p)\frac{\partial^{n-1}\hat{v}^{(s)}(p,t)}{\partial t^{n-1}} = 0,$$

whence, as in (7.4), we find

$$\begin{aligned} \left\|\frac{\partial^n\hat{v}^{(s)}(p,t)}{\partial t^n}\right\|_0 &\le c_n(1+|p|)^{\alpha_2 n}\psi^{(s)}(p)t^{-\alpha_2 n^2}e^{-\varepsilon|p|t} \\ &+ c_n|p|\sum_{l=0}^{n-1}\iint_{-\infty}^\infty (1+|p-y|)^{\alpha_2 l}(1+|y|)^{\alpha_2(n-1-l)}\psi^{(s-1)}(p-y)\psi^{(s-1)}(y) \\ &\cdot e^{-\varepsilon|p-y|t-\varepsilon|y|t}t^{-\alpha_2(n-1-l)^2-\alpha_2 l^2}\,dy \end{aligned} \tag{7.9}$$

($K(p)$ is estimated from above because the operator $\mathbb{K}$ has a finite order not exceeding α_1). All constants independent of s, p, y, t are denoted by c_n. Going to polar coordinates, we estimate the integral in (7.9) as follows:

$$\begin{aligned} \iint_{-\infty}^{\infty} &\leq \int_0^{2\pi} d\theta \Bigg\{ e^{-\varepsilon|p|t}(1+|p|)^{\alpha_2(n-1)} \int_0^{2|p|} |y|\psi^{(s-1)}(p-y)\psi^{(s-1)}(y)\,d|y| \\ &\quad + e^{-\varepsilon|p|t} \int_0^{\infty} e^{-\varepsilon|y|t}(1+|p-y|)^{\alpha_2 l}(1+|y|)^{\alpha_2(n-1-l)} \\ &\qquad \cdot \psi^{(s-1)}(p-y)\psi^{(s-1)}(y)|y|\,d|y| \\ &\leq c_n(e^{-\varepsilon|p|t}(1+|p|)^{\alpha_2(n-1)}\|\psi\|_1^2 \\ &\quad + e^{-\varepsilon|p|t}(1+|p|)^{\alpha_2(n-1)}t^{-\alpha_2(n-1)}\|\psi\|_1^2) \\ &\leq c_n e^{-\varepsilon|p|t}(1+|p|)^{\alpha_2(n-1)}t^{-\alpha_2(n-1)}\|\psi\|_1^2 \Bigg\}; \end{aligned} \tag{7.10}$$

here we have used the inequality

$$\begin{aligned} &e^{-\varepsilon|y|t}(1+|p-y|)^{\alpha_2 l}(1+|y|)^{\alpha_2(n-1-l)} \\ &\quad \leq e^{-\varepsilon|y|t}((1+|p|)(1+|y|))^{\alpha_2(n-1)} \leq ct^{-\alpha_2(n-1)}(1+|p|)^{\alpha_2(n-1)}. \end{aligned}$$

Substituting (7.10) in (7.9), we get

$$\begin{aligned} \left\| \frac{\partial^n \hat{v}^{(s)}(p,t)}{\partial t^n} \right\|_0 &\leq c_n(1+|p|)^{\alpha_2 n}\psi^{(s)}(p)t^{-\alpha_2 n^2}e^{-\varepsilon|p|t} \\ &\quad + c_n e^{-\varepsilon|p|t}(1+|p|)^{\alpha_2(n-1)+1}\sum_{l=0}^{n-1} t^{-\alpha_2 l^2-\alpha_2(n-1-l)^2-\alpha_2(n-1)} \\ &\leq c_n(1+|p|)^{\alpha_2 n}t^{-\alpha_2 n^2}e^{-\varepsilon|p|t}\psi^{(s)}(p), \end{aligned}$$

since $(1+|p|)^{-\alpha_2+1} \leq \psi^{(s)}(p)$, as $\alpha_2 - 1 = \alpha_1 \geq \beta + 1 = (\alpha-2)/2 + 1 = \frac{\alpha}{2}$. Therefore, the estimate (7.8) is proved. By (7.3) and (7.8), from the sequence $\{v^{(s)}(x,t)\}$, $s = 0, 1, 2, \ldots$, consisting of the Fourier preimages $v^{(s)}(x,t) = F^{-1}(\hat{v}^{(s)}(p,t))$ we can extract a subsequence converging to a limit $v(x,t) \in C^\infty((0,T_0];(H^\infty(R_2))^3$ in the norm of the space $C^m((0,T_0];(H^n(R_2))^3)$ with arbitrary n and m. It is obvious that $v(x,t)$ satisfies the system (6.1) with $B \equiv 0$ for all $t > 0$. In order to prove that the initial condition is satisfied, we show that some subsequence of $\{v^{(s)}(x,t)\}$ converges also in the norm of $C^0([0,T_0];(L_2(R_2))^3)$. From the sequence $\{\bar{v}^{(s)}(x)\}$ of initial vector-valued functions we choose a subsequence such that

$$\|\bar{v}^{(s)}(x) - \bar{v}^{(s-1)}(x)\|_1 \leq (c_0 e^{\tilde{c}})^{-1}2^{-s}\|\psi\|_1,$$

and prove that for all $s \geq 1$

$$\max_{t\in[0,T_0]} \|v^{(s)}(x,t) - v^{(s-1)}(x,t)\|_1 \leq 2^{5-s}\|\psi\|_1. \tag{7.11}$$

For $s = 1$ (7.11) follows from (7.3), since

$$\max_{t\in[0,T_0]} \|v^{(1)}(x,t) - v^{(0)}(x,t)\|_1 \le \max_{t\in[0,T_0]} (\|\psi^{(1)}\|_1 + \|\psi^{(0)}\|_1) \le 4\|\psi\|_1.$$

Suppose, by induction, that (7.10) has been established for all $s \le s_1$. Then for $s = s_1 + 1$, from (6.14) with $b_{ij}^k \equiv 0$ we get

(7.12)
$$\begin{aligned}
&\|\hat{v}^{(s)}(p,t) - \hat{v}^{(s-1)}(p,t)\|_0 \\
&\quad \le C_0\|\hat{\bar{v}}^{(s)}(p) - \hat{\bar{v}}^{(s-1)}(p)\|_0 e^{-\kappa(p)t} \\
&\quad\quad + 2^5\tilde{b}\int_0^t e^{-\kappa(p)(t-r)}\,d\tau|p| \iint_{-\infty}^{\infty} \|\hat{v}^{(s-1)}(p-y,\tau) - \hat{v}^{(s-2)}(p-y,\tau)\|_0 \\
&\quad\quad\quad \cdot \|\hat{v}^{(s-1)}(y,\tau) + \hat{v}^{(s-2)}(y,\tau)\|_0\,dy,
\end{aligned}$$

$$\tilde{b} = c_0 36 \max_{i,j,k} \|a_{ij}^k\|_0 .$$

We estimate the integral with respect to y in (7.12) by means of (7.3), (7.11), and the Cauchy inequality

$$\iint_{-\infty}^{\infty} \cdots\,dy \le 2^{8-s}\|\psi\|_1^2 . \tag{7.13}$$

Substituting (7.13) in (7.12), we find that for $0 \le t \le T_0$

$$\begin{aligned}
&\|\hat{v}^{(s)}(p,t) - \hat{v}^{(s-1)}(p,t)\|_0 \\
&\quad \le c_0 e^{\tilde{c}}\|\hat{\bar{v}}^{(s)}(p) - \hat{\bar{v}}^{(s-1)}(p)\|_0 + 2^{13-s}\|\psi\|_1^2\tilde{b}\int_0^t e^{-\kappa(p)(t-\tau)}|p|\,d\tau \\
&\quad \le c_0 e^{\tilde{c}}\|\hat{\bar{v}}^{(s)}(p) - \hat{\bar{v}}^{(s-1)}(p)\|_0 + 2^{4-s}\psi(p),
\end{aligned} \tag{7.14}$$

in view of (7.7) with $\varepsilon - 0$. Integrating (7.14) with respect to $p \in R_2$, we obtain (7.11). It follows from (7.11) that $v(x,t) \in C^0([0,T_0];(L_2(R_2))^3)$ and the initial condition in (6.1) is satisfied. Theorem 9 is proved. □

THEOREM 10. *Suppose the operator $\mathbb{K}$ satisfies the dissipation condition* (6.5), (6.6), *in which the function $\kappa(p) = \varepsilon(1+|p|)^\alpha$, $p \in R_2$, $\alpha > 2$, $\varepsilon > 0$, is taken in place of $\kappa(p)$, the initial perturbation $\bar{v}(x) \in (L_2(R_2))^3$ and the norm $\|\bar{v}_1\|$ is sufficiently small: $\|\bar{v}\|_1 \le \beta\varepsilon/2^{\alpha+11}\chi b_1$. Then there exists a unique solution of the Cauchy problem* (6.1) *with $B \equiv 0$ in the class $C^\infty((0,\infty);(H^\infty(R_2))^3) \cap C^0([0,\infty);(L_2(R_2))^3)$. Therefore, a nonsmooth initial perturbation becomes and remains infinitely smooth for all positive t.*

PROOF. We show that (7.3) holds for all $t > 0$ if for $\psi^{(s)}(p)$ we take the function

$$\psi^{(s)}(p) = \chi\|\hat{\bar{v}}^{(s)}(p)\|_0 + \frac{\beta}{\pi}\chi\|\bar{v}\|_1(1+|p|)^{-\beta-1}. \tag{7.15}$$

In this case (7.4) and (7.5) remain in force. For the integral in (7.5) we have

$$\begin{aligned}
\int_0^t e^{-\kappa(p)(t-\tau)-\varepsilon|p|\tau}\,d\tau &= \frac{e^{-\varepsilon|p|t} - e^{-\varepsilon(1+|p|)^\alpha t}}{\varepsilon(1+|p|)^\alpha - \varepsilon|p|} \\
&\le \frac{e^{-\varepsilon|p|t}2^\alpha}{\varepsilon(1+|p|)^\alpha} \le \frac{e^{-\varepsilon|p|t}(1+|p|)^{-\beta-2}\beta\chi\|\bar{v}\|_1}{2^7 b_1\pi\|\psi\|_1^2}
\end{aligned} \tag{7.16}$$

since

$$\alpha \geq \beta + 2 = \frac{\alpha - 2}{2} + 2 = \frac{\alpha}{2} + 1 \quad \text{and} \quad \|\psi\|_1^2 \leq 2^2\chi^2\|\bar{v}\|_1^2 \leq \frac{\varepsilon\beta\chi\|\bar{v}\|_1}{2^{\alpha+7}b_1\pi}$$

by the hypothesis of the theorem.

The bound (7.16) is analogous to (7.7), and thus the estimate (7.3) with $\psi^{(s)}(p)$ from (7.15) is proved; (7.3) in turn implies that

$$\|v\|_1^2 \leq \|\psi\|_1^2 \leq \frac{\varepsilon\beta\chi}{2^{\alpha+7}b_1\pi}\|\bar{v}\|_1$$

for all $t > 0$. Using this bound and applying Theorem 9 successively, we can extend the solution of the Cauchy problem (6.1) with $B \equiv 0$ to an arbitrary time interval. Theorem 10 is proved. □

§8. The Cauchy problem for the system of equations for surface waves with weak nonlocal interaction

We consider the Cauchy problem for the system of equations for surface waves with weak nonlocal interaction:

$$(8.1) \qquad \begin{cases} \eta_t + (\nabla, \eta\nabla\varphi) + \mu\mathbb{K}_1(\eta) + \mu\mathbb{K}_2(\nabla\varphi) = 0, \\ \varphi_t + \frac{1}{2}(\nabla\varphi)^2 + \mu\mathbb{K}_3(\eta) + \mu\mathbb{K}_4(\nabla\varphi) = 0, \\ \eta|_{t=0} = \bar{\eta}(x), \qquad \varphi|_{t=0} = \bar{\varphi}(x), \end{cases}$$

where $\mu \geq 0$ is a small parameter. Theorem 11 presented below holds under the conditions of Theorem 4. It gives the complete asymptotic expansion in powers of the small parameter μ of a solution $u_\mu(x,t) \equiv (\eta_\mu(x,t), \varphi_\mu(x,t))$ of the Cauchy problem (8.1); the remainder term is estimated in the uniform metric.

THEOREM 11. *Assume that the following conditions are satisfied*:

1) $\operatorname{Re}(i\sum_{j=1}^2 p_j K_4^{(j)}(p)) \geq 0$ *for* $|p| \geq p_0 > 0$,

2)

$$\iint_{-\infty}^{\infty} |K_1(x)|\, dx \leq c, \qquad \iint_{-\infty}^{\infty} |K_2^{(j)}(x)|^q\, dx \leq c, \qquad q = 1,2,$$

$$\iint_{-\infty}^{\infty} \left|\frac{\partial^{n+m}K_3(x)}{\partial x_1^n\, \partial x_2^m}\right| dx \leq c, \qquad 0 \leq n+m \leq 4, \qquad K_4^{(j)}(x) \in C^1(R_2\backslash 0),$$

$$|K_4^{(j)}(x)| \leq c|x|^{-1-\alpha}, \quad |\nabla K_4^{(j)}(x)| \leq c|x|^{-2-\alpha},$$

$$x \neq 0,\ c \geq 1,\ 0 < \alpha < \tfrac{2}{5} - 10\gamma,\ \gamma \in (0, \tfrac{1}{100}),$$

where $K_1(x)$, $K_2^{(j)}(x)$, $K_3(x)$, $K_4^{(j)}(x)$, $j = 1,2$, *are the kernels of the operators* $\mathbb{K}$ *in the* x*-representation*;

3) $(\bar{\eta}(x), \bar{\varphi}(x)) \in (H^\infty(R_2))^2$;

4)

$$\min_{x\in R_2} \frac{\partial^2\bar{\varphi}}{\partial x_1^2} > -\tfrac{1}{3}m_0, \quad \max_{x\in R_2}\left|\frac{\partial^2\bar{\varphi}}{\partial x_1\partial x_2}\right| \leq \gamma^3 m_0, \quad m_0 \equiv \left|\min_{x\in R_2}\frac{\partial^2\bar{\varphi}}{\partial x_2^2}\right|,$$

where m_0 *is sufficiently large* (*see* (4.6)).

Then:

1) *for any* $\mu \in [0,1]$ *there exists a unique solution*

$$u_\mu(x,t) \in C^\infty([0,T_\mu);(H^\infty(R_2))^2)$$

of the Cauchy problem (8.1), *where* T_μ *is the time of breaking*

$$(m_0(1+\mu\gamma)^2)^{-1} \le T_\mu \le (m_0(1-\mu\gamma)^2)^{-1};$$

2) $u_\mu(x,t)$ *has the asymptotic expansion* $u_\mu(x,t) = \sum_{k=0}^{\infty} \mu^k u_k(x,t)$. *Moreover, the following estimate holds for the remainder term* δ_n:

$$\max_{t\in[0,T_0-\beta]} \left\| \frac{\partial^k \delta_n}{\partial t^k} \right\|_{(l)} \le c_0 \mu^{n+1},$$

where $\beta > 0$ *and the integers* $k, l \ge 0$ *are arbitrary,* $\|\cdot\|_{(l)}$ *is the norm of the space* $C^l(R_2)$, *and* $c_0 = c_0(l,k,n,\beta)$.

PROOF. The proof of this theorem coincides word-for-word with that of Theorem 6 in Chapter 3. □

CHAPTER 5

Generalized Solutions

§1. Introduction

This chapter is devoted to the construction of various generalized solutions of the Whitham equation (§§2 through 4), the nonlinear nonlocal Schrödinger equation (§5), and the system of surface wave equations (§6). The importance of generalized solutions is well known: they appear in various applied problems including hydrodynamics and numerical modelling. As was also shown in the previous chapters, a classical solution of the Cauchy problem for the Whitham equation and the system of surface wave equations breaks in a finite time T_0, and for $t > T_0$ only generalized solutions of the corresponding Cauchy problem can exist.

We describe the content of this chapter. Sections 2 through 4 are devoted to generalized solutions of the Cauchy problem for the Whitham equation:

$$(1.1) \qquad u_t + uu_x + \mathbb{K}(u) = 0, \qquad u|_{t=0} = \bar{u}(x),$$

where

$$(1.2) \qquad \mathbb{K}(u) = \frac{1}{2\pi}\int_{-\infty}^{\infty} e^{ipx} K(p)\hat{u}(p,t)\,dp;$$

$\hat{u}(p,t) = \int_{-\infty}^{\infty} e^{-ipx}u(x,t)\,dx$ is the Fourier transform of $u(x,t)$. Suppose that the operator $\mathbb{K}$ has a finite order α_0, i.e., $\alpha_0 \in [0, +\infty)$ the numbers $\alpha \geq 0$ such that

$$(1.3) \qquad |K(p)| \leq c(1+|p|)^{\alpha}, \qquad p \in R_1,\ c > 0$$

have a minimum $\alpha_0 \in [0, +\infty)$. Along with (1.1), we consider the following regularized equation:

$$(1.4) \qquad u_t^{(\mu)} + u^{(\mu)}u_x^{(\mu)} + \mathbb{K}(u^{(\mu)}) + \mu\mathbb{R}(u^{(\mu)}) = 0, \qquad u|_{t=0} = \bar{u}(x),$$

where the dissipative operator $\mathbb{R}$ is determined by (1.2), so that its symbol $R(p)$ satisfies the condition of the theorem on smoothing presented in Chapter 2, §4:

$$(1.5) \qquad R^1(p) = \operatorname{Re} R(p) \geq \gamma|p|^{\beta}, \quad \text{for } |p| \geq p_0 > 0$$

where $\beta > \frac{3}{2}$, $\gamma > 0$. Then for an arbitrary initial perturbation $\bar{u}(x) \in L_2(R_1)$ there exists a unique classical solution $u^{(\mu)}(x,t)$ for the Cauchy problem (1.4), belonging to $C^{\infty}((0,\infty); H^{\infty}(R_1)) \cap C^0([0,\infty), L_2(R_1))$.

An interesting question arises: is there a limit as $\mu \to 0$ of the solutions $u^{(\mu)}(x,t)$ when a classical solution $u(x,t)$ of the Cauchy problem (1.1) no longer exists, so that the limiting function $u^{(0)}(x,t)$ could be considered as a generalized solution of (1.1)? Generally speaking, this generalized solution $u^{(0)}(x,t)$ of (1.1) is not unique even in the case of smooth initial conditions. It depends on how the limit is evaluated as $\mu \to 0$, i.e., on the choice of the dissipative operator $\mathbb{R}$. This question was first raised in 1950 by Hopf [**120**] in connection with the Burgers equation $u_t + uu_x - \mu u_{xx} = 0$, i.e, $\mathbb{K} = 0$ and $\mathbb{R} = -\partial e/\partial x^2$. Later, Hopf's problem was considered by many other authors [**31, 32, 36, 71, 74, 141, 170**].

Apparently, the first study of the generalized solutions for (1.1) was undertaken in 1979 in [**158**] in the case when the operator $\mathbb{R}$ is a power of the Laplace operator. In §2 of this chapter we consider the case of dissipative operators $\mathbb{K}$, whose symbol $K(p)$ satisfies the condition

$$K^1(p) = \operatorname{Re} K(p) \geq \varepsilon |p|^\delta \quad \text{for } |p| \geq p_0 > 0, \tag{1.6}$$

where δ, ε are any positive numbers. We prove the existence of a generalized solution $u^{(0)}(x,t)$ of the Cauchy problem (1.1) in the class

$$L_\infty([0,T]; L_2(R_1)) \bigcap L_2([0,T]; H^{\delta/2}(R_1))$$

for any positive T and any initial condition $\bar{u}(x) \in L_2(R_1)$. This solution is constructed as the limit as $\mu \to 0$ of the sequence of solutions $\{u^{(\mu)}(x,t)\}$ of the Cauchy problem (1.4).

Section 3 is dedicated to studying the conservative Whitham equation (i.e. $K^1(p) \equiv 0$). For any $T > 0$ we establish the existence of a generalized solution $u^{(0)}(x,t) \in L_\infty([0,T]; H^{\sigma/2}(R_1))$ of the initial value problem (1.1) by taking the limit $\mu \to 0$ of the sequence of solutions $\{u^{(\mu)}(x,t)\}$ of (1.4), in the case when the conservative component of the symbol $K^2(p) \equiv \operatorname{Im} K(p)$ has a constant sign and is increasing for large $|p|$. More precisely, when $K^3(p) \equiv K^2(p)/p$ has a finite order $\overline{\alpha}$ (and, in particular, is bounded at zero) and for $|p| \geq p_0$ satisfies one of the following conditions:

$$\text{either } K^3(p) \geq \varepsilon |p|^\sigma, \quad \text{or} \quad K^3(p) \leq -\varepsilon |p|^\sigma, \tag{1.7}$$

where ε, p_0 are positive numbers, $\sigma > \frac{1}{2}$.

As an example of application of this result we obtain the existence of generalized solutions of the Korteweg-de Vries, Benjamin-Ono, Joseph, Smith, and Kawahara equations. It should be noted, however, that since for the KdV, BO, and Joseph equations there are higher(*) conservation laws, classical solutions of the Cauchy problem for these equations exist for all times, as demonstrated in [**87**]. For the Kawahara equation $u_t + uu_x + au_{xxx} + u_{xxxx} = 0$ the existence of a global in time classical solution also follows from the result indicated above.

In §4 we construct generalized solutions $u^{(0)}(x,t)$ for the shallow water equation $u_t + uu_x = 0$ (i.e., the operator $\mathbb{K} = 0$ in (1.1)) as the limit, as $\mu \to 0$, of the solutions of (1.4) with a strongly dissipative operator $\mathbb{R}$ satisfying (1.5) and having the following form:

$$\mathbb{R}(\varphi) = \int_{-\infty}^{\infty} r(x-s)\varphi''(s)\, ds; \tag{1.8}$$

here the kernel $r(x)$ is convex up ($r''(x) \leq 0$) in some neighborhood of the singular point $x = 0$. This result generalizes the above-mentioned publication of Hopf in two aspects: first, the regularizing operator may be nonlocal, second, its order is less than two.

(*) It was shown in [**93**] that in the class of equations of type (1.1) only the KdV, BO, Joseph, and shallow water equations have higher conservation laws.

In §5 we consider the Cauchy problem for the nonlinear nonlocal Schrödinger equation

$$u_t - i|u|^2 u + \mathbb{K}(u) = 0, \qquad u|_{t=0} = \bar{u}(x), \tag{1.9}$$

where $\mathbb{K}(u) = (2\pi)^{-\mathcal{N}} \int_{-\infty}^{\infty} e^{i(p,x)} K(p)\hat{u}(p,t)\,dp$, $x, p \in R_{\mathcal{N}}$, $\mathcal{N} \geq 1$. We prove the existence of a generalized solution of the Cauchy problem (1.9) in the case when the dissipative part of the operator $\mathbb{K}$ is of order not exceeding $\mathcal{N}$, and the conservative part is of order greater than $\mathcal{N}$, i.e., when the following conditions are satisfied:

$$K^2(p) = \operatorname{Im} K(p) \geq \varepsilon |p|^\sigma \quad \text{for } |p| \geq p_0 > 0, \tag{1.10}$$

where $\sigma > \mathcal{N}$, $\varepsilon > 0$, and $K^1(p) = \operatorname{Re} K(p)$ satisfies condition (1.3) with $\alpha < \mathcal{N}$. We show that if the initial perturbation $\bar{u}(x)$ belongs to the class $H^\rho(R_{\mathcal{N}})$, where $\rho > \frac{\sigma}{2}$, then the generalized solution of the Cauchy problem (1.9) belongs to the space $L_\infty([0,T]; H^\rho(R_{\mathcal{N}}))$, for any $T > 0$. As a consequence of this result, we obtain conditions for the existence of a global in time classical solution.

In §6 we establish the existence of generalized solutions of the Cauchy problem for the system of surface wave equations (see Chapter 4)

$$\begin{cases} \eta_t + (\nabla, \eta\nabla\varphi) + \mathbb{K}_1(\eta) + \mathbb{K}_2(\varphi) = 0, \\ \varphi_t + \frac{1}{2}(\nabla\varphi)^2 + \mathbb{K}_3(\eta) + \mathbb{K}_4(\varphi) = 0 \end{cases} \tag{1.11}$$

in the conservative case, i.e., when the operators $\mathbb{K}_j$ satisfy the following conditions:

$$\mathbb{K}_1 = -\mathbb{K}_4^*, \quad \mathbb{K}_2 = \mathbb{K}_2^*, \quad \mathbb{K}_3 = \mathbb{K}_3^* \tag{1.12}$$

if the initial perturbations $\eta(x,0) = \bar{\eta}(x)$ and $\varphi(x,0) = \bar{\varphi}(x)$, $x \in R_2$ are sufficiently small. (See the conservation laws introduced under the present conditions (1.12), in §1, Chapter 4.)

The results derived in §§2 through 4 were published in **[46, 53]**, and the results of §§5 and 6 were obtained by E. I. Kaĭkina.

§2. The dissipative Whitham equation

In this section we construct a generalized solution $u(x,t)$ of the Cauchy problem (1.1) with a dissipative operator $\mathbb{K}$ obeying the condition (1.6). We say that $u(x,t)$ is a generalized solution if for any positive T: $u(x,t) \in L_\infty((0,T], L_2(R_1)) \cap L_2([0,T], H^{\delta/2}(R_1))$ and for almost all $t \in [0,T]$ the function $u(x,t)$ satisfies, for any $\varphi(x) \in H^\nu(R_1)$, the equation

$$(u_t, \varphi) - \tfrac{1}{2}(u, u\varphi_x) + (u, \mathbb{K}^*\varphi) = 0 \tag{2.1}$$

and the initial condition

$$u|_{t=0} = \bar{u}(x), \qquad x \in R_1 . \tag{2.2}$$

Here $\nu = \max(\alpha_0, \alpha_1)$, α_0, is the order of the operator $\mathbb{K}$, α_1 is the order of operator $\mathbb{R}$, $\alpha_0 \geq \delta$, $\alpha_1 \geq \beta > \frac{3}{2}$; the operator $\mathbb{K}^*$ is adjoint in $L_2(R_1)$ of the operator $\mathbb{K}$; $(\cdot,\cdot)$ in (2.1) is the scalar product in $L_2(R_1)$.

THEOREM 1. *Assume that*

1) *The condition of dissipation* (1.6) *holds with any* $\delta > 0$;

2) *The initial perturbation* $\bar{u}(x) \in L_2(R_1)$.

Then the Cauchy problem (1.1) *has a generalized solution* $u(x,t)$.

PROOF. In accordance with the results obtained in §4, Chapter 2, the solution $u^{(\mu)}(x,t)$ of the Cauchy problem (1.4) belongs to the space $C^\infty((0,\infty); H^\infty(R_1))$. Therefore, all further transformations are justified. We multiply (1.4) by $u^{(\mu)}$ and integrate with respect to x from $-\infty$ to $+\infty$ (the index μ is temporarily omitted):

$$\frac{1}{2}\frac{d}{dt}(u,u) + (u, \mathbb{K}(u)) + \mu(u, \mathbb{R}(u)) = 0, \tag{2.3}$$

since $\int_{-\infty}^{\infty} u^2 u_x\,dx = 0$. Applying the Plancherel theorem and (1.6), we find

$$\begin{aligned}(u, \mathbb{K}(u)) &= \frac{1}{2\pi}\int_{-\infty}^{\infty} |\hat{u}(p,t)|^2 K^1(p)\,dp \\ &= \int_{|p|\geq p_0} + \int_{|p|\leq p_o} \\ &\leq \frac{\varepsilon}{2\pi}\int_{-\infty}^{\infty} |\hat{u}(p,t)|^2\,|p|^\delta\,dp \\ &\quad - \left(\varepsilon|p_0|^\delta + \sup_{|p|\leq p_0}(-K^1(p),0)\right)\frac{1}{2\pi}\int_{-\infty}^{\infty}|\hat{u}(p,t)|^2\,dp \\ &\leq \varepsilon\|u\|^2_{\delta/2} - c\|u\|_0^2,\end{aligned} \tag{2.4}$$

where $\|u\|_\delta^2 = \frac{1}{2\pi}\int_{-\infty}^{\infty}|\hat{u}(p,t)|^2|p|^{2\delta}\,dp$; here and below $c > 0$ denotes various constants, independent of μ. Similarly, (1.5) implies

$$\mu(r, \mathbb{R}(u)) \geq \mu\gamma\|u\|^2_{\beta/2} - c\|u\|_0^2\,. \tag{2.5}$$

Substitution of (2.4) and (2.5) in (2.3) leads to

$$\frac{1}{2}\frac{d}{dt}\|u\|_0^2 + \varepsilon\|u\|^2_{\delta/2} + \mu\gamma\|u\|^2_{\beta/2} \leq c\|u\|_0^2,$$

whence, integrating with respect to t, we find

$$\|u\|_0^2 + 2\int_0^t (\varepsilon\|u\|^2_{\delta/2} + \mu\gamma\|u\|^2_{\beta/2})\,d\tau \leq c\int_0^t \|u\|_0^2\,d\tau + \|\bar{u}\|_0^2\,. \tag{2.6}$$

We conclude from (2.6) that for all $T > 0$

$$\|u^{(\mu)}\|_{L_2([0,T];H^{\delta/2}(R_1))} \leq c \quad \text{and} \quad \|u^{(\mu)}\|_{L_\infty([0,T];L_2(R_1))} \leq c\,. \tag{2.7}$$

Let us show that

$$\|u_t^{(\mu)}\|_{L_\infty([0,T];H^{-\nu}(R_1))} \leq c\,. \tag{2.8}$$

We multiply (1.4) by an arbitrary function $\varphi(x) \in H^\nu(R_1)$ (the index μ is omitted):

$$(u_t, \varphi) + (uu_x, \varphi) + (u, (\mathbb{K}^* + \mu\mathbb{R}^*)\varphi) = 0\,. \tag{2.9}$$

It follows from the Sobolev imbedding theorem that

$$H^{\nu-1}(R_1) \subset L_\infty(R_1), \quad \text{since } \nu > \tfrac{3}{2}, \tag{2.10}$$

so that, by (2.7),

$$|(uu_x, \varphi)| = \tfrac{1}{2}|(u, u\varphi_x)| \leq \tfrac{1}{2}\|\varphi_x\|_{L_\infty(R_1)}\|u\|_0^2 \leq c\|\varphi\|_\nu\,. \tag{2.11}$$

Applying the Plancherel theorem and the Cauchy inequality, we get

$$(2.12)\qquad \begin{aligned}|(u, (\mathbb{K}^* + \mu\mathbb{R}^*)\varphi| &= \frac{1}{2\pi}\left|\int_{-\infty}^{\infty} \hat{u}(p,t)\hat{\varphi}(p)(K^*(p) + \mu R^*(p))\,dp\right| \\ &\le c\|u\|_0\,(\|\varphi\|_{\alpha_0} + \|\varphi\|_{\alpha_1}) \le c\|\varphi\|_\nu\,.\end{aligned}$$

We conclude from (2.9)–(2.12) that $|(u_t^{(\mu)}, \varphi)| \le c\|\varphi\|_\nu$, which implies (2.8).

It follows **[67]** from the imbedding theorem, (2.7), and (2.8), that for some $\xi > 0$ the norm $\|u^{(\mu)}\|_{H^\xi(R_1\times[0,T])}$ is uniformly bounded with respect to $\mu > 0$. Denoting by Ω any compact set in R_1, we find, in view of the compactness **[37]** of the imbedding $H^\xi(\Omega \times [0, T])$ in $L_2(\Omega \times [0, T])$, that the sequence $\{u^{(\mu)}\}$ contains a subsequence (which we also denote by $\{u^{(\mu)}\}$), such that $u^{(\mu)} \to u$ weakly in $L_2([0, T]; H^{\delta/2}(R_1))$ and strongly in $L_2(\Omega \times [0, T])$, while $u_t^{(\mu)} \to u_t$ weakly in the space $L_2([0, T]; H^{-\nu}(R_1))$.

We show that $(u^{(\mu)})^2 \to u^2$ in $\mathscr{D}'(\Omega \times [0, T])$. Indeed, for any function $\varphi(x, t) \in C_0^\infty(\Omega \times [0, T])$

$$(2.13)\qquad \int_\Omega\int_0^T (u^{(\mu)})^2\varphi\,dx\,dt \to \int_\Omega\int_0^T u^2\varphi\,dx\,dt,$$

since $u^{(\mu)} \to u$ weakly in $L_2(R_1 \times [0, T])$, and $u^{(\mu)}\varphi \to u\varphi$ strongly in $L_2(\Omega \times [0, T])$. We multiply (1.4) by $\varphi(x, t) \in C_0^\infty(\Omega \times [0, T])$ and integrate with respect to x and t in the domain $R_1 \times [0, T]$:

$$\int_0^T (u_t^{(\mu)}, \varphi)\,d\tau - \tfrac{1}{2}\int_0^T (u^{(\mu)}, u^{(\mu)}\varphi_x)\,d\tau + \int_0^T (u^{(\mu)}, (\mathbb{K}^* + \mu\mathbb{R}^*)\varphi)\,d\tau = 0,$$

whence, in view of (2.13), for $\mu \to 0$ we get

$$(2.14)\qquad \int_0^T (u_t, \varphi)\,d\tau - \tfrac{1}{2}\int_0^T (u^2, \varphi_x)\,d\tau + \int_0^T (u, \mathbb{K}^*(\varphi))\,dt = 0\,.$$

The equality (2.14), owing to (2.7) and (2.10), can be extended by continuity to all functions $\varphi(x, t) = \psi_1(x)\psi_2(t)$, where $\psi_1(x) \in H^\nu(R_1)$, and $\psi_2(t) \in L_2([0, T])$. This means that $u(x, t)$ satisfies (2.1). Since $u^{(\mu)}(x, t)$ converges weakly to $u(x, t)$ in $L_2([0, T]; H^{\delta/2}(R_1))$ and $u_t^{(\mu)}(x, t)$ converges weakly to $u(x, t)$ in $L_2([0, T]; H^{-\nu}(R_1))$, the function $u(x, t)$ (possibly after changing it on a set of measure zero in the interval $[0, T]$) realizes a continuous mapping of $[0, T]$ into the space $H^{-\nu}(R_1)$. Therefore $u^{(\nu)}(x, 0)$ converges weakly to $u(x, 0)$ in $H^{-\nu}(R_1)$. Thus, the initial condition (2.2) is fulfilled in the sense of equality of functions from $H^{-\nu}(R_1)$. Theorem 1 is proved. □

§3. The conservative Whitham equation

In the conservative case, i.e.,

$$(3.1)\qquad K^1(p) \equiv 0,$$

equation (1.1) has at least two conservation laws that enable one to prove the existence of a generalized solution $u(x, t)$ of the corresponding Cauchy problem. The function $u(x, t) \in L_\infty([0, T]; H^{\sigma/2}(R_1))$ is called a generalized solution of the Cauchy problem for the conservative equation (1.1) if it satisfies conditions (2.1) and (2.2) (the constant $\sigma > \frac{1}{2}$ is taken from (1.7)).

THEOREM 2. *Suppose that*

1) *Conditions* (3.1) *and* (1.7) *are satisfied*;

2) *The initial perturbation* $\bar{u}(x) \in H^{\bar{\alpha}/2}(R_1)$.

Then there exists a generalized solution of the Cauchy problem (1.1) *for any* $T > 0$.

PROOF. We establish bounds, uniform with respect to μ, for a solution $u^{(\mu)}(x,t)$ of the Cauchy problem (1.4). As in Theorem 1, we obtain that for any $T > 0$

$$\|u^{(\mu)}\|_{L_\infty([0,T];L_2(R_1))} \le c \tag{3.2}$$

and

$$\|u_t^{(\mu)}\|_{L_\infty([0,T];H^{-\nu}(R_1))} \le c, \tag{3.3}$$

where c are various constants independent of μ, $\nu = \max(\alpha_0, \alpha_1)$. In order to obtain the bound

$$\|u^{(\mu)}\|_{L_\infty([0,T];H^{\sigma/2}(R_1))} \le c, \tag{3.4}$$

we multiply (1.4) by $\psi(u^{(\mu)}) = (u^{(\mu)})^2 + \mathbb{K}_3(u^{(\mu)}) \cdot 2$ and integrate with respect to x from $-\infty$ to $+\infty$ (the index μ is omitted):

$$\frac{d}{dt}\int_{-\infty}^{\infty}\left[\frac{u^3}{3} + u\mathbb{K}_3(u)\right] dx + \int_{-\infty}^{\infty} \psi(u)\mu\mathbb{R}(u)\, dx = 0\,. \tag{3.5}$$

Here the operator $\mathbb{K}_3$ has the symbol $K^3(p) \equiv K^2(p)/p = K(p)/ip$ and therefore is symmetric in $L_2(R_1)$ and $\mathbb{K}(\varphi) = \mathbb{K}_3(\varphi_x)$. Applying the Plancherel theorem and (1.7) we find

$$\begin{aligned}\mu\int_{-\infty}^{\infty} \mathbb{R}(u)\mathbb{K}_3(u)\, dx &= \frac{\mu}{2\pi}\int_{-\infty}^{\infty} R(p)K^3(p)|\hat{u}(p,t)|^2\, dp\\ &\ge \varepsilon\mu\gamma\int_{|p|\ge p_0} (|p|)^{\beta+\sigma}|\hat{u}(p,t)|^2\, dp - c\|u\|_0^2\\ &\ge \varepsilon\mu\gamma\|u\|_\rho^2 - c, \qquad \text{where } \rho = \frac{\beta+\sigma}{2}\,.\end{aligned} \tag{3.6}$$

Let $\kappa = \frac{1}{4}(\sigma - \frac{1}{2}) > 0$; we estimate the following expression:

$$\begin{aligned}\mu\left|\int_{-\infty}^{\infty} u^2\mathbb{R}(u)\, dx\right| &= \frac{\mu}{2\pi}\left|\int_{-\infty}^{\infty} R(p)\hat{u}^*(p,t)\, dp\int_{-\infty}^{\infty} \hat{u}(p-q,t)\hat{u}(q,t)\, dq\right|\\ &\le \mu c\int_{-\infty}^{\infty}(1+|p|)^{\rho-\frac{1}{2}-2\kappa}\,|\hat{u}(p,t)|\, dp\\ &\quad\times\int_{-\infty}^{\infty}|\hat{u}(p-q,t)\hat{u}(q,t)|\,(1+|p-q|)^\rho\, dq\\ &\le \mu c\|u\|_0\|u\|_\rho\int_{-\infty}^{\infty}(1+|p|)^{\rho-1/2-2\kappa}|\hat{u}(p,t)|\, dp\\ &\le c\mu\|u\|_0\|u\|_\rho\|u\|_{\rho-\kappa} \le \varepsilon\mu\gamma\|u\|_\rho^2 - c\,.\end{aligned} \tag{3.7}$$

Here we have taken into account the fact that the operator $\mathbb{R}$ has a finite order and that

$$|R(p)| \le c(1+|p|)^{2\rho-1/2-2\kappa}\,.$$

By the Sobolev imbedding theorem, $L_4(R_1) \supset H^{1/4}(R_1)$. Therefore, in view of (3.2), we obtain

$$\|u\|_{L_3(R_1)}^3 \le \|u\|_{L_4(R_1)}^2\|u\|_0 \le \|u\|_0\|u\|_{1/4}^2 \le \frac{\varepsilon}{2}\|u\|_{\sigma/2}^2 + c\,. \tag{3.8}$$

Application of (3.2), (1.7), and the Plancherel theorem leads to

$$
\begin{aligned}
\int_{-\infty}^{\infty} u\mathbb{K}_3(u)\,dx &= \frac{1}{2\pi}\int_{-\infty}^{\infty} K^3(p)|\hat{u}(p,t)|^2\,dp \\
&\geq \frac{\varepsilon}{2\pi}\int_{|p|\geq p_0} (|p|)^\sigma|\hat{u}(p,t)|^2\,dp - c\|u\|_0^2 \geq \varepsilon\|u\|_{\sigma/2}^2 - c\,.
\end{aligned}
\tag{3.9}
$$

Substitution of (3.6), (3.9) in (3.5) yields, after integration,

$$
\frac{\varepsilon}{2}\|u\|_{\sigma/2}^2 + 2\varepsilon\mu\gamma\int_0^t \|u\|_\rho^2\,dt \leq 2\varepsilon\|\bar{u}\|_{\sigma/2}^2 + \varepsilon\mu\gamma\int_0^t \|u\|_\rho^2\,d\tau + c,
$$

whence (3.4) is obtained. In view of the estimates (3.2)–(3.4), the proof can be completed as in Theorem 1. □

Theorem 2 implies

COROLLARY. *Suppose that*

1) *Conditions* (3.1) *and* (1.7) *are fulfilled with* $\sigma > 3$;

2) *The initial perturbation* $\bar{u}(x) \in H^\infty(R_1)$.

Then there exists a unique classical solution $u(x,t)$ *of the Cauchy problem belonging to the class* $C^\infty([0,\infty); H^\infty(R_1))$.

Thus, in the case of a conservative operator $\mathbb{K}$, satisfying condition (1.7), a classical solution of the Whitham equation can break only for $\sigma \lesssim 3$.

For the Kawahara equation we have $K(p) = -aip^3 + ip^5$, so that $K^3(p) = -ap^2 + p^4$ and condition (1.7) is satisfied for $\sigma = 4$. Therefore, a classical solution of the Cauchy problem for the Kawahara equation does not break with passage of time, i.e., exists globally.

PROOF OF COROLLARY. Since we established in Theorem 2 that for any $T > 0$ a generalized solution $u(x,t)$ of the Cauchy problem for (1.1), belonging to

$$
L_\infty([0,T]; H^{\sigma/2}(R_1)),
$$

exists, by the Sobolev imbedding theorem (see (2.10)) we find that the derivative $u_x(x,t)$ of the solution is bounded for all $T > 0$. Now the corollary follows from Theorem 2 of Chapter 2. □

Theorems 1 and 2 generalize the results of Saut **[158]** somewhat since, in contrast to **[158]**, the regularizing operator $\mathbb{R}$ in these theorems can be nonlocal and can have a lower order.

§4. The shallow water equation

We now consider the existence of a generalized solution $u(x,t)$ for the shallow water equation

$$
u_t + uu_x = 0 \tag{4.1}
$$

with the initial condition $u|_{t=0} = \bar{u}(x)$. Applying the approach presented in §§2 and 3, we construct a generalized solution of the Cauchy problem (4.1) as the limit (as $\mu \to 0$) of the sequence of solutions $\{u^{(\mu)}(x,t)\}$ of the Cauchy problem (1.4) with the operator $\mathbb{R}(u)$ satisfying the condition (1.5) and representable in the form

$$\mathbb{R}(\varphi) = \int_{-\infty}^{\infty} r(x-y)\varphi_{yy}(y)\,dy \tag{4.2}$$

where the kernel $r(x)$ is convex up in some neighborhood of the point $x = 0$. By a generalized solution we mean a function $u(x,t) \in L_2(R_1 \times [0,T]) \cap L_\infty(R_1 \times [0,T])$ for any positive T and satisfying (2.1) and (2.2) with $\mathbb{K} = 0$.

THEOREM 3. *Suppose that*

1) *The operator* $\mathbb{R}$ *satisfies* (1.5) *and is represented in the form* (4.2), *where*

$$r(x) \in L_1(R_1) \cap C^2(R_1 \backslash 0) \quad \text{and} \quad r''(x) \le 0 \quad \text{for } x \in [-a,a], \tag{4.3}$$

$$\int_{R_1\backslash[-a,a]} |r''(x)|\,dx \le b, \qquad a,b > 0,\ r^{(k)}(x) = o(x^{-k-1}), x \to 0, k = 0,1; \tag{4.4}$$

2) *The initial perturbation* $\bar{u}(x) \in L_2(R_1) \cap L_\infty(R_1)$ *and its first derivative is bounded from above.*

Then there exists a generalized solution $u(x,t)$ *of the Cauchy problem for equation* (4.1).

PROOF. For the solutions $u^{(\mu)}(x,t)$ of (1.4) we establish μ-uniform bounds, $\mu > 0$: $(0 < \mu \le 1)$

$$\sup_{x\in R_1} \left|u^{(\mu)}(x,t)\right| \le ce^{2bt} \tag{4.5}$$

and

$$u_x^{(\mu)}(x,t) < ce^{2bt}\,. \tag{4.6}$$

Here and below $c > 0$ represents various constants independent of $\mu > 0$. The derivative $u_x^{(\mu)}(x,t)$ is bounded by (4.6) only from above and no assertions are made whether it is bounded from below. We approximate the initial condition $\bar{u}(x)$ by the sequence of functions $\bar{u}^{(s)}(x) \in C_0^\infty(R_1)$, so that

$$\bar{u}_x^{(s)}(x) < c, \qquad \sup_{x\in R_1} \left|\bar{u}^{(s)}(x)\right| < c \tag{4.7}$$

and

$$\bar{u}^{(s)}(x) \to \bar{u}(x) \quad \text{in } L_2(R_1), \tag{4.8}$$

where c is independent of s. Then solutions $u^{(\mu,s)}(x,t)$ of the Cauchy problem for (1.4) with the initial data $\bar{u}^{(s)}(x)$ exist, belong to $C^\infty([0,\infty) \times R_1)$, and, together with their derivatives, tend to $u^{(\mu)}(x,t)$ and its derivatives, as $s \to \infty$. We prove that the inequalities (4.5) and (4.6) hold for $u^{(\mu,s)}(x,t)$.

We define the characteristic $y^{(\mu,s)}(\xi,t)$ as the solution of the following Cauchy problem:

$$\frac{dy^{(\mu,s)}}{dt} = u^{(\mu,s}(y^{(\mu,s)},t), \qquad y^{(\mu,s)}|_{t=0} = \xi\,. \tag{4.9}$$

From (1.4) we obtain

$$v^{(\mu,s)}(\xi,t) = u^{(\mu,s)}(y^{(\mu,s)}(\xi,t),t), \qquad v_t^{(\mu,s)}(\xi,t) + I_0(\xi,t) = 0, \tag{4.10}$$

where $I_0(\xi,t) = \mu \int_{-\infty}^{\infty} r(y^{(\mu,s)}(\xi,t) - \eta)u_{\eta\eta}^{(\mu,s)}(\eta,t)\,d\eta$.

We argue by contradiction. Let t_0 be the first moment of time when at least one of the inequalities (4.5) or (4.6) is violated. We temporarily omit the indexes μ and s. Let $v(\xi_0, t_0) = ce^{2bt_0}$ for some $\xi_0 \in R_1$. We transform the integral term $I_0(\xi, t)$ as follows:

$$\begin{aligned}
-I_0(\xi_0, t_0) &= -\mu \int_{-\infty}^{\infty} r\,(y(\xi_0, t_0) - \eta)\, u_{\eta\eta}(\eta, t_0)\, dy \\
&= -\mu \int_{-\infty}^{\infty} r''\,(y(\xi_0, t_0) - \eta)\,(u(\eta, t_0) - u(y(\xi_0, t_0), t_0))\, d\eta \\
&\le -\int_{y(\xi_0,t_0)-a}^{y(\xi_0,t_0)+a} r''\,(y(\xi_0, t_0) - \eta)\left(u(\eta, t_0) - ce^{2bt_0}\right) d\eta \\
&\quad + 2ce^{2bt_0} \int_{R_1\setminus[-a,a]} |r''(\eta)|\, d\eta < 2bce^{2bt_0}\,.
\end{aligned}$$

From this, by continuity, we have $-I_0(\xi_0, t) \le 2bce^{2bt}$ on some interval $t \in [t_1, t_0]$ $(0 \le t_1 < t_0)$. From (4.10) we therefore obtain

$$v_t(\xi_0, t) \le 2bce^{2bt}\,. \tag{4.11}$$

Integration of (4.11) with respect to t from t_1 to t_0 yields

$$v(\xi_0, t_0) \le v(\xi_0, t_1) + ce^{2bt_0} - ce^{2bt_1} < ce^{bt_0},$$

since for $t = t_1$ (4.5) is satisfied. Therefore, the bound (4.5) holds also at $t = t_0$. Similarly, analyzing the derivative $u_x^{(\mu,s)}(x, t)$ on the characteristic $x = y^{(\mu,s)}(\xi, t)$, for the function $v_1^{(\mu,s)}(\xi, t) = u_x^{(\mu,s)}(y^{(\mu,s)}(\xi, t), t)$, omitting the indexes μ and s, we obtain

$$v_{1t} + v_1^2 + I_1(\xi, t) = 0 \tag{4.12}$$

where

$$I_1(\xi, t) = \int_{-\infty}^{\infty} r\left(y^{(\mu,s)}(\xi, t) - \eta\right) u_{\eta\eta\eta}^{(\mu,s)}(\eta, t)\, dn\,.$$

Again, assuming by contradiction $[t_1, t_0]$ on some interval $t \in [t_1, t_0]$ we obtain the bound

$$-I_1(\xi_0, t) \le 2bce^{2bt} \tag{4.13}$$

for I_1. Integrating (4.12) with (4.13) taken into account we obtain (4.6) also for $t = t_0$, since (4.6) holds for $t = t_1$. The contradiction obtained proves (4.5) and (4.6) for the function $u^{(\mu)}(x, t)$.

In the same way as in Theorems 1 and 2, we prove that the sequence $\{u^{(\mu)}\}$ is bounded in $L_2(R_1)$, and that $\{u_t^{(\mu)}(x, t)\}$ is bounded in $L_2([0, T]; H^{-v}(R_1))$. Therefore, some subsequence of $\{u^{(\mu)}(x, t)\}$ converges weakly to the function $u(x, t)$ in the space $L_2([0, T]; H^{-v}(R_1))$, and a subsequence of $\{u_t^{(\mu)}(x, t)\}$ converges weakly to $u_t(x, t)$ in $L_2([0, T]; H^{-v}(R_1))$. Starting from the inequalities (4.5) and (4.6) obtained, one can establish, as in **[71]**, that $(u^{(\mu)}(x, t))^2$ converges to $u^2(x, t)$ almost everywhere on any compact set $\Omega \times [0, T]$. The proof is completed as in Theorem 1. Theorem 3 is proved. □

REMARK. In this theorem we have considered nonlocal dissipative operators of order less than 2. However, the approach presented can obviously be extended to the case of dissipative (in particular, also local) operators of any order higher than $\frac{3}{2}$. It is sufficient to consider a sequence of truncated convex kernels converging to the kernel of the operator under study.

§5. Nonlinear nonlocal Schrödinger equation

We examine here the Cauchy problem for the nonlinear nonlocal Schrödinger equation:

$$u_t - i|u|^2u + \mathbb{K}(u) = 0, \qquad u|_{t=0} = \bar{u}(x) \tag{5.1}$$

where $x \in R_{\mathcal{N}}$, $\mathcal{N} \geq 1$, and the operator $\mathbb{K}(u)$ is defined in the Introduction. We denote, as usual, $K^1(p) = \operatorname{Re} K(p)$, $K^2(p) = \operatorname{Im} K(p)$. Our goal in this section is to prove the following theorem.

THEOREM 4. *Suppose that the following conditions are satisfied*:

1) *For* $|p| \geq p_0 > 0$,

$$\begin{gathered}\lambda|p|^\theta \leq K^1(p) \leq c\lambda|p|^{\theta+\sigma-\mathcal{N}-4\kappa}, \\ \text{either } K^2(p) \geq \varepsilon|p|^\sigma, \quad \text{or} \quad K^2(p) \leq -\varepsilon|p|^\sigma,\end{gathered} \tag{5.2}$$

where $\sigma > \mathcal{N}$, $\varepsilon > 0$, $\lambda \geq 0$, $\theta \geq 0$, $0 < \kappa \leq \frac{1}{4}(\sigma - \mathcal{N})$;

2) *The initial perturbation is* $\bar{u}(x) \in H^\rho(R_i)$, $\rho \geq \sigma_0/2$.

Then there exists a generalized solution of the Cauchy problem (5.1) *such that for any* $T > 0$

$$\frac{\partial^m u}{\partial t^m} \in L_\infty\left([0,T]; H^{\rho-\nu m}(R_{\mathcal{N}})\right), \qquad m = 0, 1, 2, \dots,$$

where $\nu = \max(\alpha_0, \alpha_1)$, α_0 *and* α_1 *are the orders of the operators* $\mathbb{K}$ *and* $\mathbb{R}$, *respectively* (*see* (5.3) *below*), *and for any function* $\psi(x) \in C_0^\infty(R_{\mathcal{N}})$ *for almost all* $t \in [0,T]$ *the identity*(*)

$$(u_t, \psi) - i(|u|^2u, \psi) + (u, \mathbb{K}^*(\psi)) = 0$$

and the initial condition

$$(u, \psi)|_{t=0} = (\bar{u}, \psi)$$

hold.

PROOF. We consider the case $K^2(p) \geq \varepsilon|p|^\sigma$, for definiteness. We define the function $u^{(\mu)}(x,t)$ as the solution of the following regularized problem, $\mu \in (0,1]$:

$$u_t^{(\mu)} - i|u^{(\mu)}|^2u^{(\mu)} + \mathbb{K}(u^{(\mu)}) + \mu\mathbb{R}(u^{(\mu)}) = 0, \qquad u^{(\mu)}|_{t=0} = \bar{u}, \tag{5.3}$$

where the strongly dissipative operator $\mathbb{R}$ satisfies the conditions of Theorem 7 of §6, Chapter 4; i.e.

$$R^1(p) = \operatorname{Re} R(p) \geq \gamma|p|^\beta \quad \text{for } |p| \geq p_0, \tag{5.4}$$

where $\beta > \mathcal{N}$, $\gamma > 0$, and has a finite order:

$$|R(p) = c(1+|p|)^{\beta+\sigma-\mathcal{N}-4\kappa}, \tag{5.5}$$

Then, in accordance with that theorem, there exists a solution $u^{(\mu)}(x,t) \in$

(*) As usual, (,) is the scalar product in $L_2(R_{\mathcal{N}})$.

$C^\infty((0,\infty); H^\infty(R_{\mathcal{N}}))$ of the Cauchy problem (5.3). We omit the index μ and establish a number of bounds. From (5.3), for the function $J_1(t) = \int_{-\infty}^{\infty} uu^*\,dx = \|u\|_0^2$ we obtain

$$\dot{J}_1 + 2\,\mathrm{Re}\int_{-\infty}^{\infty} u^*\left(\mathbb{K}(u) + \mu\mathbb{R}(u)\right)\,dx = 0,$$

whence, integrating with respect to t and taking into account (5.2) and (5.4), we find that for $t \in [0, T]$, where $T > 0$ is arbitrary but fixed,

$$J_1(t) \le c\,. \tag{5.6}$$

Here and below, the constants are independent of μ (but may depend on T). In the case of a conservative operator (i.e., $K^1(p) \equiv 0$), the following integral is conserved (see §2, Chapter 4) for (5.1):

$$J_3(t) = \int_{-\infty}^{\infty}\left\{|u|^4 + 2u\mathbb{K}_2(u^*)\right\}\,dx = J_3(0),$$

where the operator $\mathbb{K}_2(u)$ has the symbol $K^2(p)$. In the presence of dissipation (brought about by the symbols $K^1(p)$ and $\mu R^1(p)$) the integral $J_3(t)$ varies in time. To estimate it, we differentiate it J_3 with respect to t and, applying (5.3), obtain

$$\begin{aligned}\dot{J}_3 = &-4\mu\int_{-\infty}^{\infty}\mathrm{Re}\left(u(u^*)^2\mathbb{R}(u)\right)\,dx - 4\mu\,\mathrm{Re}\int_{-\infty}^{\infty}\mathbb{R}(u)\mathbb{K}_2(u^*)\,dx\\ &-4\,\mathrm{Re}\int_{-\infty}^{\infty}\left\{u(u^*)^2\mathbb{K}_1(u) + \mathbb{K}_1(u)\mathbb{K}_2(u^*)\right\}\,dx\,.\end{aligned} \tag{5.7}$$

Using the Plancherel theorem, (5.5), and (5.6), for the first integral on the right-hand side of (5.7) we have

(5.8)

$$\begin{aligned}&4\mu\left|\mathrm{Re}\int_{-\infty}^{\infty}u(u^*)^2\mathbb{R}(u)\,dx\right|\\ &\quad\le c_\mu\int_{-\infty}^{\infty}dp(1+|p|)^{\beta+\sigma-\mathcal{N}/2-2\kappa}|\hat{u}(p,t)|\\ &\quad\cdot\int_{-\infty}^{\infty}dq(1+|p-q|)^{(\sigma+\beta-\mathcal{N})/2-2\kappa}|\hat{u}(p-q,t)|\int_{-\infty}^{\infty}dr\,|\hat{u}(q-r,t)\hat{u}^*(-r,t)|\\ &\quad\le c\mu\|u\|^2_{\sigma+\beta/2-\kappa}\|u\|_0^2 \le \mu\gamma\varepsilon\|u\|^2_{(\sigma+\beta)/2} + c\,.\end{aligned}$$

For the second integral on the right in (5.7), in view of (5.2), (5.4), and (5.6), we have

$$\begin{aligned}-4\mu\,\mathrm{Re}\int_{-\infty}^{\infty}\mathbb{R}(u)\mathbb{K}_2(u^*)\,dx &= -\frac{4\mu}{(2\pi)^{\mathcal{N}}}\int_{-\infty}^{\infty}dp\,R^1(p)K^2(p)|\hat{u}(p,t)|^2\\ &\le -\mu\varepsilon\gamma\|u\|^2_{(\sigma+\beta)/2} + c\,.\end{aligned} \tag{5.9}$$

The third and fourth terms on the right in (5.7) are estimated similarly:

$$-4\,\mathrm{Re}\int_{-\infty}^{\infty}\left\{u(u^*)^2\mathbb{K}_1(u) + \mathbb{K}_1(u)\mathbb{K}_2(u^*)\right\}\,dx \le c\,. \tag{5.10}$$

Substitution of (5.8)–(5.10) in (5.7) yields $\dot{J}_3(t) \le c$, whence, integrating with respect to t, we obtain $J_3(t) \leq c$ for all $t \in [0, T]$ ($T > 0$ is fixed), or, in view of the Plancherel theorem and (5.2), we have

$$\varepsilon \|u\|^2_{\sigma/2} \leq \|u\|^4_{L_4(R_{\mathcal{N}})} + c \tag{5.11}$$

since $\sigma > \mathcal{N}$, with the aid of the Sobolev imbedding theorem from (5.11) we find that

$$\|u^{(\mu)}\|_{\sigma/2} \leq c\,. \tag{5.12}$$

We now show that

$$\|u^{(\mu)}\|_{\rho} \leq c \tag{5.13}$$

since $\bar{u}(x) \in H^{\rho}(R_{\mathcal{N}})$ and $\rho \geq \sigma/2$ by condition 2) of the theorem. Applying the Fourier transform to (5.3), multiplying the result by $(1+|p|)^{2\rho}\hat{u}^*(p,t)$, and integrating with respect to $p \in R_{\mathcal{N}}$ we obtain

$$\begin{aligned}
\frac{d}{dt}\|u\|^2_{\rho} = & \frac{2\,\mathrm{Re}}{(2\pi)^{\mathcal{N}}} \int_{-\infty}^{\infty} dp (1+|p|)^{2\rho}\hat{u}^*(p,t) \\
& \times \iint_{-\infty}^{\infty} dq\, dr\, \hat{u}(p-q,t)\hat{u}(q-r,t)\hat{u}^*(-r,t) \\
& - \frac{2\mu}{(2\pi)^{\mathcal{N}}} \int_{-\infty}^{\infty} dp\, R^1(p)(1+|p|)^{2\rho}|\hat{u}(p,t)|^2 \\
& - \frac{2}{(2\pi)^{\mathcal{N}}} \int_{-\infty}^{\infty} dp\, K^1(p)(1+|p|)^{2\rho}|\hat{u}(p,t)|^2.
\end{aligned}$$

Since the operators $\mathbb{K}$ and $\mathbb{R}$ are dissipative (see (5.2) and (5.4)), using the already proved inequality (5.12), we find that

$$\begin{aligned}
\frac{d}{dt}\|u\|^2_{\rho} \leq & c\int_{-\infty}^{\infty} dr \frac{(1+|r|)^{\mathcal{N}/2+\kappa}}{(1+|r|)^{\mathcal{N}/2+\kappa}}|\hat{u}^*(-r,t)| \int_{-\infty}^{\infty} dq \frac{(1+|q|)^{\mathcal{N}/2+\kappa}}{(1+|q|)^{\mathcal{N}/2+\kappa}}|\hat{u}(q-r,t)| \\
& \cdot \int_{-\infty}^{\infty} dp\,(1+|p|)^{\rho}|\hat{u}(p,t)|(1+|p-q|)^{\rho}|\hat{u}(p-q,t)| + c\|u\|^2_0 \\
\leq & c\|u\|^2_{\rho}\|u\|^2_{\sigma/2} + c\|u\|^2_0 \leq c\|u\|^2_{\rho}\,.
\end{aligned}$$

Integration of the last inequality leads to (5.13).

The inequality

$$\frac{d^m}{dt^m}\|u^{(\mu)}(\cdot,t)\|_{\rho-\gamma m} \leq c_m, \qquad m \geq 0, \tag{5.14}$$

is proved analogously. Applying (5.13) and (5.14), as in Theorems 1 and 2, we see that some subsequence of the sequence $\{u^{(\mu)}(x,t)\}$ converges weakly in $L_2([0,T]; H^{\rho}(R_{\mathcal{N}}))$ and converges strongly in $L_2(\Omega \times [0,T])$, where Ω is any compact set in $R_{\mathcal{N}}$, and that a subsequence of $\{u_t^{(\mu)}\}$ converges weakly in $L_2([0,T], H^{\rho-\nu}(R_{\mathcal{N}}))$ to u_t. Therefore, for any $\psi \in C_0^{\infty}(R_{\mathcal{N}})$, we have

$$(|u^{(\mu)}|^2 u^{(\mu)}, \psi) \to (|u|^2 u, \psi) \quad \text{as } \mu \to 0,$$

and, thus, the function $u(x,t)$ is a generalized solution of the Cauchy problem (5.1). Theorem 4 is proved. □

REMARK 1. In the case of $\rho = +\infty$, i.e., $\bar{u}(x) \in H^{\infty}(R_{\mathcal{N}})$, the solution is classical and infinitely smooth: $u(x,t) \in C^{\infty}([0,\infty); H^{\infty}(R_{\mathcal{N}}))$.

REMARK 2. For $\rho \geq \nu$ the solution obtained is unique. Indeed, assuming the contrary, i.e., that there exist two solutions $u_1(x,t)$ and $u_2(x,t)$, for $w = u_1 - u_2$ we obtain the equation

$$\frac{d}{dt}\|w\|_0^2 - 2i\left(\left(w^2 u_1^* + 2u_2 u_1^* w + u_2^2 w^*\right), w\right) + 2\left(\mathbb{K}(w), w\right) = 0,$$

whence $\frac{d}{dt}\|w\|_0^2 \leq c\|w\|_0^2$ and, because $\|w\|_0$ is equal to zero for $t = 0$, it follows that $\|w\|_0 = 0$ for all t; that is, the solution is unique.

REMARK 3. This theorem is true for the one-dimensional ($\mathscr{N} = 1$) nonlinear Schrödinger equation, since in this case $K^1(p) \equiv 0$ (i.e., $\lambda = 0$ under the conditions of (5.2)) and $\sigma = 2 > \mathscr{N} = 1$.

The just proved Theorem 4 and Theorem 6 of §6, Chapter 4, imply

THEOREM 5. *Suppose that*

1) *The condition* (5.2) *and the following strong dissipativity condition hold*:

$$K^1(p) \geq \varepsilon|p|^\delta \quad \textit{for } |p| \geq p_0 > 0,$$

where $\delta > 0, \varepsilon > 0$;

2) *The initial perturbation* $\bar{u}(x) \in H^\rho(R_{\mathscr{N}}), \rho \geq \sigma/2$.

Then there exists a unique solution of the Cauchy problem (5.1) *belonging to* $C^0([0,\infty); H^\rho(R_{\mathscr{N}})) \cap C^\infty((0,\infty); H^\infty(R_{\mathscr{N}}))$.

This theorem provides conditions for the existence in the large of smoothed solutions of the Cauchy problem (5.1) in the case when the conservative part $K^2(p)$ of the operator is large, and the dissipative part $K^1(p)$, in contrast, is not too large, so that Theorem 7 of §6, Chapter 4 can be applied.

§6. The system of surface waves

In this section we prove a theorem on the existence of generalized solutions of the Cauchy problem for the system of equations of surface waves

$$\tag{6.1} \begin{cases} \eta_t + (\nabla, \eta\nabla\varphi) + \mathbb{K}_1(\eta) + \mathbb{K}_2(\varphi) = 0, \\ \varphi_t + \frac{1}{2}(\nabla\varphi)^2 + \mathbb{K}_3(\eta) + \mathbb{K}_4(\varphi) = 0, \\ \eta|_{t=0} = \bar{\eta}(x), \quad \varphi|_{t=0} = \bar{\varphi}(x), \quad x \in R_2 \end{cases}$$

in the conservative case, i.e., when the operators $\mathbb{K}_j$ satisfy the conditions

$$\tag{6.2} \mathbb{K}_1 = -\mathbb{K}_4^*, \quad \mathbb{K}_2 = \mathbb{K}_2^*, \quad \mathbb{K}_3 = \mathbb{K}_3^*$$

(* means "adjoint" in $L_2(R_2)$). In §1, Chapter 4, we showed that the system of equations for surface waves (6.1) has conservation laws under the condition (6.2). In particular, the energy integral is conserved:

$$J(t) = (\eta, (\nabla\varphi)^2) + (\eta, \mathbb{K}_3(\eta)) - 2(\varphi, \mathbb{K}_1(\eta)) - (\varphi, \mathbb{K}_2(\varphi)) = J(0)\,.$$

Here $(\cdot,\cdot)$ denotes the scalar product in $L_2(R_2)$. It also follows from (6.2) that the symbols $K_2(p)$ and $K_3(p)$ of the operators $\mathbb{K}_2$ and $\mathbb{K}_3$ are real.

THEOREM 6. *Suppose that*

1) *The conditions* (6.2) *are satisfied and, in addition,*(*) *for* $p \in R_2$

$$|K_1(p)| \le \varepsilon_1 M^{\alpha_1}(p), K_2(p) \le -\varepsilon_2 M^{\alpha_2}(p), K_3(p) \ge \varepsilon_3 M^{\alpha_3}(p), \tag{6.3}$$

where $\varepsilon_j > 0, \alpha_j \ge 0,\ j = 1, 2, 3$, *are such that*

$$\varepsilon_1 < \min(\varepsilon_2 \varepsilon_3), \quad 2\alpha_1 \le \alpha_2 + \alpha_3, \quad \alpha_2 \ge 2 \quad \text{and} \quad \alpha_2 > 3 - \frac{\alpha_3}{2}. \tag{6.4}$$

2) *The initial perturbations* $\bar{\eta}(x) \in \mathscr{D}(\mathbb{K}_3^{1/2})$ *and* $\bar{\varphi}(x) \in \mathscr{D}((-\mathbb{K}_2)^{1/2})$ *are sufficiently small in the sense that*

$$J(0) \le c \tag{6.5}$$

where c is a constant.

Then there exists a generalized solution (η, ϕ) *of the Cauchy problem* (6.1), $\eta \in L_\infty([0, T], H^{\alpha_3/2}(R_2))$, $\varphi \in L_\infty([0, T], H^{\alpha_2/2}(R_2))$ *satisfying for all* $t \in [0, T]$ ($T > 0$ *is arbitrary*) *the following identities*:

$$\begin{cases} (\eta_t, \psi) - (\nabla\psi, \eta\nabla\varphi) + (\eta, \mathbb{K}_1^*(\psi)) + (\varphi, \mathbb{K}_2(\psi)) = 0, \\ (\varphi_t, \psi) + \frac{1}{2}(\psi, (\nabla\varphi)^2) + (\eta, \mathbb{K}_3(\psi)) + (\varphi, \mathbb{K}_4^*(\psi)) = 0 \end{cases}$$

and the initial conditions:

$$(\varphi, \psi)|_{t=0} = (\bar{\varphi}, \psi), \qquad (\eta, \psi)|_{t=0} = (\bar{\eta}, \psi)$$

for any function $\psi(x) \in C_0^\infty(R_2)$.

PROOF. For all $\mu \in (0, 1]$ we define the functions $\eta^{(\mu)}$ and $\varphi^{(\mu)}$ as the solution of the following problem:

$$\begin{cases} \eta_t^{(\mu)} + (\nabla, \eta^{(\mu)}\nabla\varphi^{(\mu)}) + \mathbb{K}_1(\eta^{(\mu)}) + \mathbb{K}_2(\varphi^{(\mu)}) + \mu\mathbb{R}(\eta^{(\mu)}) = 0, \\ \varphi_t^{(\mu)} + \frac{1}{2}(\nabla\varphi^{(\mu)})^2 + \mathbb{K}_3(\eta^{(\mu)}) + \mathbb{K}_4(\varphi^{(\mu)}) + \mu\mathbb{R}(\varphi^{(\mu)}) = 0, \\ \eta^{(\mu)}|_{t=0} = \bar{\eta}(x), \qquad \varphi^{(\mu)}|_{t=0} = \bar{\varphi}(x), \end{cases} \tag{6.6}$$

where $R(p) = M^\beta(p)$, $\beta > \max(2, \alpha_1, \alpha_3)$. Theorem 9 of §6, Chapter 4, guarantees the existence of a solution to the Cauchy problem (6.6) for all $\mu \in (0, 1]$ from the class $C^0([0, T_0]; (L_2(R_2))^2) \cap C^\infty((0, T_0]; (H^\infty(R_2))^2)$. Indeed, after differentiating the second equation of (6.6) with respect to x_1 and x_2, and using $\mathbb{K}_2$ in the form $\mathbb{K}_2 = (\partial/\partial x_1)\mathbb{K}_2^{(1)} + (\partial/\partial x_2)\mathbb{K}_2^{(2)}$, we arrive at the system of equations (6.1) of §6, Chapter 5. The matrix-symbol $K(p)$ of the operator $\mathbb{K}$ has the following form:

$$K(p) = \begin{pmatrix} K_1(p) + \mu R(p) & K_2^{(1)}(p) & K_2^{(2)}(p) \\ ip_1 K_3(p) & K_4(p) + \mu R(p) & 0 \\ ip_2 K_3(p) & 0 & K_4(p) + \mu R(p) \end{pmatrix},$$

and its eigenvalues are

$$\lambda_1 = \mu R + K_4, \qquad \lambda_{2,3} = \mu R + i \operatorname{Im} K_1 \pm \sqrt{(\operatorname{Re} K_1)^2 + \mu K_1 R + K_2 K_3}$$

and by virtue of (6.3) and (6.4) we obtain

$$\max_{1 \le j \le 3} (\operatorname{Re} \lambda_j) \ge \varepsilon |p|^\beta \quad \text{for } |p| \ge p_0 > 0$$

with some $\varepsilon > 0$, so that the conditions of Theorem 9 of Chapter 5 are satisfied.

(*) Recall that $M(p) = \max(1, |p|)$. In this section the norm $\|\psi\|_p$ is defined by $\|\psi\|_p^2 = \int_{R_2} M^{2p}(p)|\hat{\psi}(p)|^2\,dp$.

We prove that the norms $\|\eta^{(\mu)}\|_{\alpha_3/2}$ and $\|\varphi^{(\mu)}\|_{\alpha_2/2}$ are uniformly bounded with respect to $\mu \in (0,1]$. To this end, we take scalar product of the first equation in (6.6) and $\{(\nabla\varphi)^2 - 2\mathbb{K}_1^*(\varphi) + 2\mathbb{K}_3(\eta)\}$, and of the second and $\{-2(\nabla, \eta\nabla\varphi) - 2\mathbb{K}_1(\eta) - 2\mathbb{K}_2(\varphi)\}$, and then add the results. In view of (6.2), omitting the index μ, we obtain

$$\begin{aligned}\dot{J} = &- \mu\left(\mathbb{R}(\eta), (\nabla\varphi)^2\right) + 2\mu\left(\mathbb{R}(\varphi), \nabla(\eta\nabla\varphi)\right) + 4\mu\left(\mathbb{R}(\varphi), \mathbb{K}_1(\eta)\right) \\ &- 2\mu\left(\mathbb{R}(\eta), \mathbb{K}_3(\eta)\right) + 2\mu\left(\mathbb{R}(\varphi), \mathbb{K}_2(\varphi)\right).\end{aligned} \tag{6.7}$$

We estimate the right-hand side of (6.7). Application of the Plancherel theorem gives

$$\begin{aligned}\mu\left|\left(\mathbb{R}(\eta), (\nabla\varphi)^2\right)\right| \leq & c\mu \int_{-\infty}^{\infty} dp\, M^{\beta/2+\alpha_3/2}(p)\,|\hat{\eta}(p,t)| \\ &\cdot \int_{-\infty}^{\infty} dq\, M^{\beta/2+1-\alpha_3/2}(p-q)M(q)\,|\hat{\varphi}(q,t)\hat{\varphi}(p-q,t)|\,.\end{aligned}$$

Since for any $\gamma_1 \geq 0$, $\gamma_2 \geq 0$, and $p, q \in R_2$, we have

$$M^{\gamma_1}(q)M^{\gamma_2}(p-q) \leq M^{\gamma_1+\gamma_2}(q) + M^{\gamma_1+\gamma_2}(p-q); \tag{6.8}$$

choosing $\gamma_1 = (\alpha_2+\alpha_3)/2 - 1$ and $\gamma_2 = (\alpha_2+\beta)/2 - 1$, we conclude that

$$\begin{aligned}\mu\left|\mathbb{R}(\eta), (\nabla\varphi)^2\right| \leq & c\mu \int_{-\infty}^{\infty} dq\, M^{\alpha_2/2-\kappa}(q)\,|\hat{\varphi}(q,t)| \\ &\cdot \int_{-\infty}^{\infty} dp\, M^{\beta/2+\alpha_3/2}(p)\,|\hat{\eta}(p,t)|\, M^{\beta/2+\alpha_2/2}(p-q)\,|\hat{\varphi}(p-q,t)| \\ \leq & c\mu\|\eta\|_{\beta+\alpha_3/2}\|\varphi\|_{\beta+\alpha_2/2}\|\varphi\|_{\alpha_2/2},\end{aligned} \tag{6.9}$$

where, by virtue of (6.4), $\kappa \equiv \alpha_2 - 2 + \alpha_3/2 > 1$. To estimate the second term on the right in (6.7) we proceed similarly. Using (6.4) and (6.8) and denoting $\theta = \max(0, 1 + (\beta - \alpha_2)/2)$, we obtain

$$\begin{aligned}&2\mu\left|\left(\mathbb{R}(\varphi), \nabla(\eta\nabla\varphi)\right)\right| \\ &\quad\leq c\mu \int_{-\infty}^{\infty} dp\, M^{\beta+1}(p)\,|\hat{\varphi}(p,t)| \int_{-\infty}^{\infty} dq\, M(p-q)\,|\hat{\varphi}(p-q,t)\hat{\eta}(q,t)| \\ &\quad\leq c\mu \int_{-\infty}^{\infty} dp\, M^{(\beta+\alpha_2)/2}(p)\,|\hat{\varphi}(p,t)| \int_{-\infty}^{\infty} dq\, |\hat{\varphi}(p-q,t)\hat{\eta}(q,t)| \\ &\qquad\cdot \left\{M^{\theta+1}(p-q) + M(p-q)M^{\theta}(q)\right\}.\end{aligned} \tag{6.10}$$

Now we apply the following inequality valid for all $\gamma_3 \geq 0$, $\gamma_4 \geq 0$, $\gamma_5 = \gamma_3 + \gamma_4$, $p, q \in R_2$:

$$M^{-\gamma_3}(p-q)M^{-\gamma_4}(q) \leq M^{-\gamma_5}(q) + M^{-\gamma_5}(p-q)\,. \tag{6.11}$$

We choose $\gamma_3 = (\alpha_2+\beta)/2 - 1 - \theta \geq 0$, $\gamma_4 = \alpha_3/2$ so that $\gamma_5 > 1$, and by (6.4) we obtain

$$M^{\theta+1}(p-q) \leq M^{(\alpha_2+\beta)/2}(p-q)M^{\alpha_3/2}(q)\left(M^{-\gamma_5}(q) + M^{-\gamma_5}(p-q)\right). \tag{6.12}$$

Putting in (6.11) $\gamma_3 = \alpha_2/2 - 1$ and $\gamma_4 = (\beta+\alpha_3)/2 - \theta$ with the same $\gamma_5 > 1$, we get

$$M(p-q)M^{\theta}(q) \leq M^{\alpha_2/2}(p-q)M^{(\alpha_3+\beta)/2}(q)\cdot\left(M^{-\gamma_5}(q) + M^{-\gamma_5}(p-q)\right). \tag{6.13}$$

Substitution of (6.12) and (6.13) in (6.10) yields

(6.14)
$$2\mu\,|\mathbb{R}(\varphi),\nabla(\eta\nabla\varphi)|\le c\mu\|\varphi\|_{(\beta+\alpha_2)/2}\left(\|\varphi\|_{(\alpha_2+\beta)/2}\|\eta\|_{\alpha_3/2}+\|\varphi\|_{\alpha_2/2}\|\eta\|_{(\alpha_3+\beta)/2}\right).$$

For the remaining terms on the right-hand side of (6.7), owing to the Plancherel theorem and (6.3), we obtain

(6.15)
$$(\mathbb{R}(\eta),\mathbb{K}_3(\eta))\le\varepsilon_3\|\eta\|^2_{(\alpha_3+\beta)/2},\qquad -(\mathbb{R}(\varphi),\mathbb{K}_2(\varphi))\ge\varepsilon_2\|\varphi\|^2_{(\beta+\alpha_2)/2},$$
$$|(\mathbb{R}(\varphi),\mathbb{K}_1(\eta))|\le\varepsilon_1(\|\varphi\|^2_{(\alpha_2+\beta)/2}+\|\eta\|^2_{(\alpha_3+\beta)/2}).$$

Comparison of (6.7), (6.9), (6.14), and (6.15) gives

(6.16)
$$\dot{J}(t)\le-\mu\left(2\varepsilon_4-c\|\varphi\|_{\alpha_2/2}\right)\|\eta\|^2_{(\beta+\alpha_3)/2}-\mu\left(2\varepsilon_4-c\|\eta\|_{\alpha_3/2}-c\|\varphi\|_{\alpha_2/2}\right)\|\varphi\|^2_{(\beta+\alpha_2)/2},$$

where $2\varepsilon_4=\min\{\varepsilon_2-\varepsilon_1,\varepsilon_3-\varepsilon_1\}>0$. We now estimate the terms contained in the energy integral $J(t)$. Using the inequality

$$M^{-\gamma_1}(p)M^{-\gamma_2}(p-q)M^{-\gamma_3}(q)\le M^{-\gamma_4}(p)+M^{-\gamma_4}(p-q)+M^{-\gamma_1}(q),$$

valid for any $\gamma_1,\gamma_2,\gamma_3\ge 0$, $\gamma_4=\gamma_1+\gamma_2+\gamma_3$, $p,q\in R_2$, in which we fix $\gamma_1=\alpha_3/2$ and $\gamma_2=\gamma_3=\alpha_2/2-1$, we obtain

(6.17)
$$\begin{aligned}\left|(\eta,(\nabla\varphi)^2)\right|\le&c\int_{-\infty}^{\infty}dp M^{\alpha_3/2-\kappa}(p)\,|\hat\eta(p,t)|\int_{-\infty}^{\infty}dq\,M^{\alpha_2/2}(p-q)\\&\cdot M^{\alpha_2/2}(q)\,|\hat\varphi(p-q,t)\hat\varphi(q,t)|+c\int_{-\infty}^{\infty}dq\,M^{\alpha_2/2-\kappa}(q)\,|\hat\varphi(q,t))\\&\cdot\int_{-\infty}^{\infty}dpM^{\alpha_3/2}(p)M^{\alpha_2/2}(p-q)\,|\hat\eta(p,t)\hat\varphi(p-q,t)|\\\le&c\|\eta\|_{\alpha_3/2}\|\varphi\|^2_{\alpha_2/2}.\end{aligned}$$

Further, by (6.3), we have

(6.18)
$$|((\varphi,\mathbb{K}_1(\eta))|\le\frac{\varepsilon_1}{2}(\|\varphi\|^2_{\alpha_2/2}+\|\eta\|^2_{\alpha_3/2}),$$
$$-(\mathbb{K}_2(\varphi),\varphi)\ge\varepsilon_2\|\varphi\|^2_{\alpha_2/2},\qquad(\mathbb{K}_3(\eta),\eta)\ge\varepsilon_3\|\eta\|^2_{\alpha_3/2}.$$

Substitution of (6.17) and (6.18) in the integral $J(t)$ yields

(6.19)
$$J(t)\ge\left(2\varepsilon_4-c\|\eta\|_{\alpha_3/2}\right)\|\varphi\|^2_{\alpha_2/2}+2\varepsilon_4\|\eta\|^2_{\alpha_3/2}.$$

We now prove that

(6.20)
$$\|\eta^{(\mu)}\|_{\alpha_3/2}<\frac{\varepsilon_4}{2c},\quad\|\varphi^{(\mu)}\|_{\alpha_2/2}<\frac{\varepsilon_4}{c},\quad J(t)<\frac{\varepsilon_4^2}{c}.$$

For $t=0$ the bounds (6.20) follow from the requirements (6.5) and inequality (6.19). For $t>0$, arguing by contradiction we find the first moment $T>0$ when one of the inequalities (6.20) is violated and the nonstrict inequalities (6.20) are fulfilled on $[0,T]$ by continuity. Then (6.16) gives $\dot J(t)<0$, whence $J(t)\le J(0)<\varepsilon_4^2/c$, and, by (6.19), we arrive at (6.20) on the entire interval $[0,T]$. The contradiction obtained proves (6.20) for all $t\ge 0$.

Thus, we have established the uniform boundedness (with respect to $\mu>0$) of the sequences $\{\eta^{(\mu)}\}$ and $\{\varphi^{(\mu)}\}$ in the space $L_\infty([0,T_0];H^{\alpha_3/2}(R_2))$ and

$L_\infty([0, T_0]; H^{\alpha_2/2}(R_2))$, respectively. Multiplying the equations of the system (6.6) by an arbitrary function $\psi(x) \in H^\gamma(R_2)$, $\gamma \geq \max(\bar{\alpha}_1, \bar{\alpha}_2, \bar{\alpha}_3, \beta)$, where $\bar{\alpha}_j$ is the order of the operator $\mathbb{K}_j$, and integrating with respect to κ we obtain

$$|(\eta_t^{(\mu)}, \psi)| \leq c \quad \text{and} \quad |(\varphi_t^{(\mu)}, \psi)| \leq c\,.$$

The sequences $\{\eta_t^{(\mu)}\}$ and $\{\varphi_t^{(\mu)}\}$ are thus bounded in the space $L_\infty([0,T_0]; H^{-v}(R_2))$. It follows from these bounds that one can extract subsequences $\{\eta^{(\mu)}\}$ and $\{\varphi^{(\mu)}\}$ such that $\eta^{(\mu)} \to \eta$ weakly in $L_2([0, T_0]; H^{\alpha_3/2}(R_2))$ and $\varphi^{(\mu)} \to \varphi$ weakly in $L_2([0, T_0]; H^{\alpha_2/2}(R_2))$, and $\varphi_t^{(\mu)} \to \varphi_t$ and $\eta_t^{(\mu)} \to \eta_t$ weakly in $L_2([0, T_0]; H^{-v}(R_2))$. Moreover, applying the compactness lemma [**37**], we find that $\nabla\varphi^{(\mu)} \to \nabla\varphi$ strongly in $L_2(\Omega \times [0, T_0])$ for any compact set $\Omega \subset R_2$. Therefore, for any function $\psi \in C_0^\infty(R_2 \times [0, T_0])$

$$\int_0^{T_0} (\psi, (\nabla\varphi^{(\mu)})^2)\, dt \to \int_0^{T_0} (\psi, (\nabla\varphi)^2)\, dt,$$

$$\int_0^{T_0} (\nabla\phi, \eta^{(\mu)}\nabla\varphi^{(\mu)})\, dt \to \int_0^{T_0} (\nabla\psi, \eta\nabla\psi)\, dt$$

as $\mu \to 0$. Multiplying the equations of system (6.6) by $\psi(x, t) \in C_0^\infty(R_2 \times [0, T_0])$ and integrating with respect to x and t, we pass to the limit as $\mu \to 0$. as a result, we conclude that (η, φ) is a generalized solution. It follows from (6.20) that $J(t)$ is bounded, and thus $\eta \in \mathscr{D}(\mathbb{K}_3^{1/2})$ and $\varphi \in \mathscr{D}((-\mathbb{K}_2)^{1/2})$. Therefore, continuing the solution constructed from the moment $t = T_0$, in a finite number of steps we prove the existence of a generalized solution for any interval $[0, T]$. Theorem 6 is proved. $\square$

CHAPTER 6

The Asymptotics as $t \to \infty$ of Solutions of the Generalized Kolmogorov-Petrovskiĭ-Piskunov Equation

§1. Introduction

It is well known that the problem of finding the asymptotics, as $t \to \infty$, of solutions of the Cauchy problem for a nonlinear evolution equation is one of the most important and difficult in the theory of nonlinear equations. In recent years significant progress has been achieved in solving this problem, basically on account of application of inverse scattering methods. Even in this case every nonlinear equation must be studied separately, and complete asymptotics, i.e., one for all relationships between the space variable, x and time, t, has not been constructed for any one equation except, perhaps, for the Korteweg-de Vries equation. Moreover, this method is confined to conservative nonlinear equations. However, many physical processes are described by essentially dissipative equations. Widely known examples of such equations are the nonlinear Schrödinger equation with linear dissipation (the so-called generalized Landau-Ginzburg equation) and the Burgers and Korteweg-de Vries-Burgers equations.

One of the best known model equations with dissipation is the equation suggested in 1937 by Kolmogorov, Petrovskiĭ, and Piskunov [**29**]

$$u_t - F(u) - u_{xx} = 0\,. \tag{1.1}$$

The Kolmogorov-Petrovskiĭ-Piskunov equation, KPP for short describes such phenomena as combustion (physics), evolution of dominant genes and propagation of nerve pulses (biology), propagation of concentration waves (chemical kinetics), and other (see the survey [**6**]).

The so-called step-decaying problem for the KPP equation was considered in [**29**], i.e. the Cauchy problem with the initial condition $\overline{u}(x) \to a_\pm$ as $x \to \pm\infty$, where $a_+ \neq a_-$ are constants. The authors' approach in this article was related to the possibility of applying an explicit form of the fundamental solution for the linear part of the KPP equation. The paper [**29**] gave rise to a large number of studies on the asymptotics, as $t \to \infty$, of solutions of the Cauchy problem for the KPP equation (see the surveys [**8, 99**]). As a typical nonlinearity, sources of the form $F(u) = u^2 - u$ were discussed. We shall consider the step-decaying problem of the Korteweg-de Vries-Burgers equation in Chapter 10, and here we study the asymptotics, for large times, of solutions of the Cauchy problem for the KPP equation with the initial conditions decreasing at infinity. Consider the generalized KPP equation

$$u_t - u^2 + \mathbb{K}(u) = 0 \tag{1.2}$$

where $\mathbb{K}(u)$ is the linear pseudodifferential operator

$$\mathbb{K}(u) = \frac{1}{2\pi}\int_{-\infty}^{\infty} e^{ipx} K(p)\hat{u}(p,t)\,dp, \tag{1.3}$$

$\hat{u}(p,t) = \int_{-\infty}^{\infty} e^{-ipx}u(x,t)\,dx$ is the Fourier transform, and $K(p)$ is the symbol of the operator $\mathbb{K}$. The KPP equation corresponds to the symbol $K(p) = 1 + p^2$. Our objective in this chapter is to study the asymptotics, as $t \to \infty$, of the solutions of the Cauchy problem for equation (1.2):

$$u_t - u^2 + \mathbb{K}(u) = 0, \qquad u|_{t=0} = \overline{u}(x), \quad \overline{u}(x) \in L_1(R_1)\,. \tag{1.4}$$

The requirement $\overline{u}(x) \in L_1(R_1)$ ensures stabilization to zero of a solution of the Cauchy problem, as $t \to \infty$. Indeed, the following theorem is true.

THEOREM. *Suppose that*

1) $K^1(p) \equiv \operatorname{Re} K(p) = 1 + |p|^\delta$, $|p| \le \frac{1}{2}$; $K^1(p) \ge 1 + \gamma$, $\gamma > 0$, $|p| \ge \frac{1}{2}$; $|K^2(p)| \equiv |\operatorname{Im} K(p)| \le c_1|p|^\beta$, $|p| \le \frac{1}{2}$, $\beta > \delta > 0$, $|K(p) - K(q)| \le c_2|p-q|^{\sigma_1}M^\alpha(p,q)$, $|p-q| \le 1$, $M(p,q) \equiv \max(1,|p|,|q|)$, $\sigma_1 \in (0,1]$, $\alpha \ge 0$,

2) *The Fourier transform $\hat{\overline{u}}(p)$ of the initial condition $\overline{u}(x) \in L_1(R_1)$ satisfies the following requirements* $(M(p) \equiv M(p,1))$

$$|\hat{\overline{u}}(p)| \le \varepsilon M^{-2-\alpha}(p), \quad |\hat{\overline{u}}(p) - \hat{\overline{u}}(q)| \le \varepsilon|p-q|^{\sigma_2}M^{-2}(p,q), \quad |p-q| \le 1,$$

where $\varepsilon < c$, and $c > 0$ is determined by the symbol $K(p)$, $\sigma_2 \in (0,1]$. Then, as $t \to \infty$, the following asymptotics

$$u(x,t) = Ae^{-t}t^{-1/\delta}\int_0^\infty \cos(p\xi)e^{-p^\delta}\,dp + O(e^{-t}t^{-1/\delta-\gamma_0}), \tag{1.5}$$

takes place uniformly with respect to $\xi = |x|t^{-1/\delta} \in R_1$. Here $\gamma_0 > 0$ is a constant determined by the symbol $K(p)$, and the constant A is expressed in terms of the symbol $K(p)$ and the initial function $\overline{u}(p)$ and will be indicated below (see equations (2.17) *and* (3.36)).

REMARK. The theorem is true for the KPP equation itself, and in this case we have $\delta = 2$, $\gamma_0 = \frac{1}{2}$.

§2. Proof of the theorem

The asymptotic formula (1.5) will be proved assuming the existence of a unique solution of the Cauchy problem for the generalized KPP equation (1.2) in the class $C^\infty((0,\infty); H^\infty(R_1))$. This existence itself follows readily from the a priori estimates (2.5) established below and the method of successive approximation.

Then, applying the Fourier transform to (1.4), we obtain $(p \in R_1, t > 0)$

$$\hat{u}_t(p,t) - \int_{-\infty}^{\infty} \hat{u}(p-q,t)\hat{u}(q,t)\,dq + K(p)\hat{u}(p,t) = 0, \qquad \hat{u}(p,0) = \hat{\overline{u}}(p),$$

whence, upon integration with respect to t, we find that

$$\hat{u}(p,t) = \hat{\overline{u}}(p)e^{-K(p)t} + \int_0^t e^{-K(p)(t-\tau)}\int_{-\infty}^{\infty} \hat{u}(p-q,\tau)\hat{u}(q,\tau)\,dq\,d\tau\,. \tag{2.1}$$

To express a solution of (2.1) in terms of the initial condition $\hat{\hat{u}}(p)$, we make use of perturbation theory. We set $\hat{\hat{u}}(p) = \varepsilon\overline{v}(p)$ and represent $\hat{\hat{u}}(p,t)$ as a formal power series in ε:

$$\hat{u}(p,t) = \sum_{j=0}^{\infty} \varepsilon^{j+1} v_j(p,t)\,. \tag{2.2}$$

Substituting (2.2) in (2.1) and comparing, as usual, the coefficients of like powers of ε, we obtain

$$v_0(p,t) = \overline{v}(p)e^{-K(p)t}, \tag{2.3}$$

$$v_{\mathcal{N}}(p,t) = \int_0^t e^{-K(p)(t-\tau)} \int_{-\infty}^{\infty} \sum_{j=1}^{\mathcal{N}} v_{j-1}(p-q,\tau)v_{\mathcal{N}-j}(q,\tau)\,dt\,d\tau, \qquad \mathcal{N} \geq 1\,. \tag{2.4}$$

We establish the following bound for $v_{\mathcal{N}}(p,t)$, $\mathcal{N} \geq 0$:

$$|v_i(p,t)| \leq c^{\mathcal{N}}(\mathcal{N}+1)^{-2}M^{-2}(p)\exp\left(-t - m(K^1(p)-1)\frac{t}{2}\right), \qquad c \geq 2^{10}, \tag{2.5}$$

where $m(x) = \min(1,|x|)$. For $\mathcal{N} = 0$ (2.5) is easily verified. Further, arguing by induction, from (2.4) we obtain

$$\begin{aligned}
|v_{\mathcal{N}}(p,t)| &= \left| \int_0^t e^{-K(p)(t-\tau)} \int_{-\infty}^{\infty} \sum_{j=1}^{\mathcal{N}} v_{j-1}(p-q,\tau)v_{\mathcal{N}-j}(q,\tau)\,dt\,d\tau \right| \\
&\leq \int_0^t e^{-(1+m(K^1(p)-1)/2)(t-\tau)} \\
&\qquad \times \int_{-\infty}^{\infty} \sum_{j=1}^{\mathcal{N}} c^{\mathcal{N}-1}e^{-2\tau}j^{-2}(\mathcal{N}+1-j)^{-2}M^{-2}(p-q)M^{-2}(q)\,dt\,d\tau \\
&\leq c^{\mathcal{N}-1}\sum_{j=1}^{\mathcal{N}} j^{-2}(\mathcal{N}+1-j)^{-2}e^{-t-m(K^1(p)-1)t/2} \\
&\qquad \times \int_0^t e^{-\tau/2}\,d\tau \int_{-\infty}^{\infty} M^{-2}(p-q)M^{-2}(q)\,dq \\
&\leq 2^{10}c^{\mathcal{N}-1}(\mathcal{N}+1)^{-2}M^{-2}(p)e^{-t-m(K^1(p)-1)t/2},
\end{aligned}$$

whence (2.5) follows. From the representation (2.2), inverting the Fourier transform, we find that

$$u(x,t) = \frac{1}{2\pi}\int_{-\infty}^{\infty} e^{ipx}\hat{u}(p,t)\,dp = \frac{1}{2\pi}\sum_{\mathcal{N}=0}^{\infty} \varepsilon^{\mathcal{N}+1}I_{\mathcal{N}}, \tag{2.6}$$

where

$$I_{\mathcal{N}} = \int_{-\infty}^{\infty} e^{ipx}v_{\mathcal{N}}(p,t)\,dp\,. \tag{2.7}$$

In the equality (2.6) a termwise integration of a series has been performed, which is possible because of the majorant convergence of the series of perturbation theory. The latter follows from inequalities (2.5) for $\varepsilon > 1/c$, where $c > 0$ is a constant independent of $\mathscr{N}$. Thus, computation of the asymptotics of a solution $u(x,t)$ is reduced to finding the asymptotics of the integrals $I_{\mathscr{N}}$. From (2.7) we find ($\mathscr{N}$ is fixed below):

$$I_{\mathscr{N}} = J^1_{\mathscr{N}} + J^2_{\mathscr{N}} + J^3_{\mathscr{N}},$$

where

$$\begin{aligned} J^1_{\mathscr{N}} &= \int_{|p|\geq 1/2} e^{ipx} v_{\mathscr{N}}(p,t)\,dp, \\ J^2_{\mathscr{N}} &= \int_{|p|\leq 1/2} e^{ipx-K(p)t}(v_{\mathscr{N}}(p,t)e^{K(p)t} - \Phi_{\mathscr{N}}(p))\,dp, \\ J^3_{\mathscr{N}} &= \int_{|p|\leq 1/2} e^{ipx-K(p)t}\Phi_{\mathscr{N}}(p)\,dp, \\ \Phi_{\mathscr{N}}(p) &\equiv \lim_{t\to\infty} v_{\mathscr{N}}(p,t)e^{K(p)t|}, \qquad |p|\leq \tfrac{1}{2}\,. \end{aligned}$$

We estimate $J^1_{\mathscr{N}}$. From (2.5) and condition 1) of the theorem we have

$$|J^1_{\mathscr{N}}| \leq \int_{|p|\geq 1/2} |v_{\mathscr{N}}(p,t)|\,dp \leq 3c^{\mathscr{N}} e^{-t-\gamma t/2}\,. \tag{2.8}$$

It follows from (2.4) that

$$\Phi_{\mathscr{N}}(p) = \int_0^\infty e^{K(p)\tau} \int_{-\infty}^\infty \sum_{j=1}^{\mathscr{N}} v_{j-1}(p,q,\tau)v_{\mathscr{N}-j}(q,\tau)\,dt\,d\tau\,.$$

Therefore, in view of (2.5), we obtain ($|p|\leq \frac{1}{2}$)

$$\begin{aligned} &|v_{\mathscr{N}}(p,t)e^{K(p)t} - \Phi_{\mathscr{N}}(p)| \\ &\quad= \left| \int_t^\infty e^{K(p)\tau} \int_{-\infty}^\infty \sum_{j=1}^{\mathscr{N}} v_{j-1}(p-q,\tau)v_{\mathscr{N}-j}(q,\tau)\,dt\,d\tau \right| \\ &\quad\leq \int_t^\infty e^{(1+|p|^\delta)\tau-2\tau}\,d\tau \int_{-\infty}^\infty M^{-2}(q)\,dq \sum_{j=1}^{\mathscr{N}} c^{\mathscr{N}-1} j^{-2}(\mathscr{N}+1-j)^{-2} \\ &\quad\leq c^{\mathscr{N}} e^{-\gamma_1 t}\frac{1}{\gamma_1}, \end{aligned} \tag{2.9}$$

whence ($\gamma_1 > 0$ and $c > 0$ are some constants)

$$|J^2_{\mathscr{N}}| \leq c^{\mathscr{N}} e^{-(\gamma_1+1)t}\,. \tag{2.10}$$

To avoid long delays in proving the validity of the asymptotics (1.5), let us assume(*) that it has already been established that the functions are bounded and

(*) Besides, the first bound (2.11) follows directly from (2.9).

Hölder $\Phi_{\mathcal{N}}(p)$ for $|p| \le \frac{1}{2}$, i.e.

$$|\Phi_{\mathcal{N}}(p)| \le c^{\mathcal{N}}, \quad |\Phi_{\mathcal{N}}(p) - \Phi_{\mathcal{N}}(0)| \le \mathcal{C}^{\mathcal{N}} |p|^{\sigma_3}, \quad |p| \le \tfrac{1}{2}, \tag{2.11}$$

where $c > 0$, $\sigma_3 = \min(\sigma_1, \sigma_2)$, and $\sigma_3 \in (0, 1]$, are some constants independent of $\mathcal{N}$. Now, using (2.11), for $J^3_{\mathcal{N}}$ we establish the following asymptotics as $t \to \infty$:

$$\begin{gathered} J^3_{\mathcal{N}} = 2\Phi_{\mathcal{N}}(0) t^{-1/\delta} e^{-t} \int_0^\infty \cos(p\xi) e^{-p^\delta}\, dp + c^{\mathcal{N}} O(t^{-\frac{1}{\delta}-\gamma_0} e^{-t}), \\ \gamma_0 = \frac{1}{\delta} \min\left(\sigma_3, \frac{\beta - \delta}{2}\right), \qquad \xi = |x| t^{-1/\delta}. \end{gathered} \tag{2.12}$$

Indeed, let us divide the region of integration into two parts: $t^{-v/\delta} \le |p| \le \frac{1}{2}$ and $|p| \le t^{-v/\delta}$, where $v = (\delta + \sigma_4)/\beta$, $\sigma_4 = \min(\sigma_3, \frac{\beta-\delta}{2})$, i.e. $\sigma_4 < \beta - \delta$, and, therefore, $v \in (0, 1]$. The first term can be estimated easily:

$$\begin{aligned} \left| \int_{t^{-v/\delta} \le |p| \le 1/2} e^{ipx - K(p)t} \Phi_{\mathcal{N}}(p)\, dp \right| &\le e^{-t - t^{1-v}} \int_{|p| \le 1/2} |\Phi_{\mathcal{N}}(p)|\, dp \\ &= c^{\mathcal{N}} O(e^{-t} t^{-2/\delta}). \end{aligned} \tag{2.13}$$

The integral over the region $|p| \le t^{-v/\delta}$ can be represented as the sum of the following three integrals:

$$\begin{aligned} J^4_{\mathcal{N}} &= \int_{|p| \le t^{-v/\delta}} e^{ipx - K(p)t} (\Phi_{\mathcal{N}}(p) - \Phi_{\mathcal{N}}(0))\, dp, \\ J^5_{\mathcal{N}} &= \Phi_{\mathcal{N}}(0) \int_{|p| \le t^{-v/\delta}} e^{-ipx - K^1(p)t} (e^{-K^2(p)t} - 1)\, dp, \\ J^6_{\mathcal{N}} &= \Phi_{\mathcal{N}}(0) \int_{|p| \le t^{-v/\delta}} e^{ipx - K^1(p)t}\, dp. \end{aligned}$$

Making the change of the variable of integration $\eta = tp^\delta$ and applying (2.11), we obtain

$$\begin{aligned} |J^4_{\mathcal{N}}| &\le 2c^{\mathcal{N}} \int_0^{t^{-v/\delta}} e^{-t - tp^\delta} p^{\sigma_3}\, dp \le \frac{2c^{\mathcal{N}} e^{-t}}{t\delta} \int_0^\infty e^{-\eta} \left(\frac{\eta}{t}\right)^{(\sigma_3+1)/\delta} d\eta \\ &= c^{\mathcal{N}} O(e^{-t} t^{-(1+\sigma_3)/\delta}), \end{aligned} \tag{2.14}$$

and similarly

$$\begin{aligned} |J^5_{\mathcal{N}}| &\le 2e^{-t} |\Phi_{\mathcal{N}}(0)| \int_0^{t^{1-v}} e^{-\eta} \left| \exp\left(-itK^2\left(\left(\frac{\eta}{t}\right)^{1/\delta}\right) - 1\right| \eta^{1/\delta - 1} \delta^{-1} t^{-1/\delta}\, d\eta \\ &\le 2c^{\mathcal{N}} e^{-t} \int_0^\infty e^{-\eta} c_1 t^{1 - v\beta/\delta} \eta^{(\beta+1)/\delta - 1} \delta^{-1} t^{-1/\delta}\, d\eta = c^{\mathcal{N}} O(e^{-t} t^{-(1+\sigma_4)/\delta}), \end{aligned} \tag{2.15}$$

since $1 - v\beta/\delta = -\sigma_4/\delta < 0$.

The integral $J^6_{\mathcal{N}}$ can be written in the form

$$\begin{aligned} J^6_{\mathcal{N}} &= \Phi_{\mathcal{N}}(0) \int_{|p|\le t^{-\nu/\delta}} e^{ipx-|p|^\delta t - t}\, dp \\ &= 2\Phi_{\mathcal{N}}(0)e^{-t} \int_0^t \cos(p|x|)e^{-p^\delta t}\, dp + c^{\mathcal{N}} O(e^{-t}t^{-2/\delta}), \end{aligned}$$

whence, making the change $p = t^{-1/\delta}\eta$, $\xi = |x|t^{-1/\delta}$, we find that

$$J^6_{\mathcal{N}} = 2\Phi_{\mathcal{N}}(0)e^{-t}t^{-1/\delta} \int_0^\infty \cos(\eta\xi)e^{-\eta^\delta}\, d\eta + c^{\mathcal{N}} O(e^{-t}t^{-2/\delta})\,. \tag{2.16}$$

Formula (2.12) follows from (2.13)–(2.16). Substitution of (2.8), (2.10), and (2.12) in (2.6) now yields

$$\begin{aligned} u(x,t) &= \frac{1}{\pi} \sum_{\mathcal{N}=0}^\infty \Phi_{\mathcal{N}}(0)\varepsilon^{\mathcal{N}+1} t^{-1/\delta} e^{-t} \int_0^\infty \cos(\eta\xi)e^{-\eta^\delta}\, d\eta \\ &\quad + \sum_{\mathcal{N}=0}^\infty \varepsilon^{\mathcal{N}+1} c^{\mathcal{N}} O(e^{-t}t^{-1/\delta-\gamma_0}), \end{aligned}$$

where both series converge absolutely by (2.11) if we choose $\varepsilon < 1/c$.

Thus, (1.5) is proved and

$$A = \frac{1}{\pi} \sum_{\mathcal{N}=0}^\infty \Phi_{\mathcal{N}}(0)\varepsilon^{\mathcal{N}+1}\,. \tag{2.17}$$

It remains to establish the bound (2.11) and to find a method for computing the constant $\Phi_{\mathcal{N}}(0)$ in terms of the initial condition $\hat{\overline{u}}(p)$.

§3. Computation of the functions $\Phi_{\mathcal{N}}(p)$

In this section we prove (2.11) and present a method for computing the functions $\Phi_{\mathcal{N}}(p)$ for $|p| \le \frac{1}{2}$. For this we have to go deeper into perturbation theory.

3.1. Perturbation theory. We represent the function $v_{\mathcal{N}}(p,t)$ in the form of a polilinear operator of the initial condition $\overline{v}(p)$:

$$v_{\mathcal{N}}(p,t) = \int_{-\infty}^\infty Q_{\mathcal{N}}(\check{y},t)V_{\mathcal{N}}(\check{y})\, dy, \qquad V_{\mathcal{N}}(\check{y}) = \prod_{i=0}^{\mathcal{N}} \overline{v}(y_i), \tag{3.1}$$

$$y \in R_{\mathcal{N}}, \quad \check{y} = (y_0, y), \quad \sum_{i=0}^{\mathcal{N}} y_i = p\,. \tag{3.2}$$

For $\mathcal{N} = 0$ formula (3.1) converts into (2.3), and we find that $Q_0(p,t) = e^{-K(p)t}$. In order to obtain a recurrent relation for the symbols $Q_{\mathcal{N}}(\check{y},t)$, we substitute (3.1) in both sides of (2.4), $p \in R_1$ fixed, $\mathcal{N} \geq 1$:

$$\begin{aligned}
&\int_{-\infty}^{\infty} Q_{\mathcal{N}}(\check{y}, t) V_{\mathcal{N}}(\check{y})\, dy \\
&\quad = \int_0^t e^{-K(p)(t-\tau)}\, d\tau \sum_{j=1}^{\mathcal{N}} \int_{-\infty}^{\infty} d\tau \left(\int_{-\infty}^{\infty} dz Q_{j-1}(\check{z}, \tau) V_{j-1}(\check{z}) \right) \\
&\quad\quad \times \left(\int_{-\infty}^{\infty} dq\, Q_{\mathcal{N}-j}(\check{q}, \tau) V_{\mathcal{N}-j}(\check{q}) \right),
\end{aligned} \tag{3.3}$$

where $y \in R_{\mathcal{N}}$, $z \in R_{j-1}$, $q \in R_{\mathcal{N}-j}$, $p = \sum_{k=0}^{\mathcal{N}} y_k$, $p - r = \sum_{k=0}^{j-1} z_k$, $r = \sum_{k=0}^{\mathcal{N}-j} q_k$. For each fixed j, $1 \leq j \leq \mathcal{N}$, we introduce new variables of integration η on the right-hand side of (3.3): $\eta_k = z_k$, $0 \leq k \leq j-1$, $\eta_k = q_{k-j}$, $j \leq k \leq \mathcal{N}$. The Jacobian of this transition is equal to one. Equality (3.3) takes the following form:

$$\begin{aligned}
&\int_{-\infty}^{\infty} dy V_{\mathcal{N}}(\check{y}) Q_{\mathcal{N}}(\check{y}, t) \\
&\quad = \int_0^t e^{-K(p)(t-\tau)}\, d\tau \int_{-\infty}^{\infty} d\eta \\
&\quad\quad \times V_{\mathcal{N}}(\check{\eta}) \sum_{j=1}^{\mathcal{N}} Q_{j-1}(\eta_0, \dots, \eta_{j-1}, \tau) Q_{\mathcal{N}-j}(\eta_j, \dots, \eta_{\mathcal{N}}, \tau)\,.
\end{aligned} \tag{3.4}$$

Equality (3.4) is satisfied if we set

$$\begin{aligned}
&Q_{\mathcal{N}}(\check{y}, t) = \int_0^t d\tau\, e^{-K(p)(t-\tau)} \sum_{j=1}^{\mathcal{N}} Q_{j-1}(y_0, \dots, y_{j-1}, \tau) Q_{\mathcal{N}-j}(y_j, \dots, y_{\mathcal{N}}, \tau), \\
&p = \sum_{k=0}^{\mathcal{N}} y_k\,.
\end{aligned} \tag{3.5}$$

3.2. The first formula for $Q_{\mathcal{N}}$. In order to resolve the recurrent relation (3.5) we consider a procedure for constructing a certain family of graphs by means of an algorithm for dividing the segment $(0, \mathcal{N}+1)$ of the natural sequence ($\mathcal{N} \geq 0$).

On the axis Ox we consider the segment $(0, \mathcal{N}+1)$ of the natural sequence. Along the axis Oy we mark the step of the division algorithm. We represent the vertices of the desired graph by points h_j on the xOy plane. We denote h_j^x the projections of the vertices h_j onto the axis Ox. At the first step of the algorithm we set $h_1 = (0, 1)$. We introduce temporarily the additional vertex $h_0 = (\mathcal{N}+1, 1)$. At the second step we choose an arbitrary point h_3^x on the segment (h_1^x, h_0^x) and also set $h_2^x = 0$. Thus, we obtain two new vertices $h_j = (h_j^x, 2)$, $j = 2, 3$. We draw the two edges $\overrightarrow{h_1 h_2}$ and $\overrightarrow{h_1 h_3}$, to which we assign lengths $h_3^x - h_1^x - 1$ and $h_0^x - h_3^x - 1$, respectively.

Further, by induction, suppose at some step the initial segment $(0, N+1)$ is decomposed by the vertices $h_0, h_1, \dots, h_n$ into l nonempty segments: $(h_{i_1}^x, h_{j_1}^x), \dots, (h_{i_l}^x, h_{j_l}^x)$, $h_{i_1}^x < h_{j_1}^x < \dots < h_{i_l}^x < h_{j_l}^x$, each of which does not contain in its interior points h_k^x, $k \leq n$. We choose one point one each of these segments. The collection of these points together with the left end-points of the subdivided segments gives projections of $2l$ new vertices of the graph: $h_{n+1}, \dots, h_{n+2l}$. We draw the edges $\overrightarrow{h_{i_k} h_{n+2k-1}}$ and $\overrightarrow{h_{i_k} h_{n+2k}}$, joining the new vertices h_{n+2k-1} and h_{n+2k} with the vertex h_{i_k}, determining the

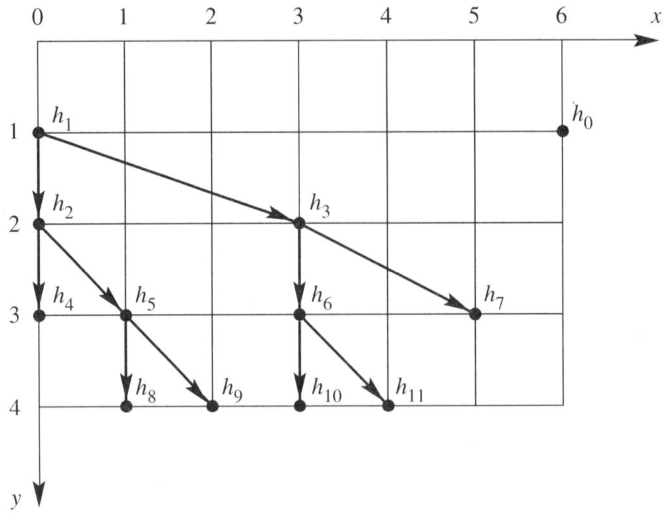

FIGURE 1. An example of a graph g of the set $G_{0,5}$.

left end-point of the segment $(h^x_{i_k}, h^x_{j_k})$, on which the point h^x_{n+2k} was chosen. We set the lengths of the edges $\overrightarrow{h_{i_k} h_{n+2k-1}}$ and $\overrightarrow{h_{i_k} h_{n+2k}}$ equal to $h^x_{n+2k} - h^x_{i_k} - 1$ and $h^x_{j_k} - h^x_{n+2k} - 1$, respectively. We continue this procedure until at some step no nonempty interval remains which can be further divided. The total number of steps of the algorithm does not exceed $\mathcal{N} + 1$ (see Figure 1).

We have thus constructed a graph g consisting of vertices h_j, $1 \le j \le 2\mathcal{N}+1$, joined by normalized and oriented edges. We do not include in g the vertex h, introduced in the construction. We denote the set of all graphs obtained by division of the segment $(0, \mathcal{N} + 1)$ by $\mathcal{G}_{0,\mathcal{N}}$. If we replace the segment $(0, \mathcal{N} + 1)$ by $(j, \mathcal{N} + j + 1)$, then the vertices of the new graph g' are shifted along the Ox axis by the number j (i.e., $h^{x'}_i = j + h^x_i$). We denote the set of such graphs by $\mathcal{G}_{j,\mathcal{N}+j}$. Vertices from which two edges exit will be called nodes. In any graph $g \in \mathcal{G}_{0,\mathcal{N}}$ there are exactly $\mathcal{N}$ nodes, while the total number of vertices is equal to $2\mathcal{N} + 1$.

We fix any two graphs $g^1 \in \mathcal{G}_{0,j-1}$ and $g^2 \in \mathcal{G}_{j,\mathcal{N}}$, $1 \le j \le \mathcal{N}$. The superscript indicates to which graph the notation refers. We define the operator of gluing two graphs g' and g^2 in the following manner:

1) We drop the graphs g^1 and g^2 one step along the Oy axis.

2) The vertices of the new graph are the vertices $h_1 = (0, 1)$ and the vertices of the graphs g^1 and g^2: $h^1_1, \dots, h^1_{2j-1}; h^2_1, \dots, h^2_{2\mathcal{N}-2j+1}$.

3) The edges of the graph g are the edges of the graphs g^1 and g^2 and the two new edges: $\overrightarrow{h_1 h^1_1}$ and $\overrightarrow{h_1 h^2_1}$ which have the lengths $|\overrightarrow{h_1 h^1_1}| = j - 1$ and $|\overrightarrow{h_1 h^2_1}| = \mathcal{N} - j$, respectively.

We denote the result of gluing of graphs by $g = g^1 \otimes g^2$. The set of glued graphs $g = g^1 \otimes g^2$ for all possible $g^1 \in \mathcal{G}_{0,j-1}$ and $g^2 \in \mathcal{G}_{j,\mathcal{N}}$ will be denoted by $\mathcal{G}_{0,j-1} \otimes \mathcal{G}_{j,\mathcal{N}}$.

We observe that the gluing operation corresponds fully to the division algorithm, i.e., any glued graph $g = g^1 \otimes g^2$ can be constructed by division of the segment $(0, \mathcal{N} + 1)$. Thus, we obtain

$$\mathcal{G}_{0,j-1} \otimes \mathcal{G}_{j,w} \subset \mathcal{G}_{0,\mathcal{N}}\,. \tag{3.6}$$

Since (3.6) holds for any j, $1 \leq j \leq \mathscr{N}$, then

$$\sum_{j=1}^{\mathscr{N}} \mathscr{G}_{0,j-1} \otimes \mathscr{G}_{j,\mathscr{N}} \subset \mathscr{G}_{0,\mathscr{N}}, \tag{3.7}$$

where there is a sum since the sets $\mathscr{G}_{0,j-1} \otimes \mathscr{G}_{j,\mathscr{N}}$ are different for different j (the graphs differ by the vertex $h_3 = (j, 2)$). We shall prove the inclusion reverse to (3.7):

$$\mathscr{G}_{0,\mathscr{N}} \subset \sum_{j=1}^{\mathscr{N}} \mathscr{G}_{0,j-1} \otimes \mathscr{G}_{j,\mathscr{N}} . \tag{3.8}$$

In order to establish (3.8) we fix any graph g and show that $g = g^1 \otimes g^2$, where $g^1 \in \mathscr{G}_{0,j-1}$, $g^2 \in \mathscr{G}_{j,\mathscr{N}}$, $j = h_3^x$. To this end, we define the operation of cutting a graph g along the vertex h_3 in the following manner.

1) Remove the vertex h_1 and the edges $\overrightarrow{h_1h_2}$ and $\overrightarrow{h_1h_3}$.

2) Divide the remaining vertices into two sets:

$$M_1 = \{h_k : h_k^x \in [0, h_3^x]\}, \qquad M_2 = \{h_k : h_k^x \in [h_3^x, \mathscr{N}]\} .$$

3) The graph g^1 consists of the set of vertices M_1 and all those edges of g which join vertices of the set M_1. The graph g^2 is defined similarly.

4) Shift the graphs g^1 and g^2 formed upward one step along the Oy axis.

Since in the construction of a graph $g \in \mathscr{G}_{0,\mathscr{N}}$ each of the segments $[0, h_3^x)$ and $[h_3^x, \mathscr{N}]$ is subdivided by means of the division algorithm, it follows that $g^1 \in \mathscr{G}_{0,j-1}$ and $g^2 \in \mathscr{G}_{j,\mathscr{N}}$. It is clear that $g = g^1 \otimes g^2$. From this we obtain (3.8). From (3.7) and (3.8) we obtain

$$\sum_{j-1}^{\mathscr{N}} \mathscr{G}_{0,j-1} \otimes \mathscr{G}_{j,\mathscr{N}} = \mathscr{G}_{0,\mathscr{N}}, \qquad \mathscr{N} \geq 1 . \tag{3.9}$$

This formula makes it possible to estimate the cardinality $|\mathscr{G}_{0,\mathscr{N}}|$ of the set $\mathscr{G}_{0,\mathscr{N}}$. It is obvious that $|\mathscr{G}_{0,\mathscr{N}}| = |\mathscr{G}_{j,\mathscr{N}+j}|$, $\forall j$. Therefore, from (3.9) we obtain the recurrent relation

$$|\mathscr{G}_{0,\mathscr{N}}| = \sum_{j=1}^{\mathscr{N}} |\mathscr{G}_{0,j-1}|\, |\mathscr{G}_{0,\mathscr{N}-j}|, \qquad \mathscr{N} \geq 1, \tag{3.10}$$

from which we easily obtain the estimate

$$|\mathscr{G}_{0,\mathscr{N}}| \leq 12^{\mathscr{N}} (\mathscr{N} + 1)^{-2}, \qquad \mathscr{N} \geq 0 . \tag{3.11}$$

Along with the natural enumeration of vertices given by the algorithm of graph construction, we also use below a special enumeration of the vertices. We first define a special succession order for the nodes. Consider the nodal part of some graph g, i.e., only the nodes and the edges joining them. For graphs of the families $\mathscr{G}_{0,0}$, $\mathscr{G}_{0,1}$, and $\mathscr{G}_{0,2}$ the special enumeration of vertices coincides with the natural one. Further, by induction we cut the graph g and obtain $g = g^1 \otimes g^2$. Suppose the special ordering of the vertices has already been established in the graphs g^1 and g^2. Then the special ordering of the vertices in g is defined as follows: the vertex h_1 remains the first vertex also in the special enumeration, and after it there follow the nodes of g' in the special ordering, and then the nodes of g^2 in the special ordering. The special enumeration of all vertices of g is now defined as follows: we first number the nodes of g in the special ordering (as described above), then the non-nodal vertices of g^1 in the natural order, and finally, we number the non-nodal vertices of g^2 also in the natural order.

With the help of the introduced family of graphs we shall solve the recurrent relation (3.5). Fix any graph $g \in \mathcal{G}_{0,\mathcal{N}}$ in which we choose a special enumeration of the vertices. Let $\xi_1 = 0$; for each node h_j, $2 \le j \le \mathcal{N}$, we determine the number ξ_j of vertices h_{ξ_j} preceding it, and also the numbers r_j^1 and r_j^2, $1 \le j \le \mathcal{N}$, of vertices $h_{r_j^1}$ and $h_{r_j^2}$ immediately following the node h_j in the sense that in g there are the edges $\overrightarrow{h_j h_{r_j^1}}$, $\overrightarrow{h_j h_{r_j^2}}$, and $\overrightarrow{h_{\xi_j} h_j}$. Let $y \in R_{\mathcal{N}}$, $\check{y} = (y_0, y) \in R_{\mathcal{N}+1}$. For each node h_j, $1 \le j \le \mathcal{N}$, we define two sums of coordinates y_k:

$$s_j^i = \sum_{k=h^x_{r_j^i}}^{h^x_{r_j^i} + |\overrightarrow{h_j h_{r_j^i}}|} y_k, \qquad 1 \le j \le \mathcal{N},\ i = 1, 2\,.$$

If the vertex $h_{r_j^i}$ is non-nodal, s_j^i is equal to $y_{h^x_{r_j^i}}$. The vertices $h_{r_j^i}$ for $1 \le j \le \mathcal{N}$, $i = 1, 2$, run through the whole set of vertices of the graph g, with the exception of the vertex h_1 only. Therefore, the collection of the sums $\{s_j^i\}_{j=1,\dots,\mathcal{N}}^{i=1,2}$ contains the subset $\{y_0, y_1, \dots, y_{\mathcal{N}}\}$. We also denote $p_j = s_j^1 + s_j^2$, $p = p_1$. For each $\mathcal{N} \ge 0$ we establish the following equality:

$$Q_{\mathcal{N}}(\check{y}, t_0) = e^{-K(p)t_0} \sum_{g \in \mathcal{G}_{\mathcal{N}}} \int_0^\infty \cdots \int_0^\infty dt_1 \cdots dt_{\mathcal{N}} \prod_{j=1}^{\mathcal{N}} e^{L_j t_j} \theta(t_{\xi_j} - t_j), \tag{3.12}$$

where

$$L_j = L(s_j^1, s_j^2) = K(s_j^1 + s_j^2) - K(s_j^1) - K(s_j^2), \tag{3.13}$$

$$\theta(x) = \begin{cases} 1, & x \ge 0, \\ 0, & x < 0, \end{cases}$$

is the Heaviside function $\mathcal{G}_{\mathcal{N}} \equiv \mathcal{G}_{0,\mathcal{N}}$, $t_{\xi_1} = t_0$. Thanks to the Heaviside function, the integration in (3.12) is taken over finite time intervals. For $\mathcal{N} = 0$ (3.12) has the following form:

$$Q_0(y, t) = e^{-K(y)t}\,. \tag{3.14}$$

Suppose formula (3.12) has been established by induction for $\mathcal{N} \le \mathcal{N}_1$. Then, substituting (3.12) in the recurrent relation (3.5) for $\mathcal{N} = \mathcal{N}_1 + 1$, we obtain

(3.15)

$$\begin{aligned} Q_{\mathcal{N}}(\check{y}, t_0) = &\int_0^{t_0} dt_1\, e^{-K(p)(t_0 - t_1)} \sum_{j=1}^{\mathcal{N}} e^{-K(p^1)t_1} \\ &\times \left(\sum_{g^1 \in \mathcal{G}_{0,j-1}} \int_0^\infty \cdots \int_0^\infty dt_1^1 \cdots dt_{j-1}^1 \prod_{k=1}^{j-1} e^{L_k^1 t_k^1} \theta(t^1_{\xi_k^1} - t_k^1) \right) \\ &\times e^{-K(p^2)t_1} \left(\sum_{g^2 \in \mathcal{G}_{j,\mathcal{N}}} \int_0^\infty \cdots \int_0^\infty dt_2^1 \cdots dt^2_{\mathcal{N}-j} \prod_{k=1}^{\mathcal{N}-j} e^{L_k^2 t_k^2} \theta(t^2_{\xi_k^2} - t_k^2) \right), \end{aligned}$$

where

$$t^1_{\xi^1_1} = t_1, \quad t^2_{\xi^2_1} = t_1, \quad p^1 = \sum_{k=0}^{j-1} y_k, \quad p^1 = \sum_{k=j}^{\mathscr{N}} y_k .$$

Superscript i indicates to which of the graphs g^i, $i = 1, 2$, the notation refers to. For $j = 1$ or $j = \mathscr{N}$ one of the factors in parentheses in (3.15) is identically equal to one. From (3.9) it follows that

$$\sum_{j=1}^{\mathscr{N}} \sum_{g^1 \in \mathscr{G}_{j-1}} \sum_{g^2 \in \mathscr{G}_{j,\mathscr{N}}} = \sum_{j=1}^{\mathscr{N}} \sum_{g = g^1 \otimes g^2 : g^1 \in \mathscr{G}_{j-1}, g^2 \in \mathscr{G}_{j,\mathscr{N}}} = \sum_{g \in \mathscr{G}_{\mathscr{N}}},$$

and, therefore, noting that $p = p^1 + p^2$, we can rewrite (3.15) in the following form:

$$\begin{aligned} Q_{\mathscr{N}}(\check{y}, t_0) = {} & e^{-K(p)t_0} \sum_{g \in \mathscr{G}_{\mathscr{N}}} \int_0^\infty dt_1\, \theta(t_0 - t_1) e^{L(p^1, p^2)t_1} \\ & \times \left(\int_0^\infty \cdots \int_0^\infty dt^1_1 \cdots dt^1_{j-1} \prod_{k=1}^{j-1} \theta(t^1_{\xi^1_k} - t^1_k) e^{L^1_k t^1_k} \right) \\ & \times \left(\int_0^\infty \cdots \int_0^\infty dt^2_1 \cdots dt^2_{\mathscr{N}-j} \prod_{k=1}^{\mathscr{N}-j} \theta(t^2_{\xi^2_k} - t^2_k) e^{L^2_k t^2_k} \right) . \end{aligned} \tag{3.16}$$

Because of the choice of the special enumeration of the vertices in the graph $g \in \mathscr{G}_{\mathscr{N}}$, the following equalities hold:

$$s^1_1 = p^1, \quad s^2_1 = p^2, \qquad t^1_{\xi^1_k} = t_{\xi_{k+1}}, \quad 1 \le k \le j-1;$$

$$t^2_{\xi^2_k} = t_{\xi_{k+j}}, \qquad 1 \le k \le \mathscr{N} - j;$$

$$t^1_k = t_{k+1}, \quad L^1_k = L_{k+1}, \qquad 1 \le k \le j-1;$$

$$t^2_k = t_{k+j}, \quad L^2_k = L_{k+j}, \qquad 1 \le k \le \mathscr{N} - j .$$

Thus, (3.16) takes the form

$$\begin{aligned} Q_{\mathscr{N}}(\check{y}, t) = {} & \sum_{g \in \mathscr{G}_{\mathscr{N}}} e^{-K(p)t} \int_0^\infty dt_1 e^{L_1 t_1} \theta(t - t_1) \\ & \times \left(\int_0^\infty \cdots \int_0^\infty dt_2 \cdots dt_j \prod_{k=2}^{j} \theta(t_{\xi_k} - t_k) e^{L_k t_k} \right) \\ & \times \left(\int_0^\infty \cdots \int_0^\infty dt_{j+1} \cdots dt_{\mathscr{N}} \prod_{k=j+1}^{\mathscr{N}} \theta(t_{\xi_k} - t_k) e^{L_k t_k} \right), \end{aligned}$$

whence we arrive at (3.12). The induction is complete.

3.3. A second formula for $Q_{\mathscr{N}}$. Let $B_0 = 1$,

$$B_1(x,t) = (x^{-1}(e^{xt} - 1),$$

$$\begin{aligned} & B_{\mathscr{N}}(x_1, x_2, \ldots, x_{\mathscr{N}}, t) \\ & \quad = x_1^{-1}(e^{x_1 t} B_{\mathscr{N}-1}(x_2, \ldots, x_{\mathscr{N}}, t) \\ & \quad -B_{\mathscr{N}-1}(x_1 + x_2, x_3, \ldots, x_{\mathscr{N}}, t)), \qquad \mathscr{N} \geq 2 . \end{aligned} \tag{3.17}$$

From the recurrent relations (3.17) we obtain the following formula for $j = 2, 3, \ldots, \mathscr{N} - 1$, $\mathscr{N} \geq 3$:

$$\begin{aligned} B_{\mathscr{N}}(x_1, \ldots, x_{\mathscr{N}}, t) = x_j^{-1}(& B_{\mathscr{N}-1}(x_1, \ldots, x_{j-1} + x_j, x_{j+1}, \ldots, x_{\mathscr{N}}, t) \\ & - B_{\mathscr{N}-1}(x_1, \ldots, x_{j-1}, x_j + x_{j+1}, \ldots, x_{\mathscr{N}}, t)) . \end{aligned} \tag{3.18}$$

We prove (3.18) first for $j = 2$, $\mathscr{N} \geq 3$. Consider the right-hand side in (3.18). In view of (3.17) we get

$$\begin{aligned} & x_2^{-1}(B_{\mathscr{N}-1}(x_1 + x_2, x_3, \ldots, x_{\mathscr{N}}, t) - B_{\mathscr{N}-1}(x_1, x_2 + x_3, \ldots, x_{\mathscr{N}}, t)) \\ & \quad = x_2^{-1}(e^{(x_1+x_2)t}(x_1 + x_2)^{-1} B_{\mathscr{N}-2}(x_3, \ldots, x_{\mathscr{N}}, t) \\ & \quad - (x_1 + x_2)^{-1} B_{\mathscr{N}-2}(x_1 + x_2 + x_3, \ldots, x_{\mathscr{N}}, t) \\ & \quad - e^{x_1 t} x_1^{-1} B_{\mathscr{N}-2}(x_2 + x_3, \ldots, x_{\mathscr{N}}, t) \\ & \qquad\qquad\qquad + x_1^{-1} B_{\mathscr{N}-2}(x_1 + x_2 + x_3, \ldots, x_{\mathscr{N}}, t)) \\ & \quad = (x_1 x_2)^{-1}(e^{(x_1+x_2)t} B_{\mathscr{N}-2}(x_3, \ldots, x_{\mathscr{N}}, t) - e^{x_1 t} B_{\mathscr{N}-2}(x_2 + x_3, \ldots, x_{\mathscr{N}}, t)) \\ & \quad - x_1^{-1}(x_1 + x_2)^{-1}(e^{(x_1+x_2)t} B_{\mathscr{N}-2}(x_3, \ldots, x_{\mathscr{N}}, t) \\ & \qquad\qquad\qquad - B_{\mathscr{N}-2}(x_1 + x_2 + x_3, \ldots, x_{\mathscr{N}}, t)) \\ & \quad = x_1^{-1}(e^{x_1 t} B_{\mathscr{N}-1}(x_2, x_3, \ldots, x_{\mathscr{N}}, t) - B_{\mathscr{N}-1}(x_1 + x_2, x_3, \ldots, x_{\mathscr{N}}, t)) \\ & \quad = B_{\mathscr{N}}(x_1, \ldots, x_{\mathscr{N}}, t) . \end{aligned}$$

Further, by induction for $3 \leq j \leq \mathscr{N} - 1$, $\mathscr{N} \geq 4$, in view of (3.17) and (3.18) we obtain

$$\begin{aligned} & x_j^{-1}(B_{\mathscr{N}-1}(x_1, \ldots, x_{j-1} + x_j, x_{j+1}, \ldots, x_{\mathscr{N}}, t) \\ & \quad - B_{\mathscr{N}-1}(x_1, \ldots, x_{j-1}, x_j + x_{j+1}, \ldots, x_{\mathscr{N}}, t)) \\ & \quad = (x_1 x_j)^{-1}(e^{x_1 t} B_{\mathscr{N}-2}(x_2, \ldots, x_{j-1} + x_j, x_{j+1}, \ldots, x_{\mathscr{N}}, t) \\ & \qquad\qquad - B_{\mathscr{N}-2}(x_1 + x_2, \ldots, x_{j-1} + x_j, x_{j+1}, \ldots, x_{\mathscr{N}}, t) \\ & \qquad\qquad - e^{x_1 t} B_{\mathscr{N}-2}(x_2, \ldots, x_{j-1}, x_j + x_{j+1}, \ldots, x_{\mathscr{N}}, t) \\ & \qquad\qquad + B_{\mathscr{N}-2}(x_1 + x_2, \ldots, x_{j-1}, x_j + x_{j+1}, \ldots, x_{\mathscr{N}}, t)) \\ & \quad = x_1^{-1}(e^{x_1 t} B_{\mathscr{N}-1}(x_2, \ldots, x_{\mathscr{N}}, t) - B_{\mathscr{N}-1}(x_1 + x_2, \ldots, x_{\mathscr{N}}, t)) \\ & \quad = B_{\mathscr{N}}(x_1, \ldots, x_{\mathscr{N}}, t) . \end{aligned}$$

The induction is complete. We now establish the following formula, $\mathscr{N} \geq 0$:

$$B_{\mathscr{N}+1}(x_1, \ldots, x_{\mathscr{N}+1}, t) = \int_0^t e^{x_1 t_1} B_{\mathscr{N}}(x_2, \ldots, x_{\mathscr{N}+1}, t_1)\, dt_1 . \tag{3.19}$$

For $\mathcal{N} = 0$ the equality (3.19) is obvious. Further, arguing by induction and using (3.17) and (3.18) we find that

$$\begin{aligned}
&\int_0^t e^{x_1 t_1} B_{\mathcal{N}}(x_2, \ldots, x_{\mathcal{N}+1}, t_1)\, dt_1 \\
&\quad = x_2^{-1} \int_0^t e^{x_1 t_1} [B_{\mathcal{N}-1}(x_3, \ldots, x_{\mathcal{N}+1}, t_1) e^{x_2 t_1} \\
&\qquad\qquad\qquad\qquad - B_{\mathcal{N}-1}(x_2 + x_3, \ldots, x_{\mathcal{N}+1}, t_1)]\, dt_1 \\
&\quad = x_2^{-1}(B_{\mathcal{N}}(x_1 + x_2, \ldots, x_{\mathcal{N}+1}, t) - B_{\mathcal{N}}(x_1, x_2 + x_3, \ldots, x_{\mathcal{N}+1}, t)) \\
&\quad = B_{\mathcal{N}+1}(x_1, x_2, \ldots, x_{\mathcal{N}+1}, t)\,.
\end{aligned}$$

We carry out an additional construction with graphs. We fix some graph $g \in \mathcal{G}_{0,\mathcal{N}}$. We consider its nodal part of it (i.e., only the nodes and the edges joining them). We define finished graphs $\tilde{g}$ as follows: all those vertices not joined by edges passing in the direction of their orientation we join by two-sided edges. We denote by the family of finished graphs by $\widetilde{\mathcal{G}}_{\mathcal{N}}$. It is obvious that the cardinality $|\widetilde{\mathcal{G}}_{\mathcal{N}}| = |\mathcal{G}_{\mathcal{N}}|$ (see Figure 2).

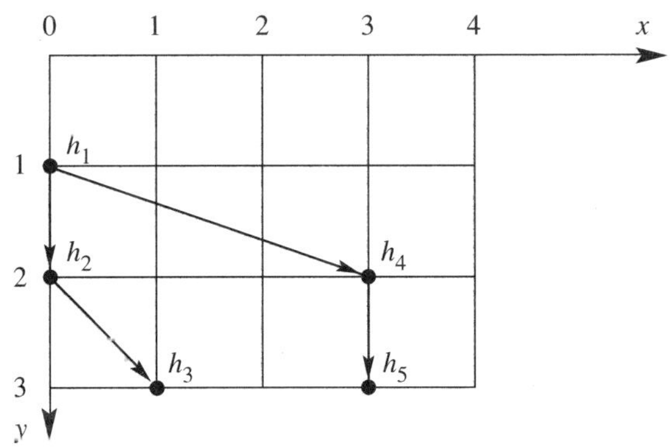

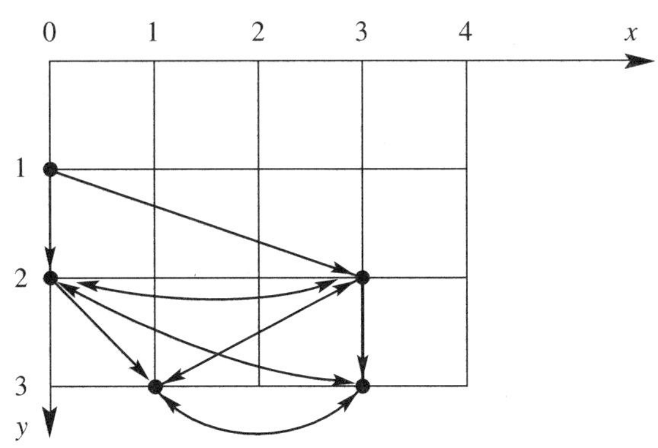

FIGURE 2. Finished graphs: (a) is the corner part of the graph g from Figure 1; (b) is the finished graph $\tilde{g}$ for the graph g from Figure 1. (Special enumeration of vertices.)

By a complete chain w we mean the sequence of all nodes of the graph $\tilde{g}$ joined together by edges so that the oriented edges, when passing from the first vertex to the last one, follow strictly their direction and subordination of the vertices is preserved. Subordination of vertices means the following: for any j, $1 \leq j \leq \mathscr{N}-1$, we consider a branch of a graph $g \in \mathscr{G}_{0,\mathscr{N}}$ beginning at the vertex h_j. All vertices of this branch we then consider subordinate to h_j, and in the complete chain w they must follow h_j. In another way subordination of vertices can be defined as follows: if we take any initial segment $(w_1, w_2, \dots, w_k)$, $k \leq \mathscr{N}$, of the chain w, then for any vertex h_j of the segment the immediately preceding vertex h_{ξ_j} (i.e., such that in the graph g the edges $\overrightarrow{h_{\xi_j} h_j}$ exists) also belongs to this segment.

Since each complete chain w determines some ordering for the $\mathscr{N}$ nodes of the graph $\tilde{g}$, and the number of all permutations of $\mathscr{N}$ vertices is equal to $\mathscr{N}!$, then in each of the finished graphs $\tilde{g}$ there are no more than $\mathscr{N}!$ complete chains w. Thus, if we denote by $W(g)$ the set of all complete chains in the finished graph $\tilde{g}$, we have

$$|W(g)| \leq \mathscr{N}! \,. \tag{3.20}$$

Below, by a complete chain we often mean the sequence of indices of the nodes of the graph $\tilde{g}$ (in the special enumeration of the vertices). We define the set of complete chains $W(w^1 \otimes w^2)$ to be the result of mixing of two complete chains $w^1 \in W(g^1)$, $g^1 \in \mathscr{G}_{j-1}$, and $w^2 \in W(g^2)$, $g^2 \in \mathscr{G}_{j,\mathscr{N}}$. This means that it is necessary to glue the chains w^1 and w^2, determine the finished graph $w^1 \otimes w^2$, and find complete chains for it. If we represent the chains w^1 and w^2 as sequences $(w^1_1, w^1_2, \dots, w^1_{j-1})$ and $(w^2_1, w^2_2, \dots, w^2_{\mathscr{N}-j})$, then the set $W(w^1 \otimes w^2)$ consists of all permutations of the sequence $(1, w^1_1, w^1_2, \dots, w^1_{j-1}, w^2_1, w^2_2, \dots, w^2_{\mathscr{N}-j})$ for which the order of succession in the chains w^1 and w^2 is not disturbed, and the first vertex always remains in its place. It is clear that for any $g^1 \in \mathscr{G}_{j-1}$ and $g^2 \in \mathscr{G}_{j,\mathscr{N}}$ we have

$$\sum_{w^1 \in W(g^1),\; w^2 \in W(g^2)} W(w^1 \otimes w^2) = W(g^1 \otimes g^2) \,. \tag{3.21}$$

The sum occurs because the sets $W(w^1 \otimes w^2)$ for different w^1 and w^2 do not intersect.

We fix any $g^1 \in \mathscr{G}_{j-1}, g^2 \in \mathscr{G}_{j,\mathscr{N}}, w^1 \in W(g^1), w^2 \in W(g^2)$, and prove the following equality, $1 \leq j \leq \mathscr{N}$:

$$\begin{aligned} &\int_0^t dt_1 e^{x_1 t_1} B_{j-1}(x_{w^1_1}, \dots, x_{w^1_{j-1}}, t_1) B_{\mathscr{N}-j}(x_{w^2_1}, \dots, x_{w^2_{\mathscr{N}-j}}, t_1) \\ &\qquad = \sum_{w \in W(w^1 \otimes w^{(2)})} B_{\mathscr{N}}(x_{w_1}, \dots, x_{w_{\mathscr{N}}}, t) \,. \end{aligned} \tag{3.22}$$

When $j = 1$ or $j = \mathscr{N}$ one of the graphs g^1 and g^2 is empty and then, since $B_0 \equiv 1$. Formula (3.22) follows from (3.19). Let $1 < j \leq \mathscr{N}$. It is sufficient to consider the case when $w^1 = (2, 3, \dots, j)$, $w^2 = (j+1, \dots, \mathscr{N})$ (the remaining cases are reduced to this by relabeling). In view of (3.17) from the left-hand side of (3.22) we have

$$
\begin{aligned}
I &\equiv \int_0^t dt_1 e^{x_1 t_1} B_{j-1}(x_2, \dots, x_j, t_1) B_{\mathcal{N}-j}(x_{j+1}, \dots, x_{\mathcal{N}}, t_1) \\
&= x_2^{-1} \int_0^t dt_1 (e^{(x_1+x_2)t_1} B_{j-2}(x_3, \dots, x_j, t_1) \\
&\qquad - e^{x_1 t_1} B_{j-2}(x_2 + x_3, \dots, x_j, t_1)) B_{\mathcal{N}-j}(x_{j+1}, \dots, x_{\mathcal{N}}, t_1) \\
&= x_2^{-1} \Bigg[\sum_{w \in W((3,\dots,j)\otimes(j+1,\dots,\mathcal{N}))} B_{\mathcal{N}-1}(x_1 + x_2, x_{w_2}, \dots, x_{w_{\mathcal{N}}}, t) \\
&\qquad - \sum_{w \in W((4,\dots,j)\otimes(j+1,\dots,\mathcal{N}))} B_{\mathcal{N}-1}(x_1, x_2 + x_3, x_{w_2}, \dots, x_{w_{\mathcal{N}-2}}, t) \\
&\qquad - \sum_{w \in W((4,\dots,j)\otimes(j+2,\dots,\mathcal{N}))} B_{\mathcal{N}-1}(x_1, x_{j+1}, x_2 + x_3, x_{w_2}, \dots, x_{w_{\mathcal{N}-3}}, t) \\
&\qquad - \dots - B_{\mathcal{N}-1}(x_1, x_{j+1}, \dots, x_{\mathcal{N}}, x_2 + x_3, x_4, x_5, \dots, x_j, t) \Bigg] .
\end{aligned} \tag{3.23}
$$

Here we have used the decomposition of the set $W((3, \dots, j) \otimes (j+1, \dots, \mathcal{N}))$:

$$
\begin{aligned}
W((3, \dots, j) \otimes (j+1, \dots, \mathcal{N})) &= \sum_{w \in W((4,\dots,j)\otimes(j+1,\dots,\mathcal{N}))} (1, 3, w_2, \dots, w_{\mathcal{N}-2}) \\
&+ \sum_{w \in W((4,\dots,j)\otimes(j+2,\dots,\mathcal{N}))} (1, j+1, 3, w_2, \dots, w_{\mathcal{N}-3}) \\
&+ \sum_{w \in W((4,\dots,j)\otimes(j+3,\dots,\mathcal{N}))} (1, j+1, j+2, 3, w_2, \dots, w_{\mathcal{N}-4}) \\
&+ \dots + (1, j+1, j+2, \dots, \mathcal{N}, 3, 4, 5, \dots, j) .
\end{aligned} \tag{3.24}
$$

Further, applying (3.24), we continue (3.23) as follows:

$$
\begin{aligned}
I = x_2^{-1} \Bigg(&\sum_{w \in W((4,\dots,j)\otimes(j+1,\dots,\mathcal{N}))} [B_{\mathcal{N}-1}(x_1 + x_2, x_3, \dots, x_{w_{\mathcal{N}-2}}, t) \\
&\qquad - B_{\mathcal{N}-1}(x_1, x_2 + x_3, \dots, x_{w_{\mathcal{N}-2}}, t)] \\
&+ \sum_{w \in W((4,\dots,j)\otimes(j+2,\dots,\mathcal{N}))} [B_{\mathcal{N}-1}(x_1 + x_2, x_{j+1}, x_3, \dots, x_{w_{\mathcal{N}-3}}, t) \\
&\qquad - B_{\mathcal{N}-1}(x_1, x_{j+1}, x_2 + x_3, \dots, x_{w_{\mathcal{N}-3}}, t)] \\
&+ \dots + [B_{\mathcal{N}-1}(x_1 + x_2, x_{j+1}, x_{j+2}, \dots, x_{\mathcal{N}}, x_3, x_4, \dots, x_j, t) \\
&\qquad - B_{\mathcal{N}-1}(x_1, x_{j+1}, x_{j+2}, \dots, x_{\mathcal{N}}, x_2 + x_3, x_4, \dots, x_j, t)] \Bigg) .
\end{aligned} \tag{3.25}
$$

By (3.18) the first square bracket in (3.25) is equal to

$$B_{\mathcal{N}}(x_1, x_2, x_3, x_{w_2}, \dots, x_{w_{\mathcal{N}-2}}, t)\,.$$

Adding and subtracting from the second bracket the term

$$B_{\mathcal{N}-1}(x_1, x_2 + x_{j+1}, x_3, x_{w_2}, \dots, x_{w_{\mathcal{N}-3}}, t)$$

we obtain

$$B_{\mathcal{N}}(x_1, x_2, x_{j+1}, x_3, x_{w_2}, \dots, x_{w_{\mathcal{N}-3}}, t) \\ + B_{\mathcal{N}}(x_1, x_{j+1}, x_2, x_3, x_{w_2}, \dots, x_{w_{\mathcal{N}-3}}, t)$$

and so on. Thus, in view of (3.24), it follows from (3.25) that

$$\begin{aligned} I = & \sum_{w\in W((4,\dots,j)\otimes(j+1,\dots,\mathcal{N}))} B_{\mathcal{N}}(x_1, x_2, x_3, x_{w_2}, \dots, x_{w_{\mathcal{N}-2}}, t) \\ & + \sum_{w\in W((4,\dots,j)\otimes(j+2,\dots,\mathcal{N}))} (\mathcal{B}_{\mathcal{N}}(x_1, x_2, x_{j+1}, x_3, x_{w_2}, \dots, x_{w_{\mathcal{N}-3}}, t) \\ & \qquad + B_{\mathcal{N}}(x_1, x_{j+1}, x_2, x_3, x_{w_2}, \dots, x_{w_{\mathcal{N}-2}}, t)) \\ & + \dots + (B_{\mathcal{N}}(x_1, x_2, x_{j+1}, \dots, x_{\mathcal{N}}, x_3, \dots, x_j, t) \\ & \qquad + B_{\mathcal{N}}(x_1, x_{j+1}, x_{j+2}, \dots, x_{\mathcal{N}}, x_3, \dots, x_j, t) \\ & \qquad + \cdots + B_{\mathcal{N}}(x_1, x_{j+1}, x_{j+2}, \dots, x_{\mathcal{N}}, x_2, x_3, \dots, x_j, t)) \\ = & \sum_{w\in W((3,\dots,j)\otimes(j+1,\dots,\mathcal{N}))} B_{\mathcal{N}}(x_1, x_2, x_{w_2}, \dots, x_{w_{\mathcal{N}-2}}, t) \\ & + \sum_{w\in W((3,\dots,j)\otimes(j+2,\dots,\mathcal{N}))} B_{\mathcal{N}}(x_1, x_{j+1}, x_2, x_{w_2}, \dots, x_{w_{\mathcal{N}-2}}, t) \\ & + B_{\mathcal{N}}(x_1, x_{j+1}, \dots, x_{\mathcal{N}}, x_2, x_3, \dots, x_j, t) \\ = & \sum_{w\in W((2,\dots,j)\otimes(j+1,\dots,\mathcal{N}))} B_{\mathcal{N}}(x_{w_1}, x_{w_2}, \dots, x_{w_{\mathcal{N}}}, t)\,. \end{aligned}$$

We have proved equality (3.22) by induction.

Now we are able to prove the second formula for the symbol $Q_{\mathcal{N}}(\check{y}, t)$:

$$\begin{aligned} Q_{\mathcal{N}}(\check{y}, t_0) &= e^{-K(p)t_0} \sum_{w\in W(g), g\in\mathcal{G}_{\mathcal{N}}} E_w(t_0), \\ E_w(t) &= B_{\mathcal{N}}(L_{w_1}, L_{w_2}, \dots, L_{w_{\mathcal{N}}}, t)\,. \end{aligned} \tag{3.26}$$

In view of (3.12) it suffices to prove that for any fixed graph $g \in \mathcal{G}_{\mathcal{N}}$

$$\int_0^\infty \cdots \int_0^\infty dt_1 \cdots dt_{\mathcal{N}} \prod_{j-1}^{\mathcal{N}} e^{L_j t_j} \theta(t_{\xi_j} - t_j) = \sum_{w\in W(g)} E_w(t_0)\,. \tag{3.27}$$

For $\mathcal{N} = 0$ and $\mathcal{N} = 1$ equality (3.27) can be verified trivially. For $\mathcal{N} > 1$ we argue by induction. We cut the graph $g \in \mathscr{G}_{\mathscr{N}}$ and obtain $g = g^1 \otimes g^2$, $g^1 \in \mathscr{G}_{j-1}$, $g^1 \in \mathscr{G}_{j-1,\mathscr{N}}$; then on the left in (3.27), in view of (3.21) and (3.22), we obtain the integral:

$$\begin{aligned} &\int_z^{t_0} dt_1 e^{L_1 t_1} \left(\int_0^\infty \cdots \int_0^\infty dt_2 \cdots dt_j \prod_{k=2}^{j} e^{L_k t_k} \theta(t_{\xi_k} - t_k \right) \\ &\quad \times \left(\int_0^\infty \cdots \int_0^\infty dt_{j+1} \cdots dt_{\mathscr{N}} \prod_{k=j+1}^{\mathscr{N}} e^{L_k t_k} \theta(t_{\xi_k} - t_k) \right) \\ &= \int_0^{t_0} dt_1 e^{L_1 t_1} \sum_{w^1 \in W(g^1)} E_{w^1}(t_1) \sum_{w^2 \in W(g^2)} E_{w^2}(t_1) \\ &= \sum_{w \in W(w^1 \otimes w^2), w^1 \in W(g^1), w^2 \in W(g^2)} E_w(t_0) = \sum_{w \in W(g)} E_w(t_0) \,. \end{aligned}$$

Thus, (3.27) and with it also (3.26) are established by induction.

3.4. A formula for $\Phi_{\mathscr{N}}(p)$. Substituting (3.26) in (3.1) and passing to the limit as $t \to \infty$, we obtain

$$\Phi_{\mathscr{N}}(p) = \sum_{w \in W(g), g \in \mathscr{G}_{\mathscr{N}}} \int_{-\infty}^{\infty} dy\, E_w(\infty) V_{\mathscr{N}}(\check{y}) \,. \tag{3.28}$$

We now show that for $|p| \leq \frac{1}{2}$

$$E_w(\infty) = (-1)^{\mathscr{N}} \prod_{k=1}^{\mathscr{N}} \left(\sum_{j=1}^{k} L_{w_j} \right)^{-1} , \tag{3.29}$$

(and $w_1 = 1$ always). For this we first establish the following inequality, $k \geq 0$ is any integer:

$$\left| \frac{d^k}{dx_1^k} B_{\mathscr{N}}(x_1, \ldots, x_{\mathscr{N}}, t) \right| \leq c_{\mathscr{N},k} t^{\mathscr{N}+k} \left(1 + \sum_{j=1}^{\mathscr{N}} \exp \left(\sum_{l=1}^{j} \operatorname{Re} x_e t \right) \right) , \tag{3.30}$$

where $c_{\mathscr{N},k} > 0$ are some constants. Applying the recurrent relation (3.17), we obtain

(3.31)

$$\begin{aligned} &\left| \frac{d^k}{dx_1^k} B_{\mathscr{N}}(x_1, \ldots, x_{\mathscr{N}}, t) \right| \\ &\quad \leq \left| \left(\frac{e^{x_1 t} - 1}{x_1} \right)^{(k)} B_{\mathscr{N}-1}(x_2, \ldots, x_{\mathscr{N}}, t) \right. \\ &\qquad \left. - (x_1^{-1}(B_{\mathscr{N}-1}(x_1 + x_2, x_3, \ldots, x_{\mathscr{N}}, t) - B_{\mathscr{N}-1}(x_2, x_3, \ldots, x_{\mathscr{N}}, t)))^{(k)} \right| . \end{aligned}$$

For any function $f(x) \in C^\infty(R_1)$ the following estimate holds:

$$\left| \left(\frac{f(x) - f(0)}{x} \right)^{(k)} \right| \leq C_k \max_{|\xi| \leq |x|} |f^{(k+1)}(\xi)|, \tag{3.32}$$

where $c_k > 0$ are some constants. In view of (3.32), from (3.31) we obtain

$$\begin{aligned}\left|\frac{d^k}{dx_1^k}B_{\mathcal{N}}\right| &\le t^{k+1}c_{\mathcal{N}-1,k+1}t^{\mathcal{N}-1}\left(1+\sum_{j=1}^{\mathcal{N}}\exp\left(\sum_{l=1}^{j}\operatorname{Re}x_l t\right)\right) + \max_{|x_1|\le 1/t}\left|\frac{d^{k+1}}{dx_1^{k+1}}B_{\mathcal{N}-1}\right| \\ &\le c_{\mathcal{N},k}t^{\mathcal{N}+k}\left(1+\sum_{j=1}^{\mathcal{N}}\exp\left(\sum_{l=1}^{j}\operatorname{Re}x_l t\right)\right).\end{aligned}$$

Fix any chain w and consider some initial segment $(w_1, w_2, \dots, w_k)$ of it. For any vertex h_{w_j}, $2 \le j \le k$, in accordance with the construction of the chain w, by the subordination of the vertices there exists a vertex $h_{\xi_{w_j}}$ also belonging to this part $(w_1, \dots, w_k)$ of the chain w. For the sum $L_{w_j} + L_{w_{\xi_j}}$, in this case we find that

$$\begin{aligned}L_{w_j} + L_{w_{\xi_j}} &= K(s^1_{\xi_{w_j}} + s^2_{\xi_{w_j}}) - K(s^1_{\xi_{w_j}}) - K(s^2_{\xi_{w_j}}) + K(s^1_{\xi_{w_j}} + s^2_{\xi_{w_j}}) - K(s^1_{w_j}) - K(s^2_{w_j}) \\ &= K(s^1_{\xi_{w_j}} + s^2_{\xi_{w_j}}) - K(z_1) - K(s^1_{w_j}) - K(s^2_{w_j}),\end{aligned}$$

where

$$z_1 = \begin{cases} s^2_{\xi_{w_j}}, & \text{if } s^1_{w_j} + s^2_{2_j} = s^1_{\xi_{w_j}}, \\ s^1_{\xi_{w_j}}, & \text{if } s^1_{w_j} + s^2_{w_j} = s^2_{\xi_{w_j}}. \end{cases}$$

Considering in this manner all vertices of the sequence $(w_1, w_2, \dots, w_k)$, we obtain

$$L_{w_1} + \cdots + L_{w_k} = K(p) - K(z_1) = \cdots = K(z_k) - K(p - z_1 - \cdots - z_k), \tag{3.33}$$

where $z_1, \dots, z_k$ coincide with certain of the sums s^i_j, $i = 1, 2$, $j = 1, \dots, \mathcal{N}$. In view of the condition 1) of the theorem for $|p| \le 1/2$, for all k, $1 \le k \le \mathcal{N}$, from (3.33) we have

$$\sum_{j=1}^{k}\operatorname{Re}L_{w_j} \le (\tfrac{1}{2})^\delta - k < 0, \quad \forall k \ge 1. \tag{3.34}$$

Formula (3.29) follows from (3.30), (3.34), and the recurrent relation (3.17). We conclude from (3.29) and (3.34) that

$$|E_w(\infty)| \le c^{\mathcal{N}}|\mathcal{N}|!. \tag{3.35}$$

3.5. Bounds for $\Phi_{\mathcal{N}}(p)$. We shall prove the inequalities (2.11). Substituting (3.29) in (3.28) we get

$$\Phi_{\mathcal{N}}(p) = \sum_{w\in W(g),\ g\in\mathcal{G}_{\mathcal{N}}} (-1)^{\mathcal{N}}\int_{-\infty}^{\infty} dy \prod_{k=1}^{\mathcal{N}}\left(\sum_{j=1}^{k}L_{w_j}\right)^{-1} V_{\mathcal{N}}(\check{y}). \tag{3.36}$$

The first of the inequalities (2.11) simply follows from (3.11), (3.20), (3.35), and the condition 2) of the theorem. Consider the difference $\Phi_{\mathscr{N}}(p) - \Phi_{\mathscr{N}}(p)$. The result of substituting $p = 0$ will be denoted by a tilde.

$$(3.37)\quad |\Phi_{\mathscr{N}} - \tilde{\Phi}_{\mathscr{N}}| \leq \sum_{w \in W(g),\, g \in \mathscr{G}_{\mathscr{N}}} \int_{-\infty}^{\infty} dy(|E_w(\infty) - \widetilde{E}_w(\infty)|\, |V_{\mathscr{N}}(\check{y})| + |\widetilde{E}_w(\infty)|\, |V_{\mathscr{N}}(\check{y}) - \widetilde{V}_{\mathscr{N}}(\check{y})|) .$$

For the first difference in (3.37) we have

$$|E_w(\infty) - \widetilde{E}_w(\infty)| \leq \sum_{k=1}^{\mathscr{N}} \prod_{j=1,\ j \neq k}^{\mathscr{N}} \left| \sum_{l=1}^{j} L_{w_l} \right|^{-1} \left| \left(\sum_{l=1}^{k} \widetilde{L}_{w_l} \right)^{-1} - \left(\sum_{l=1}^{k} L_{w_l} \right)^{-1} \right|$$
$$\leq \frac{c^{\mathscr{N}}}{\mathscr{N}!} 2\mathscr{N} |p|^{\sigma_1} Y_w^{\alpha}, \quad Y_w = \prod_{j=0}^{\mathscr{N}} M(y_j), \quad c > 0,$$

and for the second

$$|V_{\mathscr{N}}(\check{y}) - \widetilde{V}_{\mathscr{N}}(\check{y})| \leq \prod_{j=1}^{\mathscr{N}} |\overline{v}(y_j)|\, |\overline{v}(y_0) - \overline{v}(\widetilde{y}_0)| \leq y_{\overline{w}}^{-2} |p|^{\sigma_2} c^{\mathscr{N}} .$$

Hence, in view of (3.35), (3.11), (3.20), and the condition 2) of the theorem, we arrive at

$$|\Phi_{\mathscr{N}}(p) - \tilde{\Phi}_{\mathscr{N}}| \leq c^{\mathscr{N}} |p|^{\sigma_3},$$

where $c > 0$ is a constant independent of $\mathscr{N}$.

Thus, the bounds (2.11) are established and the formulas (3.36) are found for the computation of the numbers $\Phi_{\mathscr{N}}(0)$ in terms of the initial condition $\hat{\overline{u}}(p)$ and the symbol $K(p)$ are presented. The theorem is proved. In particular, we write

$$\Phi_0(0) = \frac{1}{\varepsilon} \int_{-\infty}^{\infty} \overline{u}(y)\, dy,$$
$$\Phi_1(0) = -\frac{1}{\varepsilon^2} \int_{-\infty}^{\infty} \frac{\hat{\overline{u}}(y) \hat{\overline{u}}(-y)\, dy}{K(0) - K(y) - K(-y)},$$
$$\Phi_2(0) = \frac{2}{\varepsilon^3} \iint_{-\infty}^{\infty} dy_1\, dy_2 \times \frac{\hat{\overline{u}}(y_1) \hat{\overline{u}}(y_2) \hat{\overline{u}}(-y_1 - y_2)}{(K(0) - K(-y_1) - K(y_1))(K(0) - K(y_1) - K(y_2) - K(y_1 - y_2))} .$$

CHAPTER 7

Asymptotics of Solutions of the Whitham Equation for Large Times

§1. Introduction

In this chapter we consider the behavior, as $t \to \infty$, of a solution of $u(x,t)$ of the Cauchy problem for the nonlinear nonlocal Whitham equation.(*)

$$(1.1)\qquad u_t + u_x^2 + \mathbb{K}(u) = 0, \quad u|_{t=0} = \overline{u}(x), \qquad x \in R_1,\ t \geq 0,\ \overline{u}(x) \in L_1(R_1),$$

where

$$\mathbb{K}(u) = \frac{1}{2\pi}\int_{-\infty}^{\infty} e^{ipx} K(p)\hat{u}(p,t)\,dp, \qquad \hat{u}(p,t) = \int_{-\infty}^{\infty} e^{-ipx}u(x,t)\,dx,$$

is the Fourier transform.

By now the asymptotics of solutions for the Korteweg-de Vries and Burgers equations, as $t \to \infty$, has been basically constructed (see **[1, 18, 8]**). However, for such an important particular case of the Whitham equation as the Korteweg-de Vries-Burgers equation

$$(1.2)\qquad u_t + u_x^2 + u_{xxx} - u_{xx} = 0$$

the asymptotics, as $t \to \infty$, so far has not been found. (We should note in this regard the interesting work **[91]**, where asymptotic values of various norms of solutions of the Korteweg-de Vries-Burgers equation and of the regularized equation for long waves were considered.) The answer to this question is contained, as a particular case, in Theorem 1 proved below.

The physical setting of this problem presupposes that $u(x,t)$ is a real function. Therefore, the symbol $K(p)$ should satisfy the condition $K(p) = K^*(-p)$, whence it follows that $K^1(p) = \operatorname{Re} K(p)$ is an even function and $K^2(p) = \operatorname{Im} K(p)$ is an odd function.

In the case when the operator $\mathbb{K}$ is dissipative and the initial perturbation vanishes at infinity, the solution of the Cauchy problem (1.1) sufficiently fast stabilizes at zero and this allows us to calculate the asymptotics as $t \to \infty$. Our main goal in this chapter is to prove the following theorem.

(*) The standard Whitham equation has been integrated once with respect to x.

THEOREM 1. *Suppose that*

1) *The symbol $K(p)$ satisfies the following conditions*:

$$K^1(p) = \lambda + w|p|^\delta, \qquad |K^2(p)| \leq c_1|p|^\beta, \tag{1.3}$$

$$\lambda \geq 0,\ 0 < \delta < 3,\ w > 0,\ \delta < \beta,\ |p| \leq 1;$$

$$K^1(p) \geq \lambda + c_2 m^\delta(p) M(p) \equiv \lambda + 2\kappa(p), \qquad p \in R_1; \tag{1.4}$$

$$|K(p_1) - K(p_2)| \leq c_3|\Delta p|^{\sigma_1} M^\alpha(p_1), \qquad |\Delta p| = |p_1 - p_2| \leq 1,\ p_1, p_2 \in R_1, \tag{1.5}$$

$$\sigma_1 \in (0,1],\ \alpha \geq 0,\ m(p) \equiv \min(1,|p|),\ M(p) \equiv \max(1,|p|);$$

2) *The following bounds hold for the Fourier transform $\hat{\overline{u}}(p)$ of the initial condition $\overline{u}(x) \in L_1(R_1)$*:

$$\begin{aligned} &|\hat{\overline{u}}(p)| \leq \varepsilon M^{-3-\alpha}(p), \qquad |\hat{\overline{u}}(p_1) - \hat{\overline{u}}(p_2)| \leq c_4|\Delta p|^{\sigma_2} M^{-3}(p_1), \\ &|\Delta p| \leq 1,\ p_1, p_2 \in R_1,\ \sigma_2 \in (0,1],\ c_1, c_2, c_3, c_4 > 0\,. \end{aligned} \tag{1.6}$$

Here $0 < \varepsilon < c$, where c is some positive constant determined by the operator $\mathbb{K}$ (see formula (2.3) *below).*

Then for a solution $u(x,t)$ of problem (1.1) *the following asymptotics holds, as $t \to +\infty$*:

$$u(x,t) = Ae^{-\lambda t} t^{-1/\delta} \int_0^\infty \cos(p\xi) e^{-wp^\delta}\, dp + O(e^{-\lambda t} t^{-1/\delta - \mu}), \tag{1.7}$$

uniformly with respect to $\xi = |x|t^{-1/\delta} \geq 0$. Here $mu > 0$ is some constant determined by the symbol $K(p)$, and the constant A is fully expressed through the symbol $K(p)$ and the initial condition $\overline{u}(x)$ and will be indicated in the proof (see formulae (3.24) *and* (4.36)).

REMARKS 1. Condition (1.4), the nonnegativity of the real part $K^1(p)$ of the symbol $K(p)$ for all $p \in R_1$, means that the operator $\mathbb{K}$ is dissipative.

2. The Korteweg-de Vries-Burgers equation satisfies the condition of the theorem with $\lambda = 0,\ \delta = 2,\ \beta = 3,\ \mu = \frac{1}{2}$.

3. As $\xi \to \infty$, the first term in formula (1.7) approaches zero and might not be the leading term of the asymptotics, since the remainder (the second term on the right in (1.7)) is independent of ξ.

In the last section, §5, we consider the asymptotics, as $t \to \infty$, of solutions of the Ott-Sudan-Ostrovskiĭ equation which is of great interest to physicists.

The results of this chapter were published in **[57]**, **[59]**, and **[189]**.

§2. Technical lemmas

We preface the proof of Theorem 1 with two lemmas. Above we introduced the functions $m(p)$ and $M(p)$, for which the following obvious inequalities hold, $a, b \geq 0$:

$$m(a)m(b) \leq m(ab) \leq M(a)m(b) \leq M(ab) \leq M(a)M(b), \tag{2.1}$$

for $p \in R_1,\ t \geq 0,\ \chi > \delta + 3,\ \gamma \in [0,1]$: we introduce the following notation:

$$H_{\mathcal{N}}(p,t) = (\mathcal{N}+1)^{-\chi} M^{-3}(p) \exp\left(-\lambda t - t\kappa\left(\frac{P}{\mathcal{N}+1}\right)\right), \qquad \mathcal{N} \geq 0,$$

$$V_{\mathcal{N}}(p,t,\gamma) = \int_0^t e^{-(\lambda+2\kappa(p))(t-\tau)} \times \int_{-\infty}^{\infty} \sum_{j=1}^{\mathcal{N}} |p-q|^{1-\delta} |q| H_{j-1}(p-q,\tau) H_{\mathcal{N}-j}(q,\tau)\, dq\, d\tau.$$

LEMMA 1. *If $K^1(p)$ satisfies the condition* (1.4) *with $0 < \delta < 3-\gamma$, then*

$$V_{\mathcal{N}}(p,t,\gamma) \leq c H_{\mathcal{N}}(p,t), \tag{2.2}$$

where $c > 0$ is a constant independent of $\mathcal{N}$; we may take

$$c = 2^{14+\chi}(3-\delta-\gamma)^{-1}/c_2\,. \tag{2.3}$$

PROOF. Taking into consideration the following inequalities ($p, q \in R_1$, $\mathcal{N} \geq 1$, $1 \leq j \leq \mathcal{N}$):

$$\begin{gathered} M^{-3}(p-q) M^{-3}(q) \leq 2^3 M^{-3}(p)(M^{-3}(q) + M^{-3}(p-q)), \\ j^{-\chi}(\mathcal{N}+1-j)^{-\chi} \leq 2^{\chi}(\mathcal{N}+1)^{-\chi}(j^{-\chi} + (\mathcal{N}+1-j)^{-\chi}), \end{gathered} \tag{2.4}$$

we obtain

$$\begin{aligned} V_{\mathcal{N}} \leq{}& 2^{3+\chi} M^{-3}(p)(\mathcal{N}+1)^{-\chi} e^{-\lambda t} \sum_{j=1}^{\mathcal{N}} ((\mathcal{N}+1-j)^{-\chi} + j^{-\chi}) \\ & \cdot \int_{-\infty}^{\infty} dq |p-q|^{1-\gamma} |q| (M^{-3}(q) + M^{-3}(p-q)) \int_0^t e^{-2\kappa(p)t + \mathcal{L}(p,q)\tau}\, d\tau, \end{aligned} \tag{2.5}$$

where

$$\mathcal{L}(p,q) = 2\kappa(p) - \kappa\left(\frac{p-q}{j}\right) - \kappa\left(\frac{q}{\mathcal{N}+1-j}\right).$$

We estimate a typical summand on the right hand side of (2.5) (the others are considered similarly):

$$\begin{aligned} I_1 ={}& 2^{3+\chi} e^{-\lambda t} M^{-3}(p)(\mathcal{N}+1)^{-\chi} \sum_{j=1}^{\mathcal{N}} (\mathcal{N}+1-j)^{-\chi} \\ & \cdot \int_{-\infty}^{\infty} dq |p-q|^{1-\gamma} |q| M^{-3}(q) \int_0^t e^{-2\kappa(p)t + \mathcal{L}(p,q)\tau}\, d\tau\,. \end{aligned} \tag{2.6}$$

The monotonicity of $\kappa(p)$ implies that for all $p, q \in R_1$, $\mathcal{N} \geq 1$, $1 \leq j \leq \mathcal{N}$,

$$\kappa\left(\frac{p-q}{\mathcal{N}+1-j}\right) + \kappa\left(\frac{q}{j}\right) \geq \kappa\left(\frac{p}{\mathcal{N}+1}\right). \tag{2.7}$$

Taking into account (2.7), we obtain

$$\int_0^t e^{-2\kappa t + \mathcal{L}\tau}\, d\tau \leq \frac{e^{-\kappa(p/(\mathcal{N}+1))t}}{|\mathcal{L}|}.$$

On the other hand, if $|\mathscr{L}| \le \kappa(p)$, then for $\mathscr{N} \ge 1$

$$\int_0^t e^{-2\kappa t+\mathscr{L}\tau}\,d\tau \le \kappa^{-1}(p)e^{-t\kappa(p/(\mathscr{N}+1))};$$

i.e., the following is always true:

$$\int_0^t e^{-2\kappa(p)t+\mathscr{L}(p,q)\tau}\,d\tau \le \frac{e^{-t\kappa(p/(\mathscr{N}+1))}}{\max(\kappa(p),|\mathscr{L}(p,q)|)}. \tag{2.8}$$

Observe now that

$$\begin{aligned}\max(\kappa,|\mathscr{L}|) &\ge \max\left(\kappa,\frac{1}{4}\left(\kappa\left(\frac{p-q}{j}\right)+\kappa\left(\frac{q}{\mathscr{N}+1-j}\right)\right)\right)\\ &\ge \frac{1}{4}\max\left(\kappa(p),\kappa\left(\frac{q}{\mathscr{N}+1-j}\right)\right)\\ &\ge c_2\frac{\max(m^\delta(p),m^\delta(q))\max(1,|p|,|q|)}{4(\mathscr{N}+1-j)^{\delta+1}}.\end{aligned} \tag{2.9}$$

Substituting (2.8) and (2.9) in (2.6) we get

$$\begin{aligned}I_1 &\le c_2^{-1}2^{5+\chi}H_{\mathscr{N}}(p,t)\sum_{j=1}^{\mathscr{N}}(\mathscr{N}+1-j)^{1+\delta-\chi}\\ &\quad\cdot\int_{-\infty}^{\infty}\frac{|p-q|^{1-\gamma}|q|M^{-3}(q)\,dq}{\max(m^\delta(p),m^\delta(q))\max(1,|p|,|q|)}\\ &\le c_2^{-1}2^{7+\chi}H_{\mathscr{N}}\int_{-\infty}^{\infty}m^{2-\gamma-\delta}(q)M^{-2}(q)\,dq \le C_2^{-1}2^{9+\chi}(3-\gamma-\delta)^{-1}H_{\mathscr{N}}(p,t),\end{aligned}$$

since

$$|p-q|^{1-\gamma} \le |p|^{1-\gamma}+|q|^{1-\gamma} \le 2\max(1,|p|,|q|)\max(m^{1-\gamma}(q),m^{1-\gamma}(p)).$$

Lemma 1 is proved. □

Let

$$I=\int_{-\infty}^{\infty}M^{-2}(q)m^{1-\gamma_1}(p-q)m^{1+\gamma_1-\gamma_2}(q)\int_0^t e^{K^1(p)\tau-\lambda\tau-\kappa(qa)\tau}\,d\tau\,dq,$$

$$0<a\le 1,\ |p|\le 1,\ \gamma_1,\gamma_2\in[0,1],\ \gamma_1\ge\gamma_2.$$

LEMMA 2. *Under the conditions of Lemma* 1 *we have*

$$I \le ce^{K^1(p)t-\lambda t}a^{-2-\delta-\gamma_3}|p|^{-1-\gamma_3}t^{-1/\delta-\gamma_3/\delta}, \tag{2.10}$$

where $0<\gamma_3\le m(3-\delta-\gamma_2)$, *and* $c>0$ *is some constant.*

PROOF. We split the integral I into two integrals over the domains $|q| \leq |p|$ and $|q| > |p|$. In the first domain we have

$$\int_0^t e^{K^1(p)\tau - \lambda\tau - \kappa(qa)\tau}\, d\tau \leq \kappa^{-1}(p) \exp(K^1(p)t - \lambda t - \kappa(qa)t),$$

whence, accounting for (1.4), we find

$$\begin{aligned} I_1 &\leq \int_{|q|\leq|p|} m^{1-\gamma_1}(p-q) m^{1+\gamma_1-\gamma_2}(q) \kappa^{-1}(p) e^{K^1(p)t - \lambda t - \kappa(qa)t}\, dq \\ &\leq 2(c_2|p|)^{-1} e^{K^1(p)t-\lambda t} \int_{|q|\leq|p|} |q|^{\gamma_3} e^{-c_2|q|^\delta a^\delta t}\, dq \qquad (2.11) \\ &\leq c|p|^{-1} a^{-1-\gamma_3} t^{-1/\delta - \gamma_3/\delta} \exp(K^1(p)t - \lambda t)\,. \end{aligned}$$

In the case $|q| > |p|$, we obtain

$$\begin{aligned} \int_0^t e^{K^1(p)\tau - \lambda r - \kappa(qa)\tau}\, d\tau &\leq \exp\left(K^1(p)t - \lambda t - \frac{t}{2}\kappa(pa)\right) \int_0^t e^{\kappa(pa)\tau/2 - \kappa(qa)\tau}\, d\tau \\ &\leq 2\kappa^{-1}(qa) \exp\left(K^1(p)t - \lambda t - \frac{t}{2}\kappa(pa)\right). \end{aligned}$$

Therefore, from the inequality $|p-q| \leq 2|q|$ we find that

$$\begin{aligned} I_2 &\leq 2\exp\left(K^1(p)t - \lambda t - \frac{t}{2}\kappa(pa)\right) \\ &\quad \cdot \int_{|q|>|p|} m^{1-\gamma_1}(p-q) m^{1+\gamma_1-\gamma_2}(q) M^{-2}(q) \kappa^{-1}(qa)\, dq \qquad (2.12) \\ &\leq 4a^{-1-\delta} c_2^{-1} \exp\left(K^1(p)t - \lambda t - \frac{t}{2}\kappa(pa)\right) \int_{-\infty}^{\infty} M^{-3}(q) m^{\gamma_3 - 1}(q)\, dq \\ &\leq c a^{-\delta-2-\gamma_3} |p|^{-1-\gamma_3} t^{-1/\delta - \gamma_3/\delta} e^{K^1(p)t - \lambda t}. \end{aligned}$$

By (2.11) and (2.12) we arrive at (2.10). Lemma 2 is proved. □

§3. Proof of the theorem

3.1. Perturbation theory. It is known (see Chapter 2) that there exists a unique solution of the Cauchy problem (1.1) in the class $C^\infty((0,\infty); H^\infty(R_1))$. Applying the Fourier transform to (1.1), we obtain

$$\hat{u}_t(p,t) - \int_{-\infty}^{\infty} \hat{u}(p-q,t)\hat{u}(q,t) q(p-q)\, dq + K(p)\hat{u}(p,t) = 0,$$

$$\hat{u}|_{t=0} = \hat{\bar{u}}(p), \qquad p \in R_1,\ t \geq 0,$$

whence, integrating with respect to t we get

$$\hat{u}(p,t) = \hat{\bar{u}}(p) e^{-K(p)t} + \int_0^t e^{-K(p)(t-\tau)} \int_{-\infty}^{\infty} \hat{u}(p-q,\tau)\hat{u}(q,\tau) q(p-q)\, dq\, d\tau. \qquad (3.1)$$

We now use perturbation theory. Setting $\hat{\overline{u}}(p) = \varepsilon\overline{v}(p)$, we represent $\hat{u}(p,t)$ as a formal power series in ε:

$$\hat{u}(p,t) = \sum_{j=0}^{\infty} \varepsilon^{j+1} v_j(p,t). \tag{3.2}$$

Substituting (3.2) in (3.1) and comparing, as usual, the coefficients of like powers of ε, we get ($\mathcal{N} \geq 1$):

$$v_0(p,t) = \overline{v}(p)e^{-K(p)t}, \tag{3.3}$$

$$v_{\mathcal{N}}(p,t) \int_0^t e^{-K(p)(t-\tau)} \int_{-\infty}^{\infty} \sum_{j=1}^{\mathcal{N}} v_{j-1}(p-q,\tau)v_{\mathcal{N}-j}(q,\tau)q(p-q)\,dq\,d\tau. \tag{3.4}$$

We establish the following bound for $\mathcal{N} \geq 0$, $p \in R_1$, $t \geq 0$:

$$|v_{\mathcal{N}}(p,t)| \leq c^{\mathcal{N}} H_{\mathcal{N}}(p,t)M^{-\alpha}(p), \tag{3.5}$$

where $H_{\mathcal{N}}$ was defined in Lemma 1 with a constant $\chi = \delta + 4\beta/(\beta-\delta)$.

This inequality is easily verified for $\mathcal{N} = 0$. Further, arguing by induction, by Lemma 1 with consideration of (2.5), from (3.4) we obtain

$$\begin{aligned}|v_{\mathcal{N}}(p,t)| \leq \int_0^t e^{-K^1(p)(t-\tau)} \int_{-\infty}^{\infty} \sum_{j=1}^{\mathcal{N}} c^{\mathcal{N}-1}|p-q|\,|q|M^{-\alpha}(p-q)M^{-\alpha}(q).\\ \cdot H_{j-1}(p-q,\tau)H_{\mathcal{N}-j}(q,\tau)\,dq\,d\tau\\ \leq c^{\mathcal{N}-1}2^{\alpha+1}M^{-\alpha}(p)V_{\mathcal{N}}(p,t,0) \leq c^{\mathcal{N}} H_{\mathcal{N}}(p,t)M^{-\alpha}(p),\end{aligned}$$

where $c > 0$ is a constant independent of $\mathcal{N}$.

From the representation (3.2), inverting the Fourier transform, we find that

$$u(x,t) = \frac{1}{2\pi}\int_{-\infty}^{\infty} e^{ipx}\hat{u}(p,t)\,dp = \frac{1}{2\pi}\sum_{\mathcal{N}=0}^{\infty} \varepsilon^{\mathcal{N}+1}\psi_{\mathcal{N}}(x,t), \tag{3.6}$$

where

$$\psi_{\mathcal{N}}(x,t) = \int_{-\infty}^{\infty} e^{ipx}v_{\mathcal{N}}(p,t)\,dp. \tag{3.7}$$

The series in (3.6) has been integrated termwise, which is legitimate, because of the majorant convergence of the series (3.2) of perturbation theory. This follows directly from (3.5) if we choose $\varepsilon < 1/c$. Thus, the computation of the asymptotics of $u(x,t)$ is reduced to finding the asymptotics of the integrals $\psi_{\mathscr{N}}$. We represent $\psi_{\mathscr{N}}$ in the form $(\mathscr{N} \geq 0)$: $\psi_{\mathscr{N}} = \sum_{i=1}^{5} J^i_{\mathscr{N}}$,

$$
\begin{aligned}
J^1_{\mathscr{N}} &= \int_{|p|\geq 1} e^{ipx} v_{\mathscr{N}}(p,t)\,dp,\\
J^2_{\mathscr{N}} &= \int_{|p|\leq 1} e^{ipx-K^1(p)t}(v_{\mathscr{N}}(0,t)e^{\lambda t} - \Phi_{\mathscr{N}})\,dp,\\
J^3_{\mathscr{N}} &= \int_{|p|\leq 1} e^{ipx} v_{\mathscr{N}}(p,t)(1-e^{iK^2(p)t})\,dp,\\
J^4_{\mathscr{N}} &= \int_{|p|\leq 1} e^{ipx-K^1(p)t}\Phi_{\mathscr{N}}\,dp, \qquad \Phi_{\mathscr{N}} \equiv \lim_{t\to\infty} v_{\mathscr{N}}(0,t)e^{\lambda t},\\
J^5_{\mathscr{N}} &= \int_{|p|\leq 1} e^{ipx-K^1(p)t}(v_{\mathscr{N}}(p,t)e^{K(p)t} - v_{\mathscr{N}}(0,t)e^{\lambda t})\,dp\,.
\end{aligned}
\tag{3.8}
$$

3.2. The estimation of the integrals. Everywhere below $c > 0$ is a constant independent of $\mathscr{N} \geq 0$. Let us estimate $J^1_{\mathscr{N}}$. From (3.5) we obtain

$$
\begin{aligned}
|J^1_{\mathscr{N}}| &\leq \int_{|p|\geq 1} |v_{\mathscr{N}}(p,t)|\,dp \leq c^{\mathscr{N}} e^{-\lambda t}(\mathscr{N}+1)^{-\chi} \int_{|p|\geq 1} e^{-t\kappa(p/(\mathscr{N}+1))} M^{-3}(p)\,dp\\
&\leq c^{\mathscr{N}}(\mathscr{N}+1)^{-\chi} e^{-\lambda t - c_2 t(\mathscr{N}+1)^{-\delta}} \leq c^{\mathscr{N}} O(e^{-\lambda t} t^{-2/\delta}),
\end{aligned}
\tag{3.9}
$$

since $\chi \geq 4+\delta$. From (3.4), $\mathscr{N} \geq 1$, we find that

$$
\Phi_{\mathscr{N}} = \int_0^\infty e^{\lambda\tau}\,d\tau \int_{-\infty}^{\infty} \sum_{j=1}^{\mathscr{N}} v_{j-1}(-q,\tau)v_{\mathscr{N}-j}(q,\tau)(-q^2)\,dq,
$$

and also

$$
v_{\mathscr{N}}(0,t)e^{\lambda t} - \Phi_{\mathscr{N}} = \int_t^\infty e^{\lambda\tau}\,d\tau \int_{-\infty}^{\infty} \sum_{j=1}^{\mathscr{N}} v_{j-1}(-q,\tau)v_{\mathscr{N}-j}(q,\tau)q^2\,dq\,.
\tag{3.10}
$$

Substituting (3.5) in (3.10) we get

$$
\begin{aligned}
|v_{\mathscr{N}}(0,t)e^{\lambda t} - \Phi_{\mathscr{N}}| &\leq \int_t^\infty e^{\lambda\tau}\,d\tau \int_{-\infty}^{\infty} dq \sum_{j=1}^{\mathscr{N}} j^{-\chi}(\mathscr{N}+1-j)^{-\chi} M^{-6}(q)q^2\\
&\quad \cdot \exp\left(-\kappa\left(\frac{q}{j}\right) r - k\left(\frac{q}{\mathscr{N}+1-j}\right)\tau - 2\lambda\tau\right) c^{\mathscr{N}-1}\\
&\leq c^{\mathscr{N}-1}\sum_{j=1}^{\mathscr{N}} j^{-\chi}(\mathscr{N}+1-j)^{-\chi}\int_{-\infty}^{\infty} \frac{m^2(q)e^{-\kappa(q/j)t}}{M^4(q)\kappa(q/j)}\,dq\\
&\leq 3c^{\mathscr{N}-1}\mu^{\mu} m^{-2}(c_2)M^{-\mu}(t)\sum_{j=1}^{\mathscr{N}} j^{-\chi+\delta+\delta\mu}(\mathscr{N}+1-j)^{-\chi}\\
&\quad \cdot \int_{-\infty}^{\infty} m^{2-\delta-\delta\mu}(q)M^{-2}(q)\,dq\\
&\leq c^{\mathscr{N}} M^{-\mu}(t),
\end{aligned}
\tag{3.11}
$$

since $\chi \geq 4+\delta$, $3-\delta-\delta\mu > 0$, $\mu = \frac{1}{2\delta}\min(\delta, \sigma_2, (\beta-\delta/4, \sigma_1(3-\delta)/4)$, and for all $t \geq 0$ the bound

$$e^{-\kappa(q/j)t} \leq 3\mu^\mu M^{-\mu}(t) m^{-\mu}(\kappa(q/j))$$

holds. For $\mathcal{N}=0$ we have $v_0(0,t)e^{\lambda t} = \overline{v}(0) = \Phi_0$ and by (1.6) $|\Phi_0| \leq 1$, so that (3.11), taken for $t=0$, implies

$$|\Phi_{\mathcal{N}}| \leq c^{\mathcal{N}}, \qquad \mathcal{N} \geq 0. \tag{3.12}$$

From (3.11), in view of (1.3), we have also

$$|J^2_{\mathcal{N}}| \leq c^{\mathcal{N}} M^{-\mu}(t) \int_{|p|\leq 1} e^{-K^1(p)t}\, dp = c^{\mathcal{N}} O(e^{-\lambda t} t^{-\mu-1/\delta}). \tag{3.13}$$

In order to estimate $J^3_{\mathcal{N}}$ we split the domain of integration into two parts $t^{-\nu} \leq |p| \leq 1$ and $|p| \leq t^{-\nu}$, where $\nu = (\delta+\beta)/2\delta\beta \in (0, 1/\delta)$. The first term is simply estimated by (3.5):

$$\begin{aligned}
&\left| \int_{t^{-\nu}\leq |p| \leq 1} e^{ipx} v_{\mathcal{N}}(p,t)(e^{iK^2(p)t}-1)\, dp \right| \\
&\quad \leq 2c^{\mathcal{N}} \int_{t^{-\nu}\leq |p|\leq 1} (\mathcal{N}+1)^{-\chi} M^{-3}(p) e^{-\lambda t - t\kappa(p/(\mathcal{N}+1))}\, dp \\
&\quad \leq 4c^{\mathcal{N}} e^{-\lambda t}(\mathcal{N}+1)^{-\chi} \int_{t^{-\nu}}^{\infty} e^{-c_2 p^\delta t(\mathcal{N}+1)^{-\delta}} M^{-3}(p)\, dp \\
&\quad \leq \frac{8c^{\mathcal{N}}}{(\mathcal{N}+1)^\chi} \exp(-\lambda t - c_2 t^{-1\nu\delta}(\mathcal{N}+1)^{-\delta}) \\
&\quad = c^{\mathcal{N}} O(t^{-2/\delta}) e^{-\lambda t}(\mathcal{N}n+1)^{2/(1-\nu\delta)-\chi} \\
&\quad = c^{\mathcal{N}} O(e^{-\lambda t} t^{-1/\delta-\mu}), \qquad \text{since } \chi \geq \frac{2}{1-\nu\delta}.
\end{aligned} \tag{3.14}$$

We introduce a new variable of integration $\eta = tp^\delta$ in the second term and using (3.5), we obtain

$$\begin{aligned}
&\left| \int_{|p|\leq t^{-\nu}} e^{ipx} v_{\mathcal{N}}(p,t)(e^{iK^2(p)t}-1)\, dp \right| \\
&\quad \leq 2c^{\mathcal{N}} \int_{|p|\leq t^{-\nu}} e^{-\lambda t - c_2|p|^\delta t(\mathcal{N}+1)^{-\delta}} (\mathcal{N}+1)^{-\chi} |\exp(itK^2(p))-1|\, dp \\
&\quad \leq 4c^{\mathcal{N}}(\mathcal{N}+1)^{-\chi} e^{-\lambda t} \int_0^{t^{1-\nu\delta}} \left| \exp\left(itK^2\left(\left(\frac{\eta}{t}\right)^{1/\delta}\right)\right) - 1 \right| e^{-c_2\eta(\mathcal{N}+1)^{-\delta}} \left(\frac{\eta}{t}\right)^{1/\delta} \frac{d\eta}{\eta\delta} \\
&\quad \leq \frac{8c_1 e^{-\lambda t} c^{\mathcal{N}}}{\delta(\mathcal{N}+1)^\chi} t^{1-\nu\beta-1/\delta} \int_0^\infty e^{-c_2\eta(\mathcal{N}+1)^{-\delta}} \eta^{1/\delta-1}\, d\eta \\
&\quad \leq 8c_1 c^{\mathcal{N}} \delta^{-1} e^{-\lambda t} t^{-1/\delta-\mu} (\mathcal{N}+1)^{-\chi+1} \int_0^\infty e^{-c_2\xi} \xi^{1/\delta-1}\, d\xi \\
&\quad = c^{\mathcal{N}} O(e^{-\lambda t} t^{-1/\delta-\mu}),
\end{aligned} \tag{3.15}$$

since $\mu \le \nu\beta - 1$. It follows from (3.14) and (3.15) that

$$|J^3_{\mathscr{N}}| = c^{\mathscr{N}} O(e^{-\lambda t} t^{-1/\delta - \mu}). \tag{3.16}$$

3.3. The leading term of the asymptotics. It follows from (1.3) that

$$J^4_{\mathscr{N}} = \int_{|p|\le 1} e^{ipx - K^1(p)t} \Phi_{\mathscr{N}}\, dp = 2\Phi_{\mathscr{N}} e^{-\lambda t} \operatorname{Re} \int_0^\infty e^{ipx - wtp^\delta}\, dp + 2\Phi_{\mathscr{N}} O(e^{-\lambda t - wt}),$$

whence, in view of (3.12), after the change of variables $\eta = pt^{1/\delta}$ and $\xi = |x|t^{-1/\delta}$, we get

$$J^4_{\mathscr{N}} = 2\Phi_{\mathscr{N}} e^{-\lambda t} t^{-1/\delta} \int_0^\infty \cos(\eta\xi) e^{-w\eta^\delta}\, d\eta + c^{\mathscr{N}} O(e^{-\lambda t} t^{-2/\delta}). \tag{3.17}$$

3.4. Estimation of the Hölder term. In order to estimate the remaining term $J^5_{\mathscr{N}}$ we first establish the following inequality, $|p_2| \le |p_1|$;

$$|v_{\mathscr{N}}(p_1, t) - v_{\mathscr{N}}(p_2, t)| \le c^{\mathscr{N}} H_{\mathscr{N}}(p_2, t) O(|\Delta p|^{2\gamma} t^{2\gamma/\sigma_1}), \tag{3.18}$$

where $\gamma = \delta\mu$, $\Delta p = p_1 - p_2$, $|\Delta p| \le 1$.

For $\mathscr{N} = 0$ the bound (3.18) follows from (3.3), (1.5), and (1.6). Further, arguing by induction, in view of (3.4) we write the difference on the left in (3.18) as follows:

$$v_{\mathscr{N}}(p_1, t) - v_{\mathscr{N}}(p_2, t) = \sum_{j=1}^{5} I_j,$$

where

$$I_1 = \int_0^t d\tau (e^{-K(p_1)(t-\tau)} - e^{-K(p_2)(t-\tau)}) \cdot \int_{-\infty}^{\infty} \sum_{j=1}^{\mathscr{N}} v_{j-1}(p_2 - q_2, \tau) v_{\mathscr{N}-j}(q, \tau)(p_2 - q) q\, dq,$$

$$I_2 = \int_0^t e^{-K(p_1)(t-\tau)}\, d\tau \int_{|p_1 - q| \le |p_2 - q|} dq \cdot \sum_{j=1}^{\mathscr{N}} (v_{j-1}(p_1 - q_1\tau) - v_{j-1}(p_2 - q, \tau)) v_{\mathscr{N}-j}(q, \tau)(p_1 - q) q,$$

$$I_3 = \Delta p \int_0^t e^{-K(p_1)(t-\tau)}\, d\tau \int_{|p_1 - q| \le |p_2 - q|} dq \sum_{j=1}^{\mathscr{N}} v_{j-1}(p_2 - q, \tau) v_{\mathscr{N}-j}(q, \tau) q,$$

$$I_4 = \int_0^t e^{-K(p_1)(t-\tau)}\, d\tau \int_{|p_1 - q| > |p_2 - q|} dq \cdot \sum_{j=1}^{\mathscr{N}} (v_{j-1}(p_1 - q, \tau) - v_{j-1}(p_2, q, \tau)) v_{\mathscr{N}-j}(q, \tau)(p_2 - q) q,$$

$$I_5 = \Delta p \int_0^t e^{-K(p_1)(t-\tau)}\, d\tau \int_{|p_1 - q| > |p_2 - q|} dq \sum_{j=1}^{\mathscr{N}} v_{j-1}(p_1 - q, \tau) v_{\mathscr{N}-j}(q, \tau) q.$$

First of all, we note that the inequality $|e^{-z}-1| \leq 2m^{2\gamma/\sigma_1(z)}$, $\operatorname{Re} z \geq 0$, $2\gamma < \sigma_1$, and condition (1.5) imply that for $|p_2| \leq |p_1|$

$$
\begin{aligned}
&|e^{-K(p_1)(t-\tau)} - e^{-K(p_2)(t-\tau)}| \\
&\quad \leq 2c_3^{2\gamma/\sigma_1}|\Delta p|^{2\gamma} t^{2\gamma/\sigma_1/\sigma} e^{-2\kappa(p_2)(t-\tau)} M^{\alpha}(p_2) e^{-\lambda(t-\tau)}.
\end{aligned}
\tag{3.19}
$$

Using (3.5), (3.19), and the results of Lemma 1 we obtain

$$
\begin{aligned}
|I_1| \leq & 2c_3^{2\gamma/\sigma_1} t^{2\gamma/\sigma_1} |\Delta p|^{2\gamma} M^{\alpha}(p_2) \int_0^t d\tau e^{-(\lambda+2\kappa(p_2))(t-\tau)} \\
& \cdot \int_{-\infty}^{\infty} \sum_{j=1}^{\mathcal{N}} |v_{j-1}(p_2-q,\tau) v_{\mathcal{N}-j}(q,\tau)|\, |p_2-q|\, |q|\, dq \\
= & c^{\mathcal{N}-1} M^{\alpha}(p_2) O(t^{2\gamma/\sigma_1}|\Delta p|^{2\gamma}) \int_0^t d\tau e^{-(2\kappa(p_2)+\lambda)(t-\tau)} \\
& \cdot \int_{-\infty}^{\infty} \sum_{j=1}^{\mathcal{N}} |H_{j-1}(p_2-q,\tau) H_{\mathcal{N}-j}(q,\tau)|\, |p_2-q|\, |q| M^{-\alpha}(p_2-q) M^{-\alpha}(q)\, dq \\
= & c^{\mathcal{N}-1} O(t^{2\gamma/\sigma_1}|\Delta p|^{2\gamma}) U_{\mathcal{N}}(p_2,t,0) \\
= & c^{\mathcal{N}} O(t^{2\gamma/\sigma_1}|\Delta p|^{\gamma}) H_{\mathcal{N}}(p_2,t).
\end{aligned}
$$

Now, in view of (3.5) and the induction hypothesis we have

$$
\begin{aligned}
|I_2| \leq & \int_0^t e^{-K^1(p_1)(t-\tau)}\, d\tau \int_{-\infty}^{\infty} dq \sum_{j=1}^{\mathcal{N}} c^{\mathcal{N}-1} O(\tau^{2\gamma/\sigma_1}|\Delta p|^{2\gamma}) H_{j-1}(p_1-q,\tau) \\
& \cdot H_{\mathcal{N}-j}(q,\tau)|p_1-q|\, |q| \\
= & c^{\mathcal{N}} O(t^{2\gamma/\sigma_1}|\Delta p|^{2\gamma}) H_{\mathcal{N}}(p_1,t)\,.
\end{aligned}
$$

We note that $|\Delta p| \leq 2|\Delta p|^{2\gamma}|p_2-q|^{1-2\gamma}$, in the integral I_3, and, therefore, using Lemma 1, we find that

$$
\begin{aligned}
|I_3| \leq 2c^{\mathcal{N}-1}|\Delta p|^{2\gamma} & \int_0^t e^{-K^1(p_2)(t-\tau)}\, d\tau \int_{-\infty}^{\infty} dq \sum_{j=1}^{\mathcal{N}} H_{j-1}(p_2-q,\tau) \\
& \cdot H_{\mathcal{N}-j}(q,\tau)|p_2-q|^{1-2\gamma}|q| \\
\leq 2c^{\mathcal{N}-1}|\Delta p|^{2\gamma} & V_{\mathcal{N}}(p_2,t,2\gamma) \leq 2c^{\mathcal{N}}|\Delta p|^{2\gamma} H_{\mathcal{N}}(p_2,t),
\end{aligned}
$$

since $3-\delta-2\gamma > 0$. The integrals I_4 and I_5 can be estimated in the same way as I_2 and I_3, respectively. By the same token, the bound (3.18) is established by induction.

The bound (3.18) is a preliminary; it is a crude for large t. We now proceed to proving the following estimate for $|p| \leq 1$:

$$
|v_{\mathcal{N}}(p,t)e^{K(p)t} - v_{\mathcal{N}}(0,t)e^{\lambda t}| = c^{\mathcal{N}} e^{K^1(p)t} O(e^{-\lambda t} t^{-1/\delta-\mu}|p|^{\gamma-1}).
\tag{3.20}
$$

For $\mathcal{N}=0$ (3.20) follows from (1.6). For $\mathcal{N} \geq 1$ we use induction. In view of (3.4) we represent the left-hand side of (3.20) in the following form:

$$
v_{\mathcal{N}}(p,t)e^{K(p)t} - v_{\mathcal{N}}(0,t)e^{\lambda t} = \sum_{j=1}^{5} I_j,
$$

where

$$I_1 = \int_0^t (e^{K(p)\tau} - e^{\lambda\tau})\, d\tau \int_{-\infty}^{\infty} dq \sum_{j=1}^{\mathcal{N}} v_{j-1}(-q,\tau) v_{\mathcal{N}-j}(q,\tau)(-q^2),$$

$$I_2 = \int_0^t e^{K(p)\tau}\, d\tau \int_{|p-q|\le|q|} dq$$

$$\cdot \sum_{j=1}^{\mathcal{N}} (v_{j-1}(p-q,\tau) - v_{j-1}(-q,\tau)) v_{\mathcal{N}-j}(q,\tau)(p-q)q,$$

$$I_3 = p\int_0^t e^{K(p)\tau}\, d\tau \int_{|p-q|\le|q|} dq \sum_{j=1}^{\mathcal{N}} v_{j-1}(-q,\tau) v_{\mathcal{N}-j}(q,\tau) q,$$

$$I_4 = -\int_0^t e^{K(p)\tau}\, d\tau \int_{|p-q|>|q|} dq \sum_{j=1}^{\mathcal{N}} (v_{j-1}(p-q,\tau) - v_{j-1}(-q,\tau)) v_{\mathcal{N}-j}(q,\tau) q^2,$$

$$I_5 = p\int_0^t e^{K(p)\tau}\, d\tau \int_{|p-q|>|q|} dq \sum_{j=1}^{\mathcal{N}} v_{j-1}(p-q,\tau) v_{\mathcal{N}-j}(q,\tau) q\,.$$

We have

$$\tau^{2\gamma/\delta} e^{-\kappa(q/(\mathcal{N}+1-j))\tau} \le (\mathcal{N}+1-j)^{2\gamma} \kappa^{-2\gamma/\delta}(q)\,. \tag{3.21}$$

Since for $|p| \le 1$

$$|e^{K(p)\tau} - e^{\lambda\tau}| = O(|p|^{2\gamma} \tau^{2\gamma/\delta} e^{K^1(p)\tau}),$$

by (3.21) and (3.5) we obtain

$$|I_1| \le \int_0^t d\tau\, e^{K^1(p)\tau} O(|p|^{2\gamma}\tau^{2\gamma/\delta}) \int_{-\infty}^{\infty} dq \sum_{j=1}^{\mathcal{N}} c^{\mathcal{N}-1} H_{j-1}(-q,\tau) H_{\mathcal{N}-j}(q,\tau) q^2$$

$$= c^{\mathcal{N}-1} O(|p|^{2\gamma}) \sum_{j=1}^{\mathcal{N}} j^{-\chi} \int_0^t e^{K^1(p)\tau}\, d\tau \int_{-\infty}^{\infty} \frac{e^{-\kappa(q/j)\tau - 2\lambda\tau}}{M^6(q)\kappa^{2\gamma/\delta}(q)} q^2\, dq,$$

whence, applying Lemma 2 with $\gamma_1 = 1$, $\gamma_2 = 2\gamma$, $a = 1/j$, $\gamma_3 = \gamma$, we get

$$|I_1| = c^{\mathcal{N}} O(|p|^{\gamma-1} e^{-\lambda t} t^{-1/\delta-\mu}) e^{K^1(p)t}.$$

Substitution of (3.5) and (3.18) in I_2 yields

$$|I_2| = c^{\mathcal{N}-1} O(|p|^{2\gamma}) \int_0^t e^{K^1(p)\tau} \tau^{2\gamma/\delta_1}\, d\tau$$

$$\cdot \int_{|p-q|\le|q|} \sum_{j=1}^{\mathcal{N}} H_{j-1}(p-q,\tau) H_{\mathcal{N}-j}(q,\tau) |p-q|\, |q|,$$

whence, using (3.21) and the result of Lemma 2 with $\gamma_1 = \gamma_2 = 2\gamma\delta/\sigma_1$, $a = 1/(\mathcal{N} + 1 - j)$, $\gamma_3 = \gamma$, we obtain

$$\begin{aligned}|I_2| &= c^{\mathcal{N}-1} O(|p|^{2\gamma}) \sum_{j=1}^{\mathcal{N}} (\mathcal{N} + 1 - j)^{-\chi} \int_{-\infty}^{\infty} dq M^{-2}(q) m(q) m^{1-2\delta\gamma/\sigma_1}(p-q) \\ &\quad \cdot \int_0^t e^{K^1(p)\tau - \lambda\tau - \tau\kappa(q/\mathcal{N}+1-j)}\, d\tau \\ &= c^{\mathcal{N}} e^{K^1(p)t} O(|p|^{\gamma-1} t^{-1/\delta-\mu} e^{-\lambda t}).\end{aligned}$$

Now, since for $|p - q| \le |q|$ we have $|p| \le 2|p|^{2\gamma}|q|^{1-2\gamma}$,

$$\begin{aligned}|I_3| &\le 2|p|^{2\gamma} \int_0^t e^{K^1(p)\tau}\, d\tau \int_{|p-q|\le|q|} dq \sum_{j=1}^{\mathcal{N}} |q|^{2-2\gamma} \frac{c^{\mathcal{N}-1}}{j^\chi} e^{-\lambda\tau - \tau\kappa}\left(\frac{q}{j}\right) M^{-6}(q) \\ &\le 2c^{\mathcal{N}-1}|p|^{2\gamma} \sum_{j=1}^{\mathcal{N}} j^{-\chi} \int_{-\infty}^{\infty} dq\, m^{2-2\gamma}(q) M^{-2}(q) \int_0^t e^{K^1(p)\tau - \lambda\tau - \tau\kappa(q/j)}\, d\tau .\end{aligned}$$

From this, in view of Lemma 2 with $\gamma_1 = 1$, $\gamma_2 = 2\gamma$, $\gamma_3 = \gamma$, $a = 1/j$, we arrive at (3.20). The integrals I_4 and I_5 can be estimated analogously. The estimate (3.20) has been established.

From (3.20), in turn, it follows that

$$J^5_{\mathcal{N}} = c^{\mathcal{N}} \int_{|p|\le 1} O(t^{-1/\delta-\mu} e^{-\lambda t}) |p|^{\gamma-1}\, dp = c^{\mathcal{N}} O(t^{-1/\delta-\mu} e^{-\lambda t}). \tag{3.22}$$

3.5. The asymptotics. Substitution of (3.9), (3.13), (3.16), (3.17), and (3.22) in (3.8) yields

$$\psi_{\mathcal{N}} = 2\Phi_{\mathcal{N}} e^{-\lambda t} t^{-1/\delta} \int_0^\infty \cos(\eta\xi) e^{-w\eta^\delta}\, d\eta + c^{\mathcal{N}} O(e^{-\lambda t} t^{-1/\delta-\mu}). \tag{3.23}$$

From (3.23) and (3.7) we obtain (1.7) with the constant A equal to

$$A = \frac{1}{\pi} \sum_{j=0}^{\infty} \Phi_j \varepsilon^{j+1}. \tag{3.24}$$

It remains to find a method for computing the constants Φ_j in terms of the initial condition $\bar{u}(x)$.

§4. Computation of the numbers $\Phi_{\mathcal{N}}$

The arguments in this section almost literally coincide with the constructions in §3, Chapter 6. Here we only point out the necessary minor changes.

4.1. Perturbation theory. Formulae (6.3.1)(*) and (6.3.2) remain valid. The right-hand side of (6.3.3) is multiplied by $(p-r)r$:

$$\int_{-\infty}^{\infty} dy\, Q_{\mathcal{N}}(\check{y},t)V_{\mathcal{N}}(\check{y}) \tag{4.2}$$
$$= \int_0^t e^{-K(p)(t-\tau)}\,d\tau \sum_{j=1}^{\mathcal{N}} \int_{-\infty}^{\infty} dr \left(\int_{-\infty}^{\infty} dz\, Q_{j-1}(\check{z},\tau)V_{j-1}(\check{Z})\right) \cdot \left(\int_{-\infty}^{\infty} dq\, Q_{\mathcal{N}-j}(\check{q},\tau)V_{\mathcal{N}-j}(\check{q})\right)(p-r)r\,.$$

The following two formulae are multiplied from the right by $\sum_{k=0}^{j-1} y_k \sum_{k=j}^{\mathcal{N}} y_k \equiv w(y)$, so that, for instance, (6.3.5) takes the following form:

$$Q_{\mathcal{N}}(\check{y},t) = \int_0^t d\tau\, e^{-K(p)(t-\tau)} \sum_{j=1}^{\mathcal{N}} Q_{j-1}(y_0,\dots,y_{j-1},\tau)Q_{\mathcal{N}-j}(y_j,\dots,y_{\mathcal{N}},\tau)w(y)\,. \tag{4.5}$$

4.2. First formula for the symbol $Q_{\mathcal{N}}$. All the constructions up to formula (6.3.12) remain valid. Formula (6.3.12) is multiplied from the right by $\Lambda_g = \prod_{j=1}^{\mathcal{N}} s_j^1 s_j^2$:

$$Q_{\mathcal{N}}(\check{y},t_0) = e^{-K(p)t_0} \sum_{g\in G_{\mathcal{N}}} \int_0^{\infty}\cdots\int_0^{\infty} dt_1\cdots dt_{\mathcal{N}} \prod_{j=1}^{\mathcal{N}} e^{L_j t_j}\theta(t_{\xi_j}-t_j)\Lambda_g\,. \tag{4.12}$$

Indeed, in formulae (6.3.15) and (6.3.16) on the right a new multiplier $p^1p^2\Lambda_{g^1}\Lambda_{g^2}$ appears, namely:

$$Q_{\mathcal{N}}(\check{y},t_0) = \int_0^{t_0} dt_1\, e^{-K(p)(t_0-t_1)} \sum_{j=1}^{\mathcal{N}} e^{-K(p^1)t_1}p^1p^2 \tag{4.15}$$
$$\cdot\left(\sum_{g^1\in\mathcal{G}_{0,j-1}} \int_0^{\infty}\cdots\int_0^{\infty} dt_1^1\cdots dt_{j-1}^1 \prod_{K=1}^{j-1} e^{L_k^1 t_k^1}\theta(t_{\xi_K^1}^1-t_k^1)\Lambda_{g^1}\right)e^{-K(p^2)t_1}$$
$$\cdot\left(\sum_{g^2\in\mathcal{G}_{j,\mathcal{N}}} \int_0^{\infty}\cdots\int_0^{\infty} dt_1^2\cdots dt_{\mathcal{N}-j}^2 \prod_{k=1}^{\mathcal{N}-j} \theta(t_{\xi_k^2}^2-t_k^2)e^{L_k^2t_k^2}\Lambda_{g^2}\right).$$

$$Q_{\mathcal{N}}(\check{y},t_0) = e^{-K(p)t_0} \sum_{g\in\mathcal{G}_{\mathcal{N}}} \int_0^{\infty} dt_1\theta(t_0-t_1)e^{L(p^1,p^2)t_1}p^1p^2 \tag{4.16}$$
$$\cdot\left(\int_0^{\infty}\cdots\int_0^{\infty} dt_1^1\cdots dt_{j-1}^1 \prod_{k=1}^{j-1} \theta(t_{\xi_k^1}^1-t_k^1)e^{L_k^1t_k^1}\Lambda g^1\right)$$
$$\cdot\left(\int_0^{\infty}\cdots\int_0^{\infty} dt_1^2\cdots dt_{\mathcal{N}-j}^2 \prod_{k=1}^{\mathcal{N}-j} \theta(t_{\xi_k^2}^2-t_k^2)e^{L_k^2t_k^2}\Lambda_{g^2}\right).$$

(*) References like (6.3.k) mean the formulae (k) from §3, Chapter 6.

Further, by the choice of the special enumeration of vertices in the graph $g \in \mathcal{G}_{\mathcal{N}}$, the following equalities hold:

$$\Lambda_{g^1} = \prod_{k=2}^{j} s_k^1 s_k^2, \qquad \Lambda_{g^2} = \prod_{k=j+1}^{\mathcal{N}} s_k^1 s_k^2.$$

Therefore, the formula following after (6.3.16) becomes

$$Q_{\mathcal{N}}(\check{y}, t) = \sum_{g \in \mathcal{G}_{\mathcal{N}}} e^{-K(p)t} \int_0^\infty dt_1\, e^{L_1 t_1} \theta(t - t_1) s_1^1 s_1^2$$
$$\cdot \left(\int_0^\infty \cdots \int_0^\infty dt_2 \cdots dt_j \prod_{k=2}^{j} \theta(t_{\xi_k} - t_k) e^{L_k t_k} s_k^1 s_k^2 \right)$$
$$\cdot \left(\int_0^\infty \cdots \int_0^\infty dt_{j+1} \cdots dt_{\mathcal{N}} \prod_{k=j+1}^{\mathcal{N}} \theta(t_{\xi_k} - t_k) e^{L_k t_k} s_k^1 s_k^2 \right),$$

whence (4.12) follows.

4.3. Second formula for $Q_{\mathcal{N}}$. All the constructions up to formula (6.3.26) remain valid and (6.3.26) is multiplied from the right by $\Lambda_g = \prod_{k=1}^{\mathcal{N}} s_k^1 s_k^2$, i.e.,

$$Q_{\mathcal{N}}(\check{y}, t_0) = e^{-K(p)t_0} \sum_{w \in W(g),\ g \in \mathcal{G}_{\mathcal{N}}} E_w(t_0) \Lambda_g . \tag{4.26}$$

All the subsequent formulae before (6.3.28) are also valid here.

4.4. Formula for $\Phi_{\mathcal{N}}$. We note that in the case we are considering $p = \sum_{k=0}^{\mathcal{N}} y_k = 0$. The equality (6.3.28) is also multiplied from the right by Λ_g:

$$\Phi_{\mathcal{N}} = \sum_{w \in W(g),\ g \in \mathcal{G}_{\mathcal{N}}} \int_{-\infty}^{\infty} dy\, E_w(\infty) V_{\mathcal{N}}(\check{y}) \Lambda_g. \tag{4.28}$$

But formula (6.3.29) retains the same form, as we are about to demonstrate. The bounds (6.3.30) and (6.3.31) hold, but we do not use them here. Equality (6.3.33) is also still valid, only in our case $p = 0$ and therefore ($\lambda = K(0) \geq 0$):

$$L_{w_1} + \cdots + L_{w_k} = \lambda - K(z_1) - \cdots - K(z_k) - K(z_1 - z_2 - \cdots - z_k), \tag{4.33}$$

where $z_1, \ldots, z_k$ coincide with some of the sums $\{s_j^i\}$, $i = 1, 2$, $j = 1, \ldots, \mathcal{N}$. The bounds (6.3.34), (6.3.35) are insufficient and, therefore, to prove (4.28) we act as follows. For the functions $B_{\mathcal{N}}(x_1, \ldots, x_{\mathcal{N}}, t)$ we establish the following relation:

$$B_{\mathcal{N}} = \sum_{k=0}^{\mathcal{N}} (-1)^{\mathcal{N}-k} \exp\left(t \sum_{j=1}^{k} x_j \right) \left\{ \prod_{j=1}^{k} \left(\sum_{l=j}^{k} x_l \right) \prod_{j=k+1}^{\mathcal{N}} \left(\sum_{l=k+1}^{j} x_l \right) \right\}^{-1}, \tag{4.35}$$

assuming that $\prod_{j=1}^{0} = \prod_{j=\mathcal{N}+1}^{\mathcal{N}} = 1$. We applying the recurrent formula (6.3.17), and, then, by induction:

$$B_{\mathcal{N}} = \frac{1}{x_1}(e^{x_1 t} B_{\mathcal{N}-1}(x_2, x_3, \dots, x_{\mathcal{N}}, t) - B_{\mathcal{N}-1}(x_1 + x_2, x_3, \dots, x_{\mathcal{N}}, t))$$

$$= \frac{e^{x_1 t}}{x_1} \sum_{k=0}^{\mathcal{N}-1} (-1)^{\mathcal{N}-k-1} \exp\left(t \sum_{j=1}^{k} x_{j-1}\right) \left\{ \prod_{j=1}^{k} \left(\sum_{l=j}^{k} x_{l+1} \right) \prod_{j=k+1}^{\mathcal{N}} \left(\sum_{l=k+1}^{j} x_{l+1} \right) \right\}^{-1}$$

$$- \sum_{k=1}^{\mathcal{N}-1} \frac{(-1)^{\mathcal{N}-1-k}}{x_1} \exp\left(t \sum_{j=1}^{k+1} x_j\right)$$

$$\cdot \left\{ \prod_{j=2}^{k} \left(\sum_{l=j}^{k} x_{l+1} \right) (x_1 + x_2 + \dots + x_{k+1}) \prod_{j=k+1}^{\mathcal{N}-1} \left(\sum_{l=k+1}^{j} x_{l+1} \right) \right\}^{-1}$$

$$- \frac{(-1)^{\mathcal{N}-1}}{x_1} \left\{ \prod_{j=1}^{\mathcal{N}-1} \left(\sum_{l=1}^{j+1} x_l \right) \right\}^{-1}$$

$$= (-1)^{\mathcal{N}} \left\{ \prod_{j=1}^{\mathcal{N}} \left(\sum_{l=1}^{j} x_l \right) \right\}^{-1} + (-1)^{\mathcal{N}-1} e^{x_1 t} \left\{ x_1 \prod_{j=1}^{\mathcal{N}-1} \left(\sum_{l=1}^{j} x_{l+1} \right) \right\}^{-1}$$

$$+ \sum_{k=1}^{\mathcal{N}-1} (-1)^{\mathcal{N}-k-1} \exp\left(t \sum_{j=0}^{k} x_{j+1}\right) \left\{ x_1 \prod_{j=2}^{k} \left(\sum_{l=j}^{k} x_{l+1} \right) \prod_{j=k+1}^{\mathcal{N}-1} \left(\sum_{l=k+1}^{j} x_{l+1} \right) \right\}^{-1}$$

$$\cdot \left(\frac{1}{x_2 + \dots + x_{k+1}} - \frac{1}{x_1 + \dots + x_{k+1}} \right).$$

From this we readily arrive at (4.35). Formula (4.28) follows from (4.35). Indeed, (4.33) implies that $\sum_{j=1}^{k} L_{w_j} < 0$ almost everywhere (for $s_1^1 = -s_2^1$ not equal to zero) in the measure of $R_{\mathcal{N}}$, and as $t \to \infty$ the exponential in (4.34) tends to zero if $k \neq 0$. Therefore, as $t \to \infty$, only the terms with $k = 0$ remain nonzero in (4.35). This leads to (4.28).

We substitute (6.3.29) in (4.28). For $p = \sum_{k=0}^{\mathcal{N}} y_k = 0$ we obtain

$$\Phi_{\mathcal{N}} = \sum_{w \in W(g),\, g \in \mathcal{G}_{\mathcal{N}}} (-1)^{\mathcal{N}} \int_{-\infty}^{\infty} dy \prod_{k=1}^{\mathcal{N}} \left(\sum_{j=1}^{k} L_{w_j} \right)^{-1} \Lambda_g V_{\mathcal{N}}(\check{y}), \tag{4.36}$$

$$\Phi_0 = \overline{v}(0).$$

The theorem is proved. In particular, we note that

$$\begin{aligned}
\Phi_0(0) =& \frac{1}{\varepsilon}\hat{\bar{u}}(0) = \frac{1}{\varepsilon}\int_{-\infty}^{\infty} \bar{u}(x)\,dx, \\
\Phi_1(0) =& -\frac{1}{\varepsilon^2}\int_{-\infty}^{\infty} \frac{\hat{\bar{u}}(y)\hat{\bar{u}}(-y)y(-y)\,dy}{K(0)-K(y)-K(-y)}, \\
\Phi_2(0) =& \frac{2}{\varepsilon^2}\iint_{-\infty}^{\infty} \\
&\cdot \frac{\hat{\bar{u}}(y_1)\hat{\bar{u}}(y_2)\hat{\bar{u}}(-y_1-y_2)y_1y_2(-y_1-y_2)(-y_2)\,dy_1\,dy_2}{(K(0)-K(y_2)-K(-y_2))(K(0))-K(y_2)-K(y_1)-K(-y_1-y_2)}.
\end{aligned}$$

§5. Asymptotics of solutions of the KdV equation

5.1. This section is devoted to a study of the Ott-Sudan-Ostrovskiĭ equation [**152**], [**153**], which describes nonlinear acoustic effects in the mixture of a fluid and gas bubbles; it has the form

$$u_t + u_x^2 + u_{xxx} + \lambda u + \frac{w}{(2\pi)^{3/2}}\int_{-\infty}^{\infty} \frac{\operatorname{sgn}(x-y)}{\sqrt{|x-y|}} u_y(y,t)\,dy = 0, \tag{5.1}$$

where $w \geq 0$ and $\lambda \geq 0$ are constants. This equation is a particular case of (1.1). In the case of the Ott-Sudan-Ostrovskiĭ equation (5.1) the symbol $K(p)$ of the operator $\mathbb{K}(p)$ has the following form:

$$K(p) = \lambda + w\sqrt{|p|} - ip^3\,. \tag{5.2}$$

If $\lambda = 0$ and $w = 0$, then (5.1) reduces to the well-known Korteweg-de Vries equation. Here we analyze the case when the constants λ and w do not vanish simultaneously; i.e., a weak dissipation term is added to the Korteweg-de Vries equation.

In Theorem 1 the condition

$$K^1(p) \geq c|p| \quad \text{for } |p| \geq p_0 > 0,\ c > 0$$

is required to hold. For the Ott-Sudan-Ostrovskiĭ equation (5.1) this condition is not satisfied and a special approach is needed to study the asymptotic behavior of the solutions for large times.

The case $w = 0$, $\lambda > 0$, when equation (5.1) becomes the Korteweg-de Vries equation with linear dissipation, is of particular interest. Many studies have been recently devoted to this equation [**172–176**], where the coefficient $\lambda > 0$ is considered a small parameter; the theory of perturbations in λ has been applied, and the inverse scattering method for the Korteweg-de Vries equation has been used. In this approach only soliton initial conditions and small λ have been investigated. The case of sufficiently general initial perturbations and finite λ remains outside the scope of these studies. An essential drawback in all these studies is the absence of estimates for the remainder terms in the asymptotic formulas, so that the domain of applicability of the asymptotics obtained remains undetermined. For example, in [**173**] it is stated that the asymptotic formulas (78) on p. 2337 are applicable for all large t, which is certainly erroneous without indicating the connection between t and the small parameter λ (see Theorem 3 below).

In this section we prove the following two theorems.

THEOREM 2. *Suppose that*

1) *The symbol $K(p)$ of the operator $\mathbb{K}(u)$ has the form* (5.2) *with $\lambda \geq 0$, $w > 0$;*

2) *The Fourier-image $\hat{\bar{u}}(p)$ of the initial perturbation $\bar{u}(x)$ satisfies the estimates*

$$(5.3) \qquad |\hat{\bar{u}}(p)| \leq \varepsilon M^{-6}(p), \qquad \left|\frac{d\hat{\bar{u}}(p)}{dp}\right| \leq \varepsilon M^{-3}(p),$$

where $\varepsilon < c$ and $c > 0$ is a constant dependent on the symbol $K(p)$.

Then, as $t \to \infty$, we have the following asymptotics of the solution of the Cauchy problem for (5.1):

$$(5.4) \qquad u(x,t) = At^{-2}e^{-\lambda t}\int_0^\infty \cos(p\xi)e^{-w\sqrt{p}}\,dp + O(t^{-3}e^{-\lambda t}),$$

uniformly in $\xi = xt^{-2} \in R_1$.

The constant A is expressed in terms of the symbol $K(p)$ and the Fourier transform $\hat{\bar{u}}(p)$ of the initial perturbation in the same way as in §4 (see also (5.21) below).

In the case $w = 0$, the asymptotics of the solutions as $t \to \infty$ has an oscillatory character, and to calculate it using the saddle-point method, it is necessary to impose more restrictive conditions on the behavior of the initial perturbation in the limit $x \to +\infty$.

THEOREM 3. *Suppose that*

1) *The symbol $K(p)$ has the form* (5.2) *with $\lambda > 0$ and $w = 0$, i.e., $K(p) = \lambda - ip^3$;*

2) *The Fourier transform $\hat{\bar{u}}(p)$ of the initial perturbation $\bar{u}(x)$ is analytic in the upper complex half-plane $p = \rho + i\eta$, $\eta = \operatorname{Im} p \geq 0$ and satisfies the estimate*

$$(5.5) \qquad \left|\frac{d^l\hat{\bar{u}}(p)}{dp^l}\right| \leq \varepsilon M^{-6}(p)e^{b\eta^3}, \qquad l = 0, 1,\ \eta \geq 0,$$

where $b \geq 0$ is a constant.

Then the following asymptotic formulas hold in the limit $t \to \infty$:

1) *for $x > 0$, uniformly in $\xi = x/3t^{1/3}$:*

$$(5.6) \qquad u(x,t) = Ae^{-\lambda t}t^{-1/3}\frac{2\sqrt{\xi}}{\sqrt{3}}\{\mathscr{K}_{1/3}(2\xi^{3/2}) + O(\xi^{1/2}t^{-1/3}\ln t)\};$$

2) *for $x < 0$, uniformly in $\xi = x/3t^{1/3}$:*

$$(5.7) \qquad u(x,t) = Ae^{-\lambda t}t^{-1/3}\frac{2\pi}{3}\sqrt{|\xi|}\{J_{1/3}(2|\xi|^{3/2}) + J_{1/3}(2|\xi|^{3/2}) + O(|\xi|^{1/2}t^{-1/3}\ln t)\}, \qquad A = \text{const};$$

3) *for $x > 0$, for an arbitrary fixed $B > 0$, uniformly in $\sigma = \sqrt{x/3t} \leq B$:*

$$(5.8) \qquad u(x,t) = A(\sigma i)e^{-\lambda t - 2\sigma^3 t}\left\{\sqrt{\frac{\pi}{3\sigma t}} + O(\sigma^{-2}t^{-1})\right\};$$

4) *for $x < 0$, for arbitrary fixed $B > 0$, uniformly in $\beta = \sqrt{|x|/3t} \leq B$:*

$$(5.9) \qquad u(x,t) = 4\sqrt{\frac{\pi}{3\beta t}}e^{-\lambda t}\operatorname{Re}(A(\beta)e^{-2i\beta^3 t + i\frac{\pi}{4}}) + O(e^{-\lambda t}\beta^{-2}t^{-1}\ln t).$$

REMARK. The expressions (5.6) and (5.7) have meaning (i.e., the remainder term is smaller than the principal one) for $0 \le |x| \le Bt^{1/3}$ with an arbitrary fixed $B \ge 0$ and $t \to \infty$. The expression (5.8) and (5.9) give the asymptotic behavior of the solutions of the Cauchy problem for $\frac{1}{B}t^{1/3} \le |x| \le Bt$ with an arbitrary fixed $B > 0$ and $t \to \infty$. In the region $|x| \approx t^{1/3}$ the asymptotics (5.6), (5.7), (5.8), and (5.9) interlock. In order to determine the principal term of the asymptotics in the region where $\frac{x}{t} \to +\infty$ $t \to \infty$, more detailed knowledge on the order of decrease of the initial condition at $+\infty$ is required (i.e., the behavior of $\hat{\bar{u}}(i\sigma)$ as $\sigma \to +\infty$), while in the region where $\frac{x}{t} \to -\infty$, $t \to \infty$, more detailed information about the smoothness characteristics of the initial condition are needed (i.e., the behavior of $\hat{\bar{u}}(\beta)$ as $\beta \to \pm\infty$). The analyticity requirement for the Fourier transform of the initial condition is used only in the proofs of formulas (5.8) and (5.9).

5.2.

PROOF OF THEOREM 2. Applying the Fourier transform to (5.1) and setting $\hat{\bar{u}}(p) = \varepsilon\bar{v}(p)$ and $\hat{u}(p,t) = \sum_{\mathcal{N}=0}^{\infty} \varepsilon^{\mathcal{N}+1} v_{\mathcal{N}}(p,t)$, we obtain (3.3) and (3.4) for $\mathcal{N} \ge 1$, $p \in R_1$. We introduce the notation

$$w_j(p,t) = v_j(p,t)e^{K(p)t},$$

$$\begin{aligned}\mathcal{L}(p,q) &= K(p) - K(p-q) - K(q)\\ &= -\lambda + w(\sqrt{|p|} - \sqrt{|p-q|} - \sqrt{|q|}) - 3ipq(p-q),\end{aligned}$$

and derive the following bounds, $p \in R_1 \backslash 0$:

(5.10)

$$|w_{\mathcal{N}}(p,t)| \le c^{\mathcal{N}} M^{-6}(p)(\mathcal{N}+1)^{-5}, \qquad \left|\frac{\partial w_{\mathcal{N}}}{\partial p}\right| \le c^{\mathcal{N}} M^{-2}(p)(\mathcal{N}+1)^{-5} m^{-1/2}(p).$$

Here and below we let c denote various positive constants independent of $\mathcal{N}$.

For $\mathcal{N} = 0$ (5.10) follows from condition (5.3) of the theorem. For $\mathcal{N} \ge 1$ we transform (3.4) applying integration by parts with respect to τ:

(5.11)

$$\begin{aligned}w_{\mathcal{N}}(p,t) &= \sum_{j=1}^{\mathcal{N}} \int_{-\infty}^{\infty} dq \int_0^t d\tau\, e^{\mathcal{L}(p,q)\tau} w_{j-1}(p-q,\tau) w_{\mathcal{N}-j}(q,\tau) q(p-q)\\ &= \int_{-\infty}^{\infty} dq \sum_{j=1}^{\mathcal{N}} w_{\mathcal{N}-j}(q,\tau) w_{j-1}(p-q,\tau) e^{\mathcal{L}(p,q)\tau} \frac{(p-q)q}{\mathcal{L}(p,q)}\Bigg|_{\tau=0}^{\tau=1}\\ &\quad - \sum_{j=1}^{\mathcal{N}} \int_{-\infty}^{\infty} dq \int_0^t d\tau\, e^{\mathcal{L}(-,q)\tau} \frac{2(p-q)q}{\mathcal{L}(p,q)} w_{j-1}(p-q,\tau)\\ &\quad \cdot \int_{-\infty}^{\infty} dr \sum_{l=1}^{\mathcal{N}-j} w_{\mathcal{N}-j-l}(r,\tau) w_{l-1}(q-r,\tau)(q-r) r e^{\mathcal{L}(q,r)\tau} = I_1 + I_2\,.\end{aligned}$$

Since

$$\text{(5.12)} \qquad \operatorname{Re}\mathcal{L}(p,q) \le -\frac{w}{4} m^{1/2}\left(\frac{q}{p}\right) m^{1/2}(p-q) - \frac{w}{4} m^{1/2}\left(\frac{p-q}{p}\right) m^{1/2}(q),$$

and therefore

$$|\mathscr{L}(p,q)| \geq |pq(p-q)| + \frac{w}{4}m^{1/2}\left(\frac{p}{q}\right)m^{1/2}(p-q),$$

for $p \in R_1$, $q \in R_1$ we have

$$\frac{|p-q|\,|q|}{|\mathscr{L}(p,q)|} \leq \frac{c}{M(p)}\min\{M^2(q), M^2(p-q)\}\,. \tag{5.13}$$

Suppose, by induction, inequalities (5.10) hold for indices smaller than $\mathscr{N}$. Then, taking (5.12) and (5.13) into account, we estimate I_1 as follows:

$$|I_1| \leq c^{\mathscr{N}} M^{-6}(\mathscr{N}+1)^{-5}. \tag{5.14}$$

Again by (5.12) and (5.13), for I_2 we have:

(5.15)

$$\begin{aligned}
|I_2| &\leq c^{\mathscr{N}-1}\int_{-\infty}^{\infty} dq \int_0^t d\tau \sum_{j=1}^{\mathscr{N}}\sum_{l=1}^{\mathscr{N}-j} j^{-5}\frac{\min\{M^2(q), M^2(p-q)\}}{M(p)M^6(p-q)} \\
&\quad\cdot \int_{-\infty}^{\infty} dr(\mathscr{N}-j-l+1)^{-5}l^{-5}M^{-6}(r)M^{-6}(q-r)|q-r|\,|r| \\
&\quad\cdot \exp\left\{-\frac{w}{4}\tau m^{1/2}(r)m^{1/2}\left(1-\frac{r}{q}\right) - \frac{w\tau}{4}m^{1/2}\left(\frac{r}{q}\right)m^{1/2}(q-r)\right\} \\
&\leq c^{\mathscr{N}}\int_{-\infty}^{\infty} dq \int_{-\infty}^{\infty} dr \sum_{j=1}^{\mathscr{N}}\sum_{l=1}^{\mathscr{N}-j} j^{-5}l^{-5}(\mathscr{N}-j-l-1)^{-5}M^{-6}(p-q)M^{-6}(r) \\
&\quad\cdot M^{-1}(p)M^{-6}(q-r)\frac{|q-r|\,|r|\min\{M^2(q), M^2(p-q)\}}{m^{1/2}\left(1-\frac{r}{q}\right)m^{1/2}(r) + m^{1/2}(\frac{r}{q})m^{1/2}(q-r)} \\
&\leq c^{\mathscr{N}}M^{-6}(p)(\mathscr{N}+1)^{-5}.
\end{aligned}$$

The first inequality (5.10) follows from (5.14) and (5.15). To prove the second inequality (5.10) we differentiate (3.4) with respect to p:

$$\begin{aligned}
\frac{\partial w_{\mathscr{N}}(p,t)}{\partial p} &= \int_0^t d\tau \int_{-\infty}^{\infty} dq \sum_{j=1}^{\mathscr{N}} \exp(\mathscr{L}(p,q)\tau)w_{\mathscr{N}-j}(q,\tau)q \\
&\qquad\cdot\Big\{w_{j-1}(p-q,\tau) + (p-q)\frac{\partial}{\partial p}w_{j-1}(p-q,\tau) \\
&\qquad\qquad + (p-q)w_{j-1}(p-q,\tau)\frac{\partial\mathscr{L}(p,q)}{\partial p}\tau\Big\} \\
&\equiv I_3 + I_4 + I_5.
\end{aligned} \tag{5.16}$$

In view of (5.12), for the integral I_3 we find

$$\begin{aligned}
|I_3| &\leq \int_0^t d\tau \int_{-\infty}^{\infty} dq \sum_{j=1}^{\mathscr{N}} \exp\{\operatorname{Re}\mathscr{L}(p,q)\tau\}|w_{\mathscr{N}-j}(q,\tau)|\,|w_{j-1}(p-q,\tau)|\,|q| \\
&\leq c^{\mathscr{N}}(\mathscr{N}+1)^{-5}M^{-5}(p)\int_{-\infty}^{\infty} dq\frac{M^{-5}(q)m(q)M^{-1}(p)}{m^{1/2}(\frac{q}{p})m^{1/2}(p-q)} \\
&\leq c^{\mathscr{N}}(\mathscr{N}+1)^{-5}M^{-5}(p)\,.
\end{aligned} \tag{5.17}$$

To estimate the integral I_5, observe that

$$|q|\,|p-q|\left|\frac{\partial}{\partial p}\mathscr{L}(p,q)\right| \le cm(q)m^{1/2}(p-q)m^{-1/2}(p)M^3(q)M^2(p-q)$$

and also

$$\tau e^{\operatorname{Re}\mathscr{L}(p,q)\tau} \le 2e^{(1/2)\operatorname{Re}\mathscr{L}(p,q)\tau}(\operatorname{Re}\mathscr{L}(p,q))^{-1}.$$

Therefore,

$$\begin{aligned}|I_5| &\le c^{\mathscr{N}}(\mathscr{N}+1)^{-5}M^{-2}(p)m^{-1/2}(p)\int_{-\infty}^{\infty}dq\frac{M^{-3}(q)M^{-1}(p)m(q)}{m(\frac{q}{p})m^{1/2}(p-q)} \\ &\le c^{\mathscr{N}}m^{-1/2}(p)M^{-2}(p)(\mathscr{N}+1)^{-5}.\end{aligned} \tag{5.18}$$

Integrating I_4 preliminarily by parts with respect to τ, as in (5.11), and estimating similarly to (5.14) and (5.15), we obtain

$$|I_4| \le c^{\mathscr{N}}m^{-1/2}(p)M^{-2}(p)(\mathscr{N}+1)^{-5}. \tag{5.19}$$

We arrive at the second inequality (5.10) from (5.17), (5.18), and (5.19). The induction is completed.

Inverting the Fourier transform, we represent the solution of the Cauchy problem for equation (5.1) in the form

$$u(x,t) = \frac{1}{2\pi}\sum_{\mathscr{N}=1}^{\infty}\varepsilon^{\mathscr{N}+1}\psi_{\mathscr{N}}(x,t), \tag{5.20}$$

where

$$\psi_{\mathscr{N}}(x,t) = \int_{-\infty}^{\infty}e^{ipx}v_{\mathscr{N}}(p,t)\,dp.$$

The series in (5.20) converges absolutely and uniformly in view of inequalities (5.10) if $\varepsilon < c$. We write $\psi_{\mathscr{N}}(x,t)$ in the form

$$\begin{aligned}\psi_{\mathscr{N}}(x,t) = \int_{-\infty}^{\infty}dp\,e^{ipx-K(p)t}\{&[\Phi_{\mathscr{N}} + w_{\mathscr{N}}(0,t) - \Phi_{\mathscr{N}}] \\ &+ [w_{\mathscr{N}}(p,t) - w_{\mathscr{N}}(0,t)]\} \equiv J_1 + J_2.\end{aligned} \tag{5.21}$$

Here $\Phi_{\mathscr{N}} \equiv \lim_{t\to\infty} w_{\mathscr{N}}(0,t)$. Making use of (5.10), we estimate the difference

$$\begin{aligned}|w_{\mathscr{N}}(0,t) - \Phi_{\mathscr{N}}| &= \int_t^{\infty}d\tau\,e^{\lambda\tau}\int_{-\infty}^{\infty}dq\sum_{j=1}^{\mathscr{N}}w_{j-1}(-q,\tau) \\ &\quad\cdot w_{\mathscr{N}-j}(q,\tau)q^2\exp\{-K(-q)\tau - K(q)\tau\} \\ &\le c^{\mathscr{N}}\int_{-\infty}^{\infty}dq\frac{q^2e^{-2wt\sqrt{|q|}}}{M^{12}(q)} = c^{\mathscr{N}}O(t^{-6}).\end{aligned} \tag{5.22}$$

Denoting $\xi = xt^{-2}$ and changing the variable $\eta = pt^2$, we analyze the asymptotics as $t\to\infty$ of the following integral:

$$\begin{aligned}J_3 &= \int_{-\infty}^{\infty}dp\,e^{ipx-K(p)t} = e^{-\lambda t}t^{-2}\int_{-\infty}^{\infty}d\eta\,\exp\{i\eta\xi - w\sqrt{|\eta|} + i\eta^3t^{-5}\} \\ &= 2e^{-\lambda t}t^{-2}\int_0^{\infty}d\eta\cos(\eta\xi)e^{-w\sqrt{\eta}} + O(e^{-\lambda t}t^{-7}).\end{aligned} \tag{5.23}$$

With the aid of (5.22) and (5.23) we obtain the principal term of the asymptotics

$$J_1 = 2\Phi_{\mathcal{N}} e^{-\lambda t} t^{-2} \int_0^\infty dp \cos(p\xi) e^{-w\sqrt{p}} + c^{\mathcal{N}} O(e^{-\lambda t} t^{-6}). \tag{5.24}$$

For the remainder term J_2, with the aid of the bound for the derivative in (5.10), we obtain

$$|J_2| \le c^{\mathcal{N}} \int_{-\infty}^\infty dp\, e^{-\lambda t - wt\sqrt{|p|}} \sqrt{|p|} = c^{\mathcal{N}} O(e^{-\lambda t} t^{-3}). \tag{5.25}$$

Substitution of (5.24) and (5.25) in (5.21) leads to the asymptotics (5.4). The constant A has the form

$$A = \frac{1}{\pi} \sum_{\mathcal{N}=0}^\infty \varepsilon^{\mathcal{N}+1} \Phi_{\mathcal{N}}.$$

The numbers $\Phi_{\mathcal{N}}$ are computed as in §4. Theorem 1 is proved. □

5.3.

PROOF OF THEOREM 3. *Perturbation theory.* As before, we represent the solution $u(x,t)$ of the Cauchy problem for the perturbed Korteweg-de Vries equation with $K(p) = \lambda - ip^3$, $\lambda > 0$ by applying formulas (3.3) and (3.4). From these formulas it is easy to show by induction that if $\hat{u}(p)$ is analytic in the upper half-plane $\eta \equiv \operatorname{Im} p \ge 0$ of the complex plane p, $p = \rho + i\eta$, then all functions $v_{\mathcal{N}}(p,t)$ are also analytic for $\eta \ge 0$. We shift upward by the moment $i\eta/2$ the contour of integration with respect to q in (3.4), we obtain

$$\begin{aligned} v_{\mathcal{N}}(p,t) = \int_0^t d\tau\, e^{-K(p)(t-\tau)} \int_{-\infty}^\infty dq \sum_{j=1}^{\mathcal{N}} v_{j-1}\left(p - q - \frac{i\eta}{2}, \tau\right) \\ \cdot v_{\mathcal{N}-j}\left(q + \frac{i\eta}{2}, \tau\right)\left(p - q - \frac{i\eta}{2}\right)\left(q + \frac{i\eta}{2}\right). \end{aligned} \tag{5.26}$$

Let us establish the following bounds for $\eta \ge 0$:

$$|v_{\mathcal{N}}(p,t)| \le c^{\mathcal{N}} H_{\mathcal{N}}(p,t), \tag{5.27}$$

$$\left|\frac{\partial w_{\mathcal{N}}}{\partial p}(p,t)\right| \le c^{\mathcal{N}+1} h_{\mathcal{N}}(p,t), \tag{5.28}$$

where

$$H_{\mathcal{N}}(p,t) = M^{-6}(p)(\mathcal{N}+1)^{-5} e^{-\lambda t + \eta^3(t+b) - x_{\mathcal{N}+1}(p)t},$$

$$h_{\mathcal{N}}(p,t) = M^{-3}(p) e^{3\eta\rho^2 t + b\eta^3 - \kappa_{\mathcal{N}+1}(p)t}(\mathcal{N}+1)^{-5},$$

$$w_{\mathcal{N}}(p,t) = v_{\mathcal{N}}(p,t) e^{K(p)t}, \qquad \kappa_{\mathcal{N}}(p) = \eta\rho^2 4^{-\mathcal{N}},$$

b is taken from (5.5), $c > 0$ here and below are some constants independent of $\mathcal{N}$.

For $\mathscr{N} = 0$ (5.27) and (5.28) follow from condition (5.5) of the theorem. For $\mathscr{N} \geq 1$ we argue by induction. If $|p| \leq 1$, then from (5.27) and (3.4) we obtain

$$\begin{aligned} |v_{\mathscr{N}}(p,t)| &\leq c^{\mathscr{N}}(\mathscr{N}+1)^{-5} e^{-\lambda t + \eta^3(t+b) - x_{\mathscr{N}+1}(p)t} \sum_{j=1}^{\mathscr{N}} j^{-5} \\ &\quad \cdot \int_0^t d\tau\, e^{-\lambda\tau} \int_{-\infty}^{\infty} dq\, M^{-5}\left(q + \frac{i\eta}{2}\right) \\ &\leq c^{\mathscr{N}} H_{\mathscr{N}}(p,t), \end{aligned} \tag{5.29}$$

since for all $p = \rho + i\eta$, $\eta \geq 0$,

$$\operatorname{Re} K(p) = \lambda + 3\eta\rho^2 - \eta^3 \geq \lambda - \eta^3 + \kappa_{\mathscr{N}+1}(p)$$

and

$$\kappa_{\mathscr{N}+1}(p) - \kappa_{\mathscr{N}+1-j}\left(q + \frac{i\eta}{2}\right) - \kappa_j\left(p - q - \frac{i\eta}{2}\right) \leq 0\,. \tag{5.30}$$

For $|p| \geq 1$, $\eta \geq 0$ we transform $v_{\mathscr{N}}(p,t)$ as follows:

$$\begin{aligned} v_{\mathscr{N}}(p,t) &= e^{-K(p)t} \int_0^t d\tau \int_{-\infty}^{\infty} dq \sum_{j=1}^{\mathscr{N}} e^{\widetilde{\mathscr{L}}(p,q)\tau - \lambda\tau} \\ &\quad \cdot w_{j-1}\left(p - q - \frac{i\eta}{2}, \tau\right) w_{\mathscr{N}-j}\left(q + \frac{i\eta}{2}, \tau\right)\left(p - q - \frac{i\eta}{2}\right)\left(q + \frac{i\eta}{2}\right), \end{aligned} \tag{5.31}$$

since

$$K(p) - K\left(p - q - \frac{i\eta}{2}\right) - K\left(q + \frac{i\eta}{2}\right) = -\lambda + \widetilde{\mathscr{L}}(p,q),$$

where

$$\widetilde{\mathscr{L}}(p,q) = -3ip\left(q + \frac{i\eta}{2}\right)\left(p - q - \frac{i\eta}{2}\right).$$

From (5.31) we obtain also

$$\begin{aligned} \frac{\partial w_{\mathscr{N}}}{\partial t}(p,t) &= \int_{-\infty}^{\infty} dq \sum_{j=1}^{\mathscr{N}} e^{\widetilde{\mathscr{L}}(p,q)t - \lambda t}\left(p, q - \frac{i\eta}{2}\right)\left(q + \frac{i\eta}{2}\right) \\ &\quad \cdot w_{j-1}\left(p, q - \frac{i\eta}{2}, t\right) w_{\mathscr{N}-j}\left(q + \frac{i\eta}{2}, t\right). \end{aligned} \tag{5.32}$$

Integrating (5.31) once by parts with respect to τ, in view of (5.32), we find

$$v_{\mathscr{N}}(p,t) = I_1 + I_2 + I_3,$$

where

$$I_1 = \frac{e^{-K(p)t}}{-3ip} \int_{-\infty}^{\infty} dq \sum_{j=1}^{\mathscr{N}} e^{\widetilde{\mathscr{L}}(p,q)\tau - \lambda\tau} w_{\mathscr{N}-j}\left(q + \frac{i\eta}{2}, \tau\right) w_{j-1}\left(p - q - \frac{i\eta}{2}\right)\Big|_{\tau=0}^{\tau=t}$$

$$I_2 = \lambda \frac{e^{-K(p)t}}{-3ip} \int_0^t d\tau \int_{-\infty}^{\infty} dq \sum_{j=1}^{\mathscr{N}} e^{\widetilde{\mathscr{L}}(p,q)\tau - \lambda\tau} w_{\mathscr{N}-j}\left(q + \frac{i\eta}{2}, \tau\right) w_{j-1}\left(p - q - \frac{i\eta}{2}, \tau\right)$$

$$I_3 = \frac{2}{3ip} e^{-K(p)t} \int_0^t d\tau \int_{-\infty}^{\infty} dq \sum_{j=1}^{\mathscr{N}} \sum_{l=1}^{\mathscr{N}-j} e^{\widetilde{\mathscr{L}}(p,q)\tau - \lambda\tau} w_{j-1}\left(p - q - \frac{i\eta}{2}, \tau\right)$$
$$\cdot e^{K(q+\frac{i\eta}{2})\tau} v_{\mathscr{N}-j-l-1}\left(r + \frac{i\eta}{4}, \tau\right) v_{l-1}\left(q - r + \frac{i\eta}{4}, \tau\right)\left(r + \frac{i\eta}{4}\right)\left(q - r + \frac{i\eta}{4}\right).$$

By (5.27) and (5.30) we readily obtain the relationship for I_1 and I_3

$$|I_1| + |I_2| \le c^{\mathscr{N}} H_{\mathscr{N}}(p, t)\,. \tag{5.33}$$

As in (5.29), for I_3 we find that

$$\begin{aligned} |I_3| \le & \frac{c^{\mathscr{N}}}{M(p)} e^{-\lambda t + \eta^3(t+b) - \kappa_{\mathscr{N}+1}(p)t} \int_0^t d\tau\, e^{-\lambda\tau} \int_{-\infty}^{\infty} dq \int_{-\infty}^{\infty} dr \\ & \cdot M^{-6}\left(p - q - \frac{i\eta}{2}\right) M^{-5}\left(r + \frac{i\eta}{4}\right) M^{-5}\left(q - r + \frac{i\eta}{4}\right) \\ & \cdot \sum_{j=1}^{\mathscr{N}} \sum_{l=1}^{\mathscr{N}-j} (jl(\mathscr{N} - j - l))^{-5} \\ \le & \, c^{\mathscr{N}} H_{\mathscr{N}}(p, t). \end{aligned} \tag{5.34}$$

The bound (5.27) follows from (5.33) and (5.34). The induction is completed.

To prove (5.28) we differentiate (5.26) with respect to p:

$$\begin{aligned} \frac{\partial w_{\mathscr{N}}(p, t)}{\partial p} = & \int_0^t d\tau \int_{-\infty}^{\infty} dq \sum_{j=1}^{\mathscr{N}} e^{K(p)\tau} v_{\mathscr{N}-j}\left(q + \frac{i\eta}{2}, \tau\right)\left(q + \frac{i\eta}{2}\right) \\ & \cdot \Big\{ v_{j-1}\left(p - q - \frac{i\eta}{2}, \tau\right) \\ & \times \left[1 + \left(p - q - \frac{i\eta}{2}\right)\left(K'(p) - K'\left(p - q - \frac{i\eta}{2}\right)\right)\tau\right] \\ & + \left(p - q - \frac{i\eta}{2}\right) e^{-K(p-q-i\eta/2)\tau} \frac{\partial}{\partial p} w_{j-1}\left(p - q - \frac{i\eta}{2}, \tau\right)\Big\} \\ \equiv & \, I_4 + I_5\,. \end{aligned} \tag{5.35}$$

In view of (5.27) and (5.30) we estimate the integral I_4:

$$
\begin{aligned}
|I_4| \le c^{\mathcal{N}} \int_0^t d\tau \int_{-\infty}^{\infty} dq \sum_{j=1}^{\mathcal{N}} e^{3\eta p^2 \tau} \left\{ 1 + \left(|p|^2 + \left| p - q - \frac{i\eta}{2} \right|^2 \right) \tau \right\} \\
\cdot \left| q + \frac{i\eta}{2} \right| M^{-5} \left(p - q - \frac{i\eta}{2} \right) M^{-6} \left(q + \frac{i\eta}{2} \right) (\mathcal{N} + 1 - j)^{-5} j^{-5} \\
\cdot \exp\left(-\lambda\tau + b\eta^3 - \kappa_{\mathcal{N}+1-j} \left(q + \frac{i\eta}{2} \right) \tau - \kappa_j \left(p - q - \frac{i\eta}{2} \right) \tau \right) \\
\le c^{\mathcal{N}+1} h_{\mathcal{N}}(p, t).
\end{aligned}
\tag{5.36}
$$

We estimate I_5 in the same manner as I_3 above. Owing to (5.27)–(5.30), for $|p| \le 1$ we find that

(5.37)

$$
\begin{aligned}
|I_5| \le & c^{\mathcal{N}} \int_0^t d\tau\, e^{\operatorname{Re} K(p)\tau} \int_{-\infty}^{\infty} dq \sum_{j=1}^{\mathcal{N}} j^{-5} (\mathcal{N} + 1 - j)^{-5} \left| p - q - \frac{i\eta}{2} \right| \left| q + \frac{i\eta}{2} \right| \\
& \cdot M^{-6} \left(q + \frac{i\eta}{2} \right) M^{-3} \left(p - q - \frac{i\eta}{2} \right) \\
& \cdot \exp\left\{ -2\lambda\tau + 2 \left(\frac{\eta}{2} \right)^3 (\tau + b) - \kappa_{\mathcal{N}+1-j} \left(q + \frac{i\eta}{2} \right) \tau - x_j \left(p - q - \frac{i\eta}{2} \right) \tau \right\} \\
\le & c^{\mathcal{N}} h_{\mathcal{N}}(p, t).
\end{aligned}
$$

For $|p| \ge 1$ we transform I_5 as in (5.31), i.e., preliminarily integrating once by parts and applying (5.32):

$$
I_5 = I_6 + I_7 + I_8 + I_9,
\tag{5.38}
$$

where

(5.39)

$$
\begin{aligned}
I_6 &= \int_{-\infty}^{\infty} dq \sum_{j=1}^{\mathcal{N}} \frac{e^{\widetilde{\mathcal{L}}(p,q)\tau - \lambda\tau}}{-3ip} w_{\mathcal{N}-j} \left(q + \frac{i\eta}{2}, \tau \right) \frac{\partial w_{j-1}}{\partial p} \left(p - q - \frac{i\eta}{2}, \tau \right) \Bigg|_{\tau=0}^{\tau=t}, \\
I_7 &= - \int_0^t d\tau \int_{-\infty}^{\infty} dq \sum_{j=1}^{\mathcal{N}} \frac{\lambda}{3ip} e^{\widetilde{\mathcal{L}}(p,q)\tau - \lambda\tau} w_{\mathcal{N}-j} \left(q + \frac{i\eta}{2}, \tau \right) \frac{\partial w_{j-1}}{\partial p} \left(p - q - \frac{i\eta}{2}, \tau \right), \\
I_8 &= - \int_0^t d\tau \int_{-\infty}^{\infty} dq \sum_{j=1}^{\mathcal{N}} \frac{e^{\widetilde{\mathcal{L}}(p,q)\tau - \lambda\tau}}{-3ip} \frac{\partial w_{\mathcal{N}-j}}{\partial \tau} \left(q + \frac{i\eta}{2}, \tau \right) \frac{\partial w_{j-1}}{\partial p} \left(p - q - \frac{i\eta}{2}, \tau \right), \\
I_9 &= - \int_0^t d\tau \int_{-\infty}^{\infty} dq \sum_{j=1}^{\mathcal{N}} \frac{e^{\widetilde{\mathcal{L}}(p,q)\tau - \lambda\tau}}{-3ip} w_{\mathcal{N}-j} \left(q + \frac{i\eta}{2}, \tau \right) \frac{\partial^2 w_{j-1}}{\partial p \partial \tau} \left(p - q - \frac{i\eta}{2}, \tau \right).
\end{aligned}
$$

Application of (5.27)–(5.30) leads to

$$
\begin{aligned}
|I_6| \le c^{\mathcal{N}} M^{-1}(p) \int_{-\infty}^{\infty} dq \sum_{j=1}^{\mathcal{N}} j^{-5} M^{-6} \left(q + \frac{i\eta}{2} \right) M^{-3} \left(p - q - \frac{i\eta}{2} \right) \\
\cdot \exp\left\{ \operatorname{Re} K(p) t - 2\lambda t + \eta^3 (t + b) - \kappa_{\mathcal{N}+1}(p) t \right\} \\
\le c^{\mathcal{N}} h_{\mathcal{N}}(p, t),
\end{aligned}
\tag{5.40}
$$

and similarly

$$(5.41)\quad \begin{aligned} |I_7| &\le c^{\mathcal{N}} M^{-1}(p) \int_0^t d\tau \int_{-\infty}^{\infty} dq \sum_{j=1}^{\mathcal{N}} j^{-5} M^{-6}\left(q+\frac{i\eta}{2}\right) M^{-3}\left(p-q-\frac{i\eta}{2}\right) \\ &\quad \cdot \exp\left\{\operatorname{Re} K(p)\tau - 2\lambda\tau + \eta^3(\tau+b) - \kappa_{\mathcal{N}+1}(p)\tau\right\} \\ &\le c^{\mathcal{N}} h_{\mathcal{N}}(p,t). \end{aligned}$$

As in the estimation of I_3, for I_8 we have

$$(5.42)\quad \begin{aligned} |I_8| &\le c^{\mathcal{N}} h_{\mathcal{N}}(p,t) \int_0^t d\tau\, e^{-\lambda\tau} \int_{-\infty}^{\infty} dq \int_{-\infty}^{\infty} dr\, M^{-3}\left(r+\frac{i\eta}{4}\right) \\ &\quad \cdot M^{-3}\left(q-r+\frac{i\eta}{4}\right) \sum_{j=1}^{\mathcal{N}} \sum_{l=1}^{\mathcal{N}-j} j^{-5} l^{-5} \\ &\le c^{\mathcal{N}} h_{\mathcal{N}}(p,t). \end{aligned}$$

For the second derivative $\partial^2 w_{j-1}/\partial p\,\partial\tau$ we use (5.32) and the bounds (5.27)–(5.30):

$$(5.43)\quad \begin{aligned} \left|\frac{\partial^2}{\partial p\,\partial\tau} w_{j-1}\left(p-q-\frac{i\eta}{2},\tau\right)\right| &\le c^{j-1} \int_{-\infty}^{\infty} dq \sum_{l=1}^{j-1} (j-l)^{-5} M^{-5}\left(q+\frac{i\eta}{2}\right) \\ &\quad \cdot \exp\left\{\left(\frac{\eta}{2}\right)^3 b + \frac{3}{2}\eta(p-q)^{-2} t - \kappa_j\left(p-q-\frac{i\eta}{2}\right) t\right\} \\ &\le c^j h_{j-1}\left(p-q-\frac{i\eta}{2},t\right). \end{aligned}$$

We estimate I_9 with the aid of (5.43):

$$(5.44)\quad \begin{aligned} |I_9| &\le c^{\mathcal{N}} M^{-1}(p) \int_0^t d\tau \int_{-\infty}^{\infty} dq \sum_{j=1}^{\mathcal{N}} j^{-5} M^{-6}\left(q+\frac{i\eta}{2}\right) M^{-3}\left(p-q-\frac{i\eta}{2}\right) \\ &\quad \cdot \exp(\operatorname{Re} K(p)\tau - 2\lambda\tau - \kappa_{\mathcal{N}+1}(p)\tau + \eta^3 b + \eta^3 t) \\ &\le c^{\mathcal{N}} h_{\mathcal{N}}(p,t). \end{aligned}$$

It follows from (5.40)–(5.44) that $|I_5| \le c^{\mathcal{N}} h_{\mathcal{N}}(p,t)$; therefore, owing to (5.35), (5.36), and (5.37), we arrive at the inequality (5.28). Thus, (5.27) and (5.28) are derived by induction for all $\mathcal{N}$. We proceed now with computation of the asymptotics.

Derivation of the formulas (5.6) *and* (5.7). Using (5.20), we now represent the function $\psi_{\mathcal{N}}(x,t)$ as follows (here and below $\sigma \ge 0$ and $\beta \ge 0$; if $\sigma > 0$ then $\beta = 0$, and if $\beta < 0$ then $\sigma = 0$):

$$(5.45)\quad \begin{aligned} \psi_{\mathcal{N}}(x,t) &= \int_{-\infty}^{\infty} dp\, e^{ipx-\sigma x-K(p+i\sigma)t} \\ &\quad \cdot \{[\Phi_{\mathcal{N}}(\beta+i\sigma) + (w_{\mathcal{N}}(\beta+i\sigma,t) - \Phi_{\mathcal{N}}(\beta+i\sigma))] \\ &\qquad + [w_{\mathcal{N}}(p+i\sigma,t) - w_{\mathcal{N}}(\beta+i\sigma,t)]\} \\ &= I_1 + I_2, \end{aligned}$$

where

$$\begin{aligned}\Phi_{\mathscr{N}}(\beta+i\sigma) &= \lim_{t\to\infty} w_{\mathscr{N}}(\beta+i\sigma,t)\\ &= \int_0^\infty d\tau \int_{-\infty}^\infty dq \sum_{j=1}^{\mathscr{N}} e^{K(\beta+i\sigma)\tau}\\ &\quad \cdot v_{j-1}\left(\beta+\frac{i\sigma}{2}-q,\tau\right) v_{\mathscr{N}-j}\left(q+\frac{i\sigma}{2},\tau\right)\left(\beta+\frac{i\sigma}{2}-q\right)\left(q+\frac{i\sigma}{2}\right).\end{aligned}$$

The principal term of the asymptotics can be obtained from the following integral:

$$I_3 = \int_{-\infty}^\infty dp\, e^{ipx-\sigma x-K(p+i\sigma)t} = \int_{-\infty}^\infty dp\, e^{ipx-\lambda t+ip^3 t}.$$

In this integral there are three saddle-points z corresponding to various relationships between the parameters x and t:

$$\begin{aligned} z&=0, \quad \text{if } |\xi| \equiv \left|\frac{x}{3t^{1/3}}\right| \le B,\\ z&=i\sigma, \quad \text{if } B \ge \sigma \equiv \sqrt{\frac{x}{3t}} > 0,\\ z&=\pm\beta, \quad \text{if } 0<\beta\equiv\sqrt{\frac{-x}{3t}} \le B,\end{aligned}$$

where $B>0$ is any fixed number.

After the change of variables $\xi = \frac{x}{3}t^{-1/3}$ and $z=\rho t^{1/3}$ we obtain **[177]**

$$I_3 = e^{-\lambda t} t^{-1/3} \frac{2}{\sqrt{3}} \sqrt{\xi}\mathscr{K}_{1/3}(2\xi^{3/2}) \quad \text{for } \xi>0, \tag{5.46}$$

$$I_3 = \frac{e^{-\lambda t}}{3} t^{-1/3}\sqrt{|\xi|}2\pi(J_{1/3}(2|\xi|^{3/2})) + J_{-1/3}(2|\xi|^{3/2}) \quad \text{for } \xi<0, \tag{5.47}$$

where $\mathscr{K}_{1/3}$ is the MacDonald function and $J_{1/3}$ is the Bessel function. With the aid of (5.27) we analyze the difference

(5.48)

$$\begin{aligned} &|w_{\mathscr{N}}(\beta+i\sigma,t)-\Phi_{\mathscr{N}}(\beta+i\sigma)|\\ &\le \int_t^\infty e^{\operatorname{Re} K(\beta+i\sigma)\tau}\, d\tau \int_{-\infty}^\infty dq \sum_{j=1}^{\mathscr{N}} \left|q+\frac{i\sigma}{2}\right|\\ &\quad \cdot \left|\beta-q+\frac{i\sigma}{2}\right| \left|v_{j-1}\left(\beta-q+\frac{i\sigma}{2},\tau\right) v_{\mathscr{N}-j}\left(q+\frac{i\sigma}{2},\tau\right)\right|\\ &\le c^{\mathscr{N}} \int_t^\infty d\tau\, e^{\lambda\tau-\sigma^3\tau} \int_{-\infty}^\infty dq\\ &\quad \cdot \sum_{j=1}^{\mathscr{N}} (\mathscr{N}+1-j)^{-5} j^{-5} M^{-5}\left(\beta+\frac{i\sigma}{2}-q\right) M^{-5}\left(q+\frac{i\sigma}{2}\right)\\ &\quad \cdot \exp\left\{-2\lambda\tau + \frac{\sigma^3}{4}(\tau+b) - \kappa_{\mathscr{N}+1-j}\left(q+\frac{i\sigma}{2}\right)\tau - \kappa_j\left(\beta-q+\frac{i\sigma}{2}\right)\right\}\tau\\ &\le c^{\mathscr{N}} e^{-\lambda t-(3/4)\sigma^3 t+b\sigma^3}.\end{aligned}$$

The principal term of the asymptotics in (5.6) and (5.7) can be obtained with the end of (5.46)–(5.48). And the constant A has the form

$$A = \frac{1}{2\pi}\sum_{\mathcal{N}=0}^{\infty} \varepsilon^{\mathcal{N}+1}\Phi_{\mathcal{N}}(0)\,. \tag{5.49}$$

The remainder term I_2 can be represented in the form

$$\begin{aligned} I_2 &= \int_{-\infty}^{\infty} d\rho\, e^{i\rho x-\lambda t+i\rho^3 t}(w_{\mathcal{N}}(\rho,t)-w_{\mathcal{N}}(0,t)) \\ &= \int_{|\rho|\le t^{-1/3}} + \int_{|\rho|\ge t^{-1/3}} = I_4 + I_5\,. \end{aligned}$$

Application of (5.28) leads to $(\rho_* \in (\beta,\rho))$

$$|w_{\mathcal{N}}(\rho+i\sigma,t)-w_{\mathcal{N}}(\beta+i\sigma,t)| \le |\rho-\beta| c^{\mathcal{N}+1} e^{3\sigma\rho_*^2 t + b\sigma^3 - \kappa_{\mathcal{N}+1}(\rho_*+i\sigma)t}. \tag{5.50}$$

Then $(\beta = 0,\ \sigma = 0)$:

$$|I_4| \le \int_{|\rho|\le t^{-1/3}} d\rho\, c^{\mathcal{N}+1}|\rho| e^{-\lambda t} \le c^{\mathcal{N}+1} t^{-2/3} e^{-\lambda t}. \tag{5.51}$$

In the integral I_5 we integrate by parts

$$\begin{aligned} I_5 = &\frac{e^{-\lambda t+i\rho^3 t+i\rho x}}{3i\rho^2 t}(w_{\mathcal{N}}(\rho,t)-w_{\mathcal{N}}(0,t))\Big|_{|\rho|=t^{-1/3}}^{\infty} - \frac{e^{-\lambda t}}{3it}\int_{|\rho|\ge t^{-1/3}} d\rho \\ &\times e^{i\rho^3 t+i\rho x}\left((w_{\mathcal{N}}(\rho,t)-w_{\mathcal{N}}(0,t))\left(\frac{ix}{\rho^2}-\frac{2}{\rho^3}\right)+\frac{1}{\rho^2}\frac{\partial w_{\mathcal{N}}(\rho,t)}{\partial\rho}\right). \end{aligned}$$

From this, by (5.27), (5.28), and (5.50), we obtain

$$|I_5| \le e^{-\lambda t} t^{-2/3} c^{\mathcal{N}+1}(1+|\xi|\ln t)\,. \tag{5.52}$$

The bounds (5.51) and (5.52) give the remainder term in the formulas (5.6) and (5.7).

Derivation of formula (5.8). We put $\sigma = \sqrt{x/3t} > 0$, $\beta = 0$, in (5.45). The integral I_3 now takes the form

$$\begin{aligned} I_3 &= e^{-\lambda t-2\sigma^3 t}\int_{-\infty}^{\infty} d\rho\, e^{-3\sigma\rho^2 t}(1+(e^{i\rho^3 t}-1)) \\ &= e^{-\lambda t-2\sigma^3 t}\sqrt{\frac{\pi}{3\sigma t}}\left(1+O\left(\frac{1}{\sqrt{\sigma^3 t}}\right)\right). \end{aligned} \tag{5.53}$$

Therefore, taking (5.48) in the account, we obtain the leading term of the asymptotics

$$I_1 = e^{-\lambda t-2\sigma^3 t}\sqrt{\frac{\pi}{3\sigma t}}\Phi_{\mathcal{N}}(i\sigma)\left(1+c^{\mathcal{N}} O\left(\frac{1}{\sqrt{\sigma^3 t}}\right)\right). \tag{5.54}$$

The remainder I_2 can be estimated with the aid of (5.50) as follows:

$$\begin{aligned} |I_2| &\le c^{\mathcal{N}+1}\int_{-\infty}^{\infty} d\rho\, e^{-\sigma x-\operatorname{Re} K(\rho+i\sigma)t}|\rho| e^{b\sigma^3+3\sigma\rho^2 t-\kappa_{\mathcal{N}+1}(\rho)t} \\ &\le e^{-\lambda t-2\sigma^3 t} c^{\mathcal{N}+1} e^{b\sigma^3}\int_0^{\infty} d\rho \exp(-\sigma\rho^2 4^{-\mathcal{N}} t)\rho \\ &\le c^{\mathcal{N}+1} e^{b\sigma^3-\lambda t-2\sigma^3 t}(\sigma t)^{-1}. \end{aligned} \tag{5.55}$$

From (5.54) and (5.55) we arrive at (5.8). The constant A has the form

$$A(i\sigma) = \frac{1}{2\pi}\sum_{\mathcal{N}=0}^{\infty} \varepsilon^{\mathcal{N}+1}\Phi_{\mathcal{N}}(i\sigma)\,. \tag{5.56}$$

Derivation of formula (5.9). Put $\beta = \sqrt{|x|/3t} > 0$ and $\sigma = 0$ in formula (5.45), to represent $\psi_{\mathcal{N}}(x,t)$ in the form

$$\begin{aligned}\psi_{\mathcal{N}}(x,t) = &\sum_{l=1}^{2}\int_0^\infty dp \exp((-1)^l ipx - K((-1)^l p)t)\\ &\cdot\{[\Phi_{\mathcal{N}}((-1)^l\beta) + w_{\mathcal{N}}((-1)^l\beta,t) - \Phi_{\mathcal{N}}((-1)^l\beta)]\\ &+[w_{\mathcal{N}}((-1)^l p,t) - w_{\mathcal{N}}((-1)^l\beta,t)]\} = I_1 + I_2\,.\end{aligned} \tag{5.57}$$

The integral

$$I_3 = \int_0^\infty dp\, e^{ipx-\lambda t+ip^3 t},$$

after the change of variable $z = p - \beta$ will become

$$I_3 = e^{-\lambda t - 2i\beta^2 t}\int_\beta^\infty dz\, e^{it3\beta z^2 + itz^3}$$

and by Erdelyi's lemma (see **[178]**), we obtain ($t \to \infty$)

$$I_3 = 2\sqrt{\frac{\pi}{3\beta t}}e^{-\lambda t - 2i\beta^3 t + i\pi/4} + O(e^{-\lambda t}\beta^{-2}t^{-1})\,. \tag{5.58}$$

In view of (5.48), from (5.58) we obtain the leading term of the asymptotics in formula (5.9):

$$I_1 = 2e^{-\lambda t}\sqrt{\frac{\pi}{3\beta t}}\sum_{l=1}^{2}\Phi_{\mathcal{N}}((-1)^l\beta)e^{-(-1)^l(2i\beta^3 t - i\pi/4)} + c^{\mathcal{N}}O(e^{-\lambda t}\beta^{-2}t^{-1}), \tag{5.59}$$

since $\Phi_{\mathcal{N}}(-\beta) = \lim_{t\to\infty} w_{\mathcal{N}}(-\beta,t) = \Phi^*_{\mathcal{N}}(\beta)$. The latter follows from (3.4) and the fact that $\overline{v}(p) = \overline{v}^*(p)$, since the function $\overline{u}(x)$ is real-valued.

Estimating the integral I_2, we split it into two parts, as above,

$$I_2 = \int_{|p-\beta|\le t^{-1/2}} + \int_{|p-\beta|>t^{-1/2},p>0} = I_4 + I_5\,.$$

By (5.50) we have

$$|I_4| \le c^{\mathcal{N}}\int_{|p-\beta|\le t^{-1/2}} dp\,|p-\beta|e^{-\lambda t} \le c^{\mathcal{N}}e^{-\lambda t}t^{-1}. \tag{5.60}$$

Changing in I_5 the variable of integration $z = p - \beta$, we integrate it by parts:

$$\begin{aligned}I_5 = &\sum_{l=1}^{2}\frac{e^{-\lambda t - 2i\beta^3 t(-1)^l}}{3it(-1)^l}\{(w_{\mathcal{N}}((-1)^l(z+\beta),t) - w_{\mathcal{N}}((-1)^l\beta,t)\}\\ &\cdot\left.\frac{e^{(-1)^l it(3\beta z^2+z^3)}}{2\beta z + z^2}\right|_{|z|=t^{-1/2}} - \int_{|z|\ge t^{-1/2},z\ge-\beta} dz\, e^{(-1)^l it(3\beta z + z^3)}\frac{\partial}{\partial z}\\ &\cdot\left(\frac{w_{\mathcal{N}}((-1)^l(z+\beta),t) - w_{\mathcal{N}}((-1)^l\beta,t)}{z(2\beta+z)}\right),\end{aligned}$$

whence, in view of (5.27), (5.28), and (5.50), we have

$$|I_5| \leq c^{\mathscr{N}} e^{-\lambda t} \beta^{-2} t^{-1} \ln t \,. \tag{5.61}$$

The bounds (5.60) and (5.61) give the remainder term in (5.9). Theorem 3 is proved. □

CHAPTER 8

Asymptotics as $t \to \infty$ of Solutions of the Nonlinear Nonlocal Schrödinger Equation

§1. Introduction

In this chapter we extend to the case of cubic nonlinearity the method of computing the asymptotics for large times for solutions of nonlinear nonlocal equations, developed in Chapters 6 and 7. We consider the Cauchy problem for the nonlinear nonlocal Schrödinger (NNS) equation describing the evolution in time of the envelope for an almost monochromatic wave in a weakly nonlinear dispersive medium

$$iu_t + u^2u^* + i\mathbb{K}(u) = 0, \quad u|_{t=0} = \bar{u}(x), \quad x \in R_1,\ t \geq 0, \tag{1.1}$$

where

$$\mathbb{K}(u) = \frac{1}{2\pi}\int_{-\infty}^{\infty} e^{ipx}K(p)\hat{u}(p,t)\,dp \quad \text{and} \quad \hat{u}(p,t) = \int_{-\infty}^{\infty} e^{-ipx}u(x,t)\,dx$$

is the Fourier transform of the function $u(x,t)$.

The asymptotics, as $t \to \infty$, of solutions to the Cauchy problem for the nonlinear Schrödinger equation, obtained from the NNS for $K(p) = ip^2$, was constructed in [**17**, **22**, **25**, **39**, **69**] owing to the possibility of application of the inverse scattering problem technique. In the presence of dissipation in the NNS equation, the solution rapidly stabilizes at zero, as $t \to \infty$, and this enables one to compute its asymptotics.

We introduce the notation

$$\begin{aligned} K^1(p) &= \operatorname{Re} K(p), & K^2(p) &= \operatorname{Im} K(p), \\ m(p) &= \min(1,|p|), & M(p) &= \max(1,|p|)\,. \end{aligned} \tag{1.2}$$

THEOREM 1. *Suppose that*

1) *The symbol* $K(p)$ *satisfies the conditions*

$$\begin{gathered} K^1(p) = \omega|p|^\delta + \lambda, \quad |K^2(p)| \leq c_1|p|^\beta, \quad |p| \leq 1, \\ \beta > \delta, \quad \delta \in (0,2), \quad \omega > 0, \quad \lambda \geq 0, \end{gathered} \tag{1.3}$$

$$K^1(p) \geq 2\kappa(p) + \lambda, \quad p \in R_1, \quad \kappa(p) \equiv c_2 m^\delta(p); \tag{1.4}$$

$$\begin{gathered} |K(p_1) - K(p_2)| \leq c_3|\Delta p|^{\sigma_1} M^\alpha(p_2), \qquad \alpha \geq 0, \\ \sigma_1 \in [0,1], \quad |\Delta p| = |p_1 - p_2| \leq 1, \quad p_1, p_2 \in R_1; \end{gathered} \tag{1.5}$$

2) *The Fourier transform* $\hat{\bar{u}}(p)$ *of the initial perturbation* $\bar{u}(x) \in L_1(R_1)$ *satisfies the inequalities*

$$
\begin{aligned}
&|\hat{\bar{u}}(p)| \le \varepsilon M^{-3-\alpha}(p), \qquad p \in R_1; \\
&|\hat{\bar{u}}(p_1) - \hat{\bar{u}}(p_2)| \le c_4 |\Delta p|^{\sigma_2} M^{-2}(p_2), \qquad |\Delta p| \le 1, \\
&p_1, p_2 \in R_1,\ \sigma_2 \in (0, 1];
\end{aligned} \tag{1.6}
$$

where c_1, c_2, c_3, c_4 *are positive numbers and* $\varepsilon > 0$ *is sufficiently small* (*see the bound* (2.7) *below*).

Then, the solution $u(x, t)$ *of the Cauchy problem* (1.1) *has the asymptotics,* $t \to \infty$,

$$
u(x, t) = e^{-\lambda t} t^{-1/\delta} A \int_0^\infty \cos(p\xi) e^{-wp^\delta} \, dp + O(t^{-1/\delta - \mu} e^{-\lambda t}), \tag{1.7}
$$

uniformly with respect to $\xi = |x| t^{-1/\delta} \ge 0$, *where* $\mu > 0$ *is some constant dependent on the symbol* $K(p)$. *The constant* A *is explicitly determined in terms of the symbol* $K(p)$ *and the Fourier-image* $\hat{\bar{u}}(p)$ *of the initial condition and it will be indicated in the proof* (*see* (3.22) *and* (4.26) *below*).

EXAMPLES. 1. The symbol $K(p) = c_1|p|^\delta + ic_2|p|^\beta$, $\delta \in (0, 2)$, $\beta > \delta$, $c_1, c_2 = \text{const}$, $c_1 > 0$, satisfies all conditions 1) of Theorem 1.

2. The nonlinear nonlocal Schrödinger equation of the form

$$
iu_t + u^2 u^* + u_{xx} + i \int_{-\infty}^\infty q(x - y) u(y) \, dy = 0
$$

encountered in the theory of plasma (see **[156]**) has the symbol $K(p) = ip^2 + \hat{q}(p)$. The asymptotics (1.7) holds for this equation if $\hat{q}(p) \ge \lambda + w|p|^\delta$, $\delta \in (0, 2)$, $\lambda \ge 0$, $w > 0$.

REMARK. In this theorem the symbol $K(p)$ is assumed to depend on p as a power function for small p (i.e., $w \neq 0$). This is the most interesting case for applications. In the framework of the suggested approach we can consider the general case of arbitrary dependence of the symbol $K(p)$ on p. For instance, for the generalized Landau-Ginzburg equation, whose symbol $K(p)$ has the form $K(p) = ip^2 + a$, $a > 0$, the solution $u(x, t)$ to the corresponding Cauchy problem has the following asymptotics, as $t \to \infty$:

$$
u(x, t) = A(\xi) t^{-1/2} e^{i\xi^2 t - at} + O(t^{-1} e^{-at}) \tag{1.8}
$$

uniformly in $\xi = \frac{x}{2t} \in R_1$; $A(\xi)$ can be expressed, as above, in terms of the symbol $K(p)$ and the Fourier-image of the initial perturbation (see §5 for details). In the concluding section, §6 , of this chapter we find the asymptotics for large times of spatially periodic solutions for the nonlinear nonlocal Schrödinger equation. The results of this chapter were published in **[57]**.

§2. Technical lemmas

We consider two lemmas in this section. Let $\mathcal{N} \ge 0$, $p \in R_1$, $t > 0$, $\gamma \in [0, 1]$.

$$
H_{\mathcal{N}}(p, t) = (\mathcal{N} + 1)^{-\chi} M^{-3}(p) e^{-\lambda t - t\kappa(p/(2\mathcal{N}+1))},
$$

$$
\begin{aligned}
U_{\mathcal{N}}(p, t, \gamma) = &\sum_{l=1}^{\mathcal{N}} \sum_{j=1}^{l} \int_0^t d\tau \, e^{-(\lambda + 2\kappa(p))(t-\tau)} \iint_{-\infty}^\infty dq \, dr \, H_{j-1}(p - q, \tau) \\
&\cdot H_{e-j}(q - r, \tau) H_{\mathcal{N}-l}(r, \tau) m^{-\gamma}(p), \qquad \rho = \sqrt{q^2 + r^2}\,.
\end{aligned}
$$

LEMMA 1. *Let $K(p)$ satisfy the conditions* (1.3) *and* (1.4), $0 < \delta < 2-\gamma$, *and* $\chi = \delta + \frac{4\beta}{\beta-\delta}$. *Then for all* $p \in R_1$, $t > 0$

$$U_{\mathscr{N}}(p,t,\gamma) \le cH_{\mathscr{N}}(p,t); \tag{2.1}$$

here c is a positive constant independent of $\mathscr{N}$.

PROOF. Taking account of the following bounds, valid for all p, q, r, $1 \le j \le l \le \mathscr{N}$:

$$\begin{aligned}
&(M(p-q)M(q-r)M(r))^{-3}\\
&\quad\le 4^3 M^{-3}(p)(M^{-3}(p-q)M^{-3}(r) + M^{-3}(q-r)M^{-3}(r)\\
&\qquad\qquad + M^{-3}(p-q)M^{-3}(q-r));\\
&((\mathscr{N}+1-l)(l+1-j)j)^{-\chi}) \le 4^{\chi}(\mathscr{N}+1)^{-\chi}((\mathscr{N}+1-l)^{-\chi}(l-1-j)^{-\chi}\\
&\qquad\qquad + (l+1-j)^{-\chi}j^{-\chi} + (\mathscr{N}+1-l)^{-\chi}j^{-\chi}),
\end{aligned}$$

we transform the expression for $U_{\mathscr{N}}$ as follows:

$$\begin{aligned}
U_{\mathscr{N}} \le& 4^{3+\chi}M^{-3}(p)(\mathscr{N}+1)^{-\chi}\\
&\cdot\sum_{l=1}^{\mathscr{N}}\sum_{j=1}^{l}(((\mathscr{N}+1-l)(l+1-j))^{-\chi} + ((l+1-j)j)^{-\chi}\\
&\qquad + ((\mathscr{N}+1-l)^{-\chi}\iint_{-\infty}^{\infty} dq\,dr((M(p-q)M(r))^{-3}\\
&\qquad + (M(q-r)M(r))^{-3} + (M(p-q)M(q-r))^{-3})\\
&\cdot m^{-\gamma}(p)\int_0^t d\tau\, e^{-\lambda t - 2\kappa(p)t + \mathscr{L}\tau}
\end{aligned} \tag{2.2}$$

where

$$\mathscr{L}(p,q,r) \equiv 2\kappa(p) - \kappa\left(\frac{p-q}{2j-1}\right) - \kappa\left(\frac{q-r}{2l+1-2j}\right) - \kappa\left(\frac{r}{2\mathscr{N}+1-2l}\right).$$

Consider a typical integral on the right-hand side of (2.2):

$$\begin{aligned}
I_1 =& 4^{3+\chi}M^{-3}(p)(\mathscr{N}+1)^{-\chi}\sum_{l=1}^{\mathscr{N}}\sum_{j=1}^{l}((\mathscr{N}+1-l)(l+1-j))^{-\chi}\\
&\cdot\iint_{-\infty}^{\infty} dq\,dr\, M^{-3}(p-q)M^{-3}(r)m^{-\gamma}(p)\int_0^t e^{-\lambda t-2\kappa(p)t+\mathscr{L}\tau}\,d\tau\,.
\end{aligned} \tag{2.3}$$

Since the function $\kappa(p) = c_2 m^{\delta}(p)$ is monotone nondecreasing,

$$\kappa\left(\frac{p-q}{2j-1}\right) + \kappa\left(\frac{q-r}{2l-2j+1}\right) + \kappa\left(\frac{r}{2\mathscr{N}-2l+1}\right) \ge \kappa\left(\frac{p}{2\mathscr{N}+1}\right) \tag{2.4}$$

for all $p, q, r \in R_1$, $1 \le j \le l \le \mathscr{N}$. Owing to (2.4), we obtain

$$\int_0^t e^{-\kappa(p)t+\mathscr{L}\tau}\,d\tau \le \frac{e^{-t\kappa(p/(2\mathscr{N}+1))}}{\max(\kappa(p), |\mathscr{L}(p,q,r)|)}\,. \tag{2.5}$$

It should be noted that

$$\max(\kappa, |\mathscr{L}|) \ge ((2\mathscr{N}+1-2l)(2l+1-2j))^{-\delta}4^{-3}m^{-\delta}(p)\,. \tag{2.6}$$

Substitution of (2.5) and (2.6) in (2.3) yields

$$\begin{aligned} I_1 &\le 4^{6+\chi} H_{\mathcal{N}}(p,t) \sum_{l=1}^{\mathcal{N}} \sum_{j=1}^{l} ((\mathcal{N}+1-l)(l+1-j))^{\delta-\chi} \\ &\quad \cdot \iint_{-\infty}^{\infty} dq\,dr\, \frac{m^{-3}(p-q)M^{-3}(r)}{m^{\gamma+\delta}(p)} \\ &\le \frac{4^{8+\chi+\delta}}{2-\delta-\gamma} H_{\mathcal{N}}(p,t)\,. \end{aligned}$$

The remaining integrals in (2.2) can be estimated similarly. Lemma 1 is proved. □

LEMMA 2. *Under the conditions of Lemma* 1, *for* $0 < a \le 1$, $|p| \le 1$ *we have*

$$\begin{aligned} &\int_0^{2\pi} d\varphi \int_0^{\infty} d\rho\, M^{-2}(\rho) m^{1-\gamma_1}(\rho) \int_0^t e^{K^1(p)r - \lambda\tau - \kappa(\rho a)\tau}\, d\tau \\ &\qquad \le e^{K^1(p)t-\lambda t} a^{-2-\delta} O(|p|^{-1-\gamma_2} t^{-(1+\gamma_2)/\delta)})\,, \end{aligned}$$

where $0 < \gamma_2 \le m(2-\delta-\gamma_1)$.

The proof of this lemma coincides with the arguments of §2, Chapter 7.

§3. Proof of Theorem 1

3.1. Perturbation theory. Application of the Fourier transform to the NNS equation leads to

$$i\hat{u}_t + \iint_{-\infty}^{\infty} dq\,dr\, \hat{u}^*(q-p,t)\hat{u}(q-r,t)\hat{u}(r,t) + iK(p)\hat{u}(p,t) = 0,$$

whence, upon integration with respect to t, we find

$$\begin{aligned} \hat{u}(p,t) &= \hat{\overline{u}}(p) e^{-K(p)t} \\ &\quad + i\int_0^t e^{-K(p)(t-\tau)}\, d\tau \iint_{-\infty}^{\infty} dq\,d\tau\, \hat{u}^*(q-p,\tau)\hat{u}(q-r,\tau)\hat{u}(r,\tau)\,. \end{aligned} \tag{3.1}$$

We put $\hat{\overline{u}}(p) = \varepsilon\overline{v}(p)$ and represent $\hat{u}(p,t)$ as a formal power series in ε:

$$\hat{u}(p,t) = \sum_{\mathcal{N}=0}^{\infty} \varepsilon^{2\mathcal{N}+1} v_{\mathcal{N}}(p,t)\,. \tag{3.2}$$

Substituting (3.2) in (3.1) and, as usual, comparing coefficients of like powers of ε, we obtain for $\mathcal{N} \ge 1$

$$v_0(p,t) = \overline{v}(p)e^{-K(p)t}, \tag{3.3}$$

$$\begin{aligned} v_{\mathcal{N}}(p,t) &= i\int_0^t e^{-K(p)(t-\tau)}\, d\tau \iint_{-o\infty}^{\infty} dq\,dr \sum_{l=1}^{\mathcal{N}} \sum_{j=1}^{l} v_{j-1}^*(q-p,\tau) \\ &\quad \cdot v_{l-j}(q-r,\tau) v_{\mathcal{N}-l}(r,\tau)\,. \end{aligned} \tag{3.4}$$

We prove now the following estimate, $\mathcal{N} \ge 0$:

$$|v_{\mathcal{N}}(p,t)| \le c^{\mathcal{N}} H_{\mathcal{N}}(p,t) M^{-\alpha}(p), \qquad p \in R_1,\ t > 0\,. \tag{3.5}$$

For $\mathcal{N} = 0$ this inequality (3.5) follows from (1.4) and (1.6). Next, arguing by induction, in view of the result of Lemma 1, from (3.4) we obtain

$$|v_{\mathcal{N}}(p,t)| \leq \int_0^t e^{-K^1(p)(t-\tau)} \iint_{-\infty}^{\infty} \sum_{l=1}^{\mathcal{N}} \sum_{j=1}^{l} c^{\mathcal{N}-1} M^{-\alpha}(p-q) M^{-\alpha}(q-r)$$

$$\cdot M^{-\alpha}(r) H_{j-1}(p-q,\tau) H_{l-j}(q-r,\tau) H_{\mathcal{N}-l}(r,\tau)\, dq\, dr\, d\tau$$

$$\leq c^{\mathcal{N}-1} 4^{\alpha} U_{\mathcal{N}}(p,t,0) M^{-\alpha}(p) \leq c^{\mathcal{N}} H_{\mathcal{N}}(p,t) M^{-\alpha}(p)\,.$$

From the representation (3.2), inverting the Fourier transform, we find that

$$u(x,t) = \frac{1}{2\pi} \int_{-\infty}^{\infty} e^{ipx} \hat{u}(p,t)\, dp = \frac{1}{2\pi} \sum_{\mathcal{N}=0}^{\infty} \varepsilon^{2\mathcal{N}+1} \psi_{\mathcal{N}}(x,t), \tag{3.6}$$

where

$$\psi_{\mathcal{N}}(x,t) = \int_{-\infty}^{\infty} e^{ipx} v_{\mathcal{N}}(p,t)\, dp, \qquad 0 < \varepsilon < 1/c\,. \tag{3.7}$$

The constant c is the same as in (3.5). It follows from (3.5) and (3.7) that the series of perturbation theory converges absolutely and uniformly. This allows termwise integration of the series in (3.6). From (3.7) we find ($\mathcal{N} \geq 0$ is fixed below):

$$\begin{aligned}
\psi_{\mathcal{N}}(x,t) &= \sum_{i=1}^{5} J_{\mathcal{N}}^i, \quad \text{where } J_{\mathcal{N}}^1 = \int_{|p|\geq 1} e^{ipx} v_{\mathcal{N}}(p,t)\, dp, \\
J_{\mathcal{N}}^2 &= \int_{|p|\leq 1} e^{ipx - K^1(p)t} (v_{\mathcal{N}}(o,t) e^{\lambda t} - \Phi_{\mathcal{N}})\, dp, \\
J_{\mathcal{N}}^3 &= \int_{|p|\leq 1} e^{ipx} v_{\mathcal{N}}(p,t)(1 - e^{iK^2(p)t})\, dp, \\
J_{\mathcal{N}}^4 &= \Phi_{\mathcal{N}} \int_{|p|\leq 1} e^{ipx - K^1(p)t}\, dp, \qquad \Phi_{\mathcal{N}} \equiv \lim_{t\to\infty} v_{\mathcal{N}}(0,t) e^{\lambda t}, \\
J_{\mathcal{N}}^5 &= \int_{|p|\leq 1} e^{ipx - K^1(p)t} (v_{\mathcal{N}}(p,t) e^{K(p)t} - v_{\mathcal{N}}(0,t) e^{\lambda t})\, dp\,.
\end{aligned} \tag{3.8}$$

3.2. Estimates of the integrals. Everywhere below $c > 0$ is some constant independent of $\mathcal{N}$. From (3.5), for the integral $J_{\mathcal{N}}^1$ we obtain

$$\begin{aligned}
|J_{\mathcal{N}}^1| &\leq c^{\mathcal{N}} \int_{|p|\geq 1} (\mathcal{N}+1)^{-\chi} M^{-3}(p) e^{-\lambda t - t\kappa(p/(2\mathcal{N}+1))}\, dp \\
&= c^{\mathcal{N}} O(e^{-\lambda t} t^{-2/8})\,.
\end{aligned} \tag{3.9}$$

For $\mathcal{N} \geq 1$ (3.4) leads to

$$\Phi_{\mathcal{N}} = i \int_0^{\infty} d\tau \iint_{-\infty}^{\infty} dq\, dr \sum_{l=1}^{\mathcal{N}} \sum_{j=1}^{l} v_{j-1}^*(q,\tau) v_{l-j}(q-r,\tau) v_{\mathcal{N}-l}(r,\tau),$$

and also

$$v_{\mathcal{N}}(0,t)e^{\lambda t} - \Phi_{\mathcal{N}} = -i\int_t^\infty d\tau \iint_{-\infty}^\infty dq\,dr \cdot \sum_{l=1}^{\mathcal{N}}\sum_{j=1}^{l} v_{j=1}^*(q,t)v_{l-j}(q-r,\tau)v_{\mathcal{N}-l}(r,\tau)\,. \tag{3.10}$$

Introduction of (3.5) in (3.10) yields

$$\begin{aligned}
|v_{\mathcal{N}}(0,t)e^{\lambda t} - \Phi_{\mathcal{N}} &\le c^{\mathcal{N}-1}\sum_{l=1}^{\mathcal{N}}\sum_{j=1}^{l}((\mathcal{N}+1-l)(l+1-j))^{-\chi} \\
&\quad \cdot \iint_{-\infty}^\infty dq\,dr\, M^{-3}(q)M^{-3}(r) \\
&\quad \cdot \left(\kappa\left(\frac{r}{2\mathcal{N}+1-2l}\right) + \kappa\left(\frac{q}{2j-1}\right)\right)^{-1} \\
&\quad \cdot \exp\left(-\tau\kappa\left(\frac{q}{2j-1}\right) - \tau\kappa\left(\frac{r}{2\mathcal{N}+1-2l}\right)\right) \\
&\le c^{\mathcal{N}-1}\sum_{l=1}^{\mathcal{N}}\sum_{j=1}^{l}((\mathcal{N}+1-l)j)^{\delta+\delta\mu-\chi}\mu^\mu M^{-\mu}(t)m^2(c_2) \\
&\quad \cdot \iint_{-\infty}^\infty dq\,dr\, M^{-3}(q)M^{-3}(r)(\kappa(q)+\kappa(r))^{-1-\mu} \\
&\le c^{\mathcal{N}} M^{-\mu}(t),
\end{aligned} \tag{3.11}$$

since $\chi \ge 4+\delta$, $2-\delta-\delta\mu > 0$, $\mu = \frac{1}{2\delta}\min\{\delta, \sigma_1, \sigma_2, \frac{2-\delta}{4}\sigma_1, \frac{\beta-\delta}{4}\}$ and for all $t \ge 0$, $a > 0$, $e^{-at} \le \mu^\mu M^{-\mu}(t)m^{-\mu}(a)$. For $\mathcal{N} = 0$ we have $v_0(0,t)e^{\lambda t} = \overline{v}(0) = \Phi_0$, and in view of (1.6) we conclude that $|\Phi_0| \le 1$. From this and inequality (3.11), taken for $t = 0$, we find that

$$|\Phi_{\mathcal{N}}| \le c^{\mathcal{N}}, \qquad \mathcal{N} \ge 0\,. \tag{3.12}$$

Owing to (1.3), from (3.11) we also obtain that

$$|J_{\mathcal{N}}^2| \le \int_{|p|\le 1} e^{-K^1(p)t}c^{\mathcal{N}}t^{-\mu}\,dp = c^{\mathcal{N}}O(e^{-\lambda t}t^{-1/\delta-\mu})\,. \tag{3.13}$$

The bounds for the integrals $J_{\mathcal{N}}^3$ and $J_{\mathcal{N}}^4$ coincide with inequalities (3.16) and (3.17) of §3 in the previous chapter:

$$|J_{\mathcal{N}}^3| = c^{\mathcal{N}}O(t^{-1/\delta-\mu}e^{-\lambda t}), \tag{3.14}$$

$$\begin{aligned}
J_{\mathcal{N}}^4 &= 2\Phi_{\mathcal{N}}e^{-\lambda t}t^{1/\delta}\int_0^\infty \cos(\eta\xi)e^{-w\eta^\delta}\,d\eta \\
&\quad + c^{\mathcal{N}}O(e^{-\lambda t}t^{-2/\delta}), \qquad \xi = |x|t^{-1/\delta}\,.
\end{aligned} \tag{3.15}$$

3.3. Estimation of the Hölder term $J^5_{\mathcal{N}}$. We first establish the following preliminary bound $|p_2| \le p_1|$:

$$|v_{\mathcal{N}}(p_1, t) - v_{\mathcal{N}}(p_2, t)| = c^{\mathcal{N}} O(t^{2\gamma/\sigma_1}|\Delta p|^{2\gamma}) H_{\mathcal{N}}(p_1, t), \tag{3.16}$$

where $\Delta p = p_1 - p_2$, $|\Delta p| \le 1$, $\gamma = \mu\delta$. For $\mathcal{N} = 0$ the bound (3.16) follows from (3.3), (1.5), and (1.6). Further, arguing by induction, we write the difference on the left-hand side of (3.16) as follows:

$$v_{\mathcal{N}}(p_1, t) - v_{\mathcal{N}}(p_2, t) = I_1 + I_2 + I_3,$$

where

$$I_1 = i\int_0^t (e^{-K(p_1)(t-\tau)} - e^{-K(p_2)(t-\tau)}) \iint_{-\infty}^{\infty} dq\, dr \sum_{l=1}^{\mathcal{N}} \sum_{j=1}^{l} v^*_{j-1}(q - p_2, \tau) \cdot v_{l-j}(q - r, \tau) v_{\mathcal{N}-l}(r, \tau)\, d\tau,$$

$$I_2 = i\int_0^t e^{-K(p_1)(t-\tau)}\, d\tau \iint_{|p_1-q|<|p_2-q|} dq\, dr \cdot \sum_{l=1}^{\mathcal{N}} \sum_{j=1}^{l} (v^*_{j-1}(q - p_1, \tau) - v^*_{j-1}(q - p_2, \tau)) v_{l-j}(q - r, \tau) v_{\mathcal{N}-l}(r, \tau)\,.$$

The integral I_3 is defined similarly to I_2, only the integration is taken over the domain $|p_1 - q| \ge |p_2 - q|$.

We note first of all that (1.5) implies

$$|e^{-K(p_1)(t-\tau)} - e^{-K(p_2)(t-\tau)}| \le e^{-\lambda(t-\tau)-\kappa(p_2)(t-\tau)} M^{\alpha}(p_2) O(t^{2\gamma/\sigma_1}|\Delta p|^{2\gamma})\,. \tag{3.17}$$

Therefore, applying (3.5) and the result of Lemma 1, we get

$$\begin{aligned}
|I_1| &= c^{\mathcal{N}-1} O(t^{2\gamma/\sigma_1}|\Delta p|^{2\gamma}) M^{\alpha}(p_2) \int_0^t e^{-(\lambda+\kappa(p_2))(t-\tau)}\, d\tau \\
&\quad \cdot \iint_{-\infty}^{\infty} dq\, dr \sum_{l=1}^{\mathcal{N}} \sum_{j=1}^{l} (M(p_2 - q) M(q - r) M(r))^{-\alpha} H_{j-1}(p_2 - q, \tau) \\
&\quad \cdot H_{l-j}(q - r, \tau) H_{\mathcal{N}-l}(r, \tau) \\
&= c^{\mathcal{N}-1} O(t^{2\gamma/\sigma_1}|\Delta p|^{2\gamma}) 4^{\alpha} U_{\mathcal{N}}(p_2, t, 0) \\
&= c^{\mathcal{N}} H_{\mathcal{N}}(p_2, t) O(t^{2\gamma/\sigma_1}|\Delta p|^{2\gamma})\,.
\end{aligned}$$

In view of (3.5) and the induction hypothesis, we obtain

$$\begin{aligned}
|I_2| &\le \int_0^t e^{-K^1(p)(t-\tau)}\, d\tau \iint_{-\infty}^{\infty} dq\, dr \sum_{l=1}^{\mathcal{N}} \sum_{j=1}^{l} c^{\mathcal{N}-1} O(\tau^{2\gamma/\sigma_1}|\Delta p|^{2\gamma}) \\
&\quad \cdot H_{j-1}(p_1 - q, \tau) H_{l-j}(q - r, \tau) H_{\mathcal{N}-l}(r, \tau) \\
&= c^{\mathcal{N}} O(t^{2\gamma/\sigma_1}|\Delta p|^{2\gamma}) H_{\mathcal{N}}(p_2, t)\,.
\end{aligned}$$

The integral I_3 is estimated similarly. The estimate (3.16) has been established by induction.

Now we prove the following estimate for $|p| \leq 1$:

$$|v_{\mathcal{N}}(p,t)e^{K(p)t} - v_{\mathcal{N}}(0,t)e^{\lambda t}| = c^{\mathcal{N}} e^{K^1(p)t-\lambda t} O(|p|^{\gamma-1} t^{-1/\delta-\mu}) . \tag{3.18}$$

For $\mathcal{N} = 0$ the estimate (3.18) follows from condition (1.6). For $\mathcal{N} \geq 1$ we carry out induction. By (3.4) we represent the left-hand side of (3.18) in the form

$$v_{\mathcal{N}}(p,t)e^{K(p)t} - v_{\mathcal{N}}(0,t)e^{\lambda t} = I_1 + I_2 + I_3,$$

$$I_1 = i \int_0^t (e^{K(p)\tau} - e^{\lambda\tau})\, d\tau \iint_{-\infty}^{\infty} dq\, dr \sum_{l=1}^{\mathcal{N}} \sum_{j=1}^{l} v_{j-1}^*(q,\tau) v_{l-j}(q-r,\tau) v_{\mathcal{N}-l}(r,\tau),$$

$$I_2 = i \int_0^t e^{K(p)\tau}\, d\tau \iint_{|p-q|\leq|q|} dq\, dr \sum_{l=1}^{\mathcal{N}} \sum_{j=1}^{l} (v_{j-1}^*(q-p,\tau) - v_{j-1}^*(q,\tau)) \cdot v_{l-j}(q-r,\tau) v_{\mathcal{N}-l}(r,\tau) .$$

The integral I_3 is defined similarly to I_2, only the integration is taken over the domain $|p-q| > |q|$. Further we use the following inequality:

$$\begin{aligned} &\tau^{2\gamma/\delta} e^{-\kappa(r/(2\mathcal{N}+1-2l))\tau - \kappa(q/(2j-1))\tau} \\ &\quad \leq c((2\mathcal{N}+1-2l)(2j-1))^{2\gamma} (\kappa(q)+\kappa(r))^{-2\gamma/\delta} . \end{aligned} \tag{3.19}$$

Since for $|p| \leq 1$, $|e^{K(p)\tau} - e^{\lambda\tau}| = e^{K^1(p)\tau} O(|p|^{2\gamma}\tau^{2\gamma/\delta})$, applying Lemma 2 with $\gamma_1 = 2\gamma$ and $\gamma_2 = \gamma$ in view of (3.19) and (3.5), we find that

$$\begin{aligned} |I_1| &\leq \int_0^t e^{K^1(p)\tau} O(|p|^{2\gamma} t^{2\gamma/\delta})\, d\tau \iint_{-\infty}^{\infty} dq\, dr \sum_{l=1}^{\mathcal{N}} \sum_{j=1}^{l} c^{\mathcal{N}-1} H_{j-1}(q,\tau) \\ &\quad \cdot H_{l-j}(q-r,\tau) H_{\mathcal{N}-l}(r,\tau) \\ &= c^{\mathcal{N}-1} O(|p|^{2\gamma}) \sum_{l=1}^{\mathcal{N}} \sum_{j=1}^{l} ((\mathcal{N}+1-l)j)^{2\gamma-\chi} \\ &\quad \cdot \int_0^t e^{K^1(p)\tau-\lambda\tau}\, d\tau \int_0^{2\pi} d\varphi \int_0^{\infty} \rho\, d\rho\, M^{-3}(\rho) M^{-2\gamma}(\rho) e^{-\tau\kappa(((2\mathcal{N}+1-2l)(2j-1))^{-1}\rho)} \\ &= c^{\mathcal{N}} e^{K^1(p)t} O(|p|^{\gamma-1} e^{-\lambda t} t^{-1/\delta-\mu}) . \end{aligned}$$

Substitution of (3.5) and (3.16) in I_2 yields

$$\begin{aligned} |I_2| &= \int_0^t e^{K^1(p)\tau} d\tau \iint_{|p-q|\leq|q|} dq\, dr \sum_{l=1}^{\mathcal{N}} \sum_{j=1}^{l} c^{\mathcal{N}-1} O(|p|^{2\gamma} \tau^{2\gamma/\sigma_1}) \\ &\quad \cdot H_{j-1}(p-q,\tau) H_{l-j}(q-r,\tau) H_{\mathcal{N}-l}(r,\tau) . \end{aligned}$$

From this, using (3.19) and the result of Lemma 2 for $\gamma_1 = 2\gamma\delta/\sigma_1$ and $\gamma_2 = \gamma$, we conclude that

$$
\begin{aligned}
|I_2| \leq & c^{\mathcal{N}-1} O(|p|^{2\gamma}) \sum_{l=1}^{\mathcal{N}} \sum_{j=1}^{l} ((\mathcal{N}+1-l)j)^{2\gamma\delta/\sigma_1-\chi} \\
& \cdot \int_0^{2\pi} d\varphi \int_0^{\infty} \rho \, d\rho M^{-3}(\rho) m^{-2\gamma\delta/\sigma_1}(\rho) \\
& \cdot \int_0^t \exp\{K^1(p)\tau - \lambda\tau - \tau\kappa(((2\mathcal{N}+1-2l)(2j-1))^{-1}\rho)\} \, d\tau \\
= & c^{\mathcal{N}} e^{K^1(p)t-\lambda t} O(|p|^{\gamma-1} t^{-1/\delta-\mu}) \, .
\end{aligned}
$$

The integral I_3 is estimated similarly. The bound (3.18) is established. From (3.18) in turn, it follows that

$$
(3.20) \qquad |J^5_{\mathcal{N}}| = c^{\mathcal{N}} O(t^{-1/\delta-\mu} e^{-\lambda t}) \int_{|p|\leq 1} |p|^{\gamma-1} dp = c^{\mathcal{N}} O(t^{-1/\delta-\mu} e^{-\lambda t}) \, .
$$

3.4. The asymptotics. Substitution of (3.9), (3.13), (3.14), (3.15), and (3.20) in (3.8) gives

$$
(3.21) \qquad \psi_{\mathcal{N}} = 2\Phi_{\mathcal{N}} e^{-\lambda t} t^{-1/\delta} \int_0^{\infty} \cos(\eta\xi) e^{-\eta^{\delta} w} \, d\eta + c^{\mathcal{N}} O(t^{-1/\delta-\mu} e^{-\lambda t}) \, .
$$

Introducing (3.21) into (3.6), we obtain (1.7) with the constant A equal

$$
(3.22) \qquad A = \frac{1}{\pi} \sum_{\mathcal{N}=0}^{\infty} \Phi_{\mathcal{N}} \varepsilon^{2\mathcal{N}+1} = \frac{1}{\pi} \lim_{t\to\infty} e^{\lambda t} \int_{-\infty}^{\infty} u(x,t) \, dx \, .
$$

It remains now to compute the numbers $\Phi_{\mathcal{N}}$.

§4. Computation of the numbers $\Phi_{\mathcal{N}}$

In this section we present a method for computing the $\Phi_{\mathcal{N}}$. The argument follows closely that in §3, Chapter 6.

4.1. Perturbation theory. We represent the function $v_{\mathcal{N}}(p,t)$ in the form

$$
(4.1) \qquad
\begin{gathered}
v_{\mathcal{N}}(p,t) = \int_{-\infty}^{\infty} dy \, Q_{\mathcal{N}}(\check{y},t) V_{\mathcal{N}}(\check{y}), \qquad V_{\mathcal{N}}(\check{y}) = \prod_{K=0}^{2\mathcal{N}} \overline{v}(y_k), \\
y \in R_{2\mathcal{N}}, \ \check{y} = (y_0, y) \in R_{2\mathcal{N}+1}, \ p = \sum_{k=0}^{2\mathcal{N}} y_k \, .
\end{gathered}
$$

For $\mathcal{N} = 0$ formula (4.1) becomes (2.3), whence we find that

$$
Q_0(p,t) = e^{-K(p)t} \, .
$$

Substitution of (4.1) in both sides of (2.4) leads, for all $p \in R$, and $\mathcal{N} \geq 1$, to

$$
\begin{aligned}
&\int_{-\infty}^{\infty} dy\, Q_{\mathcal{N}}(\check{y}, t) V_{\mathcal{N}}(\check{y}) \\
&\quad = i \int_0^t e^{-K(p)(t-\tau)}\, d\tau \sum_{l=1}^{\mathcal{N}} \sum_{j=1}^{l} \iint_{-\infty}^{\infty} dq\, dr \\
&\qquad \cdot \left(\int_{-\infty}^{\infty} dx\, Q_{j-1}(\check{x}, \tau) V_{j-1}(\check{x}) \right)^* \left(\int_{-\infty}^{\infty} dz\, Q_{l-j}(\check{z}, \tau) V_{l-j}(\check{z}) \right) \\
&\qquad \cdot \left(\int_{-\infty}^{\infty} d\xi\, Q_{\mathcal{N}-l}(\check{\xi}, \tau) V_{\mathcal{N}-l}(\check{\xi}) \right) .
\end{aligned} \tag{4.2}
$$

Here $y \in R_{2\mathcal{N}}$, $z \in R_{2l-2j}$, $x \in R_{2j-2}$, $\xi \in R_{2\mathcal{N}-2l}$, $p = \sum_{k=0}^{2\mathcal{N}} y_k$, $q - p = \sum_{k=0}^{2j-2} x_k$, $q - r = \sum_{k=0}^{2l-2j} z_k$, $r = \sum_{k=0}^{2\mathcal{N}-2l} \xi_k$. For each fixed l and j we introduce new variables of integration η on the right-hand side of (4.2):

$$
\eta_k = -x_k, \quad 0 \leq k \leq 2j-2; \qquad \eta_k = z_{k-2j+1}, \quad 2j-1 \leq k \leq 2l-1;
$$
$$
\eta_k = \xi_{k-2l}, \quad 2l \leq k \leq 2\mathcal{N} .
$$

As a result, (4.2) becomes

$$
\begin{aligned}
&\int_{-\infty}^{\infty} dy\, Q_{\mathcal{N}}(\check{y}, t) V_{\mathcal{N}}(\check{y}) = i \int_0^t e^{-K(p)(t-\tau)}\, d\tau \\
&\quad \cdot \int_{-\infty}^{\infty} d\eta \sum_{l=1}^{\mathcal{N}} \sum_{j=1}^{l} Q_{j-1}^*(-\eta_0, \ldots, -\eta_{2j-2}, \tau) Q_{l-j}(\eta_{2j-1}, \ldots, \eta_{2l-1}, \tau) \\
&\quad \cdot Q_{\mathcal{N}-l}(\eta_{2l}, \ldots, \eta_{2\mathcal{N}}, \tau) \prod_{k=0}^{2j-2} \overline{v}^*(-\eta_k) \prod_{k=2j-1}^{2\mathcal{N}} \overline{v}(\eta_k) .
\end{aligned} \tag{4.3}
$$

We denote by $P^-(y)$ the operator acting on a complex-valued function $f(y)$ as follows: $P^-(y) f(y) = f^*(-y)$. Then, from (4.3) we get

$$
\begin{aligned}
&\int_{-\infty}^{\infty} dy\, Q_{\mathcal{N}}(\check{y}, t) V_{\mathcal{N}}(\check{y}) \\
&\quad = i \int_0^t e^{-K(p)(t-\tau)}\, d\tau \int_{-\infty}^{\infty} dy \sum_{l=1}^{\mathcal{N}} \sum_{j=1}^{l} P^-(\eta_0) \cdots P^-(\eta_{2j-2}) \\
&\qquad \cdot Q_{j-1}(\eta_0, \ldots, \eta_{2j-2}, \tau) Q_{l-j}(\eta_{2j-1}, \ldots, \eta_{2l-1}, \tau) \\
&\qquad \cdot Q_{\mathcal{N}-l}(\eta_{2l}, \ldots, \eta_{2\mathcal{N}}, \tau) V_{\mathcal{N}}(\check{\eta}) .
\end{aligned} \tag{4.4}
$$

Interchanging the order of integration with respect to η and τ on the right-hand side of (4.4), we see that (4.4) will be satisfied if we require that

$$
\begin{aligned}
Q_{\mathcal{N}}(\check{y}, t) = {} & i \sum_{l=1}^{\mathcal{N}} \sum_{j=1}^{l} \int_0^t e^{-K(p)(t-\tau)}\, d\tau\, P^-(y_0) \cdots P^-(y_{2j-2}) \\
& \cdot Q_{j-1}(y_0, \ldots, y_{2j-2}, \tau) Q_{l-j}(y_{2j-1}, \ldots, y_{2l-1}, \tau) \\
& \cdot Q_{\mathcal{N}-l}(y_{2l}, \ldots, y_{2\mathcal{N}}, \tau), \qquad p = \sum_{k=0}^{2\mathcal{N}} y_k .
\end{aligned} \tag{4.5}
$$

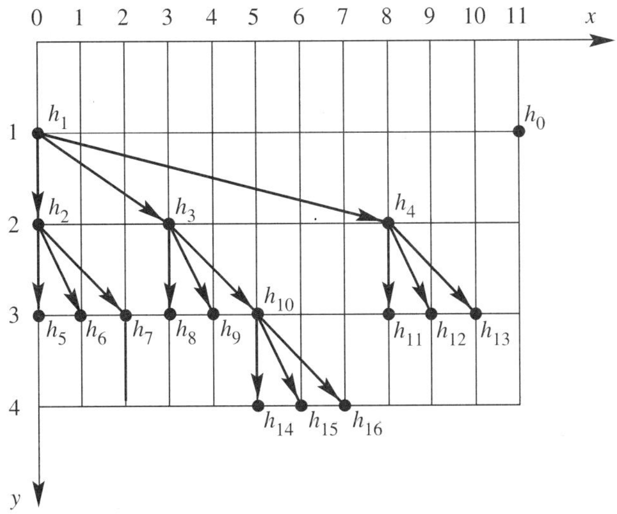

FIGURE 1. An example of graph g of the set $G_{0,5}$

4.2. First formula for $Q_{\mathcal{N}}$. In order to solve the recurrent relation (4.5), we consider a procedure of constructing a certain family of graphs by means of an algorithm for dividing the segment $[0, 2\mathcal{N}]$ of the natural sequence ($\mathcal{N} \geq 0$).

On the axis Ox we consider the segment $[0, 2\mathcal{N}]$ of the natural sequence. We represent the vertices h_j of a graph g by points on the plane xOy. We mark the step of the division algorithm along the axis Oy. We denote by h_j^x the projection of a vertex h_j on the axis Ox. At the first step of the algorithm we set $h_1 = (0, 1)$. We introduce temporarily the additional vertex $h_0 - (2\mathcal{N} + 1, 1)$. At the second step we choose two arbitrary points h_3^2 and h_4^x ($h_3^x < h_4^x$) on the segment (h_1^x, h_0^x) on the natural sequence so that the numbers $h_3^x - h_1^x$ and $h_1^x - h_3^x$ are odd. We also set $h_2^x = 0$. Thus, we obtain three new vertices $h_j = (h_j^x, 2)$, $j = 2, 3, 4$. We draw the three edges $\overrightarrow{h_1 h_2}, \overrightarrow{h_1 h_3}$ and $\overrightarrow{h_1 h_4}$ to which we assign the corresponding lengths $h_3^x - h_2^x - 1$, $h_4^x - h_3^x - 1$, and $h_0^x - h_4^x - 1$, respectively. Besides, we call the vertical edge $\overrightarrow{h_1 h_2}$ negative, and the remaining edges, oblique, positive (see Figure 1).

Next, arguing by induction, suppose at some step the initial segment $[0, 2\mathcal{N}]$ is decomposed by the vertices $h_0, h_1, \dots, h_{\mathcal{N}}$ into l nonempty segments

$$(h_{i_1}^x, h_{j_1}^x), \dots, (h_{i_l}^x, h_{j_l}^x) \qquad (h_{i_1}^x < h_{j_1}^x < \cdots < h_{i_l}^x < h_{j_l}^x),$$

each of which contains no points h_k^x, $k \leq n$, in its interior. The lengths of these segments are odd by construction. We choose two points on each of these segments in such a way that the lengths of all $3l$ newly created segments are odd as well. Then the right ends of these segments are the projections on the axis Ox of the new vertices $h_{n+1}, h_{n+2}, \dots, h_{n+3l}$ of the graph g. We draw the edges $\overrightarrow{h_{i_k} h_{n+3k+1}}, \overrightarrow{h_{i_k} h_{n+3k+2}}$, and $\overrightarrow{h_{i_k} h_{n+3k+3}}$ joining the new vertices with the vertex h_{i_k} This vertex determines the left end-point of the segment $(h_{i_k}^x, h_{j_k}^x)$, on which the corresponding points $h_{n+3k+1}^x = h_{i_k}^x, h_{n+3k+2}^x$, and h_{n+3k+3}^x are chosen. We set the lengths of these edges, to be equal to $h_{n+3k+2}^x - h_{i_k}^x - 1, h_{n+3k+3}^x - h_{n+3k+2}^x - 1$, and $h_{j_k}^x - h_{n+3k+3}^x - 1$, respectively. We call the vertical edges $\overrightarrow{h_{i_k} h_{n+3k+1}}$ negative and the remaining oblique ones positive. We continue this procedure until at some step no nonempty interval that can be further divided remains. The total number of steps of the algorithm does not exceed $\mathcal{N}+1$.

Thus, we have constructed a graph g consisting of the vertices h_j, $1 \le j \le 3\mathcal{N}+1$, joined by normalized and oriented edges. We do not include in g the vertex h_0, introduced in the construction. If we replace the segment $[0, 2\mathcal{N}]$ by $[j, 2\mathcal{N}+j]$, then the vertices of the new graph g' are shifted along the Ox axis by the number $j: h_k^{x\prime} = h_k^x + j$. We denote the set of all graphs obtained by division of the segment $[j, 2\mathcal{N}+j]$ $\mathcal{G}_{j2\mathcal{N}+j}$. Vertices from which three edges exit will be called nodes. In any graph $g \in \mathcal{G}_{0,2\mathcal{N}}$ there are exactly $\mathcal{N}$ nodes, while the total number of vertices is equal to $3\mathcal{N}+1$.

We define the sign σ_h for each vertex h of a graph $g \in \mathcal{G}_{0,2\mathcal{N}}$ as follows. For h_1 we set $\sigma_{h_1} = +$. For an arbitrary vertex h_j, $2 \le j < 3\mathcal{N}+1$, we consider the chain joining the vertices h_1 and h_j. If this chain contains an even number of negative, i.e., vertical edges, then we put $\sigma_{h_j} = +$; otherwise, we define $\sigma_{h_j} = -$.

We fix any three graphs $g^1 \in \mathcal{G}_{0,2j-2}$, $g^2 \in \mathcal{G}_{2j-1,2l-1}$, and $g^3 \in \mathcal{G}_{2l,2\mathcal{N}}$, $1 \le j \le l \le \mathcal{N}$. The set $\mathcal{G}_{0,0}$ consists of only one graph g, which, in its turn, consists of the sole vertex h_1. Further, the superscript will indicate to which of the three graphs the notation refers. We define the procedure of gluing the graphs g^1, g^2, and g^3 as follows:

1) We drop the graphs g^1, g^2, and g^3 one step along the y axis.

2) The vertices of the new graph are the vertex $h_1 = (0, 1)$ and all vertices of the graphs g^1, g^2, and g^3.

3) The edges of the graph g are the edges of the graphs g^1, g^2, and g^3, and also the three new edges $\overrightarrow{h_1h_1^1}, \overrightarrow{h_1h_1^2}$, and $\overrightarrow{h_1h_1^3}$, which have the lengths $|\overrightarrow{h_1h_1^1}| = 2j-2$, $|\overrightarrow{h_1h_1^2}| = 2l-2j$, and $|\overrightarrow{h_1h_1^3}| = 2\mathcal{N}-2l$, respectively. In this case, the vertical edge $\overrightarrow{h_1h_1^1}$ is negative, and the other two edges are positive.

In the graph g constructed by gluing, the signs of the vertices which belonged initially to the graph g^1 are changed to the opposite, while the remaining vertices preserve their signs. We denote the result of gluing by $g = g^1 \otimes g^2 \otimes g^3$. The set of all glued graphs for all possible $g^1 \in \mathcal{G}_{0,2j-2}$, $g^2 \in \mathcal{G}_{2j-1,2l-1}$, $g^3 \in \mathcal{G}_{2l,2\mathcal{N}}$ is denoted by $\mathcal{G}_{0,2j-2} \otimes \mathcal{G}_{2j-1,2l-1} \otimes \mathcal{G}_{2l,2\mathcal{N}}$. We observe that the gluing operation corresponds fully to the division algorithm, i.e., any glued graph $g = g^1 \otimes g^2 \otimes g^3$ can be constructed by division of the segment $[0, 2\mathcal{N}]$. Thus, we obtain the inclusion

$$\mathcal{G}_{0,2j-2} \otimes \mathcal{G}_{2j-1,2l-1} \otimes \mathcal{G}_{2l,2\mathcal{N}} \subset \mathcal{G}_{0,2\mathcal{N}}. \tag{4.6}$$

Since (4.6) holds for any j, l:$1 \le j \le l \in \mathcal{N}$, it follows that

$$\sum_{l=1}^{\mathcal{N}} \sum_{j=1}^{l} \mathcal{G}_{0,2j-2} \otimes \mathcal{G}_{2j-1,2l-1} \otimes \mathcal{G}_{2l,2\mathcal{N}} \subset \mathcal{G}_{0,2\mathcal{N}}. \tag{4.7}$$

In order to prove the reverse inclusion

$$\mathcal{G}_{0,2\mathcal{N}} \subset \sum_{l=1}^{\mathcal{N}} \sum_{j=1}^{l} \mathcal{G}_{0,2j-2} \otimes \mathcal{G}_{2j-1,2l-1} \otimes \mathcal{G}_{2l,2\mathcal{N}}, \tag{4.8}$$

we fix any graph $g \in \mathcal{G}_{0,2\mathcal{N}}$ and show that

$$g = g^1 \otimes g^2 \otimes g^3, \quad \text{where } g^1 \in \mathcal{G}_{0,2j-2}, g^2 \in \mathcal{G}_{2j-1,2l-1},$$
$$g^3 \in \mathcal{G}_{2l,2\mathcal{N}}, 2j-1 = h_3^x, \ 2l = h_4^x.$$

For this, we define the operation of cutting the graph g along the vertices h_3 and h_4 as follows:

1) We remove the vertex h_1 and the edges $\overrightarrow{h_1h_2}, \overrightarrow{h_1h_3}$, and $\overrightarrow{h_1h_4}$.

2) We divide the remaining vertices into three sets:

$$M_1 = \{h_k : h_k^x \in [0, h_3^x]\}, \qquad M_2 = \{h_k : h_k^x \in [h_3^x, h_4^x)\},$$
$$M_3 = \{h_k : h_k^x \in [h_4^x, 2\mathscr{N}]\}.$$

3) The graph g^1 consists of the set of vertices M_1 and all those edges of g which join vertices from the set M_1. The graphs g^2 and g^3 are defined similarly.

4) We displace the graphs g^1, g^2, and g^3 formed upward one step along the Oy axis. The signs of the vertices of the graph g^1 are changed to the opposite.

Since in the construction of a graph $g \in \mathscr{G}_{0,2\mathscr{N}}$ each of the segments $[0, h_3^x)$, $[h_3^x, h_4^x)$, and $[h_4^x, 2\mathscr{N}]$ is subdivided by means of the division algorithm, it follows that $g^1 \in \mathscr{G}_{0,2j-2}$, $g^2 \in \mathscr{G}_{2j-1,2l-1}$, $g^3 \in \mathscr{G}_{2l,2\mathscr{N}}$. It is clear from the definitions of cutting and gluing operations, that $g = g^1 \otimes g^2 \otimes g^3$, i.e., cutting and gluing are mutually inverse operations. This leads to (4.8). From (4.7) and (4.8) we obtain, $\mathscr{N} \geq 1$:

$$\sum_{l=1}^{\mathscr{N}} \sum_{j=1}^{l} \mathscr{G}_{0,2j-2} \otimes \mathscr{G}_{2j-1,2l-1} \otimes \mathscr{G}_{2l,2\mathscr{N}} = \mathscr{G}_{0,2\mathscr{N}}. \tag{4.9}$$

With the aid of (4.9), we estimate the cardinality $|\mathscr{G}_{0,2\mathscr{N}}|$ of the set $\mathscr{G}_{0,2\mathscr{N}}$. Since $|\mathscr{G}_{0,2\mathscr{N}}| = |\mathscr{G}_{j,2\mathscr{N}+j}|$, it follows from (4.9) that

$$|\mathscr{G}_{0,2\mathscr{N}}| = \sum_{l=1}^{\mathscr{N}} \sum_{j=1}^{l} |\mathscr{G}_{0,2j-2}|\, |\mathscr{G}_{0,2l-2j}|\, |\mathscr{G}_{0,2\mathscr{N}-2l}|, \tag{4.10}$$

whence it is easy to obtain the bound

$$|\mathscr{G}_{0,2\mathscr{N}}| \leq 12^{2\mathscr{N}}(\mathscr{N}+1)^{-2}, \qquad \mathscr{N} \geq 0. \tag{4.11}$$

Along with the natural enumeration of vertices given by the division algorithm, below we also use a special enumeration of the vertices. We first define the special order of succession for the nodes. Consider the nodal part of some graph g (i.e., only the nodes and the edges joining them). For graphs of the families $\mathscr{G}_{0,0}$ and $\mathscr{G}_{0,2}$ the special enumeration of vertices coincides with the natural one. Further, we argue by induction. We cut the graph g and obtain $g = g^1 \otimes g^2 \otimes g^3$. Suppose, by induction, that the special ordering of the vertices has already been established in the graphs g^1, g^2, and g^3. Then the special enumeration of nodes in g is defined as follows: the vertex h_1 remains the first also in the special enumeration and after it there follows the nodes of g^1 in the special ordering, then the nodes of g^2 in the special ordering, and, finally, the nodes of g^3 in the special ordering. The special enumeration of all vertices in g is defined as follows: we first number the nodes of g in the special ordering, as described above, then the nonnodal vertices of g^1 in the natural order, then the nonnodal vertices of g^2 in the natural order, and, finally, we number the nonnodal vertices of g^3 also in the natural order.

With the help of the family of graphs introduced above, we shall solve the recurrent relation (4.5). Fix any graph $g \in \mathscr{G}_{0,2\mathscr{N}}$, in which we choose the special enumeration of the vertices. Let $\xi_1 = 0$; for each node $h_j, 2 \leq j \leq \mathscr{N}$, we determine the number ξ_j of the vertex h_{ξ_j} preceding it, and the numbers r_j^1, r_j^2 and r_j^3, $1 \leq j \leq \mathscr{N}$, of the vertices $h_{r_j^1}$, $h_{r_j^2}$ and $h_{r_j^3}$ immediately following the vertex h_{ξ_j} in the sense that g has the edges

$\overrightarrow{h_{\xi_j}h_j}$ and $\overrightarrow{h_jh_{r_j^1}}$, $\overrightarrow{h_jh_{r_j^2}}$, $\overrightarrow{h_jh_{r_j^3}}$. Let $y \in R_{2\mathcal{N}}$, $\check{y} = (y_0, y) \in R_{2\mathcal{N}+1}$. For each node h_j, $1 \leq j \leq \mathcal{N}$, we define three sums of coordinates y_k:

$$s_j^l = \sum_{k=h^x_{r_j^l}}^{k=h^x_{r_j^l}+|\overrightarrow{h_jh_{r_j^l}}|} y_k, \qquad 1 \leq j \leq \mathcal{N},\ l = 1, 2, 3\,.$$

Moreover, let $p_j = s_j^1 + s_j^2 + s_j^3$ and $p = p_1$.

For all $\mathcal{N} \geq 0$, we establish the equality

$$Q_{\mathcal{N}}(\check{y}, t_0) = i^{\mathcal{N}} e^{-K(p)t_0} \sum_{g \in \mathcal{G}_{0,2\mathcal{N}}} \int_0^\infty \cdots \int_0^\infty \prod_{j=1}^{\mathcal{N}} e^{\mathcal{L}_j t_j} \theta(t_{\xi_j} - t_j) \Lambda_g dt_1 \cdots dt_{\mathcal{N}}, \tag{4.12}$$

where

$$\begin{aligned} &\mathcal{L}^+(s_j^1, s_j^2, s_j^3) = K(p_j) - K^*(-s_j^1) - K(s_j^2) - K(s_j^3), \\ &\mathcal{L}^-(s_j^1, s_j^2, s_j^3) = (\mathcal{L}^+(-s_j^1, -s_j^2, -s_j^3))^*, \\ &\mathcal{L}_j = \mathcal{L}^{\sigma_j}(s_j^1, s_j^2, s_j^3)), \qquad \Lambda_g = \prod_{k=\mathcal{N}+1}^{3\mathcal{N}+1} p^{\sigma_k}(y_{h_k^x}), \\ &P^+ \equiv (P^-)^2, \qquad t_{\xi_1} = t_0, \end{aligned} \tag{4.13}$$

σ_j is the sign of the vertex h_j of the graph g recall that we have fixed the special enumeration of vertices), and $\theta(t)$ is the Heaviside function.

For $\mathcal{N} = 0$ (4.12) has the following form:

$$Q_0(y, t_0) = e^{-K(y)t_0}\,. \tag{4.14}$$

Suppose, by induction, that (4.12) has been established for $\mathcal{N} \leq \mathcal{N}_1$. Substituting (4.12) in the recurrent relation (4.5) for $\mathcal{N} = \mathcal{N}_1 + 1$, we obtain

(4.15)

$$\begin{aligned} Q_{\mathcal{N}}(\check{y}, t_0) = i \int_0^{t_0} dt_1\, e^{-K(p)(t_0 - t_1)} \sum_{l=1}^{\mathcal{N}} \sum_{j=1}^{l} e^{-K^*(-p^1)t_1} \\ \cdot \Bigg(\sum_{g^1 \in \mathcal{G}_{0,2j-2}} i^{j-1} P^-(y_0) \cdots P^-(y_{2j-2}) \int_0^\infty \cdots \int_0^\infty dt_1^1 \cdots dt_{j-1}^1 \\ \cdot \prod_{k=1}^{j-1} \theta(t^2_{\xi_k^1} - t_k^1) e^{\mathcal{L}_k^1 t_k^1} \Lambda_{g^1} \Bigg) e^{-K(p^2)t_1} \\ \cdot \Bigg(\sum_{g^2 \in \mathcal{G}_{2j-1,2l-1}} i^{l-j} \int_0^\infty \cdots \int_0^\infty dt_1^2 \cdots dt_{l-j}^2 \prod_{k=2}^{l-j} \theta(t^2_{\xi_k^2} - t_k^2) e^{\mathcal{L}_k^2 t_k^2} \Lambda_{g^2} \Bigg) e^{-K(p^3)t_1} \\ \cdot \Bigg(\sum_{g^3 \in \mathcal{G}_{2l,2\mathcal{N}}} i^{\mathcal{N}-l} \int_0^\infty \cdots \int_0^\infty dt_1^3 \cdots dt_{\mathcal{N}-l}^3 \prod_{k=1}^{\mathcal{N}-l} \theta(t^3_{\xi_k^3} - t_k^3) e^{\mathcal{L}_k^3 t_k^3} \Lambda_{g^3} \Bigg), \end{aligned}$$

where

$$P^1 = \sum_{k=0}^{2j-2} y_k, \quad P^2 = \sum_{k=2j-1}^{2l-1} y_k, \quad P^3 = \sum_{k=2l}^{2\mathcal{N}} y_k,$$

$$t^1_{\xi^1_1} = t_1, \quad t^2_{\xi^2_1} = t_1, \quad t^3_{\xi^3_1} = t_1 .$$

The superscript indicates to which of the three graphs g^1, g^2, and g^3 the notation pertains. For $j = 1, l = j$ or $l = \mathcal{N}$, one of the brackets in (4.15) is identically equal to one. It follows from (4.9) that

$$\sum_{l=1}^{\mathcal{N}} \sum_{j=1}^{l} \sum_{g^1 \in \mathcal{G}_{0,2j-2}} \sum_{g^2 \in \mathcal{G}_{2j-1,2l-1}} \sum_{g^3 \in \mathcal{G}_{2l,2\mathcal{N}}}$$

$$= \sum_{l=1}^{\mathcal{N}} \sum_{j=1}^{l} \sum_{g = g^1 \otimes g^2 \otimes g^3 : g^1 \in \mathcal{G}_{0,2j-2}; g^2 \in \mathcal{G}_{2j-1,2l-1}; g^3 \in \mathcal{G}_{2l,2\mathcal{N}}} = \sum_{g \in \mathcal{G}_{0,2\mathcal{N}}} .$$

Therefore, noting that $p = p^1 + p^2 + p^3$, we write (4.15) in the form

$$\begin{aligned} Q_{\mathcal{N}}(\check{y}, t_0) = i^{\mathcal{N}} e^{-K(p)t_0} & \sum_{g \in \mathcal{G}_{0,2\mathcal{N}}} \int_0^\infty dt_1\, \theta(t_0 - t_1) e^{\mathcal{L}^+(p^1, p^2, p^3)t_1} \\ & \cdot \left(\int_0^\infty \cdots \int_0^\infty dt^1_1 \cdots dt^1_{j-1} \prod_{k=1}^{j-1} \theta(t^1_{\xi^1_k} - t^1_k) e^{\mathcal{L}^{1-}_k t^1_k} \Lambda^-_{g^1} \right) \\ & \cdot \left(\int_0^\infty \cdots \int_0^\infty dt^2_1 \cdots dt^2_{l-j} \prod_{k=1}^{l-j} \theta(t^2_{\xi^2_k} - t^2_k) e^{\mathcal{L}^2_k t^2_k} \Lambda_{g^2} \right) \\ & \cdot \left(\int_0^\infty \cdots \int_0^\infty dt^3_1 \cdots dt^3_{\mathcal{N}-l} \prod_{k=1}^{\mathcal{N}-l} \theta(t^3_{\xi^3_k} - t^3_k) e^{\mathcal{L}^3_k t^3_k} \Lambda_{g^3} \right), \end{aligned} \tag{4.16}$$

$$\Lambda^-_{g^1} = \prod_{k=j}^{3j-2} p^{-\sigma_k}(y_{h^x_k}) .$$

In view of the choice of the special enumeration of the vertices in the graph g, by the definition of the gluing operation we have

$$s^1_1 = p^1, \quad s^2_1 = p^2, \quad s^3_1 = p^3;$$

$$\begin{aligned} & t^1_k = t_{k+1}, \quad t^1_{\xi^1_k} = t_{\xi_{k+1}}, \quad -\sigma^1_k = \sigma_{k+1}, \quad \mathcal{L}^{1-}_k = \mathcal{L}_{k+1}, \qquad 1 \le k \le j-1; \\ & t^2_k = t_{k+j}, \quad t^2_{\xi^2_k} = t_{\xi_{k+j}}, \quad \sigma^2_k = \sigma_{k+j}, \quad \mathcal{L}^2_k = \mathcal{L}_{k+j}, \qquad 1 \le k \le l-j; \\ & t^3_k = t_{k+l}, \quad t^3_{\xi^3_k} = t_{\xi_{k+l}}, \quad \sigma^3_k = \sigma_{k+l}, \quad \mathcal{L}^3_k = \mathcal{L}_{k+l}, \qquad 1 \le k \le \mathcal{N}-l . \end{aligned} \tag{4.17}$$

(In the last equalities, the values on the right refer to the graph $g = g^1 \otimes g^2 \otimes g^3$.) Therefore,

$$\Lambda^-_{g_1} \Lambda_{g_2} \Lambda_{g^3} = \Lambda_g$$

and (4.16) becomes

$$
\begin{aligned}
Q_{\mathcal{N}}(\check{y}, t_0) = \sum_{g\in\mathcal{G}_{0,2\mathcal{N}}} & i^{\mathcal{N}} e^{-K(p)t_0} \int_0^\infty dt_1\, e^{\mathcal{L}_1 t_1}\theta(t_0 - t_1) \\
& \cdot\left(\int_0^\infty \cdots \int_0^\infty dt_2\cdots dt_j \prod_{k=2}^{j} \theta(t_{\xi_k} - t_k) e^{\mathcal{L}_k t_k}\right) \\
& \cdot\left(\int_0^\infty \cdots \int_0^\infty dt_{j+1}\cdots dt_l \prod_{k=j+1}^{l} \theta(t_{\xi_k} - t_k) e^{\mathcal{L}_k t_k}\right) \\
& \cdot\left(\int_0^\infty \cdots \int_0^\infty dt_{l+1}\cdots dt_{\mathcal{N}} \prod_{k=l+1}^{\mathcal{N}} \theta(t_{\xi_k} - t_k) e^{\mathcal{L}_k t_k}\right) \Lambda_g,
\end{aligned}
\tag{4.18}
$$

whence we arrive at (4.12). The induction is completed.

4.3. Second formula(*) for $Q_{\mathcal{N}}$.

The formulas (3.17), (3.18), and (3.19) of §3, Chapter 6, as well as the definition of finished graphs $\tilde{g} \in \widetilde{\mathcal{G}}_{0,2\mathcal{N}}$ and complete chains $w \in W(g)$, are carried over without any changes to the case under consideration here. Of course, Figure 2 (a), (b) now looks differently and we present it in this section.

By the set of complete chains $W(w^1 \otimes w^2 \otimes w^3)$ we mean the result of mixing of the following three complete chains:

$$
\begin{aligned}
w^1 \in W(g^1),\quad g^1 \in \mathcal{G}_{0,2j-2};\qquad & w^2 \in W(g^2),\quad g^2 \in \mathcal{G}_{2j-1,2l-1}; \\
w^3 \in W(g^3),\quad g^3 \in \mathcal{G}_{2l,2\mathcal{N}};\qquad & 1 \le j \le l \le \mathcal{N}\,.
\end{aligned}
\tag{4.19}
$$

This means that it is necessary to consider the graphs defined by the complete chains w^1, w^2, and w^3, glue them together, construct the finished graph $w^1 \otimes w^2 \otimes w^3$, and find all possible complete chains for it. If we represent the chains w^1, w^2, and w^3 as sequences of indices $(w_1^1, \dots, w_{j-1}^1)$, $(w_1^2, \dots, w_{l-j}^2)$, and $(w_1^3, \dots, w_{\mathcal{N}-l}^3)$, then the set $W(w^1 \otimes w^2 \otimes w^3)$ consists of all such permutations of sequence:

$$
(1, w_1^1, w_2^1, \dots, w_{j-1}^1, w_1^2, w_2^2, \dots, w_{l-j}^2, w_1^3, w_2^3, \dots, w_{\mathcal{N}-l}^3),
\tag{4.20}
$$

where the order of succession of the vertices in each of the chains w^1, w^2, and w^3 is not disturbed.

It is clear that for any $g^1 \in \mathcal{G}_{0,2j-2}, g^2 \in \mathcal{G}_{2j-1,2l-1}$, and $g^3 \in \mathcal{G}_{2l,2\mathcal{N}}$, we have

$$
\sum_{w^1\in W(g^1),\ w^2\in W(g^2),\ w^3\in W(g^3)} W(w^1 \otimes w^2 \otimes w^3) = W(g^1 \otimes g^2 \otimes g^3)\,.
\tag{4.21}
$$

For any $g^1 \in \mathcal{G}_{0,2j-2}$, $g^2 \in \mathcal{G}_{2j-1,2l-1}$, $g^3 \in \mathcal{G}_{2l,2\mathcal{N}}$, $1 \le j \le l \le \mathcal{N}$, $w^1 \in W(g^1)$, $w^2 \in W(g^2)$, $w^3 \in W(g^3)$, the following inequality holds:

(*) This formula and the final part of the proof of the theorem are fully analogous to those in §3, Chapter 6. Therefore, below we only note new details.

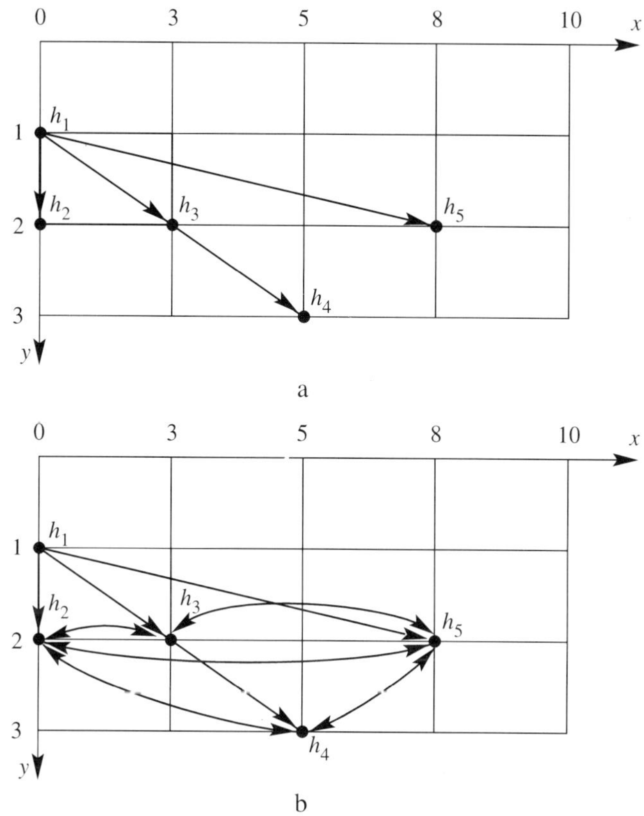

FIGURE 2. Finished graphs: (a) is the corner part of the graph g from Figure 1; (b) is the finished graph $\tilde{g}$ for the graph g from Figure 1. (Special enumeration of vertices.)

$$
\begin{aligned}
&\int_0^t dt_1\, e^{x_1 t_1} B_{j-1}(x_{w_1^1}, \dots, x_{w_{j-1}^1}, t_1) B_{l-j}(x_{w_1^2}, \dots, x_{w_{l-j}^2}, t_1) \\
&\qquad \cdot B_{\mathcal{N}-l}(x_{w_1^3}, \dots, x_{w_{\mathcal{N}-l}^3}, t_1) \\
&\quad = \sum_{w \in W(w^1 \otimes w^2 \otimes w^3)} B_{\mathcal{N}}(x_{w_1}, \dots, x_{w_{\mathcal{N}}}, t)\,.
\end{aligned}
\tag{4.22}
$$

If $j = 1$, $l = j$, or $l = \mathcal{N}$, one of the graphs g^1, g^2, or g^3 is empty, and then, owing to $B_0(t_1) = 1$, (4.22) follows from the equality (3.22) of §3, Chapter 6. However, if $1 < j < l < \mathcal{N}$, then considering complete chains

$$
w^1 = (2, 3, \dots, j), \quad w^2 = (j+1, \dots, l), \quad w^3 = (l+1, \dots, \mathcal{N}),
$$

applying (3.17) from §3, Chapter 6, and decomposing the set

$$
W((3, \dots, j) \otimes (j+1, \dots, l) \otimes (l+1, \dots, \mathcal{N}))
$$

with respect to the first vertex

$$
\begin{aligned}
&W((3,\dots,j)\otimes(j+1,\dots,l)\otimes(l+1,\dots,\mathcal{N}))\\
&\quad=\sum_{w\in W((4,\dots,j)\otimes(j+1,\dots,l)\otimes(l+1,\dots,\mathcal{N}))}(1,3,w_2,\dots,w_{\mathcal{N}-2})\\
&\quad+\sum_{w\in W((4,\dots,j)\otimes(j+2,\dots,l)\otimes(l+1,\dots,\mathcal{N}))}(1,j+1,3,w_2,\dots,w_{\mathcal{N}-2})\\
&\quad+\sum_{w\in W((4,\dots,j)\otimes(j+3,\dots,l)\otimes(l+1,\dots,\mathcal{N}))}(1,j+1,j+2,3,w_2,\dots,w_{\mathcal{N}-4})\\
&\quad+\dots+(1,j+1,j+2,\dots,l,l+1,l+2,\dots,\mathcal{N},3,4,\dots,j),
\end{aligned}
\tag{4.23}
$$

and repeating arguments of §3, Chapter 6, we obtain (4.22).

Now we prove the second formula for the symbol $Q_{\mathcal{N}}(\check{y},t)$:

$$
Q_{\mathcal{N}}(\check{y},t_0)=i^{\mathcal{N}}e^{-K(p)t_0}\sum_{w\in W(g),g\in\mathcal{G}_{0,2\mathcal{N}}}E_w(t)\Lambda_g,
\tag{4.24}
$$

where $E_w(t)\equiv B_{\mathcal{N}}(\mathcal{L}_{w_1},\mathcal{L}_{w_2},\dots,\mathcal{L}_{w_{\mathcal{N}}},t)$. In view of (4.12) it suffices to prove that for any graph $g\in\mathcal{G}_{0,2\mathcal{N}}$

$$
\int_0^\infty\cdots\int_0^\infty dt_1\cdots dt_{\mathcal{N}}\prod_{j=1}^{\mathcal{N}}e^{\mathcal{L}_jt_j}\theta(t_{\xi_j}-t_j)=\sum_{w\in W(g)}E_w(t_0)\,.
\tag{4.25}
$$

For $\mathcal{N}=0$ and $\mathcal{N}=1$ equality (4.25) can be verified trivially. For $\mathcal{N}\geq 1$ we argue by induction. We cut the graph $g\in\mathcal{G}_{0,2\mathcal{N}}$, and obtain $g=g^1\otimes g^2\otimes g^3$, $g^1\in\mathcal{G}_{0,2j-2}$, $g^2\in\mathcal{G}_{2j-1,2l-1}$, $g^3\in\mathcal{G}_{2l,2\mathcal{N}}$. Then on the left-hand side of (4.25) we have the integral

$$
\begin{aligned}
&\int_0^t dt_1e^{\mathcal{L}_1t_1}\left(\int_0^\infty\cdots\int_0^\infty dt_2\dots dt_j\prod_{k=2}^{j}e^{\mathcal{L}_kt_k}\theta(t_{\xi_k}-t_k)\right)\\
&\qquad\cdot\left(\int_0^\infty\cdots\int_0^\infty dt_{j+1}\cdots dt_l\prod_{k=j+1}^{l}e^{\mathcal{L}_kt_k}\theta(t_{\xi_k}-t_k)\right)\\
&\qquad\cdot\left(\int_0^\infty\cdots\int_0^\infty dt_{l+1}\cdots dt_{\mathcal{N}}\prod_{k=l+1}^{\mathcal{N}}e^{\mathcal{L}_kt_k}\theta(t_{\xi_k}-t_k)\right)\\
&\quad=\int_0^{t_0}dt_1e^{\mathcal{L}_1t_1}\sum_{w^1\in W(g^1)}E_{w^1}(t_1)\sum_{w^2\in W(g^2)}E_{w^2}(t_1)\sum_{w^3\in W(g^3)}E_{w^3}(t_1)\\
&\quad=\sum_{w\in W(w^1\otimes w^2\otimes w^3)}\sum_{w^1\in W(g^1)}\sum_{w^2\in W(g^2)}\sum_{w^3\in W(g^3)}E_w(t_0)\\
&\quad=\sum_{w\in W(g)}E_w(t_0),
\end{aligned}
$$

by (4.21) and (4.22). Thus , (4.25) and (4.24) with it, hold by induction.

4.4. Formula for $\Phi_{\mathscr{N}}$. Substituting (4.24) in (4.1) and passing to the limit as $t \to \infty$, we obtain

$$\Phi_{\mathscr{N}} = \sum_{w \in W(g)} \sum_{g \in \mathscr{G}_{0,2\mathscr{N}}} (-i)^{\mathscr{N}} \int_{-\infty}^{\infty} dy \prod_{k=1}^{\mathscr{N}} \left(\sum_{j=1}^{\mathscr{N}} \mathscr{L}_{w_j} \right)^{-1} \Lambda_g V_{\mathscr{N}}(\check{y}), \qquad \mathscr{N} \geq 1, \tag{4.26}$$

$$\Phi_0 = \overline{v}(0) .$$

Here $w_1 = 1$ and $p = \sum_{k=0}^{2\mathscr{N}} y_k = 0$ always. Formula (4.26) is proved as the analogous formulas in Chapters 6 and 7. Thus, the asymptotics (1.7) is established.

§5. Computation of the asymptotics for the Landau-Ginzburg equation

In this section we present a scheme for proving formula (1.8). As above, using perturbation theory, we obtain the equalities (3.1) through (3.4). We replace bound (3.5) by the following:

$$|v_{\mathscr{N}}(p,t)| \leq c^{\mathscr{N}} e^{-at} (\mathscr{N}+1)^{-4} M^{-3-\alpha}(p), \qquad p \in R_1,\ t > 0 . \tag{5.5}$$

In fact, from (3.4), arguing by induction, we obtain

$$\begin{aligned} |v_{\mathscr{N}}(p,t)| &\leq c^{\mathscr{N}-1} \int_0^t d\tau\, e^{-at+a\tau} \iint_{-\infty}^{\infty} dq\, dr \sum_{l=1}^{\mathscr{N}} \sum_{j=1}^{l} j^{-4} (\mathscr{N}-l+1)^{-4} \\ &\quad \cdot (l-j+1)^{-4} M^{-\alpha-3}(p-q) M^{-\alpha-3}(q-r) M^{-\alpha-3}(r) e^{-3a\tau} \\ &\leq c^{\mathscr{N}} e^{-at} (\mathscr{N}+1)^{-4} M^{-3-\alpha}(p) \int_0^t d\tau\, e^{-2a\tau} \\ &\quad \cdot \iint_{-\infty}^{\infty} dq\, dr\, M^{-\alpha-3}(q-r) M^{-\alpha-3}(r) \\ &\leq c^{\mathscr{N}} e^{-at} M^{-3-\alpha}(p) (\mathscr{N}+1)^{-4}, \end{aligned}$$

whence (5.5) follows for all $\mathscr{N} \geq 0$.

Next, using formulas (3.6) and (3.7), were present the function $\psi_{\mathscr{N}}(x,t)$ as follows:

$$\psi_{\mathscr{N}}(x,t) = J^1_{\mathscr{N}} + J^2_{\mathscr{N}} + J^3_{\mathscr{N}},$$

where

$$\begin{aligned} J^1_{\mathscr{N}} &= \int_{-\infty}^{\infty} dp\, e^{ipx-K(p)t} \Phi_{\mathscr{N}}(p) \rho(p-\xi), \\ J^2_{\mathscr{N}} &= \int_{-\infty}^{\infty} dp\, e^{ipx-K(p)t} (v_{\mathscr{N}}(p,t) - \Phi_{\mathscr{N}}(p)) \rho(p-\xi), \\ J^3_{\mathscr{N}} &= \int_{-\infty}^{\infty} dp\, e^{ipx-K(p)t} v_{\mathscr{N}}(p,t) (1-\rho(p-\xi)), \end{aligned}$$

$$\Phi_{\mathscr{N}}(p) = \lim_{t\to\infty} v_{\mathscr{N}}(p,t) e^{at}, \qquad \xi = \frac{x}{2t},$$

$$\rho(y) \in C_0^{\infty}(R_1), \qquad \rho(y) = \begin{cases} 1, & |y| \leq 1, \\ 0, & |y| \geq 2 . \end{cases}$$

After the change of variable of integration $y = p - \xi$ in the integral $J^1_{\mathcal{N}}$, we have

$$(5.9)\qquad \begin{aligned} J^1_{\mathcal{N}} &= \int_{-2}^{2} dy\, e^{-at+i\xi^2 t - iy^2 t}\Phi_{\mathcal{N}}(y+\xi)\rho(y) \\ &= \Phi_{\mathcal{N}}(\xi)\sqrt{\frac{2\pi}{t}}e^{-i\pi/4+i\xi^2 t-at} + O(t^{-1}e^{-at}). \end{aligned}$$

Further, introducing (5.5) into (3.10), we readily obtain the estimate

$$(5.11)\qquad |v_{\mathcal{N}}(p,t) - \Phi_{\mathcal{N}}(p)| = c^{\mathcal{N}}O(e^{-at}).$$

Therefore, owing to (5.11), we find

$$(5.13)\qquad J^2_N = c^{\mathcal{N}}O(e^{-2at}).$$

Finally, for the integral $J^3_{\mathcal{N}}$, integrating by parts, we obtain

$$(5.20)\qquad J^3_{\mathcal{N}} = c^{\mathcal{N}}O(e^{-at}t^{-1}).$$

The estimates (5.9), (5.13), and (5.20) lead to the asymptotics (1.8), where

$$A(\xi) = \frac{1}{\sqrt{2\pi}}e^{-i\pi/4}\sum_{\mathcal{N}=0}^{\infty}\varepsilon^{2\mathcal{N}+1}\Phi_{\mathcal{N}}(\varepsilon).$$

The function $\Phi_{\mathcal{N}}(\varepsilon)$ is computed by formula (4.26), where one needs to put $p = \sum_{k=0}^{2\mathcal{N}} y_k = \xi$. The asymptotics (1.8) is established. In particular, we note that

$$\begin{aligned}
\Phi_0(\xi) &= \frac{1}{\varepsilon}\hat{\bar{u}}(\xi) = \frac{1}{\varepsilon}\int_{-\infty}^{\infty} e^{i\xi x}\bar{u}(x)\,dx, \\
\Phi_1(\xi) &= \frac{1}{\varepsilon^2}\iint_{-\infty}^{\infty}\frac{dy_1\,dy_2\,\hat{\bar{u}}^*(y_1+y_2-\xi)\hat{\bar{u}}(y_1)\hat{\bar{u}}(y_2)}{K(\xi)-K(y_1)-K(y_2)-K^*(y_1+y_2-\xi)}, \\
\Phi_2(\xi) &= \frac{1}{\varepsilon^3}\iiiint_{-\infty}^{\infty} dy_1\,dy_2\,dy_3\,dy_4\left\{\frac{\mathcal{F}^1_u}{\mathcal{F}^1_k} + 2\varepsilon\frac{\mathcal{F}^2_u}{\mathcal{F}^2_k}\right\} \\
\mathcal{F}^1_u &= \hat{\bar{u}}(\xi-y_1-y_2-y_3-y_4)\hat{\bar{u}}^*(-y_1)\hat{\bar{u}}^*(-y_2)\hat{\bar{u}}(y_3)\hat{\bar{u}}(y_4), \\
\mathcal{F}^2_u &= \hat{\bar{u}}^*(y_1+y_2+y_3+y_4-\xi)\hat{\bar{u}}^*(-y_1)\hat{\bar{u}}(y_2)\hat{\bar{u}}(y_3)\hat{\bar{u}}(y_4), \\
\mathcal{F}^1_k &= (K(\xi)-K^*(y_3+y_4-\xi)-K(y_3)-K(y_4)) \\
&\quad\cdot(K(\xi)-K(\xi-y_1-y_2-y_3-y_4)-K^*(-y_1)-K^*(-y_2)-K(y_3)-K(y_4)), \\
\mathcal{F}^2_k &= (K(\xi)-K(y_1+y_2+y_3)-K^*(-\xi+y_1+y_2+y_3+y_4)-K(y_4)) \\
&\quad\cdot(K(\xi)-K^*(-y_1)-K^*(y_1+y_2+y_3+y_4-\xi)-K(y_2)-K(y_3)-K(y_4)),
\end{aligned}$$

where in the case of the Landau-Ginzburg equation for the symbol $K(\rho)$ of the operator $\mathbb{K}(u)$ we have $K(p) = a + ip^2$, $a > 0$.

§6. Asymptotics of solutions for periodic problem of the nonlinear Schrödinger equation for large times

We consider a periodic problem for the following model equation:

$$u_t - i\left(-i\frac{\partial}{\partial x}\right)^{\beta}(|u|^2u) + \mathbb{K}(u) = 0, \tag{6.1}$$

where $\beta = 0, 1$, and the pseudodifferential operator $\mathbb{K}$ is defined on a 2π-periodic function $u(x)$ as follows:

$$\mathbb{K}(u) = \frac{1}{2\pi}\sum_{p=-\infty}^{\infty} e^{ipx}K_p\hat{u}_p,$$

where $\hat{u}_p = \int_0^{2\pi} e^{-ipx}u(x)\,dx$ are the Fourier coefficients. We thus assume that a solution $u(x,t)$ to the Cauchy problem for (6.1) satisfies the condition $u(x+2\pi,t) = u(x,t)$ for all $x \in R_1$ and $t \geq 0$.

Equation (6.1) unites a significant number of model nonlinear equations known in physics. For instance, for $\beta = 0$ it becomes the nonlinear nonlocal Schrödinger equation **[57]**, and, in particular, the nonlinear Schrödinger equation with dissipation

$$u_t - i|u|^2u + biu_{xx} - au_{xx} = 0, \tag{6.2}$$

for $K_p = ap^2 - ibp^2$: and the generalized Landau-Ginzburg equation

$$u_t - i|u|^2u + au + biu_{xx} = 0, \tag{6.3}$$

for $K_p = a - bip^2$, $a > 0$, $b \in R_1$. On the other hand, for $\beta = 1$, for the real-valued function $u(x,t)$ we arrive at the modified Korteweg-de Vries-Burgers equation

$$u_t - 3u^2u_x - au_{xx} + bu_{xxx} = 0 \tag{6.4}$$

for $Kp = ap^2 - ibp^3$, $a > 0$, $b \in R_1$; the modified Kuramoto-Sivashinsky equation

$$u_t - 3u^2u_x - au_{xx} + bu_{xxxx} = 0 \tag{6.5}$$

for $Kp = ap^2 + bp^4$, $b > 0$, $a + b > 0$; the modified Ott-Sudan-Ostrovskiĭ equation

$$u_t = 3u^2u_x - \frac{a}{4\pi}\int_{-\pi}^{\pi}\coth h\frac{x-y}{2}u_y(y,t)\,dy + bu_{xxx} = 0 \tag{6.6}$$

for $Kp = a|p| - ibp^3$, $a > 0$, $b \in R_1$, and many others, Thus, the model equation (6.1) contains in itself an entire class of nonlinear equations and is suitable for describing various wave processes in dissipative and conservative media, including media with energy pumping. In §§2 through 5 of this chapter we develop a method of constructing the asymptotics, for large times, of solutions to the nonlinear nonlocal Schrödinger equation in the case of a Cauchy problem with decaying at infinity initial data $\overline{u}(x)$. In this section we apply this method to the periodic problem for equation (6.1). We shall show that the leading term of the asymptotics, as $t \to \infty$, of a solution $u(x,t)$ to the periodic problem of equation (6.1) can be determined by the behavior of the zeroth harmonics.

THEOREM 2. *Suppose that*

1) *The symbol K_p in (6.1) satisfies the following conditions:*

$$\operatorname{Re} K_p \geq \lambda + 3\alpha|p|^\beta, \qquad p \neq 0,\ p \in z, \qquad \lambda = \operatorname{Re} K_0 \geq 0,\ \alpha > 0,\ \beta = 0, 1; \tag{6.7}$$

2) *The Fourier coefficients $\hat{\overline{u}}_p$ of the 2π-periodic initial function $\overline{u}(x)$ have the bound*

$$|\hat{\overline{u}}_p| \leq \varepsilon(1+|p|)^{-4}, \qquad p \in z,$$

where $0 < \varepsilon < c$, c a positive constant.

Then, as $t \to \infty$, we have following asymptotic formula:

$$u(x,t) = Ae^{-K_0 - 4\pi^2 i\nu|A|^2 t} + O(e^{-\lambda t - \gamma t}), \tag{6.8}$$

where γ is some positive constant, $\nu = 0$ for $\beta = 1$ or $\lambda = \operatorname{Re} K_0 > 0$, but $\nu = 1$ for $\beta = 0$ and $\lambda = 0$. The number A is expressed explicitly in terms of the initial data $\overline{u}(x)$ and the symbol K_p of the operator $\mathbb{K}$.

REMARK. Theorem 2 is true, in particular, for equations (6.2) through (6.6). In equations (6.4) through (6.6) the parameter $\nu = 0$ because $\beta = 1$, and in (6.3) ν is also zero since $\lambda = a > 0$. In (6.2) $\beta = 0$ and, therefore, $\nu = 1$ in the asymptotics (6.5). The analogous phase rotation in the asymptotics of a solution also occurs for the nonlinear Schrödinger equation without dissipation (i.e., $a = 0$ in (6.2)) in the case of the Cauchy problem with initial data decreasing at infinity (see, e.g., **[1]**, **[17]**).

PROOF. The existence of a classical solution for the periodic problem of equation (6.1) in the class $C^\infty([0,\infty); C^\infty_{2\pi}(R_1))$ can be proved the same way as in Chapter 3. We represent the solution in the an absolutely and uniformly form of convergent Fourier series $u(x,t) = \frac{1}{2\pi}\sum_{p=-\infty}^{\infty} \hat{u}_p(t)e^{ipx}$. Then, from (6.1) we obtain an infinite system of ordinary differential equations for the coefficients $\hat{u}_p(t)$:

$$\frac{d\hat{u}_p}{dt} + \sum_{n=-\infty}^{\infty}\sum_{q=-\infty}^{\infty} ip^\beta \hat{u}^*_{q-p}(t)\hat{u}_{q-n}(t)\hat{u}_n(t) + K_p\hat{u}_p(t) = 0\,.$$

From this, making the change

$$\hat{u}_p(t) = w_p(t)\exp\left(-i\nu\int_0^t |w_0(\tau)|^2\,d\tau\right), \tag{6.9}$$

where ν is defined in the theorem, and integrating with respect to t, we find

$$w_p(t) = \hat{\overline{u}}_p e^{-K_p(t)} - i\sum_{\substack{r=-\infty \\ \|r|+|q|\geq\nu}}^{\infty}\sum_{q=-\infty}^{\infty}\int_0^t d\tau\, e^{-K_p(t-\tau)}p^\beta w_{p-q}(\tau)w_{q-r}(\tau)w^*_{-r}(\tau)\,. \tag{6.10}$$

We apply perturbation theory as in §3. Setting $\hat{\overline{u}}_p = \varepsilon\overline{v}_p$ we represent $w_p(t)$ in the following form:

$$w_p(t) = \sum_{\mathcal{N}=0}^{\infty} \varepsilon^{2\mathcal{N}+1} v_p^{(\mathcal{N})}(t)\,. \tag{6.11}$$

Substitution of (6.11) in (6.10) yields

$$v_p^{(0)}(t) = \overline{v}_p e^{-Kpt}; \tag{6.12}$$

$$\begin{aligned} v_p^{(\mathscr{N})}(t) = ip^\beta \sum_{\substack{l=-\infty \\ |l|+|q|\geq\nu}}^{\infty} \sum_{q=-\infty}^{\infty} \sum_{r=1}^{\mathscr{N}} \sum_{j=1}^{r} \int_0^t d\tau\, e^{-K_p(t-\tau)} \\ \cdot v_l^{(\mathscr{N}-r)}(\tau) v_{q-l}^{(r-j)}(\tau) v_{q-p}^{(j-1)^*}(\tau), \qquad \mathscr{N} \geq 1. \end{aligned} \tag{6.13}$$

We prove the following estimate:

$$|v_p^{(\mathscr{N})}(t)| \leq c^{\mathscr{N}}(\mathscr{N}+1)^{-2}(1+|p|)^{-4} e^{-\lambda t - \kappa_p t}, \qquad \mathscr{N} \geq 0,\ p \in Z, \tag{6.14}$$

where $c > 0$ is independent of $\mathscr{N}$ and p, $\kappa_0 = 0$, $\kappa_p = 2|p|^\beta$, $p \neq 0$. For $\mathscr{N} = 0$ (6.14) readily follows from (6.12) under the condition of the theorem. We carry out induction on $\mathscr{N}$ for $\mathscr{N} > 0$. From (6.13) we get

$$\begin{aligned} |v_p^{(\mathscr{N})}(t)| \leq & c^{\mathscr{N}-1}|p|^\beta \sum_{\substack{q=-\infty \\ |q|+|l|\geq\nu}}^{\infty} \sum_{l=-\infty}^{\infty} (1+|q-p|)^{-4}(1+|q-l|)^{-4}(1+|l|)^{-4} \\ & \cdot \sum_{r=1}^{\mathscr{N}} \sum_{j=1}^{r} (\mathscr{N}+1-r)^{-2}(r+1-j)^{-2} j^{-2} \int_0^t d\tau \exp(-\lambda t - 2\lambda\tau) \\ & \cdot \exp(-3\kappa_p(t-\tau) - \kappa_{q-p}\tau - \kappa_{q-l}\tau - \kappa_l\tau). \end{aligned} \tag{6.15}$$

Since $|q| + |l| \geq \nu$, we have for $\lambda = 0$

$$3\kappa_p - \kappa_{q-p} - \kappa_{q-l} - \kappa_l \leq \begin{cases} -2\alpha|p|^\beta, & \beta = 0, \\ 2\alpha|p|^\beta, & p \neq 0 \end{cases}$$

(in the case $p = 0$, $\beta = 0$, $|p|^\beta$ is replaced by 1) and therefore,

$$|p|^\beta \int_0^t d\tau \exp(-3\kappa_p t + (3\kappa_p - \kappa_{q-p} - \kappa_{q-l} - \kappa_l)\tau) \leq \frac{1}{2\alpha} e^{-\kappa_p t}.$$

We thus arrive at (6.14) form (6.15). The case $\lambda > 0$ is even simpler. Introducing the notation $\Phi^{(\mathscr{N})} = \lim_{t\to\infty} v_0^{(\mathscr{N})}(t) e^{K_0 t}$, by (6.14) we obtain

(6.16)

$$\begin{aligned} & |v_0^{(\mathscr{N})}(t) - \Phi^{(\mathscr{N})} e^{-K_0 t}| \\ & = \left| e^{-K_0 t} \int_t^\infty d\tau\, e^{K_0\tau}(1-\beta) \cdot \sum_{\substack{q=-\infty \\ |q|+|l|\geq\nu}}^{\infty} \sum_{l=-\infty}^{\infty} \sum_{r=1}^{\mathscr{N}} \sum_{j=1}^{r} v_l^{(\mathscr{N}-r)}(\tau) v_{q-l}^{(r-j)}(\tau) v_q^{(j-1)^*}(\tau) \right| \\ & \leq (1-\beta) c^{\mathscr{N}} e^{-\lambda t} \\ & \quad \cdot \sum_{\substack{q=-\infty \\ |q|+|l|\geq\nu}}^{\infty} \sum_{l=-\infty}^{\infty} (1+|q|+|l|)^{-4} \int_t^\infty d\tau \exp(-2\lambda\tau - \kappa_q\tau - \kappa_{q-l}\tau - \kappa_l\tau) \\ & \leq c^{\mathscr{N}}(1-\beta) e^{-\lambda t - \gamma t}, \end{aligned}$$

where $\gamma > 0$. Thus, it follows from (6.16) that

$$w_0(t)\sum_{\mathcal{N}=0}^{\infty}\varepsilon^{2\mathcal{N}+1}v_0^{(\mathcal{N})}(t) = Fe^{-K_0 t} + O(e^{-\lambda t-\gamma t}), \tag{6.17}$$

$$|w_0(t)|^2 = |F|^2 e^{-2\operatorname{Re}K_0 t} + O(e^{-2\lambda t-\gamma t}). \tag{6.18}$$

Here we have denoted $F = \sum_{\mathcal{N}=0}^{\infty}\varepsilon^{2\mathcal{N}+1}\Phi^{(\mathcal{N})}$. We use representation (6.9) for the solution $u(x,t)$:

$$\begin{aligned} u(x,t) =& \frac{1}{2\pi}\sum_{p=-\infty}^{\infty}\hat{u}_p(t)e^{ipx} \\ =& \frac{1}{2\pi}\sum_{p=-\infty}^{\infty}w_p(t)e^{ipx}\exp\left(-iv\int_0^t|w_0(\tau)|^2\,d\tau\right) \\ =& \frac{1}{2\pi}Fe^{-K_0 t}\exp\left(-iv|F|^2 t + iv\int_0^{\infty}(|w_0(\tau)|^2 - |F^2)\,d\tau\right) \\ &\cdot\left\{1 + \left(-1+\exp\left(iv\int_t^{\infty}(|w_0(\tau)|^2-|F|^2)\,d\tau\right)\right)\right\} \\ &+ \frac{1}{2\pi}\sum_{p\neq 0}w_p(t)e^{ipx}\exp\left(-iv\int_0^t|w_0(\tau)|^2\,d\tau\right) + O(e^{-\lambda t-\gamma t}). \end{aligned} \tag{6.19}$$

We denote $A = \frac{1}{2\pi}Fe^{-i\psi}$, $\psi = v\int_0^{\infty}(|w_0(\tau)|^2 - |F|^2)\,d\tau$. Therefore, since $|A| = \frac{1}{2\pi}|F|$, the first summand in (6.19) gives the main contribution to the asymptotics of (6.8). By (6.18),

$$\left|-1+\exp\left(iv\int_t^{\infty}(|w_0(\tau)|^2-|F|^2)\,d\tau\right)\right| = O(e^{-\lambda t-\gamma t}). \tag{6.20}$$

And applying the estimate (6.14), we find

$$\begin{aligned} &\left|\sum_{p\neq 0}w_p(t)e^{ipx}\exp\left(iv\int_0^t|w_0(\tau)|^2 d\tau\right)\right| \\ &\leq \sum_{\mathcal{N}=0}^{\infty}\varepsilon^{2\mathcal{N}+1}c^{\mathcal{N}}\sum_{p\neq 0}(1+|p|)^{-4}e^{-\kappa_p t-\lambda t} = O(e^{-\lambda t-\gamma t}). \end{aligned} \tag{6.21}$$

The asymptotics (6.8) is obtained from (6.19) through (6.21). The value of $\Phi^{(\mathcal{N})}$ is computed in the same was as in §4. Here we only note some minor changes. Moreover, we need to compute the phase value ψ which is included in the expression for the constant A. As in §4, we represent $v_p^{(\mathcal{N})}(t)$ in the form of a polilinear operator acting on the initial conditions

$$v_p^{(\mathcal{N})}(t) = \sum_{m=-\infty}^{\infty}Q_{\check{m}}^{(\mathcal{N})}(t)V_{\check{m}}^{(\mathcal{N})}, \tag{6.22}$$

where $V_{\check{m}}^{(\mathcal{N})} = \prod_{j=0}^{2\mathcal{N}}\bar{v}_{m_j}$, $m_j \in Z$, $m = (m_1, m_2, \ldots, m_{2\mathcal{N}}) \in Z^{2\mathcal{N}}$, $p = \sum_{j=0}^{2\mathcal{N}}m_j$. (The role of the variables of integration $y \in R_n$ of §4 is now played by the indices of $\check{m}$

summation over the Fourier harmonics.) For $\mathscr{N}=0$ it follows from (6.12) and (6.22) that

$$\Phi^{(0)}(t)=e^{-K_p t}.$$

From (6.13) we obtain the recurrent relations for the symbols $Q^{\mathscr{N}}_{\check m}(t)$, as in §4,

$$\begin{aligned}Q^{(\mathscr{N})}_{\check m}(t)&=ip^{\beta}\sum_{l=1}^{\mathscr{N}}\sum_{j=1}^{l}\int_0^\infty d\tau\,\exp(-K_p(t-\tau))P^-_{m_0}\cdots P^-_{m_{2j-2}}\\ &\quad\cdot Q^{(j-1)}_{m_0,\dots,m_{2j-2}}(\tau)Q^{(l-j)}_{m_{2j-1},\dots,m_{2l-1}}(\tau)Q^{(\mathscr{N}-l)}_{m_{2l},\dots,m_{2\mathscr{N}}}(\tau),\end{aligned}\tag{6.24}$$

where $p=\sum_{k=0}^{2\mathscr{N}}m_k$, $q=-\sum_{k=0}^{2j-2}m_k$, $r=\sum_{k=0}^{2l-1}m_k$. The operators P^-_m act on the complex coefficients f_m by $P^-_m f_m=f^*_{-m}$ (i.e., by complex conjugation and inversion of the index). Using the graphs $\mathscr{G}_{0,2\mathscr{N}}$ from §4, we obtain

$$Q^{(\mathscr{N})}_{\check m}(t_0)=i^{\mathscr{N}}e^{-K_p t_0}\sum_{g\in\mathscr{G}_{0,2\mathscr{N}}}\int_0^\infty\cdots\int_0^\infty dt_1\cdots dt_{\mathscr{N}}\prod_{j=1}^{\mathscr{N}}e^{\mathscr{L}_j t_j}\theta(t_{\xi_j}-t_j)\Lambda_g,\tag{6.25}$$

where

$$\begin{aligned}L^+(s^1_j,s^2_j,s^3_j)&=K_{p_j}-K^*_{-s^1_j}-K_{s^2_j}-K_{s^3_j},\\ L^-(s^1_j,s^2_j,s^3_j)&=(L^+(-s^1_j,-s^2_j,-s^3_j))^*,\end{aligned}$$

$$\mathscr{L}_j=L^{i\sigma_j}(s^1_j,s^2_j,s^3_j),\qquad \Lambda_g=\prod_{k=1}^{\mathscr{N}}(s^1_k+s^2_k+s^3_k)\prod_{k=\mathscr{N}+1}^{3\mathscr{N}+1}P^{\sigma_k}_{m_{h^x_k}}.$$

The remaining quantities are determined as in §4. From (6.22) and (6.25) we find the numbers $\Phi^{(\mathscr{N})}$:

(6.26)
$$\begin{aligned}\Phi^{(\mathscr{N})}&=\lim_{t\to\infty}v_0^{(\mathscr{N})}(t)e^{K_0 t}\\ &=i^{\mathscr{N}}\sum_{m=-\infty}^{\infty}\sum_{g\in\mathscr{G}_{0,2\mathscr{N}}}\int_0^\infty\cdots\int_0^\infty dt_1\cdots dt_{\mathscr{N}}e^{\mathscr{L}_1 t_1}\prod_{j=2}^{\mathscr{N}}e^{\mathscr{L}_j t_j}\theta(t_{\xi_j}-t_j)\Lambda_g V^{(\mathscr{N})}_{\check m}\end{aligned}$$

for $\sum_{k=0}^{2\mathscr{N}}m_k=0$.

With the aid of (6.26) we compute $F=\sum_{\mathscr{N}=0}^{\infty}\varepsilon^{2\mathscr{N}+1}\Phi^{(\mathscr{N})}$ and then the phase ψ:

$$\begin{aligned}\psi&=v\int_0^\infty(|w_0(\tau)|^2-|F|^2)\,d\tau\\ &=v\int_0^\infty\Bigg(\Bigg|\sum_{\mathscr{N}=0}^{\infty}\varepsilon^{2\mathscr{N}+1}\sum_{m=-\infty}^{\infty}\sum_{g\in\mathscr{G}_{0,2\mathscr{N}}}\int_0^\infty\cdots\int_0^\infty dt_1\cdots dt_{\mathscr{N}}\\ &\qquad\cdot\prod_{j=1}^{\mathscr{N}}e^{\mathscr{L}_j t_j}\theta(t_{\xi_j}-t_j)\Lambda_g V^{(\mathscr{N})}_{\check m}\Bigg|^2-|F|^2\Bigg)\,d\tau\end{aligned}$$

where $\sum_{k=0}^{2\mathscr{N}}m_k=0$, $t_{\xi_1}=\tau$. Theorem 2 is proved. □

REMARK. Analogous to (6.8) asymptotics of solutions for the periodic problem can be considered for other nonlinear nonlocal equations as well. For instance, we can analyze a model equation of the form

$$u_t + \left(\frac{\partial^\beta u}{\partial x^\beta}\right)^2 + \mathbb{K}(u) = 0, \tag{6.27}$$

where $\beta = 0, 1$, and $\mathbb{K}$ is a pseudodifferential operator; this equation unites in itself the Kolmogorov-Petrovskiĭ-Piskunov equation and the Whitham equation.

THEOREM 3. *Suppose that*

1) *The symbol K_p of the operator $\mathbb{K}$ satisfies the following conditions*:

$$\operatorname{Re} K_p \geq \max(0, \operatorname{Re} K_0) + \alpha|p|^\beta \quad \textit{for } p \neq 0,\ \beta = 0, 1,$$
$$\operatorname{Re} K_0 > 0 \quad \textit{for } \beta = 0;$$

2) *The Fourier coefficients $\hat{\overline{u}}_p$ of the 2π-periodic initial data $\overline{u}(x)$ satisfy the estimates*

$$|\hat{\overline{u}}_p| \leq \varepsilon(1 + |p|)^{-4}, \qquad p \in Z,$$

where $0 < \varepsilon < c$, $c > 0$ some constant.

Then, as $t \to \infty$, for a solution $u(x, t)$ of the Cauchy problem for equation (6.27) *with the initial condition $\overline{u}(x)$, the following asymptotic formula is valid*:

$$u(x, t) = Ae^{-K_0 t} + O(e^{-\operatorname{Re} K_0 t - \gamma t}) \tag{6.28}$$

where $\gamma > 0$, and A can be expressed explicitly in terms of the initial condition and the symbol of the operator $\mathbb{K}$.

It should be noted that in the case $\beta = 1$ the value $\operatorname{Re} K_0$ can become negative, and then the leading term of the asymptotics (6.28) increases for large times. The possibility of such growth is what distinguishes the periodic problem from the corresponding Cauchy problem with decaying at infinity initial data.

CHAPTER 9

Asymptotics of Solutions for a System of Equations of Surface Waves for Large Times

§1. Introduction

In Chapter 4 the following model system of equations, describing surface waves on water, was proposed:

$$(1.1)\qquad \begin{cases} \eta_t + (\nabla, \eta\theta) + \mathbb{K}_1(\eta) + \mathbb{K}_2(\theta) = 0, \\ \theta_t + (\theta, \nabla)\theta + \mathbb{K}_3(\eta) + \mathbb{K}_4(\theta) = 0 \,. \end{cases}$$

Here $\eta(x,t)$ is the free water surface, $\theta(x,t)$ is the two-dimensional velocity vector, $x \in R_2$, $t \geq 0$, and the $\mathbb{K}_j$ are linear pseudodifferential operators defined as follows:

$$\mathbb{K}_j(\psi) = (2\pi)^{-2} \int_{-\infty}^{\infty} e^{i(p,x)} k_j(p) \hat{\psi}(p)\, dp, \qquad j = 1,2,3,4,$$

where $\hat{\psi}(p) = \int_{-\infty}^{\infty} e^{-i(p,x)} \psi(x)\, dx$ is the Fourier transform of the function $\psi(x)$. The symbols $k_j(p)$ ($k_2(p)$ is a two-dimensional row, $k_3(p)$ is a two-dimensional column, and $k_4(p)$ is a 2×2 matrix) of the operators $\mathbb{K}_j$ give the dispersion relation, and for different choices of them the operators $\mathbb{K}_j$ correspond to various wave propagation processes: conservative, dissipative, or processes associated with energy pumping. The system of equations (1.1) for surface waves on water includes the standard nonlinearity of the theory of shallow water and complete linear dispersion, so that it is rather general and contains many equations and systems of equations that are important from a physical point of view. In particular, the system (1.1) includes the following systems of equations: the shallow water [**82,** p. 439], Boussinesq [**81,** p. 15], Dobrokhotov [**13**], Broer [**100**], Kaup [**130**], and other equations.

In this chapter we investigate the asymptotic behavior, as $t \to \infty$, of the solutions of the Cauchy problem for the system of equations (1.1). The asymptotic behavior of solutions of various evolution equations has recently been the subject of many studies. The main attention has been focused on conservative equations solved by the inverse scattering method (see [**1,** p. 85] and the literature cited there). However, friction, thermal conductivity, viscosity, and other phenomena typical in nature lead to dissipative equations.

We are able to extend the ideas developed in Chapters 6 through 8 to the Cauchy problem for the system of equations of surface waves (1.1):

$$(1.2)\qquad \begin{cases} \eta_t + (\nabla, \eta\theta) + \mathbb{K}_1(\eta) + \mathbb{K}_2(\theta) = 0, \\ \theta_t + (\theta, \nabla)\theta + \mathbb{K}_3(\eta) + \mathbb{K}_4(\theta) = 0, \\ \eta|_{t=0} = \overline{\eta}(x), \qquad \theta|_{t=0} = \overline{\theta}(x) \,. \end{cases}$$

We assume that the initial perturbation of the surface $\overline{\eta}(x)$ satisfies the condition

$$\int_{-\infty}^{\infty} \overline{\eta}(x)\,dx = 0\,.$$

It follows from the mass conservation law (see Chapter 4) that $\int_{-\infty}^{\infty} \eta(x,t)\,dx = 0$ for all $t \geq 0$, provided the condition $k_2(0) = 0$ is satisfied. We assume below that this requirement is satisfied. We define the vector

$$\hat{u}(p,t) = \begin{pmatrix} \hat{\eta}(p,t)|p|^{-1} \\ \hat{\theta}(p,t) \end{pmatrix}$$

and the solution vector $u(x,t)$ as the inverse Fourier transform:

$$u(x,t) = (2\pi)^{-2} \int_{-\infty}^{\infty} e^{i(p,x)} \hat{u}(p,t)\,dp\,.$$

We introduce the matrix

$$K(p) = \begin{pmatrix} K_2, & K_2 \\ K_3, & K_4 \end{pmatrix} = \begin{pmatrix} k_1(p), & k_2(p)|p|^{-1} \\ |p|k_3(p), & k_4(p) \end{pmatrix}$$

and denote by $\lambda_1(p)$, $\lambda_2(p)$, $\lambda_3(p)$ its eigenvalues. We assume that the $\lambda_j(p)$ are distinct(*) for $p \in R_2\backslash 0$ and we order them according to the increasing real parts. We also consider the fundamental Cauchy matrix

$$e^{-K(p)t} \equiv \sum_{j=1}^{3} e^{-\lambda_j(p)t} \mathscr{D}_j(p) \tag{1.3}$$

of the system of ordinary differential equations with constant coefficients, depending on the parameter $p \in R_2$:

$$\frac{d\hat{v}(p,t)}{dt} + K(p)\hat{v}(p,t) = 0,$$

where $\mathscr{D}_j(p)$ is a 3×3 matrix. Everywhere below we denote $|h| = (\sum_{i=1}^{\mathscr{N}} (h^{(i)})^2)^{1/2}$ for a vector $h + (h^{(1)}, h^{(2)}, \dots, h^{(\mathscr{N})}) \in R_{\mathscr{N}}$. Our goal in this chapter is to prove the following theorem.

THEOREM. *Suppose that*

1) *the matrix $K(p)$ is such that*

$$\lambda_j^1 \equiv \operatorname{Re}\lambda_j = a_j + b_j|p|^{\delta_j} + O(|p|^{\delta+\sigma_1}) \quad \textit{for } |p| \leq 1, \tag{1.4}$$

$$\lambda_j^1 \geq a_1 + 2\kappa(p) \quad \textit{for } p \in R_2, \tag{1.5}$$

$$|\operatorname{Im}\lambda_j(p)| \leq c_2|p|^{\delta+\sigma_2} \quad \textit{for } |p| \leq 1, \tag{1.6}$$

$$\|\mathscr{D}_j(p)\| \leq c_3(**) \quad \textit{for } p \in R_2, \tag{1.7}$$

$$|\lambda_j(p_1) - \lambda_j(p_2)| + \|\mathscr{D}_j(p_1) - \mathscr{D}_j(p_2)\| \leq c_4|\Delta p|^{\sigma_3} M^{\alpha}(p_1)\,. \tag{1.8}$$

(*) The case of equal roots introduces additional technical complications, but the nature of the asymptotic limit remains the same.

(**) Here and below $\|A\|$ is understood to be the norm of the linear operator A.

2) *The following estimates hold for the Fourier transform of the initial perturbation* $\hat{\overline{u}}(p)$:

$$|\hat{\overline{u}}(p)| \leq \frac{\varepsilon}{3c_3} M^{-4-\alpha}(p) \quad \text{for } p \in R_2, \tag{1.9}$$

$$|\hat{\overline{u}}(p_1) - \hat{\overline{u}}(p_2)| \leq c_5 |\Delta p|^{\sigma_4} M^{-4}(p_1), \tag{1.10}$$

where

$$0 \leq a_1 \leq a_2 \leq a_3, \; b_j > 0, \; \delta_j > 0, \; j = 1, 2, 3;$$
$$\delta = \max(\delta_1, \delta_2, \delta_3), \; \delta \in (0, 3), \; \sigma_1, \sigma_2, \sigma_3, \sigma_4 \in (0, 1];$$
$$\alpha \geq 0, \; c_1, c_2, c_3, c_4, c_5 > 0, \; |\Delta p| = |p_1 - p_2| \leq 1,$$
$$p_1, p_2 \in R_2 \backslash 0, \; m(p) = \min(1, |p|), \; M(p) = \max(1, |p|),$$
$$\kappa(p) = c_1 m^{\delta}(p) M(p) \,.$$

Here $0 < \varepsilon < c$, *where* c *is a positive constant, determined by the matrix* $K(p)$ (*see Lemma* 1 *below*).

Then the solution $u(x,t)$ *of the Cauchy problem* (1.2) *has the uniform with respect to* $\xi = xt^{-1/\delta} \in R_2$ *asymptotics as* $t \to \infty$,

$$u(x,t) = (2\pi)^{-2} A(\xi) e^{-a_1 t} t^{-2/\delta_1} + O(e^{-a_1 t} t^{-2/\delta_1 - \mu}), \tag{1.11}$$

where $\mu > 0$ *is a constant, and the function* $A(\xi)$ *has the form*

1) *if* $0 \leq a_1 = a_2 = a_3$ *and* $\delta_1 = \delta_2 = \delta_3 > 0$:

$$A(\xi) = \sum_{\mathcal{N}=0}^{\infty} \xi^{\mathcal{N}+1} \sum_{j=1}^{3} \int_0^{2\pi} d\varphi \int_0^{\infty} d|q| \, |q| e^{i(q,\xi) - b_j |q|^{\delta_1}} \widetilde{\mathscr{D}}_j(\varphi) \Phi_{\mathcal{N}}(\varphi), \tag{1.12}$$

2) *if* $0 \leq a_1 = a_2 = a_3$ *and* $\delta_1 > \delta_2 \geq \delta_3 > 0$ *or* $0 \leq a_1 = a_2 < a_3$ *and* $\delta_1 > \delta_2 > 0$, $\delta_3 > 0$, *or* $0 \leq a_1 < a_2 \leq a_3$ *and* $\delta_1 > 0, \delta_2 > 0, \delta_3 > 0$:

$$A(\xi) = \sum_{\mathcal{N}=0}^{\infty} \xi^{\mathcal{N}+1} \int_0^{2\pi} d\varphi \int_0^{\infty} d|q| \, |q| e^{i(q,\xi) - b_1 |q|^{\delta_1}} \Phi_{\mathcal{N}}(\varphi), \tag{1.13}$$

3) *if* $0 \leq a_1 = a_2 = a_3$ *and* $\delta_1 = \delta_2 > \delta_3$, *or* $0 \leq a_1 = a_2 < a_3$ *and* $\delta_1 = \delta_2$:

$$A(\xi) = \sum_{\mathcal{N}=0}^{\infty} \xi^{\mathcal{N}+1} \sum_{j=1}^{2} \int_0^{2\pi} d\varphi \int_0^{\infty} d|q| \, |q| e^{i(q,\xi) - b_j |q|^{\delta_1}} \widetilde{\mathscr{D}}_j(\varphi) \Phi_i(\varphi),$$

where

$$\widetilde{\mathscr{D}}_j(\varphi) = \lim_{p \to 0: \; \frac{p}{|p|} = (\cos\varphi, \sin\varphi) = \text{const}} \mathscr{D}_j(p),$$

and the vectors $\Phi_{\mathcal{N}}(\varphi)$ *can be expressed in terms of the symbol* $K(p)$ *and the initial perturbation* $\overline{u}(x)$ *and will be indicated in the course of the proof* (*see* (4.28) *below*).

EXAMPLES. 1) Consider the system of Boussinesq equations with viscosity

$$\begin{cases} \eta_t + (\nabla, \eta\theta) - g_1\Delta\eta + g_2\Delta(\nabla, \theta) = 0, \\ \theta_t + (\theta, \nabla)\theta + g_3\nabla\eta - g_4\Delta\theta = 0, \end{cases}$$

where g_1, g_2, g_3, g_4 are constants: $g_1g_4 > g_2g_3 > 0, g_4 > 0$. The eigenvalues of the matrix $K(p)$ are

$$\lambda_1(p) = \frac{|p|^2}{2}(g_1 + g_4 - g_5), \qquad \lambda_2(p) = g_4|p|^2,$$

$$\lambda_3(p) = \frac{|p|}{2}(g_1 + g_4 + g_5), \qquad g_5 = \sqrt{(g_1 - g_4)^2 + 4g_2g_3}\,.$$

The asymptotics has the form

$$u(x,t) = A(\xi)t^{-1} + O(t^{-3/2}), \tag{1.14}$$

and the vector $A(\xi)$ is given by (1.12), where $\delta_1 = 2$, $b_j = \lambda_j(p)/|p|^2$.

2) Consider the system of Dobrokhotov equations with viscosity

$$\begin{cases} \eta_t & +(\nabla, \eta\theta) - g_1\Delta\eta \\ & +\sum_{j=1}^{2} \frac{g_2}{(2\pi)^2}\int_{-\infty}^{\infty} e^{i(p,x)}\left(\frac{\tanh|p|}{|p|} - 1\right) ip^{(j)}\hat{\theta}^{(j)}(p,t)\,dp = 0, \\ \theta_t & +(\theta, \nabla)\theta + g_3\nabla\eta - g_4\Delta\theta = 0, \end{cases}$$

where g_1, g_2, g_3, g_4 are constants: $g_1g_4 > g_2g_3 > 0$, $g_4 > 0$. The eigenvalues of the matrix $K(p)$ are

$$\lambda_1(p) = \tfrac{1}{2}((g_1 + g_4)|p|^2 - g_5), \qquad \lambda_2(p) = g_4|p|^2,$$

$$\lambda_3(p) = \tfrac{1}{2}((g_1 + g_4)|p|^2 + g_5), \qquad g_5 = \sqrt{(g_1 - g_4)^2|p|^4 + 4g_2g_3(|p|^2 - |p|\tanh|p|)}\,.$$

The asymptotics has the form (1.14), (1.12).

The results of this chapter were published in **[62]**, **[65]**.

§2. Lemmas

We note that for the functions $m(p)$ and $M(p)$ the following inequalities hold, $p, q \in R_2$:

$$m(p)m(q) \le m(|p|\,|q|) \le M(p)m(q) \le M(|p|\,|q|) \le M(p)M(q)\,. \tag{2.1}$$

Denote, for all $p \in R_2$, $t \ge 0$, $\chi \ge \delta + 5$, $\gamma \in [0, 1]$:

$$H_{\mathcal{N}}(p,t) = (\mathcal{N}+1)^{-\chi}M^{-4}(p)\exp\left(-at - t\kappa\left(\frac{p}{\mathcal{N}+1}\right)\right), \qquad \mathcal{N} \ge 0,$$

$$U_{\mathcal{N}}(p,t,\gamma) = \int_0^t dt\, e^{-(a+2\kappa(p))(t-\tau)}\int_{-\infty}^{\infty} dq \sum_{j=1}^{\mathcal{N}} |p-q|^{1-\gamma}$$

$$\cdot H_{\mathcal{N}-j}(q,\tau)H_{j-1}(p-q,\tau), \qquad \mathcal{N} \ge 1,$$

where $\kappa(p) = c_1 m^{\delta}(p)M(p)$.

LEMMA 1. *Let $0 < \delta < 3 - \gamma$, $a \ge 0$. Then the following estimates holds*

$$U_{\mathcal{N}}(p,t,\gamma) \le cH_{\mathcal{N}}(p,t), \tag{2.2}$$

where $c > 0$ is a constant.

LEMMA 2. *If* $0 < \gamma_2 \le \frac{1}{2}m(3-\delta-\gamma_1)$, *then for* $|p| \le 1$ *the following estimate holds*:

$$\int_{-\infty}^{\infty} dq\, M^{-3}(q)m^{1-\gamma_1}(q) \int_0^t d\tau\, e^{-2\kappa(p)(t-\tau)-\kappa(gr)\tau} = r^{3-\delta-\gamma_2}|p|^{-2-\gamma_2}O(t^{-2/\delta-\gamma_2/\delta}), \tag{2.3}$$

where $0 \le \gamma_1 \le \frac{1}{2}$, $0 < r \le 1$.

The proof of these two lemmas is a word for word repetition of the proof of the lemmas in §2, Chapter 7.

§3. Proof of the theorem

3.1. Perturbation theory. In the proof we confine ourselves to only two cases: 1) $a_1 = a_2 = a_3$ and 2) $a_1 < a_2 \le a_3$ (the remaining case $a_1 = a_2 < a_3$ can be considered similarly). It was proved in Chapter 4 that there exists a unique solution of the Cauchy problem for the system of surface wave equations in the class $C^\infty((0,\infty); H^\infty(R_2))$. Applying the Fourier transform to (1.1), we find that

$$\hat{\eta}_t + i\int_{-\infty}^{\infty} dq\, \hat{\eta}(p-q,t)\sum_{l=1}^{2} p^{(l)}\hat{\theta}^{(l)}(q,t) + k_1(p)\hat{\eta}(p,t) + k_2(p)\hat{\theta}(p,t) = 0,$$

$$\hat{\theta}_t + i\int_{-\infty}^{\infty} dq\, \hat{\theta}(p-q,t)\sum_{l=1}^{2} \hat{\theta}^{(l)}(q,t)(p^{(l)}-q^{(l)}) + k_3(p)\hat{\eta}(p,t) + k_4(p)\hat{\theta}(p,t) = 0,$$

or for the vector

$$\hat{u}(p,t) = \begin{pmatrix} \hat{\eta}(p,t)|p|^{-1} \\ \hat{\theta}(p,t) \end{pmatrix}$$

we obtain

$$\hat{u}_t(p,t) - \int_{-\infty}^{\infty} dq\, V(p,q,t) + K(p)\hat{u}(p,t) = 0, \tag{3.1}$$

where

$$V(p,q,t) = -i\left\{ e_1|p-q|\hat{u}^{(1)}(p-q,t)\sum_{n=1}^{2}\frac{p^{(n)}}{|p|}\hat{u}^{(n+1)}(q,t) + \sum_{l=2}^{3} e_l\hat{u}^{(l)}(p-q,t)\sum_{n=1}^{2}\hat{u}^{(n+1)}(q,t)(p^{(n)}-q^{(n)})\right\};$$

e_1, e_2, e_3 are the (canonical) basis unit vectors in R_3. Using the fundamental Cauchy matrix, we integrate (3.1) with respect to t:

$$\hat{u}(p,t) = e^{-K(p)t}\hat{\bar{u}}(p) + \int_0^t d\tau\, e^{-K(p)(t-\tau)}\int_{-\infty}^{\infty} dq\, V(p,q,\tau)\,. \tag{3.2}$$

We make use of perturbation theory. We set $\hat{\bar{u}}(p) = \varepsilon\bar{v}(p)$ and represent $\hat{u}(p,t)$ as a formal series in ε:

$$\hat{u}(p,t) = \sum_{j=0}^{\infty}\varepsilon^{j+1}v_j(p,t)\,. \tag{3.3}$$

Substituting (3.3) in (3.2) and comparing coefficients of like powers of ε, we obtain, $\mathcal{N} \geq 1$:

$$v_0(p,t) = e^{-K(p)t}\overline{v}(p), \tag{3.4}$$

$$v_{\mathcal{N}}(p,t) = \int_0^t d\tau\, e^{-K(p)(t-\tau)} \int_{-\infty}^{\infty} dq\, V_{\mathcal{N}}(p,q,\tau), \tag{3.5}$$

where

$$V_{\mathcal{N}}(p,q,\tau) = -i\sum_{j=1}^{\mathcal{N}} \Bigg\{ e_1|p-q|v_{j-1}^{(1)}(p-q,\tau)\sum_{n=1}^{2}\frac{p^{(n)}}{|p|}v_{\mathcal{N}-j}^{(n+1)}(q,\tau)$$
$$+\sum_{l=2}^{3} e_l v_{j-1}^{(e)}(p-q,\tau)\sum_{n=1}^{2} v_{j-1}^{(n+1)}(q,\tau)(p^{(n)}-q^{(n)})\Bigg\}\,.$$

3.2. Estimation of the function $v_{\mathcal{N}}$. We shall establish the following bound for all $\mathcal{N} \geq 0, p \in R_2, t \geq 0$:

$$|v_{\mathcal{N}}(p,t)| \leq c^{\mathcal{N}} H_{\mathcal{N}(}(p,t)M^{-\alpha}(p), \tag{3.6}$$

where the function $H_{\mathcal{N}}(p,t)$ is defined in Lemma 1 with $\delta = \max(\delta_1,\delta_2,\delta_3)$ and $a = a_1$. For $\mathcal{N} = 0$ (3.6) is readily verified. Indeed, from (1.5) and (1.7) we obtain

$$\|e^{-K(p)t}\| \leq 3c_3 e^{-at-2\kappa(p)t}\,. \tag{3.7}$$

Thus, by (1.9) we have

$$\begin{aligned}|v_0(p,t)| &\leq e^{-at-2\kappa(p)t}3c_3|\overline{v}(p)|\\ &\leq e^{-at-2\kappa(p)t}M^{-4-\alpha}(p)\,.\end{aligned}$$

Next, arguing by induction, from (3.5), by virtue of Lemma 1 and taking into account (2.1) and (3.7), we obtain

$$\begin{aligned}|v_{\mathcal{N}}(p,t)| &\leq c\int_0^t e^{-a(t-\tau)-2\kappa(p)(t-\tau)}\,d\tau \sum_{j=1}^{\mathcal{N}}\int_{-\infty}^{\infty} dq\,|v_{j-1}(p-q,\tau)||v_{w-j}(q,\tau)|\,|p-q|\\
&\leq c^{\mathcal{N}}\int_0^t e^{-a(t-\tau)-2\kappa(p)(t-\tau)}\,d\tau\\
&\quad\cdot\sum_{j=1}^{\mathcal{N}}\int_{-\infty}^{\infty} dq\,H_{\mathcal{N}-j}(q,\tau)H_{j-1}(p-q,\tau)|p-q|M^{-\alpha}(p-q)M^{-\alpha}(q)\\
&\leq c^{\mathcal{N}}2^{\alpha+1}M^{-\alpha}(p)V_{\mathcal{N}}(p,t,0) \leq c^{\mathcal{N}}H_{\mathcal{N}}(p,t)M^{-\alpha}(p);\end{aligned}$$

here and below $c > 0$ denotes various constants independent of $\mathcal{N}$. From the representation (3.3), inverting the Fourier transform, we find

$$u(x,t) = (2\pi)^{-2}\int_{-\infty}^{\infty} e^{i(p,x)}\hat{u}(p,t)\,dp = \sum_{\mathcal{N}=0}^{\infty}\varepsilon^{\mathcal{N}+1}\psi_{\mathcal{N}}(x,t), \tag{3.8}$$

where

$$\psi_{\mathcal{N}}(x,t) = (2\pi)^{-2}\int_{-\infty}^{\infty} e^{i(p,x)}v_{\mathcal{N}}(p,t)\,dp\,.$$

The termwise integration of the series performed in obtaining (3.8) is possible because of the majorant convergence of the series of perturbation theory (3.3) for $\varepsilon < 1/c$ (see (3.6)). The computation of the asymptotics of $u(x,t)$ reduces to finding the asymptotics of the integrals $\psi_{\mathcal{N}}$. We represent $\psi_{\mathcal{N}}(x,t)$ in the following form:

$$(2\pi)^2\psi_{\mathcal{N}}(x,t) = \sum_{j=1}^{5} J_j, \qquad J_1 = \int_{|p|\geq 1} dp\, e^{i(p,x)} v_{\mathcal{N}}(p,t)\,.$$

In case 1) we set

$$\begin{aligned}
\int_{|p|\leq 1} & e^{i(p,x)} v_{\mathcal{N}}(p,t)\,dp \\
&= \int_0^{2\pi} d\varphi \int_0^1 d|p|\cdot|p|e^{i(p,x)}\left\{\left[\sum_{j=1}^{3}\widetilde{\mathscr{D}}_j e^{-\overline{\lambda}_j(p)t}(\tilde{v}_{\mathcal{N}}(\varphi,t)e^{at}-\Phi_{\mathcal{N}})\right]\right. \\
&\qquad + \left[\sum_{j=1}^{3}\widetilde{\mathscr{D}}_j\Phi_{\mathcal{N}}e^{-\overline{\lambda}_j(p)t}\right] \\
&\qquad + \left[\sum_{j=1}^{3}\widetilde{\mathscr{D}}_j\tilde{v}_{\mathcal{N}}(\varphi,t)e^{at}(e^{-\lambda_j(p)t}-e^{-\overline{\lambda}_j(p)t})\right] \\
&\qquad \left. + \left[v_{\mathcal{N}}(p,t) - \sum_{j-1}^{3} e^{-\lambda_j(p)t}\widetilde{\mathscr{D}}_j e^{at}\tilde{v}_{\mathcal{N}}(\varphi,t)\right]\right\} \\
&= J_2 + J_3 + J_4 + J_5,
\end{aligned}$$

where $\Phi_{\mathcal{N}}(\varphi) = \lim_{t\to\infty} e^{at}\tilde{v}_{\mathcal{N}}(\varphi,t)$, $\overline{\lambda}_j(p) = a + b_j|p|^{\delta_j}$. Here and below a tilde over a function denotes the result of taking the limit as $p\to 0$, so that $p/|p| = (\cos\varphi,\sin\varphi) = \text{const}$; in particular

$$\tilde{v}_{\mathcal{N}}(\varphi,t) = \lim_{p\to 0, p/|p|=(\cos\varphi,\sin\varphi)=\text{const}} v_{\mathcal{N}}(p,t)\,.$$

In case 2) we write

$$\begin{aligned}
\int_{|p|\leq 1} & e^{i(p,x)} v_{\mathcal{N}}(p,t)\,dp \\
&= \int_0^{2\pi} d\varphi \int_0^1 d|p|\cdot|p|e^{i(p,x)}\{[e^{-\overline{\lambda}_1(p)t}\cdot(\tilde{v}_{\mathcal{N}}(\varphi,t)e^{at}-\Phi_{\mathcal{N}})] \\
&\qquad + [\Phi_i e^{-\overline{\lambda}_1(p)t}] + [\tilde{v}_{\mathcal{N}}(\varphi,t)e^{at}(e^{-\lambda_1(p)t}-e^{-\overline{\lambda}_1(p)t})] \\
&\qquad + [v_{\mathcal{N}}(p,t) - \tilde{v}_{\mathcal{N}}(\varphi,t)e^{-\lambda_1(p)t+at}]\} \\
&= J_2 + J_3 + J_4 + J_5\,.
\end{aligned}$$

3.3. Estimates of the integrals J_1, J_2, J_3. Let us estimate J_1. From (3.6) we obtain

$$
\begin{aligned}
|J_1| &\le \int_{|p|\ge 1} |v_{\mathcal{N}}(p,t)|\,dp \le c^{\mathcal{N}} e^{-at}(\mathcal{N}+1)^{-\chi} \int_{|p|\ge 1} dp\, M^{-4-\alpha}(p) e^{-t\kappa(p/(\mathcal{N}+1))} \\
&\le c^{\mathcal{N}}(\mathcal{N}+1)^{-\chi} e^{-at-c_1 t(\mathcal{N}+1)^{-\delta}} \int_{|p|\ge 1} dp\, M^{-4-\alpha}(p) = c^{\mathcal{N}} O(e^{-at} t^{-3/\delta}),
\end{aligned} \tag{3.9}
$$

since $\chi \ge 5+\delta > 3$. We introduce new matrices: in case 1) $\overline{\mathscr{D}}_1 = E$, $\overline{\mathscr{D}}_2 = \overline{\mathscr{D}}_3 = 0$, and in case 2) $\overline{\mathscr{D}}_1 = \widetilde{\mathscr{D}}_1$, $\overline{\mathscr{D}}_2 = \widetilde{\mathscr{D}}_2$, $\overline{\mathscr{D}}_3 = \widetilde{\mathscr{D}}_3$ (E is the 3×3 identity matrix). Then, for $\mathcal{N} \ge 1$ from (3.5) we obtain

$$
\Phi_{\mathcal{N}} = \int_0^\infty d\tau\, e^{a\tau}\overline{\mathscr{D}}_1 \int_{-\infty}^\infty dq\, \widetilde{V}_{\mathcal{N}}(\varphi,q,\tau) + \lim_{t\to\infty} I_1(\varphi,t),
$$

where

$$
I_1(\varphi,t) = e^{at}\int_0^t d\tau \sum_{j=2}^3 \overline{\mathscr{D}}_j e^{-\lambda_j(0)(t-\tau)} \int_{-\infty}^\infty dq \widetilde{V}_{\mathcal{N}}(\varphi,q,\tau),
$$

$$
\begin{aligned}
\widetilde{V}_{\mathcal{N}}(\varphi,q,\tau) = -i\sum_{j=1}^{\mathcal{N}} \Bigg\{ & e_1|q| v_{j-1}^{(1)}(-q,\tau)(\cos\varphi v_{\mathcal{N}-j}^{(2)}(q,\tau) + \sin\varphi v_{\mathcal{N}-j}^{(3)}(q,\tau)) \\
& + \sum_{p=n}^3 e_l v_{j-1}^{(l)}(-q,\tau) \sum_{n=1}^2 v_{\mathcal{N}-j}^{(n+1)}(q,\tau)(-q^{(n)}) \Bigg\}.
\end{aligned}
$$

Also,

$$
\begin{aligned}
&\tilde{v}_{\mathcal{N}}(\varphi,t)e^{at} - \Phi_{\mathcal{N}} \\
&\quad = -\int_t^\infty d\tau\, e^{a\tau}\overline{\mathscr{D}}_1 \int_{-\infty}^\infty dq\, \widetilde{V}_{\mathcal{N}}(\varphi,q,\tau) + I_1(\varphi,t) - \lim_{t\to\infty} I_1(\varphi,t).
\end{aligned} \tag{3.10}
$$

Applying (3.6) and (1.7), we estimate I_1:

$$
\begin{aligned}
|I_1| &\le c\int_0^t e^{(a-a_2)(t-\tau)} e^{a\tau}\, d\tau \sum_{j=1}^{\mathcal{N}} \int_{-\infty}^\infty dq\,|q|\,|v_{\mathcal{N}-j}(q,\tau)|\,|v_{j-1}(-q,\tau)| \\
&\le c^{\mathcal{N}}\left(\int_0^{t/2} + \int_{t/2}^t\right) d\tau\, e^{(a-a_2)(t-\tau)-a\tau} \\
&\quad \cdot \sum_{j=1}^{\mathcal{N}} \int_{-\infty}^\infty e^{-\tau\kappa(q/j)} e^{-\tau\kappa(q/(\mathcal{N}+1-j))} M^{-8}(q) j^{-\chi}(\mathcal{N}+1-j)^{-\chi}|q|\,dq \\
&= c^{\mathcal{N}} O(e^{(-a_2+a)t/2}) + c^{\mathcal{N}} \sum_{j=1}^{\mathcal{N}} j^{-\chi} \int_{-\infty}^\infty \frac{e^{-\kappa(q/j)t/2}|q|\,dq}{\kappa(q/j)M^8(q)} \\
&= c^{\mathcal{N}} O(M^{-3/\delta+1}(t)).
\end{aligned} \tag{3.11}
$$

Therefore, $\lim_{t\to\infty} I_1(t) = 0$ since $\delta < 3$. Substitution of (3.6) and (3.11) in (3.10) yields

$$
\begin{aligned}
&|\tilde{v}_{\mathcal{N}}(\varphi, t)e^{at} - \Phi_i(\varphi)| \\
&\quad = c\int_t^\infty e^{a\tau}\,d\tau \sum_{j=1}^{\mathcal{N}} \int_{-\infty}^{\infty} dq\,|q| c^{\mathcal{N}-1} \cdot e^{-2a\tau-\tau\kappa(q/j)-\tau\kappa(q/(\mathcal{N}+1-j))} \\
&\qquad \cdot j^{-\chi}(\mathcal{N}+1-j)^{-\chi} M^{-8}(q) + c^{\mathcal{N}} O(M^{1-3/\delta}(t)) \qquad (3.12)\\
&\quad \le c^{\mathcal{N}} \sum_{j=1}^{\mathcal{N}} j^{-\chi} \int_{-\infty}^{\infty} \frac{M(q)m(q)e^{-\kappa(q/j)t}}{\kappa(q/j)M^8(q)}\,dq + c^{\mathcal{N}} O(M^{1-3/\delta}(t)) \\
&\quad = c^{\mathcal{N}} O(M^{1-3/\delta}(t)) .
\end{aligned}
$$

And for $\mathcal{N} = 0$ we have

$$
\begin{aligned}
\tilde{v}_0(\varphi, t)e^{at} &= \overline{\mathscr{D}}_1 \tilde{\bar{v}}(\varphi) + \sum_{j=2}^{3} e^{(a-a_j)t} \overline{\mathscr{D}}_j \tilde{\bar{v}}(\varphi) \\
&= \overline{\mathscr{D}}_1 \tilde{\bar{v}}(\varphi) + O(M^{1-3/\delta}(t)),
\end{aligned}
$$

since for $a_1 \neq a_2 \le a_3$ the second term decays even exponentially, and for $a_1 = a_2 = a_3$ it is equal to zero by the definition of $\overline{\mathscr{D}}_j$. Hence

$$
\Phi_0(\varphi) \equiv \lim_{t\to\infty} e^{at}\tilde{v}_0(\varphi, t) = \overline{\mathscr{D}}_1 \tilde{\bar{v}}(\varphi),
$$

so that by (1.9) we have $|\Phi_0(\varphi)| \le 1$. This shows, together with (3.12) taken at $t = 0$, that for all $\mathcal{N} \ge 0$

$$
|\Phi_{\mathcal{N}}(\varphi)| \le c^{\mathcal{N}} . \tag{3.13}
$$

Further, we consider the integral

$$
I_2 = \int_{|p|\le 1} d|p| \cdot |p| e^{i(p,x)-\bar{\lambda}_j(p)t},
$$

where $j = 1, 2, 3$ in case 1) and $j = 1$ in case 2). By the definition of $\bar{\lambda}_j(p)$, for $t \ge 1$ we have

$$
\begin{aligned}
I_2 &= \int_{|p|\le 1} d|p|\,|p| e^{i(p,x)-at-b_j t|p|^{\delta_j}} \\
&= \int_0^\infty d|p|\,|p| e^{-at-b_j t|p|^{\delta_j}+i(p,x)} + O(e^{-at}t^{-3/\delta}),
\end{aligned}
$$

whence, after the change $\eta = pt^{1/\delta_j}$ and $\xi = xt^{-1/\delta_j}$ we obtain

$$
I_2 = e^{-at}t^{-2/\delta_j} \int_0^\infty d|\eta| e^{i(\eta,\xi)-b_j|\eta|^{\delta_j}} |\eta| + O(e^{-at}t^{-3/\delta}) . \tag{3.14}
$$

Substituting (3.12), (3.13), and (3.14) in the integrals J_2 and J_3, we obtain

$$
|J_2| = c^{\mathcal{N}} O(e^{-at}t^{1-5/\delta}) . \tag{3.15}
$$

For the leading term of the asymptotics of J_3 in case 1) we have

$$J_3 = \int_0^{2\pi} d\varphi \int_0^{\infty} d|\eta|\,|\eta| \sum_{j=1}^{3} \widetilde{\mathscr{D}}_j \Phi_{\mathscr{N}} e^{-at} t^{-2/\delta_j} e^{i(\eta,\xi)-b_j|\eta|^{\delta_j}} + c^{\mathscr{N}} O(e^{-at} t^{-3/\delta}), \tag{3.16}$$

and, similarly, in case 2)

$$J_3 = \int_0^{2\pi} d\varphi \int_0^{\infty} d|\eta|\,|\eta| \Phi_{\mathscr{N}} e^{-at-b_1|\eta|^{\delta_1}+i(\eta,\xi)} + c^{\mathscr{N}} O(e^{-at} t^{-3/\delta}) .$$

3.4. Estimation of the integral J_4. We first consider the integral

$$I_3 = \int_{|p|\le 1} d|p| e^{i(p,x)-\overline{\lambda}_j(p)t}(e^{-s(p)t}-1)|p|,$$

where $s = \lambda_j - \overline{\lambda}_j$ is set. The index j assumes the values 1, 2, 3 in case 1) and $j = 1$ in case 2). We split the domain of integration into two parts: $0 < t^{-\nu} \le |p| \le 1$ and $0 \le |p| \le t^{-\nu}$, where $\nu = (\delta+\beta)/2\delta\beta \in (0, 1/\delta)$, $\beta = \delta + \sigma$, $\sigma = \min(\sigma_1, \sigma_2, \sigma_3, \sigma_4, \delta)$. For the first term we have

$$\left| \int_{t^{-\nu}\le|p|\le 1} d|p|\,|p| e^{i(p,x)-\overline{\lambda}_j(p)t}(e^{-s(p)t}-1) \right| \le 1 \int_{t^{-\nu}\le|p|\le 1} e^{-at-\overline{b}|p|^{\delta_t}} |p| d\,|p| \le 2e^{-at-\overline{b}t^{1-\delta\nu}} = O(e^{-at} t^{-3/\delta}), \tag{3.17}$$

since $1-\delta\nu > 0$, $\overline{b} = \min\{b_j, 2c_1\}$. In the second term we introduce the new variable of integration $\eta = |p|^{\delta_t}$, and find that

$$\begin{aligned} &\left| \int_{|p|\le t^{-\nu}} e^{i(p,x)-\overline{\lambda}_j(p)t}(e^{-s(p)t}-1)|p| d\,|p| \right| \\ &\quad \le \frac{1}{\delta} \int_0^{t^{1-\nu\delta}} d\eta e^{-at-b_j\eta} t^{-2/\delta} \eta^{2/\delta-1} \left| \exp\left(-ts\left(\left(\frac{\eta}{t}\right)^{1/\delta}\right)-1\right| \right. \\ &\quad = e^{-at} O(t^{-2/\delta+1-\nu\beta}) \int_0^{\infty} d\eta\, e^{-b_j\eta} \eta^{2/\delta-1} \\ &\quad = O(e^{-at} t^{-2/\delta-\sigma/2\delta}) . \end{aligned} \tag{3.18}$$

From (3.17), (3.18), and (3.6) we obtain

$$|J_4| = c^{\mathscr{N}} O(e^{-at} t^{-2/\delta-\sigma/2\delta}) . \tag{3.19}$$

3.5. Estimation of the Hölder term J_5. We first establish the following bound for $0 < |p_2| \leq |p_1|$:

$$|v_{\mathcal{N}}(p_1,t) - v_{\mathcal{N}}(p_2,t)| = c^{\mathcal{N}} H_{\mathcal{N}}(p_2,t) O(|\Delta p|^{2\gamma}) m^{-2\gamma^{\delta}/\gamma}(p_2), \tag{3.20}$$

where

$$|\Delta p| = |p_1 - p_2| \leq 1, \quad \gamma = \delta\mu, \quad \mu = \frac{\sigma}{4\delta} m\left(\frac{3-\delta}{4}\right) m(1/\delta). \tag{3.21}$$

For $\mathcal{N} = 0$ the estimate (3.20) follows from (3.4) and (1.7)–(1.10):

$$\begin{aligned}
&|v_0(p_1,t) - v_0(p_2,t)| \\
&\quad \leq \sum_{j=1}^{3} \{|e^{-\lambda_j(p_1)t}\mathcal{D}_j(p_1) - e^{-\lambda_j(p_2)t}\mathcal{D}_j(p_2)|\overline{v}(p_1)| + e^{-\lambda_j(p_2)t} c_3 |\overline{v}(p_1) - \overline{v}(p_2)|\} \\
&\quad = e^{-at-2\kappa(p_2)t}(M^{-\alpha-4}(p_2) O(|\Delta p|^{\sigma_3}) M^{\alpha}(p_1) + O(|\Delta p|^{\sigma_4}) M^{-4}(p_1)) \\
&\quad = H_0(p_2,t) O(|\Delta p|^{\sigma}).
\end{aligned}$$

Next, for $N \geq 1$, arguing by induction, in view of (3.5), we write the difference on the left-hand side of (3.20) in the following form:

$$\begin{aligned}
&v_{\mathcal{N}}(p_1,t) - v_{\mathcal{N}}(p_2,t) \\
&\quad - \int_0^t d\tau\,(e^{-K(p_1)(t-\tau)} - e^{-K(p_2)(t-\tau)}) \int_{-\infty}^{\infty} dq\, V_{\mathcal{N}}(p_2,q,\tau) \\
&\quad + \int_0^t d\tau e^{-K(p_1)(t-\tau)} \Bigg\{ \int_{|p_1-q|\leq|p_2-q|} dq\, F_1(p_1,p_2,q,\tau) \\
&\qquad + \int_{|p_1-q|\leq|p_2-q|} dq\, F_2(p_1,p_2,q,\iota) \\
&\qquad + \int_{|p_2-q|<|p_1-q|} dq\, F_1(p_2,p_2,q,\tau) \\
&\qquad + \int_{|p_2-q|<|p_1-q|} dq\, F_2(p_2,p_1,q,\tau) \Bigg\} \\
&\quad \equiv I_4 + I_5 + I_6 + I_7 + I_8,
\end{aligned}$$

where

$$\begin{aligned}
F_1(p_1,p_2,q,\tau) = i \sum_{j=1}^{\mathcal{N}} \Bigg\{ & e_1|p_1 - q| \Bigg[v_{j-1}^{(1)}(p_1 - q,\tau) \sum_{n=1}^{2} \frac{p_1^{(n)}}{|p_1|} v_{\mathcal{N}-j}^{(n+1)}(q,\tau) \\
& - v_{j-1}^{(1)}(p_2 - q,\tau) \sum_{n=1}^{2} \frac{pj_2^{(n)}}{|p_2|} v_{\mathcal{N}-j}^{(n+1)}(q,\tau) \Bigg] \\
& + \sum_{l=2}^{3} e_l [v_{j-1}^{(l)}(p_1 - q,\tau) - v_{j-1}^{(l)}(p_2 - q,\tau)] \\
& \cdot \sum_{n=1}^{2} v_{\mathcal{N}-j}^{(n+1)}(q,\tau)(p_1^{(n)} - q^{(n)}) \Bigg\},
\end{aligned}$$

$$F_2(p_1,p_2,q,\tau)=i\sum_{j=1}^{n}\Big\{e_1[|p_1-q|-|p_2-q|]v_{j-1}^{(1)}(p_2-q,\tau)$$

$$\cdot\sum_{n=1}^{2}\frac{p_2^{(n)}}{|p_2|}v_{\mathcal{N}-j}^{(n+1)}(q,\tau)$$

$$+\sum_{l=2}^{3}e_l v_{j-1}^{(l)}(p_2-q,\tau)\sum_{n=1}^{2}v_{\mathcal{N}-j}^{(n+1)}(q,\tau)\Delta p^{(n)}\Big\}.$$

First of all, we note that from the estimate

$$|e^{-z}-1|\leq 2m^{2\gamma/\sigma}(z),\qquad \operatorname{Re}z\geq 0,\ 2\gamma<\sigma,$$

and condition (1.8), it follows that for $|p_2|\leq|p_1|$

$$\|e^{-K(p_1)(t-\tau)}-e^{-K(p_2)(t-\tau)}\|\leq\sum_{j=1}^{3}|e^{-\lambda_j(p_1)(t-\tau)}-e^{-\lambda_j(p_2)(t-\tau)}|\cdot\|\mathscr{D}_j(p_1)\|$$

$$+\sum_{j=1}^{3}|e^{-\lambda_j(p_2)(t-\tau)}|\,\|\mathscr{D}_j(p_1)-\mathscr{D}_j(p_2)\|$$

$$\leq ce^{-a(t-\tau)-2\kappa(p_2)(t-\tau)}|\Delta p|^{2\gamma}m^{-2\gamma\delta/\sigma}(p_2)M^{2\gamma\alpha/\sigma}(p_1).$$

Using (3.6), (3.21), and the result of Lemma 1, we obtain ($2\gamma<\sigma$):

$$|I_4|=O(\Delta p|^{2\gamma})m^{-2\gamma\delta/\sigma}(p_2)M^{2\gamma\alpha/\sigma}(p_1)\int_0^t d\tau\, e^{-(a+2\kappa(p_2))(t-\tau)}$$

$$\cdot\sum_{j=1}^{\mathcal{N}}\int_{-\infty}^{\infty}dq\,|v_{j-1}(p_2-q,\tau)|\,|v_{\mathcal{N}-j}(q,\tau)|\,|p_2-q|$$

$$=c^{\mathcal{N}}M^{2\gamma\alpha/\sigma}(p_1)O(m^{-2\gamma\delta/\delta}(p_2)|\Delta p|^{2\gamma})\int_0^t d\tau\, e^{-(a+2\kappa(p_2))(t-\tau)}$$

$$\cdot\sum_{j=1}^{\mathcal{N}}\int_{-\infty}^{\infty}dq\,H_{j-1}(p_2-q,\tau)H_{\mathcal{N}-j}(q,\tau)|p_2-q|M^{-\alpha}(p_2-q)M^{-\alpha}(q)$$

$$=c^{\mathcal{N}}O(m^{-2\gamma\delta/\sigma}(p_2)|\Delta p|^{2\gamma})V_{\mathcal{N}}(p_2,t,0)$$

$$=c^{\mathcal{N}}H_{\mathcal{N}}(p_2,t)O(m^{-2\gamma\delta/\sigma}(p_2)|\Delta p|^{2\gamma}).$$

Further, by (3.56), (3.7), and the induction hypothesis we have

$$\begin{aligned}|I_5| &\le c\int_0^t \|e^{-K(p_1)(t-\tau)}\|d\tau \int_{|p_1-q|\le|p_2-q|} dq \sum_{j=1}^{\mathscr{N}} |v_{\mathscr{N}-j}(q,\tau)|\,|p_1-q| \\ &\quad\cdot\left(|v_{j-1}(p_1-q,\tau)-v_{j-1}(p_2-q,\tau)| + |v_{j-1}(p_1-q,\tau)|\left|\frac{p_1}{|p_1|}-\frac{p_2}{|p_2|}\right|\right) \\ &= O(|\Delta p|^{2\gamma})\int_0^t d\tau\, e^{-(a+2\kappa(p_1))(t-\tau)}\int_{|p_1-q|\le|p_2-q|} dq\sum_{j=1}^{\mathscr{N}} c^{\mathscr{N}} H_{\mathscr{N}-j}(q,\tau)|p_1-q| \\ &\quad\cdot H_{j-1}(p_1-q,\tau)(m^{-2\gamma\delta/\sigma}(p_1-q)+m^{-2\gamma\delta/\sigma}(p_2))|p_1-q| \\ &= c^{\mathscr{N}} O(|\Delta p|^{2\gamma}m^{-2\gamma\delta/\sigma}(p_2))\,V_{\mathscr{N}}\left(p_1,t,\frac{2\gamma\delta}{\sigma}\right) \\ &= c^{\mathscr{N}} H_{\mathscr{N}}(p_2,t)O(|\Delta p|^{2\gamma}m^{-2\gamma\delta/\sigma}(p_2)),\end{aligned}$$

since $3-\delta-2\gamma\delta/\sigma>0$ and $\delta/\sigma\ge 1$. In the integral I_6,

$$\big||p_1-q|-|p_2-q|\big| \le |\Delta p| \le 2|\Delta p|^{2\gamma}|p_2-q|^{1-2\delta};$$

thus, using Lemma 1, we find

$$\begin{aligned}|I_6| &\le \int_0^t e^{-(a+2\kappa(p_1))(t-\tau)}\,d\tau\int_{|p_1-q|\le p_2-q|} dq\sum_{j=1}^{\mathscr{N}} c^{\mathscr{N}}|\Delta p|^{2\gamma}|p_2-q|^{1-2\gamma} \\ &\qquad\qquad\qquad\qquad\cdot H_{j-1}(p_2-q,\tau)H_{\mathscr{N}-j}(q,\tau) \\ &= c^{\mathscr{N}} O(|\Delta p|^{2\gamma})V_{\mathscr{N}}(p_2,t,2\gamma) = c^{\mathscr{N}} H_{\mathscr{N}}(p_2,t)O(|\Delta p|^{2\gamma}).\end{aligned}$$

The integrals I_7 and I_8 can be estimated similarly to I_5 and I_6, and the estimate (3.20) is established by induction. We now prove a more precise Hölder estimate for $|p|\le 1$, $p\ne 0$, $p=|p|(\cos\varphi,\sin\varphi)$. For case 1);

$$\left|v_{\mathscr{N}}(p,t)-\sum_{j=1}^{3} e^{-\lambda_j(p)t}\widetilde{\mathscr{D}}_j e^{at}\tilde{v}_{\mathscr{N}}(\varphi,t)\right| = c^{\mathscr{N}}|p|^{\gamma-2}O(e^{-at}t^{-2/\delta-\mu}), \tag{3.22}$$

and for case 2):

$$|v_{\mathscr{N}}(p,t)-\tilde{v}_{\mathscr{N}}(\varphi,t)e^{at-\lambda_1(p)t}| = c^{\mathscr{N}}|p|^{\gamma-2}O(e^{-at}t^{-2/\delta-\mu}). \tag{3.23}$$

For $\mathscr{N}=0$ the estimates (3.22) and (3.23) follow from conditions (1.8) and (1.10). Indeed, since $\sum_{j=1}^{3}\mathscr{D}_j(p)=E$, in case 1) we have

$$\begin{aligned}&\left|v_0(p,t)-\sum_{j=1}^{3} e^{-\lambda_j(p)t}\widetilde{\mathscr{D}}_j e^{at}\tilde{v}_0(\varphi,t)\right| \\ &\quad=\left|\sum_{j=1}^{3} e^{-\lambda_j(p)t}\left\{\mathscr{D}_j(p)\overline{v}(p)-\widetilde{\mathscr{D}}_j e^{at}\sum_{l=1}^{3} e^{-at}\widetilde{\mathscr{D}}_l\tilde{\overline{v}}(\varphi)\right\}\right| \\ &\quad=\left|\sum_{j=1}^{3} e^{-\lambda_j(p)t}(\mathscr{D}_j(p)\overline{v}(p)-\widetilde{\mathscr{D}}_j\tilde{\overline{v}}(\varphi))\right| \\ &\quad= e^{-at-2\kappa(p)t}O(|p|^{\sigma}) = O(e^{-at}t^{-2/\delta-\mu})|p|^{\gamma-2},\end{aligned}$$

since $\sigma - \delta(2/\delta + \mu) \geq \gamma - 2$. Case 2) is considered analogously.

For $\mathscr{N} \geq 1$ we carry out induction. In view of (3.5) we represent the left-hand side of (3.22) (case 1) in the form

$$
\begin{aligned}
v_{\mathscr{N}}(p,t) &- \sum_{j=1}^{3} e^{-\lambda_j(p)t} \widetilde{\mathscr{D}}_j e^{at} \tilde{v}_{\mathscr{N}}(\varphi, t) \\
&= \sum_{l=1}^{3} e^{-\lambda_l(p)t} \left\{ \mathscr{D}_l(p) \int_0^t d\tau\, e^{\lambda_l(p)\tau} \int_{-\infty}^{\infty} dq\, V_{\mathscr{N}}(p,q,\tau) \right. \\
&\qquad\qquad \left. - \widetilde{\mathscr{D}}_l \int_0^t d\tau\, e^{a\tau} \int_{-\infty}^{\infty} dq\, \widetilde{V}_{\mathscr{N}}(\varphi,q,\tau) \right\} \\
&= \int_0^t d\tau \sum_{l=1}^{3} e^{-\lambda_l(p)t} (\mathscr{D}_l(p) e^{\lambda_l(p)\tau} - \widetilde{\mathscr{D}}_l(\varphi) e^{a\tau}) \int_{-\infty}^{\infty} dq\, \widetilde{V}_{\mathscr{N}}(\varphi,q,\tau) \\
&\quad + \int_0^t d\tau\, e^{-K(p)(t-\tau)} \left\{ \int_{|p-q|\leq|q|} dq\, F_3(p-q,-q,q,\tau) \right. \\
&\qquad\qquad + \int_{|p-q|\leq|q|} dq\, F_4(p-q,-q,q,\tau) \\
&\qquad\qquad + \int_{|q|<|p-q|} dq\, F_3(-q,p-q,q,\tau) \\
&\qquad\qquad \left. + \int_{|q|<|p-q|} dq\, F_4(-q,p-q,q,\tau) \right\} \\
&= I_9 + I_{10} + I_{11} + I_{12} + I_{13},
\end{aligned}
$$

where

$$
\begin{aligned}
F_3(x,y,q,\tau) = -i \sum_{j=1}^{\mathscr{N}} \Bigg\{ & e_1 |x| (v_{j-1}^{(1)}(x,\tau) - v_{j-1}^{(1)}(y,\tau)) \\
& \cdot (\cos\varphi v_{\mathscr{N}-j}^{(2)}(q,\tau) + \sin\varphi v_{\mathscr{N}-j}^{(3)}(q,\tau)) \\
& + \sum_{l=2}^{3} e_l (v_{j-1}^{(l)}(x,\tau) - v_{j-1}^{(l)}(y,\tau) \sum_{n=1}^{2} v_{\mathscr{N}-j}^{(n+1)}(q,\tau) x^{(n)} \Bigg\},
\end{aligned}
$$

$$
\begin{aligned}
F_4(x,y,q,\tau) = -i \sum_{j=1}^{\mathscr{N}} \Bigg\{ & e_1 [|x| - |y|] v_{j-1}^{(1)}(y,\tau) \\
& \cdot (\cos\varphi v_{\mathscr{N}-j}^{(2)}(q,\tau) + \sin\varphi v_{\mathscr{N}-j}^{(3)}(q,\tau)) \\
& + \sum_{l=2}^{3} e_l v_{j-1}^{(l)}(y,\tau) \sum_{n=1}^{2} v_{\mathscr{N}-j}^{(n+1)}(q,\tau)(x^{(n)} - y^{(n)}) \Bigg\}.
\end{aligned}
$$

And in case 2) we write

$$
\begin{aligned}
&v_{\mathcal{N}}(p,t) - \tilde{v}_{\mathcal{N}}(\varphi,t)e^{at-\lambda_1(p)t} \\
&\quad = \sum_{l=2}^{3}\int_0^t d\tau\, e^{-\lambda_l(p)(t-\tau)}\mathscr{D}_l(p)\cdot\int_{-\infty}^{\infty} dq\, V_{\mathcal{N}}(p,q,\tau) \\
&\qquad - e^{at-\lambda_1(p)t}\sum_{l=2}^{3}\int_0^t d\tau\, e^{-a_l(t-\tau)}\widetilde{\mathscr{D}}_l\cdot\int_{-\infty}^{\infty} dq\, \widetilde{V}_{\mathcal{N}}(\varphi,q,\tau) \\
&\qquad + \int_0^t d\tau\, e^{-\lambda_1(p)(t-\tau)}\mathscr{D}_1(p)\int_{-\infty}^{\infty} dq\, V_{\mathcal{N}}(p,q,\tau) \\
&\qquad - \int_0^t d\tau\, e^{-\lambda_1(p)t+a\tau}\widetilde{\mathscr{D}}_1\int_{-\infty}^{\infty} dq\, \widetilde{V}_{\mathcal{N}}(\varphi,q,\tau) \\
&\quad = \Bigg[\int_0^t d\tau\sum_{l=2}^{3} e^{-\lambda_l(p)t-\tau)}\mathscr{D}_l(p)\int_{-\infty}^{\infty} dq\, V_{\mathcal{N}}(p,q,\tau) \\
&\qquad\qquad - e^{at-\lambda_1(p)t}\sum_{l-2}^{3}\int_0^t d\tau\, e^{-a_l(t-\tau)}\widetilde{\mathscr{D}}_l(\varphi)\int_{-\infty}^{\infty} dq\, \widetilde{V}_{\mathcal{N}}(\varphi,q,\tau)\Bigg] \\
&\qquad + \int_0^t d\tau\, e^{-\lambda_1(p)(t-\tau)}\mathscr{D}_1(p)\Bigg\{\int_{|p-q|\le|q|} dq\, F_3(p-q,-q,q,\tau) \\
&\qquad\qquad + \int_{|p-q|\le|q|} dq\, F_4(p-q,-q,q,\tau) \\
&\qquad\qquad + \int_{|q|<|p-q|} dq\, F_3(-q,p-q,q,\tau) \\
&\qquad\qquad + \int_{|q|\le|p-q|} dq\, F_4(-q,p-q,q,\tau)\Bigg\} \\
&\qquad + \left[\int_0^t d\tau\, e^{-\lambda_1(p)t}(e^{\lambda_1(p)\tau}\mathscr{D}_1(p) - e^{a\tau}\widetilde{\mathscr{D}}_1)\int_{-\infty}^{\infty} dq\, \widetilde{V}_{\mathcal{N}}(\varphi,q,\tau)\right] \\
&\quad \equiv I_{14} + \cdots + I_{19},
\end{aligned}
$$

we have

$$
M^{2\gamma/\sigma}(\tau)e^{-\tau\kappa(q/(\mathcal{N}+1-j))} \le (\mathcal{N}+1-j)^{\delta+1}\kappa^{-2\kappa/\sigma}(q)\,. \tag{3.24}
$$

Since

$$
\left\|\sum_{l=1}^{3}(e^{-\lambda_l(p)(t-\tau)}\mathscr{D}_l(p) - e^{-\lambda_l(p)t+a\tau}\widetilde{\mathscr{D}}_l(\varphi))\right\| = O(|p|^{2\gamma}M^{2\gamma/\sigma}(\tau))e^{-a(t-\tau)-2\kappa(p)(t-\tau)},
$$

for $|p| \leq 1$, in view of (3.6) and (3.24) we obtain

$$|I_9| = \int_0^t d\tau\, O(|p|^{2\gamma} M^{2\gamma/\sigma}(\tau)) e^{-(a+2\kappa(p))(t-\tau)}$$

$$\cdot \int_{-\infty}^{\infty} dq \sum_{j=1}^{\mathcal{N}} c^{\mathcal{N}-1} H_{\mathcal{N}-j}(q,\tau) H_{j-1}(-q,\tau)|q|$$

$$= c^{\mathcal{N}} O(|p|^{2\gamma}) \cdot \sum_{j=1}^{\mathcal{N}} j^{-\chi} \int_0^t d\tau \int_{-\infty}^{\infty} dq\, |q| \frac{e^{-at-2\kappa(p)(t-\tau)-\kappa}(q/j)^{\tau}}{\kappa^{2\gamma/\sigma}(q) M^8(q)} .$$

Whence, using Lemma 2 with $\gamma_1 = 2\gamma\delta/\sigma$, $\gamma_2 = \gamma$, $r = 1/j$, we obtain

$$|I_9| = c^{\mathcal{N}} O(|p|^{\gamma-2} t^{-2/\delta-\mu} e^{-at}) .$$

Substitution of (3.6) and (3.20) in I_{10} yields

$$|I_{10}| = c^{\mathcal{N}} O(|p|^{2\gamma}) \int_0^t d\tau\, e^{-(a+2\kappa(p))(t-\tau)}$$

$$\cdot \sum_{j=1}^{\mathcal{N}} \int_{|p-q| \leq |q|} dq\, |p-q| m^{-2\gamma\delta/\sigma}(p-q) H_{j-1}(p-q,\tau) H_{\mathcal{N}-j}(q,\tau) .$$

We make the change $p - q = z$; then, applying the result of Lemma 2 with $\gamma_1 = 2\gamma\delta/\sigma$, $r = 1/j$, $\gamma_2 = \gamma$ we find that

$$|I_{10}| = c^{\mathcal{N}} O(|p|^{2\gamma}) \sum_{j=1}^{\mathcal{N}} j^{-\chi} \int_{-\infty}^{\infty} dq\, M^{-3}(q) m^{1-2\gamma\delta/\sigma}(q) \int_0^t d\tau\, e^{-at-2\kappa(p)(t-\tau)-\kappa(q/j)\tau}$$

$$= c^{\mathcal{N}} O(|p|^{\gamma-2} t^{-2/\delta-\mu} e^{-at}),$$

since $\gamma < (\sigma/3\delta)M(3-\delta)$. Further, using the estimate $||p-q| - |q|| \leq |p| \leq 2|p|^{2\gamma}|q|^{1-2\gamma}$, valid for $|p-q| \leq |q|$, we obtain the following estimate for I_{11}:

$$|I_{11}| = O(|p|^{2\gamma}) c^{\mathcal{N}} \int_0^t d\tau\, e^{-2\kappa(p)(t-\tau)-a(t-\tau)} \sum_{j=1}^{\mathcal{N}} \int_{|p-q| \leq |q|} dq$$

$$\cdot H_{\mathcal{N}-j}(q,\tau) H_{j-1}(q,\tau) |q|^{1-2\delta}$$

$$= c^{\mathcal{N}} \sum_{j=1}^{\mathcal{N}} j^{-\chi} O(|p|^{2\gamma}) \int_{-\infty}^{\infty} dq\, M^{-4}(q) m^{1-2\gamma}(q) \int_0^t d\tau\, e^{-2\kappa(p)(t-\tau)-at-\kappa(q/j)\tau},$$

whence, by the result of Lemma 2 with $\gamma_1 = 2\gamma$, $\gamma_2 = \gamma$, $r = 1/j$, we arrive at (3.22).

The integrals I_{12}, I_{13}, and I_{15} through I_{19} can be estimated analogously. It remains to estimate I_{14}:

$$
\begin{aligned}
|I_{14}| &\le c\int_0^t d\tau\, e^{-a_2(t-\tau)}\int_{-\infty}^{\infty} dq \sum_{j=1}^{\mathcal{N}} c^{\mathcal{N}} H_{\mathcal{N}-j}(q,\tau) \\
&\qquad\qquad \cdot \left(H_{j-1}(p-q,\tau)|p-q| + H_{j-1}(q,\tau)|q|e^{-\kappa(p)t}\right) \\
&= \int_0^{t/2} + \int_{t/2}^t \\
&= c^{\mathcal{N}} O(e^{-at-(a_2-a)t/2}) + \int_{t/2}^t d\tau\, e^{-a(t-\tau)} c^{\mathcal{N}} \int_{-\infty}^{\infty} dq \sum_{j=1}^{\mathcal{N}} M^{-4}(q) \\
&\quad \cdot j^{-\chi}(\mathcal{N}+1-j)^{-\chi}\left\{|p-q|\exp\left(-\tau\kappa\left(\frac{p-q}{j}\right) - \tau\kappa\left(\frac{q}{\mathcal{N}+1-j}\right)\right)\right. \\
&\qquad\qquad \left. + |q|\exp\left(-\tau\kappa\left(\frac{q}{j}\right) - t\kappa(p)\right)\right\} e^{-2a\tau} \\
&= c^{\mathcal{N}} O(e^{-at}t^{-3/\delta}) + c^{\mathcal{N}} e^{-at}\sum_{j=1}^{\mathcal{N}} j^{-\chi}(\mathcal{N}+1-j)^{-\chi} \\
&\quad \cdot \int_{-\infty}^{\infty} dq\, M^{-4}(q)\left\{\frac{\exp(-\kappa(\frac{p}{\mathcal{N}+1})\frac{t}{4} - \kappa(\frac{p-q}{j})\frac{t}{4})}{\kappa(\frac{q}{\mathcal{N}+1-j}) + \kappa(\frac{p-q}{j})}|p-q|\right. \\
&\qquad\qquad \left. + |q|\exp\left(-\kappa(p)t - \kappa\left(\frac{q}{j}\right)\frac{t}{2}\right)\kappa^{-1}\left(\frac{q}{j}\right)\right\} \\
&= c^{\mathcal{N}}|p|^{\delta-2} O(e^{-at}t^{-2/\delta-\mu}),
\end{aligned}
$$

by (3.24) and the inequality

$$
\kappa\left(\frac{p-q}{j}\right) + \kappa\left(\frac{q}{N+1-j}\right) \ge \kappa\left(\frac{p}{\mathcal{N}+1}\right).
$$

Thus, (3.22) and (3.23) hold. From them we now obtain

$$
\begin{aligned}
|J_5| &= c^{\mathcal{N}} O(e^{-at}t^{-2/\delta-\mu})\int_0^{2\pi} d\varphi \int_0^1 d|p|\cdot|p|^{\gamma-1} \\
&= c^{\mathcal{N}} O(e^{-at}t^{-2/\delta-\mu}).
\end{aligned}
\tag{3.25}
$$

3.6. The asymptotics. Substituting the estimates (3.9), (3.15)–(3.17), (3.19), and (3.25) in (3.8), we obtain in case 1) (i.e., $a_1 = a_2 = a_3$):

$$
\begin{aligned}
(2\pi)^2\psi_{\mathcal{N}} &= \sum_{j=1}^{3} e^{-at}t^{-2/\delta_j}\int_0^{2\pi} d\varphi\int_0^{\infty} d|\eta|\cdot|\eta|\widetilde{\mathscr{D}}_j\Phi_{\mathcal{N}}e^{i(\eta,\xi)-b_j|\eta|^{\delta_j}} \\
&\quad + c^{\mathcal{N}} O(t^{-2/\delta-\mu}e^{-at});
\end{aligned}
\tag{3.26}
$$

in case 2) (i.e., $a_1 < a_2 \le a_3$):

$$
\begin{aligned}
(2\pi)^2\psi_{\mathcal{N}} &= e^{-at}t^{-2/\delta_1}\int_0^{2\pi} d\varphi\int_0^{\infty} d|\eta|\,|\eta|\,\Phi_{\mathcal{N}}e^{i(\eta,\xi)-b_1|\eta|^{\delta_1}} \\
&\quad + c^{\mathcal{N}} O(e^{-at}t^{-2/\delta-\mu}).
\end{aligned}
$$

From (3.26) and (3.8) we obtain the representation (1.1) and the formulas (1.12)–(1.13). It remains to find a method for computing the functions $\Phi_{\mathcal{N}}(\varphi)$ in terms of the initial condition.

§4. Computation of the vectors $\Phi_{\mathcal{N}}$

The arguments in this section follow closely the constructions in §3, Chapter 6, and §4, Chapter 7.

4.1. Perturbation theory. We represent the function $v_{\mathcal{N}}(p,t)$ in the following form:

$$v_{\mathcal{N}}(p,t) = \int_{-\infty}^{\infty} dy\, Q_{\mathcal{N}}(\check{y},t) \tag{4.1}$$

where $y \in (R_2)^{\mathcal{N}}$, $\check{y} = (y_0, y) \in (R_2)^{\mathcal{N}+1}$, $\sum_{j=0}^{\mathcal{N}} y_j = p$. For $\mathcal{N} = 0$ formula (4.1) transforms into (3.4), whence we find

$$Q_0(p,t) = e^{-K(p)t}\overline{v}(p)\,. \tag{4.2}$$

In order to obtain a recurrent relation for the functions $Q_N(\check{y},t)$, we substitute (4.1) in both sides of (3.5), where $p \in R_2$ is fixed and $\mathcal{N} \ge 1$:

$$\begin{aligned}
&\int_{-\infty}^{\infty} dy\, Q_{\mathcal{N}}(\check{y},t) = -i\int_0^t d\tau\, e^{-K(p)(t-\tau)} \\
&\quad\cdot \sum_{j=1}^{\mathcal{N}} \int_{-\infty}^{\infty} dr \Bigg\{ e_1|p-r|\mathbb{P}_1 \left(\int_{-\infty}^{\infty} dq\, Q_{\mathcal{N}-j}(\check{q},\tau)\right) (\cos\varphi \mathbb{P}_2 + \sin\varphi \mathbb{P}_3) \\
&\qquad\cdot \left(\int_{-\infty}^{\infty} dz\, Q_{j-1}(\check{z},\tau) + \sum_{l=2}^{3} e_l \mathbb{P}_l \left(\int_{-\infty}^{\infty} dq\, Q_{\mathcal{N}-j}(\check{q},\tau)\right)\right. \\
&\qquad\qquad \left.\left.\cdot \sum_{n=1}^{2} (p^{(n)} - r^{(n)}) \mathbb{P}_{n+1} \left(\int_{-\infty}^{\infty} dz\, Q_{j-1}(\check{z},\tau)\right)\right)\right\}.
\end{aligned} \tag{4.3}$$

Here $\mathbb{P}_j$ is the projection onto the basis unit vector e_j, $y \in (R_2)^{\mathcal{N}}$, $q \in (R_2)^{\mathcal{N}-j}$, $z \in (R_2)^{j-1}$, $p = \sum_{k=0}^{\mathcal{N}} y_k$, $p - r = \sum_{k=0}^{j-1} z_k$, $r = \sum_{k=0}^{\mathcal{N}-j} q_k$. For each fixed j: $1 \le j \le \mathcal{N}$ we introduce new variables of integration η on the right-hand side of (4.3):

$$\eta_k = z_k, \quad 0 \le k \le j-1; \qquad \eta_k = q_{k-j}, \quad j \le k \le \mathcal{N}\,.$$

Then we obtain

$$\begin{aligned}
&\int_{-\infty}^{\infty} dy\, Q_{\mathcal{N}}(\check{y},\tau) = -i\int_0^t d\tau\, e^{-K(p)(t-\tau)} \int_{-\infty}^{\infty} d\eta \\
&\quad\cdot \sum_{j=1}^{\mathcal{N}} \Bigg\{ e_i \left|\sum_{k=0}^{j-1} \eta_k\right| \mathbb{P}_1 Q_{\mathcal{N}-j}(\eta_j,\dots,\eta_{\mathcal{N}},\tau) \sum_{n=1}^{2} \frac{p^{(n)}}{|p|} \mathbb{P}_{n+1} Q_{j-1}(\eta_0,\dots,\eta_{j-1},\tau) \\
&\qquad + \sum_{l=2}^{3} e_l \mathbb{P}_l Q_{\mathcal{N}-j}(\eta_j,\dots,\eta_{\mathcal{N}},\tau) \sum_{n=1}^{2} \left(\sum_{k=0}^{j-1} y_k^{(n)}\right) \mathbb{P}_n Q_{j-1}(\eta_0,\dots,\eta_{j-1},\tau) \Bigg\}.
\end{aligned} \tag{4.4}$$

The equality (4.4) is satisfied for

(4.5)

$$Q_{\mathscr{N}}(\check{y},t) = -i\sum_{\nu=1}^{3}\mathscr{D}_\nu(p)\int_0^t d\tau\, e^{-\lambda_\nu(p)(t-\tau)}$$
$$\cdot\sum_{j=1}^{\mathscr{N}}\left\{e_1\left|\sum_{k=0}^{j-1}y_k\right|\mathbb{P}_1 Q_{\mathscr{N}-j}(y_j,\dots,y_{\mathscr{N}},\tau)\sum_{n=1}^{2}\frac{p^{(n)}}{|p|}\mathbb{P}_{n+1}Q_{j-1}(y_0,\dots,y_{j-1},\tau)\right.$$
$$\left.+\sum_{l=2}^{3}e_l\mathbb{P}_l Q_{\mathscr{N}-j(y_j,\dots,y_{\mathscr{N}},\tau)}\sum_{n=1}^{2}\left(\sum_{k=0}^{j-1}y_k^{(n)}\right)\mathbb{P}_n Q_{j-1}(y_0,\dots,y_{j-1},\tau)\right\}$$

where $p=\sum_{k=0}^{\mathscr{N}}y_k$.

4.2. First formula for $Q_{\mathscr{N}}$. In order to solve the recurrent relation (4.5), we consider a procedure of constructing a certain family of graphs $\mathscr{G}_{0,\mathscr{N}}$ described in §3, Chapter 6, with one minor modification: to each vertex h_j of a graph $g\in\mathscr{G}_{0,\mathscr{N}}$ we assign a weight ρ_j, which can take one of the three values: 1, 2, or 3.

The operations of cutting and gluing graphs changes slightly. We fix any two graphs $g^1\in\mathscr{G}_{0,j-1}$ and $g^2\in\mathscr{G}_{j,\mathscr{N}}$, $1\le j\le\mathscr{N}$. The superscript indicates to which graph the designation refers. We define the operation of gluing two graphs g^1 and g^2 as follows:

1) We shift down the graphs g^1 and g^2 one step along the Oy axis;

2) The verticles of the new graph g are the vertex $h_1=(0,1)$ with weight of ρ_1 equal 1, 2, or 3, and all vertices of the graphs g^1 and g^2: $h_1^1,\dots,h_{2j-1}^1$, and $h_1^2,\dots,h_{2\mathscr{N}+1-2j}^2$, with their weights.

3) The edges of the new graph g are the edges of the graphs g^1 and g^2, and also the two new edges $\overrightarrow{h_1h_1^1}$ and $\overrightarrow{h_1h_1^2}$; these have the lengths $|\overrightarrow{h_1h_1^1}|=j-1$, $|\overrightarrow{h_1h_1^2}|=\mathscr{N}-j$.

As a result of gluing we obtain three new graphs g differing by the weight ρ_1 of the first vertex h_1. We denote these graphs as follows: $g=(g^1\otimes g^2)_{\rho_1}$. The set of all glued graphs $g=(g^1\otimes g^2)_{\rho_1}$ with fixed weight ρ_1 of the first vertex for all possible $g^1\in\mathscr{G}_{0,j-1}$ and $g\in\mathscr{G}_{j,\mathscr{N}}$ will be denoted by $(g_{0,j-1}\otimes\mathscr{G}_{j,\mathscr{N}})_{\rho_1}$. We note that the gluing operation corresponds fully to the division algorithm, i.e., any glued graph $g=(g^1\otimes g^2)_{\rho_1}$ can be constructed by division of the segment $(0,\mathscr{N}+1)$ of the natural sequence. Thus, we obtain

$$(\mathscr{G}_{0,j-1}\otimes\mathscr{G}_{j,\mathscr{N}})_{\rho_1}\subset\mathscr{G}_{0,\mathscr{N}}\,. \tag{4.6}$$

Since (4.6) is valid for any j: $1\le j\le\mathscr{N}$ and $\rho_1=1,2,3$,

$$\sum_{\rho_1=1}^{3}\sum_{j=1}^{\mathscr{N}}(\mathscr{G}_{0,j-1}\otimes\mathscr{G}_{j,\mathscr{N}})_{\rho_1}\subset\mathscr{G}_{0,\mathscr{N}}, \tag{4.7}$$

where the symbol of sum is due to the fact that the sets $(\mathscr{G}_{0,j-1}\otimes\mathscr{G}_{j,\mathscr{N}})_{\rho_1}$ are distinct for different j and ρ_1: the graphs differ by the vertex h_3 or the weight ρ_1. Let us prove the reverse inclusion

$$\mathscr{G}_{0,\mathscr{N}}\subset\sum_{\rho_1=1}^{3}\sum_{j=1}^{\mathscr{N}}(\mathscr{G}_{0,j-1}\otimes\mathscr{G}_{j,\mathscr{N}})_{\rho_1}\,. \tag{4.8}$$

In order to establish (4.8) we fix any graph g and show that $g=(g^1\otimes g^2)_{\rho_1}$, where

$g^1 \in \mathscr{G}_{0,j-1}$, $g^2 \in \mathscr{G}_{j,\mathscr{N}}$, $j = h_3^x$, and ρ_1 is the weight of the first vertex h_1 of g. To this end, we define the operation of cutting the graph g as follows:

1) Remove the vertex h_1 and the edges $\overrightarrow{h_1h_2}$ and $\overrightarrow{h_1h_3}$;

2) Divide the remaining vertices into two sets

$$\mathscr{M}_1 = \{h_k\colon h_k^x \in [0, h_3^x)\} \quad \text{and} \quad \mathscr{M}_2 = \{h_k\colon h_k^x \in [h_3^x, \mathscr{N}]\};$$

3) Let the graph g^1 consist of the set of vertices $\mathscr{M}_1$ and all those edges of g which join vertices from the set $\mathscr{M}_1$. The graph g^2 is defined analogously.

4) Shift the graphs g^1 and g^2 formed upward one step along the Oy axis.

Since in the construction of the graph $g \in \mathscr{G}_{0,\mathscr{N}}$ each of the segments $[0, h_3^x)$ and $[h_3^x, \mathscr{N}]$ is subdivided with the help of the division algorithm, it follows that $g^1 \in \mathscr{G}_{0,j-1}$ and $g^2 \in \mathscr{G}_{j,\mathscr{N}}$. It is also clear that $g = (g^1 \otimes g^2)_{\rho_1}$, whence (4.8) follows, while (4.7) and (4.8) yield

$$\sum_{\rho_1=1}^{3}\sum_{j=1}^{\mathscr{N}}(\mathscr{G}_{0,j-1} \otimes \mathscr{G}_{j,\mathscr{N}})_{\rho_1} = \mathscr{G}_{0,\mathscr{N}}, \qquad \mathscr{N} \geq 1. \tag{4.9}$$

Formula (4.9) makes it possible to estimate the cardinality $|\mathscr{G}_{0,\mathscr{N}}|$ of the set $\mathscr{G}_{0,\mathscr{N}}$. It is obvious that $|\mathscr{G}_{0,\mathscr{N}}| = |\mathscr{G}_{j,\mathscr{N}+j}|$ for all j. Therefore, from (4.9) we get the recurrence relation

$$|\mathscr{G}_{0,\mathscr{N}}| = 3\sum_{j=1}^{\mathscr{N}} |\mathscr{G}_{0,j-1}|\,|\mathscr{G}_{0,\mathscr{N}-j}|, \qquad \mathscr{N} \geq 1, \tag{4.10}$$

from which we easily obtain

$$|\mathscr{G}_{0,\mathscr{N}}| \leq (36)^{2\mathscr{N}}(\mathscr{N}+1)^{-2}, \qquad \mathscr{N} \geq 0. \tag{4.11}$$

Along with the natural enumeration of vertices given by the division algorithm, we also use below a special enumeration of vertices, defined as in Chapter 6.

With the help of the family of graphs introduced we solve the recurrent relation (4.5). Fix any graph $g \in \mathscr{G}_{0,\mathscr{N}}$, in which we choose the special enumeration of the vertices. Let $\xi_1 = 0$. For each node h_j, $2 \leq j \leq \mathscr{N}$, we determine the number ξ_j of the vertex h_{ξ_j} preceding it, and also the numbers r_j^1 and r_j^2, $1 \leq j \leq \mathscr{N}$, of the verticles $h_{r_j^1}$ and $h_{r_j^2}$ immediately following the node h_j in the sense that $\overrightarrow{h_jh_{r_j^1}}$, $\overrightarrow{h_jh_{r_j^2}}$, and $\overrightarrow{h_{\xi_j}h_j}$ are edges in g. Let $y \in (R_2)^{\mathscr{N}}$, $\check{y} = (y_0, y) \in (R_2)^{\mathscr{N}+1}$. For each node h_j, $1 \leq j \leq \mathscr{N}$, we define two sums of coordinates J_k, $1 \leq j \leq \mathscr{N}$, $i = 1, 2$; $s_j^i = \sum_k y_k$, where the index k runs through values from $h_{r_j^i}^x$ to $h_{r_j^i}^x + |\overrightarrow{h_jh_{r_j^i}}|$ inclusively. We also set $p_j = s_j^1 + s_j^2$, $p = p_1 = \sum_{k=0}^{\mathscr{N}} y_k$.

For all $\mathscr{N} \geq 0$ we establish the following equality:

$$Q_{\mathscr{N}}(\check{y}, t_0) = \sum_{g\in\mathscr{G}_{\mathscr{N}}} e^{-\lambda_{\rho_1}(p)t_0}\int_0^\infty \cdots \int_0^\infty dt_1 \cdots dt_{\mathscr{N}}\Lambda_g(\check{y})\prod_{j=1}^{\mathscr{N}} e^{\mathscr{L}_j t_j}\theta(t_{\xi_j} - t_j), \tag{4.12}$$

where $\mathscr{L}_j = \mathscr{L}(s_j^1, s_j^2) = \lambda_{\rho_j}(s_j^1 + s_j^2) - \lambda_{\rho_{r_j^1}}(s_j^1) - \lambda_{\rho_{r_j^2}}(s_j^2)$,

$$\theta(x) = \begin{cases} 1, & x > 0, \\ 0, & x < 0, \end{cases}$$

is the Heaviside function. $\mathscr{G}_{\mathscr{N}} \equiv \mathscr{G}_{0,\mathscr{N}}$, $t_{\xi_1} = t_0$, and the $\Lambda_g(\check{y})$ are defined recurrently as follows. If $\mathscr{N} = 0$, then a graph $g \in \mathscr{G}_{0,0}$ consists of one vertex h_1 with weight ρ_1;

then $p = y_0$ and we set

$$\Lambda_g \equiv \mathscr{D}_{\rho_1}(y_0)\overline{v}(y_0)\,.$$

Further, by induction, for $g = (g^1 \otimes g^2)_{\rho_1}$, where $g^1 \in \mathscr{G}_{0,j-1}$, $g^2 \in \mathscr{G}_{j,\mathscr{N}}$, we define Λ_g by the formula

(4.13)

$$\begin{aligned}\Lambda_g(\check{y}) = -i\mathscr{D}_{\rho_1}(p)\Bigg\{ e_1 \left|\sum_{k=0}^{j-1} y_k\right| \mathbb{P}_1 \Lambda_{g^2}(y_j, \dots, y_{\mathscr{N}}) \\ \cdot \frac{1}{|p|}(p^{(1)}\mathbb{P}_2 + p^{(2)}\mathbb{P}_3)\Lambda_{g^1}(y_0, \dots, y_{j-1}) + (e_2\mathbb{P}_2 + e_3\mathbb{P}_3) \\ \cdot \Lambda_{g^2}(y_j, \dots, y_{\mathscr{N}})\left(\sum_{k=0}^{j-1}(y_k^{(1)}\mathbb{P}_2 + y_k^{(2)}\mathbb{P}_3)\Lambda_{g^1}(y_0, \dots, y_{j-1})\right)\Bigg\}\,.\end{aligned}$$

For $\mathscr{N} = 0$, (4.12) has the form

$$Q_0(y_0, t_0) = \sum_{g\in\mathscr{G}_0} e^{-\lambda_{\rho_1}(y_0)t_0}\Lambda_g(y_0) = \sum_{\rho_1=1}^{3} e^{-\lambda_{\rho_1}(y_0)t_0}\mathscr{D}_{\rho_1}(y_0)\overline{v}(y_0), \tag{4.14}$$

which coincides with (4.2). Suppose, by induction, that (4.12) has been established for $\mathscr{N} \le \mathscr{N}_1$. Substituting (4.12) in the recurrent relation (4.5) for $\mathscr{N} = \mathscr{N}_1 + 1$, we obtain

$$\begin{aligned}Q_{\mathscr{N}}(\check{y}, t_0) = -i\sum_{\rho_1=1}^{3}\mathscr{D}_{\rho_1}(p)\int_0^{t_0} dt_1\, e^{-\lambda_{\rho_1}(p)(t_0-t_1)} \\ \cdot\sum_{j=1}^{\mathscr{N}}\Bigg\{ e_1\left|\sum_{k=0}^{j-1} y_k\right| \cdot \mathbb{P}_1\Bigg(\sum_{g^2\in\mathscr{G}_{j,\mathscr{N}}} e^{-\lambda_{\rho_1}(p^2)t_1}\int_0^\infty\cdots\int_0^\infty dt_1^2\cdots dt_{\mathscr{N}-j}^2 \\ \cdot\prod_{k=1}^{\mathscr{N}-j}\theta(t_{\xi_k^2}^2 - t_k^2)e^{-\mathscr{L}_k^2 t_k^2}\Lambda_{g^2}(y_j, \dots, y_{\mathscr{N}})\Bigg) \\ \cdot\left(\sum_{k=0}^{\mathscr{N}}(y_k^{(1)}\mathbb{P}_2 + y_k^{(2)}\mathbb{P}_3)\right)\left(\sum_{k=0}^{\mathscr{N}} y_k\right)^{-1} \\ \cdot\Bigg(\sum_{g^1\in\mathscr{G}_{0,j-1}} e^{-\lambda_{\rho_1^1}(p^1)t_1}\int_0^\infty\cdots\int_0^\infty dt_1^1\cdots dt_{j-1}^1 \\ \cdot\prod_{k=1}^{j-1} e^{\mathscr{L}_k^1 t_k^1}\theta(t_{\xi_k^1}^1 - t_k^1)\Lambda_{g^1}(y_0, \dots, y_{j-1}) + (e_2\mathbb{P}_2 + e_3\mathbb{P}_3)\Bigg) \\ \cdot\Bigg(\sum_{g^2\in\mathscr{G}} e^{-\lambda_{\rho_1^2}(p^2)t_1}\int_0^\infty\cdots\int_0^\infty dt_1^2\cdots dt_{\mathscr{N}-j}^2\prod_{k=1}^{\mathscr{N}-j}\theta(t_{\xi_k^2}^2 - t_k^2)e^{-\mathscr{L}_k^2 t_k^2}\Lambda_{g^2}(y_j, \dots, y_{\mathscr{N}})\Bigg) \\ \cdot\left(\sum_{k=0}^{j-1}(y_k^{(1)}\mathbb{P}_2 + y_k^{(2)}\mathbb{P}_3)\right)\cdot\Bigg(\sum_{g^1\in\mathscr{G}_{0,j-1}} e^{-\lambda_{\rho_1^1}(p^1)t_1}\int_0^\infty\cdots\int_0^\infty dt_1^1\cdots dt_{j-1}^1 \\ \cdot\prod_{k=1}^{j-1} e^{\mathscr{L}_k^2 t_k^1}\theta(t_{\xi_k^1}^1 - t_k^1)\Lambda_{g^1}(y_0, \dots, y_{j-1})\Bigg)\Bigg\}\,.\end{aligned}$$

Here $t^1_{\xi^1_1} = t_1$, $t^2_{\xi^2_1} = t_1$, $p^1 = \sum_{k=0}^{j-1} y_k$, $p^2 = \sum_{k=j}^{\mathscr{N}} y_k$. Superscript j indicates to which of the graphs g^j, $j = 1, 2$, the notation refers. Using the recurrent relation (4.13), we obtain

(4.15)

$$Q_{\mathscr{N}}(\check{y}, t_0) = \sum_{\rho_1=1}^{3} \int_0^{t_0} dt_1\, e^{-\lambda_{\rho_1}(p)(t_0-t_1)} \sum_{g^2 \in \mathscr{G}_{j,\mathscr{N}}} e^{-\lambda_{\rho_1^2}(p^2)t_1}$$
$$\cdot \left(\int_0^\infty \cdots \int_0^\infty dt_1^2 \cdots dt^2_{\mathscr{N}-j} \prod_{k=1}^{\mathscr{N}-j} \theta(t^2_{\xi^2_k} - t^2_k) e^{\mathscr{L}^2_k t^2_k} \right) \sum_{g^1 \in \mathscr{G}_{j-1}} e^{-\lambda_{\rho_1^1}(p^1)t_1}$$
$$\cdot \left(\int_0^\infty \cdots \int_0^\infty dt_1^1 \cdots dt^1_{j-1} \prod_{k=1}^{j-1} \theta(t^1_{\xi^1_k} - t^1_k) e^{\mathscr{L}^1_k t^1_k} \right) \Lambda_(g^1 \otimes g^2)_{\rho_1} .$$

It follows from (4.9) that

$$\sum_{\rho_1=1}^{3} \sum_{j=1}^{\mathscr{N}} \sum_{g^1 \in \mathscr{G}_{j-1}} \sum_{g^2 \in \mathscr{G}_{j,\mathscr{N}}} = \sum_{\rho_1=1}^{3} \sum_{j=1}^{\mathscr{N}} \sum_{g \in (\mathscr{G}_{j-1} \otimes \mathscr{G}_{j,\mathscr{N}})_{\rho_1}} = \sum_{g \in \mathscr{G}_{\mathscr{N}}} .$$

Therefore, noting that $p = p^1 + p^2$ we have

$$Q_{\mathscr{N}}(\check{y}, t_0) = \sum_{g \in \mathscr{G}_N} e^{-\lambda_{\rho_1}(p)t_0} \int_0^\infty dt_1\, \theta(t_0 - t_1) e^{\mathscr{L}(p^1, p^2)t_1}$$
$$\cdot \left(\int_0^\infty \cdots \int_0^\infty dt_1^1 \cdots dt^1_{j-1} \prod_{k=1}^{j-1} \theta(t^1_{\xi^1_k} - t^1_k) e^{\mathscr{L}^1_k + t^1_k} \right) \tag{4.16}$$
$$\cdot \left(\int_0^\infty \cdots \int_0^\infty dt_1^2 \cdots dt^2_{\mathscr{N}-j} \prod_{k=1}^{\mathscr{N}-j} \theta(t^2_{\xi^2_k} - t^2_k) e^{\mathscr{L}^2_k t^2_k} \right) \Lambda_g(\check{y}) .$$

In view of the choice of the special enumeration of vertices in a graph $g \in \mathscr{G}_{\mathscr{N}}$, the following equalities hold:

$$s^1_1 = p^1, s^2_1 = p^2;$$
$$t^1_k = t_{k+1}, \quad t^1_{\xi^1_k} = t_{\xi_{k+1}}, \quad \mathscr{L}^1_k = \mathscr{L}_{k+1}, \qquad 1 \le k \le j-1;$$
$$t^2_k = t_{k+j}, \quad t^2_{\xi^2_k} = t_{\xi_{k+j}}, \quad \mathscr{L}^2_k = \mathscr{L}_{k+j}, \qquad 1 \le k \le \mathscr{N} - j .$$

Therefore, (4.16) takes the following form:

$$Q_{\mathscr{N}}(\check{y}, t_0) = \sum_{g \in \mathscr{G}_{\mathscr{N}}} e^{-\lambda_{\rho_1}(p)t_0} \int_0^\infty dt_1\, e^{\mathscr{L}_1 t_1} \theta(t_0 - t_1)$$
$$\cdot \left(\int_0^\infty \cdots \int_0^\infty dt_2 \cdots dt_j \prod_{k=2}^{j} \theta(t_{\xi_k} - t_k) e^{\mathscr{L}_k t_k} \right)$$
$$\cdot \left(\int_0^\infty \cdots \int_0^\infty dt_{j+1} \cdots dt_{\mathscr{N}} \prod_{k=j+1}^{\mathscr{N}} \theta(t_{\xi_k} - t_k) e^{\mathscr{L}_k t_k} \right) \Lambda_g(\check{y}),$$

whence we arrive at (4.12). The induction is completed.

4.3. Second formula for $Q_{\mathscr{N}}$. We set

$$
\begin{aligned}
&B_0 = 1, \qquad B_1(x,t) = x^{-1}(e^{xt}-1),\\
&B_{\mathscr{N}}(x_1,x_2,\dots,x_{\mathscr{N}},t) = x_1^{-1}(e^{x_1 t}B_{\mathscr{N}-1}(x_2,\dots,x_{\mathscr{N}},t)\\
&\qquad\qquad - B_{\mathscr{N}-1}(x_1+x_2,x_3,\dots,x_{\mathscr{N}},t)).
\end{aligned} \tag{4.17}
$$

From the recurrent relations (4.17) we obtain the following three formulas (see the proof in §3, Chapter 6, and §4, Chapter 7):

$$
\begin{aligned}
B_{\mathscr{N}}(x_1,\dots,x_{\mathscr{N}},t) = x_j^{-1}(&B_{\mathscr{N}-1}(x_1,\dots,x_{j-1}+x_j,x_{j+1},\dots,x_{\mathscr{N}},t)\\
&- B_{\mathscr{N}-1}(x_1,\dots,x_{j-1},x_j+x_{j+1},\dots,x_{\mathscr{N}},t))\\
&j = 2,3,\dots,\mathscr{N}-1,\ \mathscr{N}\ge 3.
\end{aligned} \tag{4.18}
$$

$$
B_{\mathscr{N}+1}(x_1,\dots,x_{\mathscr{N}+1},t) = \int_0^t e^{x_1 t_1}B_{\mathscr{N}}(x_2,\dots,x_{\mathscr{N}+1},t_1)\,dt_1, \qquad \mathscr{N}\ge 0. \tag{4.19}
$$

(4.20)

$$
B_{\mathscr{N}} = \sum_{k=1}^{\mathscr{N}}(-1)^{\mathscr{N}-k}\exp\left(t\sum_{j=1}^{k}x_j\right)\left\{\prod_{j=1}^{k}\left(\sum_{l=j}^{k}x_l\right)\prod_{n=k+1}^{\mathscr{N}}\left(\sum_{i=k+1}^{n}x_i\right)\right\}^{-1}.
$$

As in Chapter 6, we carry out additional constructions with graphs and define finished graphs and complete chains in them.

We note that for any $g^1\in\mathscr{G}_{0,j-1}$ and $g^2\in\mathscr{G}_{j,\mathscr{N}}$ the following equality holds:

$$
\sum_{w^1\in W(g^1),\ w^2\in W(g^2)} W(w^1\otimes w^2) = W(g^1\otimes g^2). \tag{4.21}
$$

Fix any $g^1\in\mathscr{G}_{j-1}$, $g^2\in\mathscr{G}_{j,\mathscr{N}}$, $w^1\in W(g^1)$, $w^2\in W(g^2)$. Then the following equality holds for $1\le j\le\mathscr{N}$ (see the proof in §3, Chapter 6):

$$
\begin{aligned}
&\int_0^t dt_1\, e^{x_1 t_1}B_{j-1}(x_{w_1^1},\dots,x_{w_{j-1}^1},t_1)B_{\mathscr{N}-j}(x_{w_1^2},\dots,x_{w_{\mathscr{N}-j}^2},t_1)\\
&\quad = \sum_{w\in W(w^1\otimes w^2)}(x_{w_1},\dots,x_{w_{\mathscr{N}}},t).
\end{aligned} \tag{4.22}
$$

We prove the second formula for the symbol $Q_{\mathscr{N}}(\check{y},t)$:

$$
Q_{\mathscr{N}}(\check{y},t_0) = \sum_{w\in W(g)}\sum_{g\in\mathscr{G}_{\mathscr{N}}} e^{-\lambda_{\rho_1}(p)t_0}E_w(\check{y},t_0)\Lambda_g(\check{y}), \tag{4.23}
$$

where

$$
E_w(\check{y},t_0) = B_{\mathscr{N}}(\mathscr{L}_{w_1},\dots,\mathscr{L}_{w_{\mathscr{N}}},t_0).
$$

In view of (4.12), it is sufficient to prove that for any fixed graph $g\in\mathscr{G}_{\mathscr{N}}$

$$
\int_0^\infty\cdots\int_0^\infty dt_1\cdots dt_{\mathscr{N}}\prod_{j=1}^{\mathscr{N}}e^{\mathscr{L}_j t_j}\theta(t_{\xi_j}-t_j) = \sum_{w\in W(g)}E_w(\check{y},t_0). \tag{4.24}
$$

For $\mathscr{N}=0$ and $\mathscr{N}=1$ equality (4.24) can be verified trivially. For $\mathscr{N}>1$ we argue by induction. Cut the graph $g \in \mathscr{G}_{\mathscr{N}}$ to obtain $g = (g^1 \otimes g^2)_{\rho_1}$, $g^1 \in \mathscr{G}_{0,j}$, $g^2 \in \mathscr{G}_{j,\mathscr{N}}$. Then on the left-hand side of (4.24), by (4.22) and (4.21), we obtain the integral

$$\int_0^t dt_1 e^{\mathscr{L}_1 t_1} \left(\int_0^\infty \cdots \int_0^\infty dt_2 \cdots dt_j \prod_{k=2}^{j} e^{\mathscr{L}_k t_k} \theta(t_{\xi_k} - t_k) \right)$$
$$\cdot \left(\int_0^\infty \cdots \int_0^\infty dt_{j+1} \cdots dt_{\mathscr{N}} \prod_{k=j+1}^{\mathscr{N}} e^{\mathscr{L}_k t_k} \theta(t_{\xi_k} - t_k) \right)$$
$$= \int_0^{t_0} dt_1\, e^{\mathscr{L}_1 t_1} \sum_{w^1 \in W(g^1)} E_{w^1}(y_0, \dots, y_{j-1}, t_1) \sum_{w^2 \in W(g^2)} E_{w^2}(y_j, \dots, y_{\mathscr{N}}, t_1)$$
$$= \sum_{w \in W(w^1 \otimes w^2),\ w^1 \in W(g^1),\ w^2 \in W(g^2)} E_w(t_0) = \sum_{w \in W(g)} E_w(t_0)\,.$$

Thus, (4.23) and (4.24) are established by induction.

4.4. A formula for $\Phi_{\mathscr{N}}$. Substituting (4.24) in (4.21) and passing to the limit as $t \to \infty$ and $|p| \to 0$ in such a way that $p = |p|(\cos\varphi, \sin\varphi)$, $\varphi = \text{const}$, we obtain

$$\Phi_{\mathscr{N}} = \sum_{w \in W(g)} \sum_{g \in \mathscr{G}_{\mathscr{N}}} \int_{-\infty}^{\infty} dy \lim_{t \to \infty} e^{at - a_{\rho_1} t} \widetilde{E}_w(\check{y}, t) \widetilde{\Lambda}_g(\check{y}), \tag{4.25}$$

where the tilde denotes the result of taking the limit as $p \to 0$ with $p/|p| = (\cos\varphi, \sin\varphi) = \text{const}$. We show that

$$\lim_{t \to \infty} e^{at - a_{\rho_1} t} \widetilde{E}_w(\check{y}, t) = \begin{cases} (-1)^{\mathscr{N}} \prod_{k=1}^{\mathscr{N}} \left(\sum_{j=1}^{k} \mathscr{L}_{w_j} \right)^{-1}, \\ 0, \quad \text{if } \rho_1 \neq 1 \text{ in case 2)}\,. \end{cases} \tag{4.26}$$

Fix any chain w and consider some initial segment $(w_1, \dots, w_k)$ of it. For any vertex h_{w_j}, $z \leq j \leq k$, in accordance with the construction of the chain (by subordination of vertices), there exists a vertex $h_{\xi_{w_j}}$ also belonging to this part $(w_1, \dots, w_k)$ of the chain w. For the sum $\mathscr{L}_{w_j} + \mathscr{L}_\Delta$, $\Delta = \xi_{w_j}$, in this case we find

$$\begin{aligned} \mathscr{L}_{w_j} + \mathscr{L}_\Delta =& \lambda_{\rho_\Delta}(s_\Delta^1 + s_\Delta^2) - \lambda_{\nu_1}(s_\Delta^1) - \lambda_{\nu_2}(s_\Delta^2) \\ &+ \lambda_{\rho_{w_j}}(s_{w_j}^1 + s_{w_j}^2) - \lambda_{\nu_3}(s_{w_j}^1) - \lambda_{\nu_4}(s_{w_j}^2) \\ =& \lambda_{\rho_\Delta}(s_\Delta^1 + s_\Delta^2) - \lambda_{\nu_3}(s_{w_j}^1) - \lambda_{\nu_4}(s_{w_j}^2) - \lambda_{\nu_5}(z), \end{aligned}$$

where $\nu_1 = \rho_{r_\Delta^1}$, $\nu_2 = \rho_{r_\Delta^2}$, $\nu_3 = \rho_{r_{w_j}^1}$, $\nu_4 = \rho_{r_{w_j}^2}$, $z = s_\Delta^2$, $\nu_5 = \nu_2$, if $s_{w_j}^1 + s_{w_j}^2 = s_\Delta^1$, and $z = s_\Delta^1$, $\nu_5 = \nu_1$, if $s_{w_j}^1 + s_{w_j}^2 = s_\Delta^2$. We consider in this manner all vertices of the sequence $(w_1, \dots, w_k)$ and obtain

$$\begin{aligned} &\mathscr{L}_{w_1} + \mathscr{L}_{w_2} + \cdots + \mathscr{L}_{w_k} \\ &\quad = \lambda_{\rho_1}(p) - \lambda_{\bar{\rho}_1}(z_1) - \cdots - \lambda_{\bar{\rho}_k}(z_k) - \lambda_{\bar{\rho}_{k+1}}(p - z_1 - z_2 - \cdots - z_k), \end{aligned} \tag{4.27}$$

where the $z_1, z_2, \dots, z_k$ coincide with some of the sums s_j^i, $i = 1, 2$, $j = 1, \dots, \mathscr{N}$, and

$\overline{\rho}_1, \ldots, \overline{\rho}_{k+1}$ take the values 1, 2, or 3. It follows from (4.27) that almost everywhere in $(R_2)^{\mathscr{N}}$ (for $s_1^1 = -s_1^2$ not equal to zero), we have

$$a - \lambda_{\rho_1}(0) + \sum_{j=1}^{k} \mathscr{L}_{w_j} < 0$$

and as $t \to \infty$ the exponential in (4.20) tends to zero, if $k \neq 0$, or if $a < a_{\rho_1}$ for case 2). This means that in (4.20) it is necessary to leave $k = 0$, and in case 2) it is necessary to put $\rho_1 = 1$ in the graphs $g \in \mathscr{G}_{\mathscr{N}}$. This leads to (4.26). We substitute (4.26) in (4.25) and obtain

$$\Phi_{\mathscr{N}} = \sum_{w \in W(g)} \sum_{g \in \mathscr{G}_{\mathscr{N}}} (-1)^{\mathscr{N}} \int_{-\infty}^{\infty} dy \prod_{k=1}^{\mathscr{N}} \left(\sum_{j=1}^{k} \widetilde{\mathscr{L}}_{w_j} \right)^{-1} \widetilde{\Lambda}_g(\check{y}), \tag{4.28}$$

where $\rho_1 = 1$ in case 2). The theorem is proved. In particular, we have

$$\Phi_0 = \frac{\overline{\mathscr{D}}_1}{\varepsilon} \int_{-\infty}^{\infty} \overline{u}(x)\,dx,$$

$$\begin{aligned}\Phi_1 = &-\frac{i}{\varepsilon^2} \overline{\mathscr{D}}_1 \sum_{\rho_2, \rho_3 = 1}^{3} \int_{-\infty}^{\infty} dy\,(a - \lambda_{\rho_2}(-y) - \lambda_{\rho_3}(y))^{-1} \\ &\cdot \{e_1 |y| \mathbb{P}_1 \mathscr{D}_{\rho_3}(y) \hat{\overline{u}}(y)(\cos\varphi \mathbb{P}_2 + \sin\varphi \mathbb{P}_3) \mathscr{D}_{\rho_2}(-y) \hat{\overline{u}}(-y) \\ &\qquad - (e_2 \mathbb{P}_2 + e_3 \mathbb{P}_3) \mathscr{D}_{\rho_3}(y) \hat{\overline{u}}(y)(y^{(1)} \mathbb{P}_2 + y^{(2)} \mathbb{P}_3) \mathscr{D}_{\rho_2}(-y) \hat{u}(-y)\},\end{aligned}$$

where $\overline{\mathscr{D}}_1 = \widetilde{\mathscr{D}}_1(\varphi)$ in case 2) and $\overline{\mathscr{D}}_1 = E$ in case 1).

CHAPTER 10

The Step-Decaying Problem for the Korteweg-de Vries-Burgers Equation

§1. Introduction

The Korteweg-de Vries-Burgers (KdVB) equation

$$u_t + 2uu_x + au_{xxx} - bu_{xx} = 0, \qquad a, b = \text{const}, \tag{1.1}$$

presents considerable mathematical interest, since it combines a typical hydrodynamic nonlinearity, the simplest nontrivial dispersion, and dissipation. This equation describes physical phenomena like the distribution of weak shock waves in plasma, the evolution of gas-fluid mixtures, propagation of waves in a liquid-filled elastic tube, etc.

We consider the Cauchy problem for equation (1.1) with initial condition $\bar{u}(x)$ and investigate the asymptotic behavior of a solution as $t \to \infty$. When the initial perturbation $\bar{u}(x)$ decreases at infinity sufficiently fast, the asymptotic behavior of a solution to the Cauchy problem for (1.1) for large times was found in Chapter 7; namely,

$$u(x,t) = At^{-1/2}e^{-\xi^2} + O(t^{-1}), \tag{1.2}$$

uniformly with respect to $\xi = \frac{|x|}{2\sqrt{t}} \geq 0$, where A is a constant, depending explicitly on $\bar{u}(x)$.

An extremely interesting problem for the KdVB equation is the step-decaying problem, i.e., the study of the asymptotic behavior, as $t \to \infty$, of the solution to the Cauchy problem for (1.1) with a sufficiently smooth initial condition $\bar{u}(x)$ tending to ∓ 1 as $x \to \pm\infty$. This problem has been discussed by several authors: Jeffrey **[123]** found two exact solutions of the traveling wave type; Bona and Schonbek **[97]** proved the existence of solutions of the traveling wave type and considered their limits for a, $b \to 0$; Novikov, Avilov, and Krichever **[2, 3]** investigated the step-decaying problem by numerical methods.

In this chapter we consider the step-decaying problem in general setting and, in particular, we give an affirmative answer to a problem, posed in **[97]** concerning the stability of traveling-wave type solutions. In §3 we briefly examine the step-decaying problem for the Kuramoto-Sivashinsky equation. The results of this chapter were published in **[66]** and **[190]**.

§2. First theorem

We consider the step-decaying problem for the KdVB equation:

$$\begin{cases} u_t + 2uu_x + au_{xxx} - u_{xx} = 0, \\ u(x,0) = \bar{u}(x), \qquad \bar{u}(x) \to \mp 1 \quad \text{as } x \to \mp\infty\,. \end{cases} \tag{2.1}$$

The somewhat more general problem

$$\begin{cases} \psi_t + 2\psi\psi_x + \alpha\psi_{xxx} - \beta\psi_{xx} = 0, \\ \psi(x,0) = \bar{\psi}(x), \quad \bar{\psi}(x) \to b_\pm \quad \text{as } x \to \pm\infty, \\ b_+ < b_-, \qquad \beta > 0 \end{cases} \tag{2.2}$$

is reduced to problem (2.1) with the aid of the substitution

$$u(x,t) = \frac{1}{\eta}\left[\psi\left(\frac{\beta}{\eta}\left(x + \frac{2\xi t}{\eta}\right), \frac{\beta t}{\eta^2}\right) - \xi\right],$$

where

$$\xi = \tfrac{1}{2}(b_- + b_+),\ \eta = \tfrac{1}{2}(b_- - b_+),$$
$$a = \alpha\eta/\beta^2,\ \bar{u}(x) = \tfrac{1}{\eta}[\bar{\psi}(\tfrac{\beta x}{\eta}) - \xi]\,.$$

Without loss of generality, we can assume that the constant a in the problem (2.1) is positive, which is attained by the change of variable

$$w(x,t) = -u(-x,t).$$

For problem (2.1) we consider traveling-wave type solutions $\varphi(x - \lambda t)$, satisfying the following conditions at infinity:

$$\varphi \to \pm 1, \qquad \varphi_x, \varphi_{xx} \to 0 \quad \text{as } x \to \mp\infty\,.$$

The function φ satisfies the equation

$$-\lambda\varphi_x + 2\varphi\varphi_x + a\varphi_{xxx} - \varphi_{xx} = 0\,. \tag{2.3}$$

Integrating (2.3) with respect to x, we obtain

$$c - \lambda\varphi + \varphi^2 + a\varphi_{xx} - \varphi_x = 0, \tag{2.4}$$

where c is the constant of integration. We introduce a new function $\psi = \frac{1}{2} - \frac{\varphi}{2}$; ψ satisfies the equation

$$c - \lambda + 1 + (2\lambda - 4)\psi + 4\psi^2 + 2\psi_x - 2a\psi_{xx} = 0\,. \tag{2.5}$$

From the conditions at infinity

$$\psi \to \begin{Bmatrix} 0 \\ 1 \end{Bmatrix}, \qquad \psi_x, \psi_{xx} \to 0 \quad \text{as } x \to \mp\infty, \tag{2.6}$$

it follows that $\lambda = 0$ and $c = -1$. Thus, for $\psi(x - \lambda t) = \psi(x)$ we have the equation

$$-a\psi_{xx} + \psi_x + 2\psi(\psi - 1) = 0. \tag{2.7}$$

This equation was investigated in the well-known paper of Kolmogorov, Petrovskiĭ, and Piskunov **[29]**, where it was shown that there exists a unique solution $\psi(x)$ of the problem (2.6), (2.7), if the condition

$$1 \geq 8|a| \tag{2.8}$$

is satisfied. Moreover, this solution has the property $\psi'(x) > 0$, $x \in R_1$, i.e.,

$$\varphi'(x) < 0, \qquad x \in R_1 . \tag{2.9}$$

We note that the indicated solution $\varphi(x)$ of the stationary equation (2.1) is determined uniquely to within a shift of the argument. We fix this shift by the condition $\varphi(0) = 0$.

THEOREM 1. *Assume that the following conditions are satisfied*:

1) $0 \leq a \leq \frac{1}{8}$;
2) *There exists a number* α *such that*

$$\bar{w}(x) = \int_{-\infty}^{x} (\bar{u}(x+\alpha) - \varphi(x))\, dx \in H^{\infty}(R_1),$$

$$\int_{-\infty}^{\infty} x^2 \bar{w}^2(x)\, dx < \infty . \tag{2.10}$$

3) *The initial condition* $\bar{u}(x+\alpha)$ *is sufficiently close to* $\varphi(x)$:

$$\|\bar{w}(x)\|_{H^3(R_1)} \leq c_1 \tag{2.11}$$

(c_1 *is given below, see* (2.22)).

Then the Cauchy problem (2.1) *has a unique solution* $u(x,t)$, *and for all* $t \geq 0$ *we have the estimate*

$$\|w(x,t)\|^2_{H^3(R_1)} \leq c_2(t+1)^{-1/2}, \tag{2.12}$$

where $c_2 > 0$ *is some constant, and*

$$w(x,t) \equiv \int_{-\infty}^{x} (u(x+\alpha,t) - \varphi(x))\, dx . \tag{2.13}$$

REMARK. It follows from this estimate that for $t \geq 0$ we have

$$\sup_{x \in R_1} |u(x+\alpha,t) - \varphi(x)|^2 \leq c_2(t+1)^{-1/2}, \tag{2.14}$$

i.e., the solution of the problem (2.1) exists, is unique, and tends to the stationary solution $\varphi(x)$, as $t \to \infty$, uniformly with respect to $x \in R_1$, with an estimate of the rate of convergence. The asymptotic stability of the traveling (stationary) wave follows from inequality (2.14).

PROOF. Denote $\mathbb{K}(u) = au_{xxx} - u_{xx}$. Subtracting (2.3) from (2.1), we obtain the following equation for the function $w(x,t)$, defined in (2.13):

$$w_{xt} + 2w_x w_{xx} + 2(\varphi w_x)_x + \mathbb{K}(w_x) = 0 .$$

Integrating this equation with respect to x from $-\infty$ to x, we find

$$\dot{w} + w_x^2 + 2\varphi w_x + \mathbb{K}(w) = 0, \quad w|_{x=\pm\infty} = 0 . \tag{2.15}$$

According to the first assertion of the lemma proved in §4, a solution of the Cauchy problem for (2.15) exists and is unique. We differentiate (2.15) n times with respect to x. Denoting $w_{(n)} = \partial^n w(x,t)/\partial x^n$: we obtain

$$\dot{w}_{(n)} + \sum_{i=0}^{n} c_n^i (w_{(i+1)} + 2\varphi_{(i)}) w_{(n+1-i)} + \mathbb{K}(w_{(n)}) = 0 . \tag{2.16}$$

We multiply (2.16) by $w_{(n)}$ and integrate with respect to x on R_1:

$$\tfrac{1}{2}\dot{J}_n + \sum_{i=0}^{n} c_n^i \int_{-\infty}^{\infty} w_{(n)} w_{(n+1-i)} (w_{(i+1)} + 2\varphi_{(i)})\, dx + \int_{-\infty}^{\infty} w_{(n)} \mathbb{K}(w_{(n)})\, dx = 0, \tag{2.17}$$

where

$$I_n \equiv \int_{-\infty}^{\infty} w_{(n)} \mathbb{K}(w_{(n)})\, dx = \int_{-\infty}^{\infty} w_{(n+1)}^2\, dx \equiv J_{n+1} .$$

Integration by parts yields

$$2\int_{-\infty}^{\infty} w_{(n)} w_{(n+1)} w_{(1)}, dx = -\int_{-\infty}^{\infty} w_{(n)}^2 w_{(2)}\, dx,$$

$$2\int_{-\infty}^{\infty} w_{(n)} w_{(n+1)} \varphi_{(0)}\, dx = -\int_{-\infty}^{\infty} w_{(n)}^2 \varphi_{(1)}\, dx .$$

Therefore, from (2.17) we find

$$\tfrac{1}{2}\dot{J}_0 + \int_{-\infty}^{\infty} w_{(1)}^2 w_{(0)}\, dx - \int_{-\infty}^{\infty} w_{(0)}^2 \varphi_{(1)}\, dx + J_1 = 0,$$

$$\begin{aligned} \tfrac{1}{2}\dot{J}_3 &+ 5\int_{-\infty}^{\infty} w_{(3)}^2 w_{(2)}\, dx + 5\int_{-\infty}^{\infty} w_{(3)} \varphi_{(1)}\, dx \\ &+ 6\int_{-\infty}^{\infty} w_{(3)} w_{(2)} \varphi_{(2)}\, dx + 2\int_{-\infty}^{\infty} w_{(3)} w_{(1)} \varphi_{(3)}\, dx + J_4 = 0 . \end{aligned} \tag{2.18}$$

Let $b = \max_{x\in R_1}(|\varphi_{(1)}| + |\varphi_{(2)}| + |\varphi_{(3)}|)$. Since $\sup_{x\in R_1} |w_{(2)}| \le (4J_2J_3)^{1/4}$, we have

$$\tfrac{1}{2}\dot{J}_3 \le -J_4 + 10J_2^{1/4}J_3^{5/4} + 6bJ_3^{1/2}(\sqrt{J_1} + \sqrt{J_2} + \sqrt{J_3}) .$$

And since $J_1 \le J_0^{2/3}J_3^{1/3}$, $J_2 \le J_0^{1/3}J_3^{2/3}$, it follows that

$$J_i \le J, \quad i = 0, 1, 2, 3, \quad J = J_0 + J_3 . \tag{2.19}$$

Therefore,

$$\tfrac{1}{2}\dot{J}_3 \le -J_4 + (10\sqrt{J} + 18b)J . \tag{2.20}$$

According to the Plancherel theorem, we have

$$\begin{aligned} J_4 &= \int_{-\infty}^{\infty} w_{(4)}^2\, dx = \frac{1}{2\pi}\int_{-\infty}^{\infty} |\hat{w}(p,t)|^2 p^8\, dp \\ &= \frac{1}{2\pi}\left(\int_{|p|\ge\delta} + \int_{|p|\le\delta}\right) \ge \frac{1}{2\pi}\int_{|p|\ge\delta} \\ &\ge \frac{\delta^2}{2\pi}\int_{-\infty}^{\infty} |\hat{w}(p,t)|^2 p^6\, dp - \frac{\delta^2}{2\pi}\int_{|p|\le\delta} |\hat{w}(p,t)|^2 p^6\, dp \ge \delta^2 J_3 - \delta^8 J_0 . \end{aligned}$$

Thus, inequality (2.20) takes the form

$$\begin{aligned}\tfrac{1}{2}\dot{J}_3 &\le -\delta^2 J_3 + \delta^8 J_0 + (10\sqrt{J} + 18b)J \\ &= -\delta^2 J + (\delta^2 + \delta^8)J_0 + (10\sqrt{J} + 18b)J\,.\end{aligned} \tag{2.21}$$

We choose δ such that $\delta^2 = 11 + 18b$ and show that for all $t \ge 0$

$$J(t) < \frac{1}{16}, \quad J_0(t) < \frac{1}{16(\delta^2 + \delta^8)}\,. \tag{2.22}$$

Arguing by contradiction, let $T > 0$ be the first moment of time when the inequalities (2.22) are violated (by the assumption of the theorem, the inequalities (2.22) are satisfied at the initial moment). Then on $[0, T]$ we have

$$J(t) \le \frac{1}{16}, \qquad J_0(t) \le \frac{1}{16(\delta^2 + \delta^8)}\,. \tag{2.23}$$

From (2.23) and the first of the relations (2.18), using (2.9) we obtain

$$\tfrac{1}{2}\dot{J}_0 \le (4J_0J_1)^{1/4} \cdot J_1 - J_1 \le -J_1(1 - \sqrt{2J}) \le -\frac{J_1}{2}\,. \tag{2.24}$$

Thus, J_0 decreases and equality in (2.22) is not possible. Adding (2.21) and (2.24), and taking into account that $J_0 \le 1$, we obtain

$$\dot{J} \le -2(J - \tfrac{1}{16}),$$

whence it follows that $J(t) < \frac{1}{16}$ for all $t \in [0, T]$. The contradiction obtained proves the estimates (2.22) for all $t \ge 0$. Thus, the norm $\|w(\cdot, t)\|_{H^3(R_1)}$ is bounded for all t and, in accordance with the second assertion of the lemma, $w(x, t) \in C^\infty([0, \infty); H^\infty(R_1))$.

Multiplying equation (2.15) by $x^2 w$ and integrating over R_1, we obtain

$$\dot{V}_0 + 2\int_{-\infty}^{\infty} x^2 w w_x^2\,dx + 4\int_{-\infty}^{\infty} x^2 w w_x \varphi\,dx + 2\int_{-\infty}^{\infty} x^2 w \mathbb{K}(w)\,dx = 0, \tag{2.25}$$

where we have introduced the notation

$$V_0 = \int_{-\infty}^{\infty} x^2 w^2(x, t)\,dx\,.$$

We also set

$$V_1(t) = \int_{-\infty}^{\infty} x^2 w_x^2(x, t)\,dx\,.$$

From (2.19) we have

$$\begin{aligned}2\left|\int_{-\infty}^{\infty} x^2 w w_x^2\,dx\right| &\le 2\sup_{x \in R_1} |w| V_1(t) \\ &\le 2(4J_0J_1)^{1/4} V_1(t) \le 4J^{1/2}(t) V_1(t)\,.\end{aligned} \tag{2.26}$$

Since $-\varphi_x \ge 0$ and $\varphi(0) = 0$, it follows that $-x\varphi \ge 0$. Integrating by parts, we obtain

$$\begin{aligned}4\int_{-\infty}^{\infty} x^2 w w_x \varphi\,dx &= -4\int_{-\infty}^{\infty} x w^2 \varphi\,dx - 2\int_{-\infty}^{\infty} x^2 w^2 \varphi_x\,dx \\ &\ge -\int_{|x| \ge A} x\varphi w^{-2}\,dx = 2B\int_{-\infty}^{\infty} w^{-2}\,dx - 2B\int_{|x| \le A} w^2\,dx \\ &\ge 2BJ_0 - 8BA\sqrt{J_0J_1} \ge BJ_0 - (4A)^2 BJ_1,\end{aligned} \tag{2.27}$$

where $2B \equiv \inf_{|x|\geq A}(-x\varphi(x)) = A|\varphi(A)|$. (The value of B increases monotonically from zero to infinity for $A \in [0,\infty)$; we choose A such that $B = 4$.) Then, taking into account (2.22), from (2.25)–(2.27) we find that

$$\dot{V}_0 \leq V_1 - 4J_0 + 64A^2 J_1 - 2\int_{-\infty}^{\infty} x^2 w \mathbb{K}(w)\,dx\,. \tag{2.28}$$

We estimate the last term on the right-hand side of (2.28):

$$\begin{aligned} -2\int_{-\infty}^{\infty} x^2 w\mathbb{K}(w)\,dx &= -2\int_{-\infty}^{\infty} x^2 w(aw_{xxx} - w_{xx})\,dx \\ &= -6a\int_{-\infty}^{\infty} xw_x^2\,dx - 2V_1 + 2J_0 \leq 6a\sqrt{V_1 J_1} - 2V_1 + 2J_0 \\ &\leq -V_1 + 9a^2 J_1 + 2J_0\,. \end{aligned} \tag{2.29}$$

Denoting $A_1 = 9a^2 + 64A^2$, from (2.28) and (2.29) we obtain

$$\dot{V}_0 \leq A_1 J_1\,. \tag{2.30}$$

Setting $W = V_0 + A_1 J_0$, from (2.24) and (2.30) we find $\dot{W} \leq 0$, from which

$$V_0(t) \leq V_0(0) + A_1 J_0(0) \equiv A_2\,. \tag{2.31}$$

Now we show that $J_0(t)$ not only does not increase with time, but tends to zero as $t \to \infty$. We rewrite (2.24) in the form

$$\begin{aligned} \dot{J}_0 \leq -J_1 &= -\frac{1}{2\pi}\int_{|p|\leq\delta} - \frac{1}{2\pi}\int_{|p|\geq\delta} \leq -\frac{\delta^2}{2\pi}\int_{|p|\geq\delta} |\hat{w}|^2\,dp \\ &= -\delta^2 J_0(t) + \frac{\delta^2}{2\pi}\int_{|p|\leq\delta} |\hat{w}|^2\,dp \leq -\delta^2 J_0(t) + \frac{2\delta^3}{2\pi}\max_{|p|\leq\delta}|\hat{w}|^2 \\ &\leq -\delta^2 J_0(t) + 4\delta^3(J_0 V_0)^{1/2} \leq -\delta^2 J_0(t) + 4\delta^3(J_0(0)A_2)^{1/2}\,. \end{aligned} \tag{2.32}$$

Choose $\delta = (t+1)^{-1/2}$; then

$$\dot{J}_0(t) \leq -\frac{J_0(t)}{t+1} + A_3(t+1)^{-3/2}, \qquad A_3 = 4(J_0(0)A_2)^{1/2}\,. \tag{2.33}$$

Integrating (2.33), we obtain

$$J_0(t) \leq A_4(t+1)^{-1/2}\,. \tag{2.34}$$

From (2.34) and (2.21) we find the inequality

$$\dot{J} \leq -J + A_5(t+1)^{-1/2},$$

integration of which yields

$$J(t) \leq A_6(t+1)^{-1/2}\,.$$

Theorem 1 is proved. □

§3. Second theorem

In §2 we have considered the case when the parameter a in the KdVB equation satisfies the condition $|a| \leq \frac{1}{8}$. It is known that in this particular case, a running wave $\varphi(x)$ decreases monotonically for all $x \in R_1$. But if $|a| > \frac{1}{8}$, then, as shown in **[97]**, $\varphi(x)$ remains monotonically decreasing for $x \geq x_0$, while it oscillates for $x < x_0$, where x_0 is a maximum point of the function $\varphi(x)$.

In this section we analyze the behavior as $t \to \infty$ of a solution $u(x, t)$ of equation (2.1) is the case when $|a| < \frac{1}{8}$. We will show that the function $u(x, t)$ tends, uniformly with respect to $x \in R_1$, as $t \to \infty$ to a standing wave $\varphi(x)$, when $|a| > \frac{1}{8}$. By the same token, the standing wave is asymptotically stable in the case $|a| < \frac{1}{8}$. We also find an estimate of the rate of convergence.

THEOREM 2. *Suppose the following conditions hold*:

1) *There exists a number α such that the initial perturbation $\bar{u}(x + \alpha)$ of the Cauchy problem for* (2.1) *is sufficiently close to a stationary wave $\varphi(x)$ in the sense that*

$$\bar{w}(x) \equiv \int_{-\infty}^{x} (\bar{u}(x + \alpha) - \varphi(x))\, dx \in H^{\infty}(R_1),$$

$$\int_{-\infty}^{\infty} x^4 \bar{w}^2(x)\, dx < \infty, \tag{3.1}$$

$$\|\bar{w}\|^2_{H^3(R_1)} + \int_{-\infty}^{\infty} x^2 \bar{w}^2(x)\, dx \leq h,$$

where $h > 0$ is determined below;

2) *For the standing wave $\varphi(x)$ the following inequality holds*:

$$\sup_{x \leq x_0} \varphi'(x) \leq \lambda, \tag{3.2}$$

where $\lambda > 0$ is a positive constant (*see* (3.19) *below*).

Then there exists a unique classical solution $u(x, t)$ of the Cauchy problem for equation (2.1), *and the following inequality holds for it for all $t \geq 0$*:

$$\|w\|^2_{H^3(R_1)} \leq \eta(t + 1)^{-1/2}, \tag{3.3}$$

where $\eta > 0$ is some constant and

$$w(x, t) = \int_{-\infty}^{x} (u(x + \alpha, t) - \varphi(x))\, dx\,. \tag{3.4}$$

REMARKS. 1) This theorem is a generalization of Theorem 1.

2) When $|a| > \frac{1}{8}$, (3.2) is certainly satisfied if $|a|$ is sufficiently close to $\frac{1}{8}$, due to the continuous dependence of the solutions $\varphi(x)$ of equation (2.3) on the parameter a, and the fact that for $|a| \leq \frac{1}{8}$ the inequality $\varphi' < 0$ holds for all $x \in R_1$.

PROOF. We use the notation $\mathbb{K}(u) = a u_{xxx} - u_{xx}$. The function $w(x, t)$, defined by (3.4), satisfies the following equation and boundary conditions:

$$w_t + w_x^2 + 2\varphi w_x + \mathbb{K}(w) = 0, \qquad w|_{x \to \pm\infty} = 0\,. \tag{3.5}$$

According to the lemma proved below in §4, there is a unique classical solution $w(x, t)$

of the Cauchy problem (3.5). This justifies all subsequent calculations. By (3.5) for the function

$$J_n(t) = \int_{-\infty}^{\infty} (w_{(n)}(x,t))^2\,dx, \qquad w_{(n)} = \frac{\partial^n w}{\partial x^n},\ n = 0, 1, 2, \dots,$$

we find (see §2):

$$\dot{J}_n + 2\int_{-\infty}^{\infty} w_{(1)}^2 w\,dx - 2\int_{-\infty}^{\infty} w^2\varphi_{(1)}\,dx + 2J_1 = 0, \tag{3.6}$$

$$\begin{aligned} &\dot{J}_3 + 10\int_{-\infty}^{\infty} w_{(3)}^2 w_{(2)}\,dx + 10\int_{-\infty}^{\infty} w_{(3)}^2\varphi_{(1)}\,dx \\ &\quad + 12\int_{-\infty}^{\infty} w_{(3)}w_{(2)}\varphi_{(2)}\,dx + 4\int_{-\infty}^{\infty} w_{(3)}w_{(1)}\varphi_{(3)}\,dx + 2J_4 = 0\,. \end{aligned} \tag{3.7}$$

Introducing the constant $b = \max_{x\in R_1}(|\varphi_{(1)}| + |\varphi_{(2)}| + |\varphi_{(3)}|)$ and taking into consideration the inequalities

$$\sup_{x\in R_1}|w_{(2)}| \le (4J_1J_3)^{1/4}, \qquad J_i \le J \equiv J_0 + J_3, \quad i = 0, 1, 2, 3, \tag{3.8}$$

from (3.7) we obtain

$$\dot{J}_3 \le -2J_4 + 2J(10\sqrt{J} + 18b)\,. \tag{3.9}$$

By the Plancherel theorem, for any $\delta \ge 0$ we have

$$J_4 \ge \delta^2 J_3 - \delta^8 J_0\,.$$

Therefore, choosing $\delta^2 = 10 + 18b$, from (3.9) we obtain

$$\dot{J}_3 \le -2\delta^2 J_3 + 2gJ_0 + 2(10\sqrt{J} + 18b)J, \tag{3.10}$$

where $g = \delta^2 + \delta^8$.

We show that for all $t \ge 0$

$$J_3(t) < \frac{1}{16} \quad\text{and}\quad I(t) \equiv J_0 + \sigma V_0 < \frac{1}{16g}, \tag{3.11}$$

where $V_0 \equiv \int_{-\infty}^{\infty} x^2 w^2(x,t)\,dx$, and the number $\sigma \in (0,1]$ will be specified below. We argue by contradiction. Suppose $T > 0$ is the first moment of time when one of the inequalities (3.11) is violated. (Initially, (3.11) is satisfied by the condition of the Theorem, if $h \le \frac{1}{16g}$.) Then, on the interval $[0,T]$ we have the nonstrict inequalities

$$J_3(t) \le \frac{1}{16} \quad\text{and}\quad I(t) \le \frac{1}{16g}\,. \tag{3.12}$$

By (3.12), from (3.10) we obtain

$$\dot{J}_3 \le -2(J_3 - \tfrac{1}{16}) \le 0,$$

whence $J_3(t) < \frac{1}{16}$ for all $t \in [0,T]$.

In view of (3.12), from (3.6) we find that

$$\dot{J}_0 \le -J_1 + \chi J_0, \tag{3.13}$$

on the interval $[0,T]$ where $\chi = 2\sup_{x\in R_1}\varphi'(x)$.

We multiply (2.1) by $x^2 w$ and integrate with respect to x:

$$\dot{V}_0 + 2\int_{-\infty}^{\infty} x^2 w w_{(1)}^2\, dx + 4\int_{-\infty}^{\infty} x^2 w w_{(1)}\varphi\, dx + 2\int_{-\infty}^{\infty} x^2 w \mathbb{K}(w)\, dx = 0\,. \tag{3.14}$$

On $[0, T]$ we have

$$\left| 2\int_{-\infty}^{\infty} x^2 w w_x^2\, dx \right| \le 2 \sup_{x\in R_1} |w| V_1 \le \tfrac{1}{2} V_1, \tag{3.15}$$

where $V_1 \equiv \int_{-\infty}^{\infty} x^2 w_{(1)}^2\, dx$. Further,

$$\begin{aligned} Q \equiv 4\int_{-\infty}^{\infty} x^2 w w_{(1)}\varphi\, dx &= -2\int_{-\infty}^{\infty} (\varphi_{(1)}x^2 + 2x\varphi) w^2\, dx \\ = \left(\int_{|x|\ge A} + \int_{|x|\le A}\right) &\ge 2B\int_{|x|\ge A} w^2(x,t)\, dx - \sup_{x\in R_1} w^2 \cdot \sup_{|x|\le A} (|\varphi_{(1)}|x^2 + |x\varphi|)4A; \end{aligned}$$

here

$$B = \min_{|x|\ge A} (-\varphi_{(1)}x^2 - 2x\varphi)\,.$$

We note that B is positive and arbitrarily large if A is sufficiently large, because $\varphi_{(1)} \to 0$ exponentially as $x \to \pm\infty$ (see **[97]**), while $\varphi \to \pm 1$ as $x \to \mp\infty$. Fix $A > 0$ so that the inequality $B > 4$ holds. Then

$$Q \ge 2BJ_0 - c_1 \sup_{x\in R_1} w^{-2}(x,t) \ge BJ_0 - c_2 J_1, \tag{3.16}$$

since $\sup_{x\in R_1} w^2(x,t) \le (4J_0J_1)^{1/2}$; here c_1 and c_2 are some positive constants.

The remaining term in (3.14) can be estimated as follows (see §2):

$$-2\int_{\infty}^{\infty} x^2 w\mathbb{K}(w)\, dx \le -V_1 + 9a^2 J_1 + 4J_0\,.$$

By (3.15) and (3.16), from (3.14) we obtain

$$\dot{V}_0 \le (4 - B)J_0 - \tfrac{1}{2}V_1 + c_3 J_1, \tag{3.17}$$

while (3.13) and (3.17) yield

$$\dot{I} \le -(B\sigma - \chi - 4\sigma)J_0 - (1 - \sigma c_3)J_1 - \xi V_1, \tag{3.18}$$

where $\xi = \frac{\sigma}{2}$. We impose the following requirements: $B\sigma - \chi - 4\sigma \equiv \chi\kappa > 0$ and $1 - \sigma c_3 \equiv \mu > 0$. Since $\sigma > 0$ is at our disposal, then $\chi > 0$ has to satisfy the following condition:

$$\chi < \frac{(B-4)\sigma}{1+\kappa} \equiv \lambda\,. \tag{3.19}$$

From (3.18) we now get

$$\dot{I} \le -\mu J_1 - \xi V_1 - \chi\kappa J_0 \le 0\,. \tag{3.20}$$

This implies $I(t) \le I(0)$ and (3.11) becomes valid also at $t = T$. The contradiction obtained proves (3.11) for all times $t > 0$.

Next we establish the boundedness of the functions $W_0 \equiv \int_{-\infty}^{\infty} x^4 w^2(x,t)\,dx$ for all $t \geq 0$. We multiply (2.1) by $x^4 w$ and integrate with respect to $x \in R_1$:

$$\dot{W}_0 + 2\int_{-\infty}^{\infty} x^4 w w_{(1)}^2\,dx + 4\int_{-\infty}^{\infty} x^4 w w_{(1)}\varphi\,dx + 2\int_{-\infty}^{\infty} x^4 w \mathbb{K}(w)\,dx = 0\,. \tag{3.21}$$

We have

$$\left|2\int_{-\infty}^{\infty} x^4 w w_{(1)}^2\,dx\right| \leq 4J^{1/2}W_1 < W_1, \tag{3.22}$$

where $W_1 \equiv \int_{-\infty}^{\infty} x^4 w_{(2)}^2\,dx$. Applying the same calculations as in the derivation of (3.16) above, we obtain

$$4\int_{-\infty}^{\infty} x^4 w w_{(1)}\varphi\,dx \geq 2BV_0 - c_4(J_0 + J_1), \tag{3.23}$$

where this time we may consider that $B = 6(|a|+1) > 0$. Further, integrating by parts we find that

$$-2\int_{-\infty}^{\infty} x^4 w\mathbb{K}(w)\,dx \leq -W_1 + 36a^2 V_1 + 12(|a|+1)V_0 + 12|a|J_0\,. \tag{3.24}$$

Taking into account (3.22)–(3.24), from (3.21) we obtain

$$\dot{W}_0 \leq \nu(J_0 + J_1 + V_1), \tag{3.25}$$

where $\nu > 0$ is a constant. If we introduce the function $Y \equiv W_0 + \nu I/\min(\mu, \xi, \chi\kappa)$, then $\dot{Y} \leq 0$, by (3.25) and (3.20), and, thus, $Y(t) < Y(0)$. This implies the boundedness of $W_0(t)$ for all $t \geq 0$.

It remains to prove that $I(t)$ not only does not increase, but tends to zero as $t \to \infty$. By the Plancherel theorem, for the right-hand side of (3.20) we have

(3.26)

$$\begin{aligned}
&\mu J_1 + \xi V_1 + \chi\kappa J_0 \\
&\quad \geq \frac{1}{2\pi}\int_{-\infty}^{\infty} p^2\left(\mu|\hat{w}(p,t)|^2 + \frac{\xi}{2}|\hat{w}_p'(p,t)|^2\right)dp + \frac{\chi\kappa - \xi}{2\pi}\int_{-\infty}^{\infty} |\hat{w}(p,t)|^2\,dp \\
&\quad \geq \int_{|p|\geq\delta} \\
&\quad \geq \frac{\delta^2}{2\pi}\int_{|p|\geq\delta}\left(|\hat{w}(p,t)|^2\mu + |\hat{w}_p'(p,t)|^2\frac{\xi}{2}\right)dp + \frac{\chi\kappa - \xi}{2\pi}\int_{|p|\geq\delta} |\hat{w}(p,t)|^2\,dp \\
&\quad \geq \delta^2\min(\mu, \tfrac{1}{4})I - c_5\delta^3(\sqrt{J_0 V_0} + \sqrt{V_0 W_0})\,.
\end{aligned}$$

In view of the boundedness of the functions J_0, V_0, W_0, from (3.20) and (3.26) we obtain

$$\dot{I} \leq -\tilde{\delta}^2 I + c_6\tilde{\delta}^3 \quad \text{where } \tilde{\delta}^2 = \delta^2\min(\mu, \tfrac{1}{4})\,. \tag{3.27}$$

Put $\tilde{\delta}^2 = (t+1)^{-1}$; then,

$$\dot{I} \leq -\frac{I}{t+1} + \frac{c_6}{(t+1)^{3/2}},$$

whence

$$I(t) \leq c_7(t+1)^{-1/2}. \tag{3.28}$$

From (3.28) and (3.10) we obtain

$$\dot{J}_3 \leq -2J_3 + c_8(t+1)^{-1/2}.$$

Integrating this inequality, with the help of (3.28), we obtain (3.3). Theorem 2 is proved. □

Hence, the value $a = \frac{1}{8}$, being critical for the behavior of the running wave for the Korteweg-de Vries-Burgers equation, is not critical for the stability of the running wave, since by (3.3) the convergence of a solution of the Cauchy problem for the Korteweg-de Vries-Burgers equation to the stationary wave is uniform with respect to x. Indeed, the following estimate holds, $t \geq 0$;

$$\sup_{x \in R_1} |(u(x+\alpha, t) - \varphi(x)|^2 \leq c(t+1)^{-1/2}.$$

Thus, we have obtained a definitive answer to the question posed in **[97]** concerning the stability of traveling wave type solutions of the Korteweg-de Vries-Burgers equation.

§4. A lemma

The purpose of this section is the proof of the following lemma.

LEMMA. 1) *If $\bar{w}(x) \in H^\infty(R_1)$, then for some $T > 0$ there exists a unique solution $w(x,t) \in C^\infty([0,T]; H^\infty(R_1))$ of the following Cauchy problem*:

$$w_t + w_x^2 + 2\varphi w_x + a w_{xxx} - w_{xx} = 0, \qquad w|_{t=0} = \bar{w}(x), \tag{4.1}$$

where(*) $\varphi(x) \in C_b^\infty(R_1)$;

2) *If the norm $\|w(\cdot,t)\|_{H^3(R_1)}$ of the solution remains bounded with the passage of time, then the solution $w(x,t) \in C^\infty([0,\infty); H^\infty(R_1))$, i.e., it exists globally in time.*

For the sake of convenience, we divide the proof into several steps.

4.1. ASSERTION 1. *Let $\bar{u}(x) \in H^\infty(R_1)$, and let $\varphi(x) \in C_b^\infty(R_1)$. Then, for some $T > 0$ there exists a solution $u(x,t) \in C^\infty([0,T]; H^\infty(R_1))$ of the following Cauchy problem*:

$$u_t + 2uu_x + 2\varphi u_x + \mathbb{K}(u) = 0, \quad u|_{t=0} = \bar{u}(x), \tag{4.2}$$

where

$$\mathbb{K}(u) = \int_{-\infty}^{\infty} k(x-y)u(y,t)\,dy + \xi(x)u(x,t), \tag{4.3}$$

$\xi(x) \in C_b^\infty(R_1)$, and $k(x)$, is absolutely integrable on R_1.

(*) $C_b^\infty(R_1)$ is the class of infinitely differentiable functions that are bounded together with all of their derivatives.

PROOF(*). We apply the method of successive approximations. We set $v^{(0)}(x,t) = \bar{u}(x)$ and define $v^{(s)}(x,t)$, $s \geq 1$, as the solution of the following linear problem:

$$v_t^{(s)} + 2(v^{(s-1)} + \varphi)v_x^{(s)} + \mathbb{K}(v^{(s-1)}) = 0, \qquad v^{(s)}|_{t=0} = \bar{u}(x)\,. \tag{4.4}$$

Omitting for the moment the index s, we rewrite (4.4) in the form

$$v_t + gv_x + f = 0, \qquad v|_{t=0} = \bar{u}(x)\,. \tag{4.5}$$

We assume, by induction, that $v^{(s-1)}(x,t)$ is already constructed and belongs to $C^\infty([0,\infty); H^\infty(R_1))$. Then

$$g \in C^\infty([0,\infty); C_b^\infty(R_1)), \qquad f \in C^\infty([0,\infty); H^\infty(R_1))$$

and (4.5) is a Cauchy problem for a linear equation, whose solution is constructed in the following manner. We define the characteristic $y(\xi,t)$ as the solution of the Cauchy problem for the ordinary differential equation

$$\dot{y} = g(y,t), \quad y|_{t=0} = \xi, \quad \xi \in R_1\,. \tag{4.6}$$

It is well known (see **[73]**) that there exists a unique solution $y(\xi,t) \in C^\infty(R_1 \times [0,T])$ of the problem (4.6). This solution realizes a diffeomorphism $\xi \in R_1 \leftrightarrow y \in R_1$ for each $t \in [0,T]$ and all of its derivatives are bounded. With the help of the function $y(\xi,t)$, the solution of the problem (4.5) is given by the formula

$$v(y(\xi,t),t) = \bar{u}(\xi) + \int_0^t f(y(\xi,\tau),\tau)\,d\tau\,.$$

From here it follows that $v(x,t) \in C^\infty([0,\infty); H^\infty(R_1))$. Thus, we have determined all the successive approximations $v^{(s)}(x,t)$.

We prove that for some $T > 0$ the sequence $\{v^{(s)}\}$ is bounded uniformly with respect to $s \geq 0$ in the class $C^\infty([0,T]; H^\infty(R_1))$. For this we differentiate (4.4) n times with respect to x:

$$\frac{\partial}{\partial t}v_{(n)}^{(s)} + 2\sum_{j=0}^{n} C_j^n(v_{(j)}^{(s-1)} + \varphi_{(j)})v_{(n+1-j)}^{(s)} + \mathbb{K}(v_{(n)}^{(s-1)}) = 0, \tag{4.7}$$

where $\varphi_{(n)} \equiv \frac{\partial^n \varphi}{\partial x^n}$, and C_j^n are the binomial coefficients. Multiplying (4.7) by $v_{(n)}^{(s)}$ and integrating with respect to x from $-\infty$ to ∞, we obtain

$$\dot{V}_n^s + 4\sum_{j=0}^{n}\int_{-\infty}^{\infty} v_{(n)}^{(s)}v_{(n+1-j)}^{(s)}(v_{(j)}^{(s-1)} + \varphi_{(j)})\,dx + 2\int_{-\infty}^{\infty} v_{(n)}^{(s)}\mathbb{K}(v_{(n)}^{(s-1)})\,dx = 0, \tag{4.8}$$

where $V_n^s = \int_{-\infty}^{\infty}(v_{(n)}^{(s)})^2\,dx$. For each t the function $v_{(n)}^{(s)}(x,t) \to 0$ as $|x| \to \infty$, since it belongs to the class $H^\infty(R_1)$. Therefore, integrating by parts we get

$$\int_{-\infty}^{\infty} v_{(n)}^{(s)}v_{(n+1)}^{(s)}v_{(0)}^{(s-1)}\,dx = -\frac{1}{2}\int_{-\infty}^{\infty}(v_{(n)}^{(s)})^2 v_{(1)}^{(s-1)}\,dx, \tag{4.9}$$

and also

$$\sup_{x\in R_1}|v_{(n)}^{(s)}(x,t)| \leq (4V_n^s V_{n+1}^s)^{1/4}\,. \tag{4.10}$$

(*) Conceptually, the proof of this statement is known **[74]**. The new feature here is the introduction of a nonlocal term into the equation and the assertion regarding the membership of the solution in the class $H^\infty(R_1)$.

Estimating the last term in (4.8) with the aid of the Cauchy inequality and taking into account (4.9), (4.10), for the function $J_n^s = 1 + \sum_{j=0}^n V_j^s$ from (4.8) we obtain

$$\dot{J}_2^s \le c_2 J_2^{s-1} J_2^s, \tag{4.11}$$

$$\dot{J}_n^s \le c_n (J_n^s J_{n-1}^{s-1} + J_s^s \sqrt{J_n^s J_n^{s-1}}), \qquad n \ge 3, \tag{4.12}$$

where $c_n \ge 0$ are some numbers. Assume, by induction, that for $J_2^{s-1}(t)$ we have the inequality

$$J_2^{s-1}(t) \le J_2(0)(1 - J_2(0)c_2 t)^{-1}. \tag{4.13}$$

Introducing it into (4.11) and integrating, we obtain (4.13) also for $J_2^s(t)$, since $J_2^{-s}(0) = J_2(0)$. (For $s = 0$, inequality (4.13) is valid by the definition of $J_2^0(t)$.) From (4.13) it follows that if T satisfies the condition $0 < T < (J_2(0)c_2)^{-1}$, then on the segment $[0, T]$ for all $s \ge 0$ we have $J_2^s(t) \le \mathcal{D}_2$, where $\mathcal{D}_2$ is some constant not depending on s.

Now, arguing by induction for $n \ge 2$ we prove that

$$J_n^s(t) \le \mathcal{D}_n, \qquad t \in [0, T],\ s = 0, 1, 2, \dots. \tag{4.14}$$

Indeed, we assume that (4.14) has been established for $n \le n_1$; then for $n = n_1 + 1$ from (4.12) we obtain

$$\dot{J}_n^s \le \tilde{c}_n (J_n^s + \sqrt{J_n^s J_n^{s-1}}). \tag{4.15}$$

We prove that for all s:

$$J_n^s(t) < 2J_n(0)e^{4\tilde{c}_n t}. \tag{4.16}$$

We carry out induction on $s \ge 0$. Suppose that the estimate (4.16) has been proved for $s \le s_1$; then for $s = s_1 + 1$, assuming the contrary, we find the first moment of time $T_1 > 0$ when (4.16) is violated. Then, by continuity, on $[0, T_1]$ we obtain

$$J_n^s(t) \le 2J_n(0)e^{4\tilde{c}_n t}. \tag{4.17}$$

Introducing (4.17) and (4.16) into (4.12), we find

$$\dot{J}_n^s(t) \le 4\tilde{c}_n J_n(0)e^{4\tilde{c}_n t},$$

whence we obtain (4.16) also for $t = T_1$. The contradiction obtained proved (4.16) for $s = s_1 + 1$ on the entire segment $[0, T]$. The induction on s has been carried out. From (4.16) we obtain (4.14), which concludes the induction on $n \ge 2$.

Similarly, one proves the boundedness of the quantities

$$\int_{-\infty}^{\infty} \left(\frac{\partial^j v_{(n)}^{(s)}(x, t)}{\partial t^j} \right)^2 dx$$

uniformly with respect to s. Thus, the sequence $\{v^{(s)}\}$ is bounded in

$$C^\infty([0, T]; H^\infty(R_1))$$

uniformly with respect to $s \ge 0$. Therefore, some subsequence of $\{v^{(s)}\}$ converges to the solution $u(x, t) \in C^\infty([0, T]; H^\infty(R_1))$ of the problem (4.2). Assertion 1 is proved. □

4.2. Suppose $\mathbb{K}$ in problem (4.2) is represented in the form of a pseudodifferential operator:

$$\mathbb{K}(u) = \frac{1}{2\pi}\int_{-\infty}^{\infty} e^{ipx}K(p)\hat{u}(p,t)\,dp + \xi(x)u(x,t),$$

where $\hat{u}(p,t) = \int_{-\infty}^{\infty} e^{ipx}u(x,t)\,dx$ is the Fourier transform, and the symbol $K(p) \in L_\infty^{\mathrm{loc}}(R_1)$, $\xi(x) \in C_b^\infty(R_1)$.

ASSERTION 2. *Suppose that*

1) *the operator $\mathbb{K}$ is dissipative, i.e.,*

$$K^1(p) = \operatorname{Re} K(p) \geq 0\,. \tag{4.18}$$

2) *The initial condition $\bar{u}(x) \in H^\infty(R_1)$.*

Then on some segment $[0, T]$ there exists a unique solution $u(x,t)$ of the Cauchy problem (4.2), *belonging to $C^\infty([0,T]; H^\infty(R_1))$. The time $T > 0$ is determined by the norm $\|u\|_{H^2(R_1)}$ of the solution.*

PROOF. We carry out the regularization of the operator $\mathbb{K}$. We define the sequence of symbols $K^l(p)$, $l = 1, 2, \dots$, in the following manner:

$$K^l(p) = (1+|p|)^\beta \int_{-l}^{l} \rho_l(p-q)(1+|q|)^{-\beta}K(q)\,dq,$$

where $\beta = 2 + \alpha$, and α is the order of the operator $\mathbb{K}$, i.e., the infimum of all α for which the following inequality holds:

$$|K(p)| \leq c(1+|p|)^\alpha\,.$$

The function $\rho_l(p) = l\rho_1(pl)$, $\rho_1(p) \in C_0^\infty(R_1)$, is such that $\rho_1(p) \geq 0$; $\rho_1(p) = 0$ for $|p| \geq 1$, $\int_{-\infty}^{\infty} \rho_1(p)\,dp = 1$. The operators $\mathbb{K}^l$ with the symbols $K^l(p)$ can be represented in the form (4.3) with the kernels $k^{(l)}(x)$ belonging to the Schwartz space. From (4.18) it follows that

$$K^{1,l}(p) \geq 0, \tag{4.19}$$

$$K^l(p)(1+|p|)^{-\beta} \to K(p)(1+|p|)^{-\beta} \tag{4.20}$$

in the form of $L_2(R_1)$ as $l \to \infty$. From Assertion 1 it follows that there exists a solution $u^l(x,t) \in C^\infty([0,T^l]; H^\infty(R_1))$ of the Cauchy problem (4.2) with the operator $\mathbb{K}^{(l)}$. We establish the boundedness of the sequence $\{u^{(l)}\}$ in the space $C^\infty([0,T]; H^\infty(R_1))$ for some $T > 0$. For the moment we omit the index l. We multiply (4.2) and its derivatives by $u_{(n)}(x,t)$ and integrate with respect to x from $-\infty$ to $+\infty$:

$$\dot{J}_n + 4\sum_{j=0}^{n} C_j^n \int_{-\infty}^{\infty} u_{(n)}u_{(n+1-j)}(u_{(j)} + \varphi_{(j)})\,dx + 2I_n = 0, \tag{4.21}$$

where $J_n = \int_{-\infty}^{\infty} u_{(n)}^2\,dx$, $I_n = \int_{-\infty}^{\infty} u_{(n)}\mathbb{K}(u_{(n)})\,dx$. We estimate the integral I_n by using (4.19) and the Plancherel theorem:

$$\begin{aligned} I_n &= \int_{-\infty}^{\infty} u_{(n)}\mathbb{K}(u_{(n)})\,dx \\ &= \frac{1}{2\pi}\int_{-\infty}^{\infty} K^{1,l}(p)|p|^{2n}|\hat{u}(p,t)|^2\,dp + \int_{-\infty}^{\infty} \xi(x)(u_{(n)}(x,t))^2\,dx \\ &\geq -\xi_0 J_n, \qquad \xi_0 = \sup_{x\in R_1}|\xi(x)|\,. \end{aligned}$$

Applying the identity

$$2\int_{-\infty}^{\infty} u_{(n)}u_{(n+1)}u_{(0)}\,dx = -\int_{-\infty}^{\infty} u_{(n)}^2 u_{(1)}\,dx$$

and the inequality $u_n(t) = \sup_{x\in R_1}|u_{(n)}| \leq (4J_nJ_{n+1})^{1/4}$, we transform (4.21) to the form

$$\dot{J}_0 \leq 4MJ_0, \tag{4.22}$$

$$\dot{J}_2 \leq 20J_2^{5/4}(J_1)^{1/4} + 16MJ_2 + 4\sqrt{J_1J_2}\varphi_2, \tag{4.23}$$

$$\begin{aligned} \dot{J}_n &\leq 2(2n+1)J_n(M + (4J_1J_2)^{1/4}) \\ &\quad + 4\sum_{j=2}^{n-1} C_j^n\sqrt{J_nJ_{n+1-j}}(\varphi_j + (4J_jJ_{j+1})^{1/4}), \qquad n\geq 3 \end{aligned} \tag{4.24}$$

where $M = \varphi_1 + \varphi_2 + \xi_0$. Put $J = J_0 + J_2$. Then, by virtue of the inequality $J_1 \leq \sqrt{J_0J_2} \leq J$, from (4.22) and (4.23) we obtain

$$\dot{J} \leq 20(J^{3/2} + MJ)\,.$$

From here, integrating with respect to t, we find that

$$J(t) \leq J(0)e^{20Mt}\{1 - M^{-1}\sqrt{J(0)}(e^{10Mt} - 1)\}^{-2}. \tag{4.25}$$

It follows from (4.25) that if we choose T from the condition

$$0 < T < (10M)^{-1}\ln(1 + M/\sqrt{J(0)}),$$

then there exists a number c_1, independent of l, such that for all $t \in [0, T]$ we have $J(t) \leq c_1$. Now, integration of the linear inequalities (4.24) with respect to t yields $J_n(t) \leq c_n$, $t \in [0, T]$, where $c_n > 0$ does not depend on l.

Similarly, one proves the boundedness of the quantities

$$\int_{-\infty}^{\infty}\left(\frac{\partial^{n+m}u^l}{\partial x^n\,\partial t^m}\right)^2 dx,$$

uniformly with respect to l.

From the estimates obtained there follows that some subsequence of $\{u^l\}$ converges to $u(x,t) \in C^\infty([0,T]; H^\infty(R_1))$ in the norm of the spaces $C^m([0,T]; H^n(R_1))$ for arbitrary n and m. Moreover, by virtue of (4.20), $\mathbb{K}^l(u^l) \to \mathbb{K}(u)$ uniformly with respect to $x \in R_1$ and $t \in [0, T]$, so that $u(x,t)$ is a solution of the problem (4.2).

Now we prove uniqueness. Assuming the contrary, suppose that there exist two solutions: $u^1(x,t)$ and $u^2(x,t)$. For the difference $v = u^1 - u^2$ we obtain

$$v_t + 2(u^1+\varphi)v_x + 2vu_x^2 + \mathbb{K}(v) = 0, \qquad v|_{t=0} = 0\,. \tag{4.26}$$

Multiplying (4.26) by v and integrating with respect to x from $-\infty$ to $+\infty$, for $V = \int_{-\infty}^{\infty} v^2(x,t)\,dx$ we find

$$\dot{V} + 2\int_{-\infty}^{\infty} v^2(-u_x^1 + 2u_x^2 - \varphi_x)\,dx + 2\int_{-\infty}^{\infty} v\mathbb{K}(v)\,dx = 0\,. \tag{4.27}$$

Transforming the last term in (4.27) with the aid of the Plancherel theorem, we have

$$\dot{V} \le 4V(u_1^1 + u_1^2 + M), \quad V(0) = 0\,.$$

Hence, it follows that $V(t) \equiv 0$ for all $t \in [0,T]$. Assertion 2 is proved. □

4.3.

PROOF OF THE LEMMA. The solution $w(x,t)$ of the problem (4.1) is obtained from the solution $u(x,t)$ of the problem (4.2) in the following way:

$$w(x,t) = \int_{-\infty}^{x} u(x,t)\,dx,$$

and, consequently, it exists, it is unique, and it belongs to $C^\infty([0,T];C^\infty(R_1))$. We show that $w(x,t) \in L_(R_1)$ for all $t \in [0,T]$. For this, we multiply (4.1) by $\rho(x)w(x,t)$ and integrate over all x ($\rho(x) \in C_0^\infty(R_1)$, $1 \ge \rho(x) \ge 0$ is an arbitrary function). For $W(t) \equiv \int_{-\infty}^{\infty} w^2(x,t)\rho(x)\,dx$ we obtain

$$\tfrac{1}{2}\dot{W} + \int_{-\infty}^{\infty} \rho w(w_x^2 + 2\varphi w_x)\,dx + \int_{-\infty}^{\infty} \rho w(aw_{xxx} - w_{xx})\,dx = 0,$$

when, by virtue of the Cauchy inequality, we find

$$\dot{W} \le \sqrt{WJ_0}\left(\sup_{x\in R_1}|w_x| + 2\sup_{x\in R_1}|\varphi|\right) + 2\sqrt{W(aJ_2+J_2)} < c\sqrt{W}, \tag{4.28}$$

where $c > 0$ is some constant. The boundedness of $W(t)$ follows from (4.28), independently of the function $\rho(x) \ge 0$. Hence, $w(x,t) \in L_2(R_1)$ for all $t \ge 0$. Thus $w(x,t) \in C^\infty([0,T],H^\infty(R_1))$. The time of existence of $w(x,t)$ is determined by the norm $\|w\|_{H^3(R_1)}$, as shown in Assertion 2. Thus, the assertion of the lemma follows from this. The lemma is proved. □

§5. The step-decaying problem for the Kuramoto-Sivashinsky equation

In this section we discuss the step-decaying problem for the Kuramoto-Sivashinsky equation

$$u_t + 2uu_x + au_{xxxx} - bu_{xx} = 0, \tag{5.1}$$

with constants $a, b > 0$, i.e., we study the behavior, as $t \to \infty$, of the solution $u(x,t)$ of (5.1) with the initial condition $\bar{u}(x)$ satisfying the following requirement:

$$\bar{u}(x) \to \pm 1 \quad \text{as } x \to \mp\infty \tag{5.2}$$

(the initial perturbation has the form of a shock wave propagating to the right).

The model equation (5.1) arises in various problems of physics: in the theory of long waves in thin films **[179]**–**[181]**, and also of long waves at an interface between two viscous liquids [**182**], in systems of the reaction-diffusion type [**139**], in the description of the instability of a flame front [**183**], etc. Of greatest interest is the case of $a > 0$ and $b < 0$, when the linear terms describe the interaction of long-wavelength pumping and short-wavelength dissipation, and the nonlinear term characterizes energy redistribution between various modes.

If the initial perturbation $\bar{u}(x)$ decays at infinity, then the asymptotic behavior, as $t \to \infty$, of a solution of the Cauchy problem for (5.1) for $a > 0$ and $b > 0$ (strict dissipation) was found in Chapter 7, where we showed that the asymptotic relationship formula

$$u(x,t) = At^{-1/2}e^{-\xi^2} + O(t^{-1}) \tag{5.3}$$

is valid uniformly with respect to $\xi = |x|/2\sqrt{bt} \geq 0$ as $t \to \infty$, and the constant A is expressed explicitly in terms of the Fourier transform of the initial perturbation $\bar{u}(x)$.

If the initial perturbation $\bar{u}(x)$ does not decay at infinity, but satisfies condition (5.2), then the solution $u(x,t)$ of the Cauchy problem for (5.1) behaves differently as $t \to \infty$. Here we show that the solution $u(x,t)$ does not tend to zero as $t \to \infty$, but to a solution of (5.1) of the traveling-wave type $\varphi(x - ct)$. For the function φ in view of (5.1) we obtain

$$-c\varphi' + 2\varphi\varphi' + a\varphi^{(4)} - b\varphi^{(2)} = 0,$$

from which, integrating with respect to x, we find

$$d - c\varphi + \varphi^2 + a\varphi''' - b\varphi' = 0, \tag{5.4}$$

where d is the constant of integration. Assuming the conditions, $\varphi^2(x) \to 1$, $\varphi'(x) \to 0$, $\varphi'''(x) \to 0$, for the function φ to be satisfied as $x \to \pm\infty$, from (5.4) we find that $c = 0$ and $d = -1$. Thus, the traveling (standing) wave $\varphi(x)$ satisfies the equation

$$a\varphi''' - b\varphi' = 1 - \varphi^2 . \tag{5.5}$$

Equation (5.5) has been the subject of many studies. The existence of a unique (to within a shift of the argument) solution $\varphi(x)$ of (5.5) for $a > 0$ and $b > 0$, satisfying the boundary conditions (5.2), was proved in [**184**]. That solution is odd and has a single zero at $x = 0$ (this condition eliminates the translational nonuniqueness). Moreover, for $0 < a < b^2/27$ the function $\varphi(x)$ decreases monotonically over the entire real line, and for $a > b^2/27$ it oscillates at infinity. Considerably less is known about the existence of solutions for equation (5.5) if $a > 0$ and $b < 0$. First of all, it is easy to notice that for $a = 0$ the function $\varphi(x) = -\tanh\frac{x}{b}$ is a solution of (5.5) with the boundary conditions opposite to those for (5.2):

$$\varphi(x) \to \mp 1 \quad \text{as } x \to \mp\infty \tag{5.6}$$

(this boundary condition corresponds to a rarefaction wave propagating to the right). With the special choice of the parameters $a = a_0 = 11 \cdot (15)^2 \cdot b^2/(19)^3$ and $b < 0$, an exact solution of (5.5) was found in [**139**]:

$$\varphi(x) = \frac{9}{2}\tanh\left(\frac{19x}{30b}\right) - \frac{11}{2}\tanh^3\left(\frac{19x}{30b}\right) \tag{5.7}$$

which satisfies a boundary condition of the type (5.6). By the way, the exact solution analogous to (5.7) can also be found easily for the case $a > 0$ and $b > 0$. Indeed, for $a = 25 \cdot (3b)^2/(19)^3$, $b > 0$ it has the form

$$\varphi(x) = -\frac{3}{2}\tanh\left(\frac{19x}{30b}\right) + \frac{1}{2}\tanh^3\left(\frac{19x}{30b}\right).$$

Furthermore, in **[182]**, **[185]** numerical calculations were used to demonstrate the existence of a solution $\varphi(x)$ of equation (5.5) with the boundary conditions (5.6) for small $a > 0$ (i.e., close to the function $-\tanh\frac{x}{b}$), as well as for a close to a_0. It was later shown analytically in **[186]** that no solution $\varphi(x)$ of (5.5) with the condition (5.6) exists for $b < 0$ and small $a > 0$. It was also shown in **[184]** that no monotonic solution of (5.5) with the condition (5.6) exists for $b < 0$ and $a \geq 2b^2/9$. Finally, the existence of solutions of equation (5.5) with the boundary condition (5.2) for $b < 0$ and sufficiently large $a > 0$ was shown in **[187]**, and for $b = 0$ and $a > 0$ in **[188]**.

In this monograph we confine our consideration to the case of positive a and b, for which there are exhaustive results on the behavior of traveling waves **[184]** and we present the following two theorems on asymptotic stability of traveling waves.

THEOREM 3. *Suppose the following conditions are satisfied*:

1) $0 < a \leq b^2/27$, $b > 0$;

2) *The initial data $\bar{u}(x)$ are sufficiently close to a standing wave $\varphi(x)$ in the sense that a number α exists such that*

$$\bar{w}(x) \equiv \int_{-\infty}^{x} (\bar{u}(y+\alpha) - \varphi(y))\, dy \in H^{\infty}(R_1),$$
$$\int_{-\infty}^{\infty} x^2\bar{w}^2(x)\, dx < \infty, \qquad \|\bar{w}(x)\|_{H^3(R_1)} \leq h, \tag{5.8}$$

$h > 0$ being sufficiently small. Then there exists a unique solution of the Cauchy problem for equation (5.1) *and for all $t \geq 0$ the following bound holds*:

$$\|w(x,t)\|^2_{H^3(R_1)} \leq \eta(t+1)^{-1/2}, \tag{5.9}$$

where $\eta > 0$ is a constant, and

$$w(x,t) = \int_{-\infty}^{x} (u(y+\alpha,t) - \varphi(y))\, dy\,. \tag{5.10}$$

THEOREM 4. *Suppose the following conditions are satisfied*:

1) *A number α exists such that the initial perturbation $\bar{u}(x)$ is sufficiently close to a standing wave $\varphi(x)$ in the sense that*

$$\bar{w}(x) \in H^{\infty}(R_1),$$
$$\int_{-\infty}^{\infty} x^4\bar{w}^2(x)\, dx < \infty, \tag{5.11}$$
$$\|\bar{w}\|^2_{H^3(R_1)} + \int_{-\infty}^{\infty} x^2\bar{w}^2(x)\, dx \leq h$$

where $h > 0$ is a constant;

2) *For the standing wave* $\varphi(x)$ *we have the bound*

$$\sup_{x \in R_1} \varphi'(x) \leq \lambda, \tag{5.12}$$

where $\lambda > 0$ *is a constant.*(*)

Then there exists a unique classical solution $u(x, t)$ *of the Cauchy problem for equation* (5.1), *and the bound* (5.9) *holds for it for all* $t \geq 0$.

REMARKS. 1) The bound (5.9) means that the traveling wave $\varphi(x)$ is asymptotically stable, since it follows from (5.9), by virtue of the imbedding theorem, that

$$\sup_{x \in R_1} |u(x + \alpha, t) - \varphi(x)|^2 \leq c(t + 1)^{-1/2}.$$

2) The condition (5.12) is certainly satisfied for $a > b^2/27$, $b > 0$, if a is sufficiently close to $b^2/27$, because of the continuous dependence of the solution $\varphi(x)$ of (5.5) on a and because for $a \leq b^2/27$ the inequality $\varphi'(x) < 0$ is satisfied for all $x \in R_1$.

The proof of Theorems 3 and 4 literally coincides with the proof of Theorems 1 and 2 presented above.

(*) The constants h and λ can be expressed explicitly through the parameters a and b.

References

1. M. J. Ablowitz and H. Segur, *Solutions and the inverse scattering transform*, SIAM Stud. Appl. Math., vol. 4, SIAM, Philadelphia, PA, 1981.
2. V. V. Avilov, I. M. Krichever, and S. P. Novikov, *Evolution of the Whitham zone in the Korteweg-de Vries theory*, Dokl. Akad. Nauk SSSR **295** (1987), no. 2, 345–349; English transl. in Soviet Phys. Dokl. **32** (1987).
3. V. V. Avilov and S. P. Novikov, *Evolution of the Whitham zone in KdV theory*, Dokl. Akad. Nauk SSSR **294** (1987), no. 2, 327–329; English transl. in Soviet Phys. Dokl. **32** (1987).
4. N. Bloembergen, *Nonlinear optics*, Benjamin, New York, 1965.
5. V. S. Buslaev, *Application of determinant representation of a solution to the Korteweg-de Vries equation for analysis of its asymptotic behavior for large times*, Uspekhi Mat. Nauk **36** (1981), no. 4, 217–218. (Russian)
6. V. A. Vasilev, Yu. M. Romanovskiĭ, and V. G. Yakhno, *Autowave processes in distributed kinetic systems*, Uspekhi Fiz. Nauk **128** (1979), no. 4, 625–666; English transl. in Soviet Phys. Uspekhi **22** (1979).
7. A. B. Vasil′eva and V. F. Butuzov, *Asymptotic expansions of the solutions of singularly perturbed equations*, "Nauka", Moscow, 1973. (Russian)
8. A. I. Vol′pert, Commentary, *I. G. Petrovskiĭ's Selected Works. Differential Equations. Probability Theory* (P. S. Aleksandrov and O. A. Oleĭnik, eds.), "Nauka", Moscow, 1987, pp. 333–358. (Russian)
9. S. A. Gabov, *On Whitham's equation*, Dokl. Akad. Nauk SSSR **242** (1978), no. 5, 993–996; English transl. in Soviet Math. Dokl. **19** (1978).
10. ———, *On the property of annihilation of solitary waves described by the Whitham equation*, Dokl. Akad. Nauk SSSR **246** (1979), no. 6, 1292–1295; English transl. in Soviet Math. Dokl. **20** (1979).
11. ———, *Introduction to the theory of nonlinear waves*, Izdat. Moskov. Gos. Univ., Moscow, 1988. (Russian)
12. V. N. Goldberg, I. G. Zarnitsyna, T. N. Fedoseeva, and V. E. Fridman, *Effects of relaxation for weak shock waves propagating in ocean*, Akust. Zh. **27** (1981), no. 1, 88–92.
13. S. Yu. Dobrokhotov, *Nonlocal analogues of the nonlinear Boussinesq equation for surface waves over an uneven bottom and their asymptotic solutions*, Dokl. Akad. Nauk **292** (1987), no. 1, 63–67; English transl. in Soviet Phys. Dokl. **32** (1987).
14. E. A. Zabolotskaya and R. V. Khokhlov, *Quasiplane waves in the nonlinear acoustics of confined beams*, Akust. Zh. **15** (1969), no. 1, 40–47; English transl. in Soviet Phys. Acoustics **15** (1969).
15. A. A. Zaitsev, *Stationary Whitham waves and their dispersion relation*, Dokl. Akad. Nauk SSSR **286** (1986), no. 6, 1364–1369; English transl. in Soviet Phys. Dokl. **31** (1986).
16. V. E. Zakharov, *Collapse of the Langmuir waves*, Zh. Èksper. Teoret. Fiz. **62** (1972), no. 5, 1745–1759; English transl. in Soviet Phys. JETP **35** (1972).
17. V. E. Zakharov and S. V. Manakov, *Asymptotic behavior of non-linear wave systems integrated by the inverse scattering method*, Zh. Èksper. Teoret. Fiz. **71** (1976), no. 1, 203–215; English transl. in Soviet Phys. JETP **44** (1976).
18. V. E. Zakharov, S. V. Manakov, S. P. Novikov, and L. P. Pitaevskiĭ, *Theory of solutions. The inverse scattering method*, "Nauka", Moscow, 1980; English transl., Plenum, New York, 1984.

19. V. E. Zakharov and A. B. Shabat, *Exact theory of the two-dimensional self-focusing and one-dimensional self-modulation of waves in nonlinear media*, Zh. Èksper. Teoret. Fiz. **61** (1971), no. 1, 118–134; English transl. in Soviet Phys. JETP **34** (1972).

20. A. M. Il′in and O. A. Oleĭnik, *Asymptotic behavior of solutions of the Cauchy problem for some quasilinear equations for large values of the time*, Mat. Sb. **51** (1960), no. 2, 191–216. (Russian)

21. A. T. Il′ichev and A. B. Marchenko, *Propagation of the long nonlinear waves in a ponderable fluid beneath an ice sheet*, Izv. Akad. Nauk SSSR Mekh. Zhidk. Gaza **1989**, no. 1, 88–95; English transl. in Fluid Dynamics **24** (1989).

22. A. P. Its, *Asymptotics behavior of the solutions to the nonlinear Schrödinger equation, and isomonodromic deformations of systems of linear differential equations*, Dokl. Akad. Nauk SSSR **261** (1981), no. 1, 14–18; English transl. in Soviet Math. Dokl. **24** (1982).

23. B. B. Kadomtsev and V. I. Petiashvilli, *On the stability of solitary waves in weakly dissipative media*, Dokl. Akad. Nauk SSSR **192** (1970), no. 4, 753–756; English transl. in Soviet Phys. Dokl. **15** (1970).

24. F. Calogero and A. Degasperis, *Spectral transform and solitons*. Vol. I. *Tools to solve and investigate nonlinear evolution equations*, North-Holland, Amsterdam and New York, 1982.

25. V. G. Kamenskiĭ and S. V. Manakov, *Formation of stable-state domains from unstable states in nonlinear dissipative sytems*, Pys′ma Zh. Èksper. Teoret. Fiz. **45** (1987), no. 10, 499–502; English transl. in JETP Letters **45** (1987).

26. V. I. Karpman, *Nonlinear waves in dispersive media*, "Nauka", Moscow, 1973; English transl., Pergamon Press, New York, 1975.

27. Yu. L. Klimontovich, *Statistical theory of non-equilibrium processes in plasma* (1964), Izdat. Moskov. Gos. Univ., Moscow; English transl. (1967), MIT Press, Cambridge, MA.

28. Yu. A. Kobelev and L. A. Ostrovskiĭ, *Nonlinear acoustics. Theoretical and experimental studies*, Gorky, 1980, pp. 143–160. (Russian)

29. A. N. Kolmogorov, I. G. Petrovskiĭ, and N. S. Piskunov, *Étude de l'équation de la chaleur avec croissance de la quantité de matière et son application à un problème biologique*, Bull. Moskov. Gos. Univ. Mat. Mekh. **1** (1937), no. 6, 1–25.

30. A. A. Samarskiĭ (ed.), *Computers and nonlinear phenomena. Information science and contemporary natural science*, "Nauka", Moscow, 1987. (Russian)

31. S. N. Kruzhkov and N. S. Petrosyan, *Asymptotic behaviour of the solutions of the Cauchy problem for non-linear first order equations*, Uspekhi Mat. Nauk **42** (1987), no. 5, 3–40; English transl. in Russian Math. Surveys **42** (1987).

32. O. A. Ladyzhenskaya, *On construction of discontinuous solutions of quasilinear hyperbolic equations as the viscosity coefficient approaching zero*, Dokl. Akad. Nauk **3** (1956), no. 2, 291–294. (Russian)

33. ______, *Mathematical problems in the dynamics of a viscous incompressible fluid*, 2nd rev. aug. ed., "Nauka", Moscow, 1970; English transl. of 1st ed., *The mathematical theory of viscous incompressible flow*, Gordon and Breach, New York, 1963; rev., 1969.

34. P. D. Lax, *Integrals of nonlinear equations of evolution and solitary waves*, Comm. Pure Appl. Math. **21** (1968), 467–490.

35. L. D. Landau and E. M. Lifshitz, *Fluid mechanics*, Pergamon Press, London, 1989.

36. J. L. Lions, *Quelques méthodes de résolution des problèmes aux limites non linéaires*, Dunod and Gauthier-Villars, Paris, 1969.

37. J. L. Lions and E. Magenes, *Problèmes aux limites non homogènes et applications*. Vol. 1, Travaux et Recherches Mathématiques, no. 17, 1968; Vol. 2, Travaux et Recherches Mathématiques, no. 18, 1968; Vol. 3, Travaux et Recherches Mathématiques, no. 20, Dunod, Paris, 1970.

38. H. Lamb, *Hydrodynamics*, Dover, New York, 1945.

39. S. V. Manakov, *Nonlinear Fraungofer's diffraction*, Zh. Èksper. Teoret. Fiz. **65** (1973), no. 10, 1392–1396; English transl. in Soviet Phys. JETP **38** (1973).

40. S. Mandelbrojt, *Séries adhérentes, régularisation des suites. Applications*, Gauthier-Villars, Paris, 1952.

41. V. A. Marchenko, *The periodic Korteweg-de Vries problem*, Mat. Sb. **95** (1974), no. 3, 331–356; English transl. in Math. USSR-Sb. **24** (1974).

42. V. P. Maslov, *Asymptotic methods for solving pseudodifferential equations*, "Nauka", Moscow, 1987. (Russian)

43. ______, *Perturbation theory and asymptotic methods*, "Nauka", Moscow, 1988. (Russian)

44. V. P. Maslov and G. A. Omelyanov, *Soliton-like asymptotics of internal waves in a stratified fluid with small dispersion*, Differentsial′nye Uravneniya **21** (1985), no. 10, 1766–1775; English transl. in Differential Equations **21** (1985).

45. V. E. Nakoryakov and I. P. Schreiber, *A model for propagation of distribution in a vapor-liquid mixture*, Teplofizika Vysokikh Temperatur. **17** (1979), no. 4, 798–803; English transl. in High Temperature **17** (1979).

46. P. I. Naumkin, *The Whitham equation with a singular kernel*, Zh. Vychisl. Mat. i Mat. Fiz. **27** (1987), no. 4, 633–636; English transl. in Comput. Math. and Math. Phys. **27** (1987).

47. P. I. Naumkin and I. A. Shishmarev, *The wave break for the Whitham equation*, Dokl. Akad. Nauk SSSR **265** (1982), no. 4, 809–811; English transl. in Soviet Math. Dokl. **26** (1982).

48. ______, *On the Cauchy problem for the Whitham equation*, Dokl. Akad. Nauk SSSR **273** (1983), no. 4, 804–807; English transl. in Soviet Math. Dokl. **28** (1983).

49. ______, *Breaking of waves for the Whitham equation with a singular kernel.* I, Differentsial′nye Uravneniya **21** (1985), no. 3, 499–508; English transl. in Differential Equations **21** (1985).

50. ______, *Breaking of waves for the Whitham equation with a singular kernel.* II, Differentsial′nye Uravneniya **21** (1985), no. 10, 1775–1790; English transl. in Differential Equations **21** (1985).

51. ______, *The Whitham equation with a singular kernel and small interaction*, Differentsial′nye Uravneniya **21** (1985), no. 10, 1818–1819; English transl. in Differential Equations **21** (1985).

52. ______, *On the existence and destruction of waves that can be described by the Whitham equation*, Dokl. Akad. Nauk SSSR **288** (1986), no. 1, 90–95; English transl. in Soviet Phys. Dokl. **31** (1986).

53. ______, *On the Cauchy problem for the nonlinear Whitham equation*, Allstate Conference "Differential Equations and their Applications", Ashkhabad, 1988, pp. 114–115. (Russian)

54. ______, *A periodic problem for the Whitham equation*, Dokl. Akad. Nauk SSSR **299** (1988), no. 5, 1063–1065; English transl. in Soviet Math. Dokl. **37** (1989).

55. ______, *A system of equations of surface waves*, Dokl. Akad. Nauk SSSR **301** (1988), no. 4, 788–793; English transl. in Soviet Math. Dokl. **38** (1989).

56. ______, *Asymptotic behavior as $t \to \infty$ of solutions of the generalized Kolmogorov-Petrovskiĭ-Piskunov equation*, Mat. Model. **1** (1989), no. 6, 109–125. (Russian)

57. ______, *Asymptotic behavior as $t \to \infty$ of solutions of nonlinear evolution equations with dissipation*, Mat. Zametki **45** (1989), no. 4, 118–121; English transl. in Math. Notes **45** (1989).

58. ______, *A periodic problem for the Whitham equation*, Mat. Sb. **180** (1989), no. 7, 946–968; English transl. in Math. USSR-Sb. **67** (1990).

59. ______, *Asymptotic behavior of the solutions of the Whitham equation for large time*, Mat. Model. **2** (1990), no. 3, 75–88. (Russian)

60. ______, *A system of equations that describe surface waves*, Izv. Akad. Nauk SSSR Ser. Mat. **54** (1990), no. 4, 774–809; English transl. in Math. USSR-Izv. **37** (1991).

61. ______, *Nonlinear nonlocal equations in wave theory.* I. *The Whitham equation*, Vestnik Moskov. Univ. Ser. III Fiz. Astronom. **31** (1990), no. 5, 3–16; English transl. in Moscow Univ. Phys. Bull. **45** (1990).

62. ______, *Nonlinear nonlocal equations in wave theory.* II. *A system of equations of surface wave. Asymptotics of dissipative equations*, Vestnik Moskov. Univ. Ser. III Fiz. Astronom. **31** (1990), no. 6, 3–17; English transl. in Moscow Univ. Phys. Bull..

63. ______, *The Cauchy problem for the Whitham equation.* Part I, Mat. Model. **2** (1990), no. 9, 80–87. (Russian)

64. ______, *The Cauchy problem for the Whitham equation.* Part II, Mat. Model. **2** (1990), no. 9, 88–104. (Russian)

65. ______, *On the asymptotic behavior for large time values of solutions of a system of equations of surface wave for long time scales*, Dokl. Akad. Nauk SSSR **315** (1990), no. 6, 1357–1360; English transl. in Soviet Phys. Dokl. **35** (1990).

66. ———, *A problem on the decay of step-like data for the Korteweg-de Vries-Burgers equation*, Funktsional. Anal. i Prilozhen. **25** (1991), no. 1, 21–32; English transl. in Functional Anal. Appl. **25** (1991).

67. S. M. Nikol′skiĭ, *Approximation of functions of several variables and imbedding theorems*, "Nauka", Moscow, 1969; English transl., Grundlehren Math. Wiss., vol. 205, Springer-Verlag, New York and Heidelberg, 1975.

68. S. P. Novikov, *A periodic problem for the Korteweg-de Vries equation*. I, Funktsional. Anal. i Prilozhen. **8** (1974), no. 3, 54–66; English transl. in Functional Anal. Appl. **8** (1974).

69. V. Yu. Novokshenov, *Asymptotic behavior as $t \to \infty$ of the solution to the Cauchy problem for a two-dimensional generalization of a Toda chain*, Izv. Akad. Nauk SSSR Ser. Mat. **48** (1984), no. 2, 372–410; English transl. in Math. USSR-Izv. **24** (1985).

70. A. Newell, *Solitons in mathematics and physics*, CBMS-NSF Regional Conf. Ser. in Appl. Math., vol. 48, SIAM, Philadelphia, PA, 1985.

71. O. A. Oleĭnik, *Discontinious solutions for non-linear differential equations*, Uspekhi Mat. Nauk **12** (1957), no. 3, 3–73; English transl. in Amer. Math. Soc. Transl. Ser. 2 **26** (1963).

72. M. A. Petrova and I. A. Shishmarev, *On periodic problems with weak interaction*, Abstracts for USSR Seminar on Small Parameter Methods, Nal′chik, 1987, p. 118. (Russian)

73. L. S. Pontryagin, *Ordinary differential equations*, 3rd ed., "Nauka", Moscow, 1970; English transl. of 1st ed., Addison-Wesley, Reading, MA, 1962.

74. B. L. Rozhdestvenskiĭ and N. N. Yanenko, *Systems of quasilinear equations and their applications to gas dynamics*, 2nd ed., "Nauka", Moscow, 1978; English transl., Transl. Math. Monographs, vol. 55, Amer. Math. Soc., Providence, RI, 1983.

75. O. V. Rudenko and S. T. Soluyan, *Theoretical foundations of nonlinear acoustics*, "Nauka", Moscow, 1975; English transl., Consultants Bureau, New York and London, 1977.

76. L. N. Sretenskiĭ, *Theory of wave motion of fluids*, "Nauka", Moscow, 1977. (Russian)

77. J. J. Stoker, *Water Waves: The mathematical theory with applications*, Pure Appl. Math., vol. 4, Interscience, New York, 1957.

78. V. V. Sukhanov, *Asymptotic behavior of solutions of the Cauchy problem for a system of KdV type for large times*, Dokl. Akad. Nauk SSSR **269** (1983), no. 5, 1091–1094; English transl. in Soviet Phys. Dokl. **28** (1983).

79. L. A. Takhtadzhyan and L. D. Faddeev, *Hamiltonian approach in soliton theory*, "Nauka", Moscow, 1986; English transl., *Hamiltonian methods in the theory of solitons*, Springer-Verlag, Berlin and New York, 1987.

80. A. N. Tikhonov, *Systems of differential equations containing small parameters in the derivatives*, Mat. Sb. **31** (1952), no. 3, 575–586. (Russian)

81. G. B. Whitham, *Variational methods and applications to water waves*, Hyperbolic Equations and Waves (Rencontres, Battelle Res. Inst., Seattle, WA, 1968), Springer, Berlin, 1970, pp. 153–172.

82. ———, *Linear and nonlinear waves*, Pure Appl. Math., Wiley, New York, 1974.

83. L. Hörmander, *The analysis of linear partial differential operators*, Grundlehren Math. Wiss., vol. 256, Springer-Verlag, Berlin and New York, 1983.

84. A. B. Shabat, *The Korteweg-de Vries equation*, Dokl. Akad. Nauk SSSR **211** (1973), no. 6, 1310–1313; English transl. in Soviet Math. Dokl. **14** (1973).

85. I. A. Shishmarev, *On the system of surface wave equations with weak nonlocal interaction*, Abstract for the USSR School on Functional Methods in Applied Mathematics and Mathematical Physics, Tashkent, 1988. (Russian)

86. ———, *On the smoothing of solutions of the Cauchy problem for a system of equations of surface waves*, Mat. Zametki **45** (1989), no. 1, 136–138; English transl. in Math. Notes **45** (1989).

87. L. Abdelouhab, J. L. Bona, M. Felland, and J.-C. Saut, *Nonlocal models for nonlinear, dispersive waves*, Phys. D **40** (1989), 360–392.

88. M. J. Ablowitz, D. J. Kaup, A. C. Newell, and H. Segur, *The inverse scattering transform–Fourier analysis for nonlinear problems*, Stud. Appl. Math. **53** (1974), 249–315.

89. M. J. Ablowitz and Y. C. Newell, *The decay of the continuous spectrum for solutions of the Korteweg-de Vries equation*, J. Math. Phys. **14** (1973), 1277–1284.

90. A. Acrivos and R. E. Davis, *Solitary internal waves in deep water*, J. Fluid Mech. **29** (1967), 593–607.

91. C. J. Amick, J. L. Bona, and M. E. Schonbek, *Decay of solutions of some nonlinear wave equations*, J. Differential Equations **81** (1989), 1–49.

92. J. M. Ball, *Remarks on blow up and nonexistence theorems for nonlinear evolution equations*, Quart. J. Math. Oxford Ser. (2) **28** (1977), 473–486.

93. R. Benguria and M. Depassier, *Equations of the Korteweg-de Vries type with nontrivial conserved quantities*, J. Phys. A **22** (1989), 4135–4142.

94. T. B. Benjamin, *Internal waves of permanent form in fluids of great depth*, J. Fluid Mech. **29** (1967), 559–592.

95. T. B. Benjamin, J. L. Bona, and J. J. Mahony, *Model equations for long waves in nonlinear dispersive systems*, Philos. Trans. Roy. Soc. London Ser. A **272** (1972), 47–78.

96. B. Birnir, *An example of blow-up for the complex KdV equation and existence beyond the blow-up*, SIAM J. Appl. Math. **47** (1987), 710–725.

97. J. L. Bona and M. E. Schonbek, *Travelling-wave solutions to the Korteweg-de Vries-Burgers equation*, Proc. Roy. Soc. Edinburgh Sect. A **101** (1985), 207–226.

98. J. Boussinesq, *Theorie de l'intumescence liquide appelee onde solitaire on de translation se propageant dans un canal rectangulaire*, Compt. Rend. **72** (1871), 755–759.

99. M. Bramson, *Convergence of solutions of the Kolmogorov equation to travelling waves*, Mem. Amer. Math. Soc. **285** (1983), 1–190.

100. L. J. F. Broer, *Approximate equations for long water waves*, Appl. Sci. Res. **31** (1975), 377–395.

101. J. M. Burgers, *A mathematical model illustrating the theory of turbulence*, Adv. Appl. Mech. **1** (1948), 171–199.

102. K. M. Case, *The Benjamin-Ono equation: a remarkable dynamical system*, Ann. Nuclear Energy **7** (1980), 273–277.

103. H. H. Chen and Y. C. Lee, *Internal wave solitons of fluids with finite depth*, Phys. Rev. Lett. **43** (1979), 264–266.

104. D. Christodoulou, *Global solution of nonlinear hyperbolic equations for small initial data*, Comm. Pure Appl. Math. **39** (1986), 267–282.

105. I. D. Cole, *On a quasilinear parabolic equation occuring in aerodynamic*, Quart. Appl. Math. **9** (1951), 226–236.

106. R. Courant, *Methods of mathematical physics*. Vol. II: *Partial differential equations*, Interscience, London, 1962.

107. S. Takeno (ed.), *Dynamical problems in soliton systems*, Springer Ser. Synergetics, vol. 30, Springer-Verlag, Berlin and New York, 1985.

108. J. Engelbrecht, *Nonlinear wave processes of deformation in solids*, Monographs and Studies in Mathematics, vol. 16, Pitman, Boston and London, 1983.

109. A. S. Fokas and M. J. Ablowitz, *The inverse scattering transform for the Benjamin-Ono equation. A pivot to multidimensional problems*, Stud. Appl. Math. **68** (1983), 1–10.

110. H. Fujita, *On the blowing up of solution of the Cauchy problem* $u_t = \Delta u + u^{1+\alpha}$, J. Fac. Sci. Univ. Tokyo Sect. I **13** (1966), 109–124.

111. C. S. Gardner, J. M. Greene, M. D. Kruskal, and R. M. Miura, *Method for solving the Korteweg-de Vries equation*, Phys. Rev. Lett. **19** (1967), 1095–1097.

112. R. T. Glassey, *On the blowing up of solutions to the Cauchy problem for nonlinear Schrödinger equation*, J. Math. Phys. **18** (1977), 1794–1797.

113. ______, *Blow up theorems for nonlinear wave equations*, Math. Z. **132** (1973), 182–203.

114. J. Glimm, *The interaction of nonlinear hyperbolic waves*, Comm. Pure Appl. Math. **41** (1988), 569–590.

115. ______, *Solution in the large for nonlinear hyperbolic systems of equations*, Comm. Pure Appl. Math. **18** (1965), 697–715.

116. J. Glimm and P. Lax, *Decay of solution of systems of hyperbolic conservation laws*, Mem. Amer. Math. Soc. **101** (1970).

117. M. V. Goldman, *Langmuir wave solitons and spatial collapse in plasma physics*, Phys. D **18** (1986), 67–76.

118. H. Grad and P. N. Hu, *Unified shock profile in a plasma*, Phys. Fluids **10** (1967), 2596–2602.

119. R. Hirota, *Direct method of finding exact solutions of nonlinear evolution equations*, Bäcklund Transformations, the Inverse Scattering Method, Solitons, and their Applications, Lecture Notes in Math., vol. 515, Springer-Verlag, Berlin and New York, 1976, pp. 40–68.

120. E. Hopf, *The partial differential equation* $u_t + uu_x = \mu u_{xx}$, Comm. Pure. Appl. Math. **3** (1950), 201–230.

121. L. Hörmander, *On global existence of solutions of nonlinear hyperbolic equations in* R^{1+3}, Report No. 9, Institut Mittag Leffler, 1985, pp. 1–22.

122. A. Jeffrey and J. Engelbrecht, *Nonlinear dispersive waves in a relaxing medium*, Wave Motion **2** (1980), 255–266.

123. A. Jeffrey and S. Xu, *Exact solutions to the Korteweg-de Vries-Burgers equation*, Wave Motion **11** (1989), 559–564.

124. F. John, *Existence for large times of strict solution of nonlinear wave equations in three space dimensions for small initial data*, Comm. Pure Appl. Math. **40** (1987), 79–109.

125. ———, *Blow up for quasilinear wave equations in three dimensions*, Comm. Pure Appl. Math. **34** (1981), 29–51.

126. K. Jörgens, *Nonlinear wave equations*, Lecture Notes, Univ. Colorado, 1970.

127. R. J. Joseph, *Solitary waves in a finite depth fluid*, J. Phys. A **10** (1977), 225–227.

128. T. Kakutani and K. Matsuuchi, *Effect of viscosity in long gravity waves*, J. Phys. Soc. Japan **39** (1975), 237–246.

129. S. Kaplan, *On the growth of solutions of quasilinear parabolic equations*, Comm. Pure Appl. Math. **16** (1963), 327–330.

130. D. J. Kaup, *A higher-order water wave equation and the method for slowing it*, Progr. Theoret. Phys. **54** (1975), 396–408.

131. T. Kawahara, *Oscillatory solitary waves in dispersive media*, J. Phys. Soc. Japan **33** (1972), 260–264.

132. J. B. Keller, *On solutions of nonlinear wave equations*, Comm. Pure Appl. Math. **10** (1957), 523–530.

133. P. L. Kelley, *Self-focusing of pitical beams*, Phys. Rev. Lett. **15** (1965), 1005–1008.

134. J. U. Kim, *On the model equation which describes nonlinear wave motions in a rotating fluid*, Trans. Amer. Math. Soc. **287** (1985), 403–417.

135. S. Klainerman and G. Ponce, *Global, small amplitude solutions to nonlinear evolution equations*, Comm. Pure Appl. Math. **36** (1983), 133–141.

136. Y. Kodama, M. J. Ablowitz, and J. Satsuma, *Direct and inverse scattering problems of the nonlinear intermediate long-wave equation*, J. Math. Phys. **23** (1982), 564–576.

137. D. J. Korteweg and G. de Vries, *On the change of form of long waves advancing in a new type of long stationary waves*, Philos. Mag. **5** (1895), 422–443.

138. T. Kubota, D. R. Ko, and D. Dobbs, *Weakly nonlinear long stratified fluids of finite depth*, AIAA J. Hydronautics **12** (1978), 157–165.

139. Y. Kuramoto, *Chemical oscillations, waves and turbulence*, Springer Ser. Synergetics, vol. 19, Springer-Verlag, Berlin and New York, 1984.

140. P. D. Lax, *Periodic solutions of the KdV equation*, Comm. Pure Appl. Math. **28** (1975), 141–188.

141. P. D. Lax and C. D. Levermore, *The small dispersion limit of the Korteweg-de Vries equation*. I, Comm. Pure Appl. Math. **36** (1983), 253–290; II, 571–593; III, 809–829.

142. S. J. Leibovich, *Weakly non-linear waves in rotating fluids*, J. Fluid Mech. **42** (1970), 803–822.

143. V. S. Manoranjan, T. Ortega, and J. M. Sanz-Serna, *Soliton and antisoliton interactions in the "good" Boussinesq equation*, J. Math. Phys. **29** (1988), 1964–1968.

144. J. Miles, *KdV equation modified by viscosity*, Phys. Fluids **19** (1976), 1063.

145. ______, *The asymptotic solution of the Korteweg-de Vries equation in the absence of solitons*, Stud. Appl. Math. **60** (1979), 59–72.

146. C. S. Morawetz and W. A. Strauss, *Decay and scattering of solutions of a nonlinear relativisitc wave equation*, Comm. Pure Appl. Math. **25** (1972), 1–31.

147. I. Nakata, *Long nonlinear waves on a liquid layer adjacent to a gas stream*, J. Phys. Soc. Japan **41** (1976), 1387–1393.

148. G. A. Nariboli and A. Sedov, *Burgers-Korteweg-de Vries equation for viscoelastic rods and paltes*, J. Math. Anal. Appl. **32** (1970), 661–677.

149. A. Novick-Cohen and G. I. Sivashinsky, *On the solidification front of a dilute binary alloy: thermal diffusivity effects and breathing solutions*, Phys. D **20** (1986), 237–258.

150. K. Nozaki and N. Bekki, *Exact solutions of the generalized Ginzburg-Landau equation*, J. Phys. Soc. Japan **53** (1984), 1581–1582.

151. H. Ono, *Algebraic solitary waves in stratified fluids*, J. Phys. Soc. Japan **39** (1975), 1082–1091.

152. L. A. Ostrovsky, *Short-wave asymptotics for weak shock waves and solitons in mechanics*, Internat. J. Non-Linear Mech. **11** (1976), 401–416.

153. E. Ott and R. N. Sudan, *Nonlinear theory of ion acoustic waves with Landau damping*, Phys. Fluids **12** (1969), 2388–2394.

154. L. E. Payne, *Improperly posed problems in partial differential equations*, CBMS-NSF Regional Conf. Ser. in Appl. Math., vol. 22, SIAM, Philadelphia, PA, 1975.

155. D. Pfirsch and R. N. Sudan, *Conditions for the existence of shock-like solutions of Korteweg-de Vries equation with dissipation*, Phys. Fluids **14** (1971), 1033–1035.

156. G. Pelletier, M. Goldman, H. T. Moon, and W. Merryfield, *Autonomous dynamical systems arising from self-similar parametrization of damped/driven NLS equations*, Phys. D **18** (1986), 154–156.

157. W. G. Pritchard, *Solitary waves in rotating fluids*, J. Fluid Mech. **42** (1970), 61–83.

158. J.-C. Saut, *Sur quelques généralisations de l'équation de Korteweg-de Vries*, J. Math. Pures Appl. **58** (1979), 21–61.

159. H. Segur and M. J. Ablowitz, *Asymptotic solutions and conservation laws for the nonlinear Schrödinger equation*. I, J. Math. Phys. **17** (1976), 710–713.

160. R. L. Seliger, *On the breaking of waves*, Proc. Roy. Soc. **303** (1968), 493–496.

161. T. C. Sideris, *Nonexistence of global solutions to semilinear wave equations in high dimensions*, J. Differential Equations **52** (1984), 378–406.

162. R. S. Smith, *Nonlinear Kelvin and continental-shelf waves*, J. Fluid Mech. **52** (1972), 379–391.

163. D. Spehler and G. C. Marques, *Classical solutions of nonrelativisitc model exhibiting spontaneous symmetry breakdown*, J. Math. Phys. **30** (1989), 464–469.

164. B. Straughan, *Further global nonexistence theorems for abstract nonlinear wave equation*, Proc. Amer. Math. Soc. **48** (1975), 381–390.

165. J. Swift and P. C. Hohenberg, *Hydrodynamic fluctuations at the convective instability*, Phys. Rev. A **15** (1977), 319–334.

166. S. Tanaka, *Korteweg-de Vries equation; asymptotic behavior of solutions*, Publ. Res. Inst. Math. Sci. **10** (1975), 367–379.

167. M. Toda, *Coupled nonlinear waves*, Phys. D **33** (1988), 317–322.

168. M. Tsutsumi, *On global solutions of the generalized Korteweg de Vries equation*, Publ. Res. Inst. Math. Sci. **7** (1971), 329–344.

169. ______, *Nonexistence of global solutions to the Cauchy problem for the damped nonlinear Schrödinger equation*, SIAM J. Math. Anal. **15** (1984), 357–366.

170. S. Venakides, *The Korteweg-de Vries equation with small dispersion: higher order Lax-Levermore theory*, Comm. Pure Appl. Math. **43** (1990), 335–361.

171. J. A. Zufira, *Symmetry breaking in periodic and solitary gravity-capillary waves on water of finite depth*, J. Fluid Mech. **184** (1987), 183–206.

172. R. L. Herman, *Resolution of the motion of a perturbed KdV soliton*, Inverse Problems **6** (1990), 43–54.

173. ______, *A direct approach to studying soliton perturbations*, J. Phys. A **23** (1990), 2327–2362.

174. D. J. Kaup and A. C. Newell, *Soliton as particles oscillators, and in slowly changing media: a singular perturbation theory*, Proc. Roy. Soc. **361** (1978), 413–446.

175. V. I. Karpman and E. M. Maslov, *Structure of tails produced under the action of perturbations on solitons*, Soviet Phys. JETP **48** (1978), 252–258.

176. K. Ko and H. H. Kuehl, *Energy loss of Korteweg-de Vries solitary wave in a slowly varying medium*, Phys. Fluids **23** (1980), 834–836.

177. A. Erdélyi, W. Magnus, F. Oberhettinger, and F. G. Tricomi, *Tables of integral transform*. Vols. I–II, based, in part, on notes left by Harry Bateman, McGraw-Hill, New York, Toronto, and London, 1954.

178. M. B. Fedoryuk, *Asymptotics: integrals and series*, "Nauka", Moscow, 1987. (Russian)

179. D. J. Berney, *Long waves on liquid films*, J. Math. Phys. **45** (1966), 150–155.

180. R. W. Atherton, Chem. Eng. Comm. **2** (1976), 57–77.

181. J. Topper and T. Kawahara, *Approximate equations for long nonlinear waves on a viscous fluid*, J. Phys. Soc. Japan **44** (1978), 663–666.

182. A. P. Hooper and R. Grimshaw, *Nonlinear instability at the interface between two viscous fluids*, Phys. Fluids **28** (1985), 37–45.

183. G. I. Sivashinsky, *Instabilities pattern formation, and turbulence in flames*, Ann. Rev. Fluid Mech. **15** (1983), 179–199.

184. J. F. Toland, *Existence and uniqueness of heteroclinic orbits for the equation $\lambda u''' + u' = f(u)$*, Proc. Roy. Soc. Edinburgh Sect. A **109** (1988), 23–36.

185. A. P. Hooper and R. Grimshaw, *Travelling wave solutions of the Kuramoto-Sivashinsky equation*, Wave Motion **10** (1988), 405–420.

186. ______, *The nonexistence of a certain class of travelling wave solutions of the Kuramoto-Sivashinsky equation*, Phys. D **50** (1991), 231–238.

187. D. Michelson, *Steady solutions of the Kuramoto-Sivashinsky equation*, Phys. D **19** (1986), 89–111.

188. C. K. McCord, *Uniqueness of connecting orbits in the equation $Y^{(3)} = Y^2 - 1$*, J. Math. Anal. Appl. **114** (1986), 584–592.

189. P. I. Naumkin and I. A. Shishmarev, *On the asymptotic behavior as $t \to \infty$ of solutions of some nonlinear equations*, Dokl. Akad. Nauk SSSR **321** (1991), no. 2, 290–293; English transl. in Soviet Phys. Dokl. **36** (1991).

190. ______, *On stability of running wave solutions for the Kuramoto-Sivashinskiĭ equation*, Russian Acad. Sci. Dokl. Math. **323** (1992), no. 2, 266–269; English transl. in Soviet Phys. Dokl. **37** (1992).

SUPPLEMENTARY REFERENCES(*)

1*. P. Biler, *Partition of energy in strongly damped generalized wave equations*, Math. Methods Appl. Sci. **12** (1990), 95–103.

2*. T. Cazenave, *An introduction to nonlinear Schrödinger equations*, Textos de Métodos Matematicos, vol. 22, Rio de Janeiro, 1989.

3*. F. M. Christ and M. I. Weinstein, *Dispersion of small amplitude solutions of the generalized Korteweg-de Vries equation*, J. Funct. Anal. **100** (1991), 87–109.

4*. Ph. Clément and J. A. Nohel, *Asymptotic behavior of solutions of nonlinear Volterra equations with completely positive kernels*, SIAM J. Math. Anal. **12** (1981), 514–535.

5*. P. Constantin and J.-C. Saut, *Local smoothing properties of Schrödinger equations*, Indiana Univ. Math. J. **38** (1989), 791–810.

6*. D. B. Dix, *Temporal asymptotic behavior of solutions of the Bejamin-Ono-Burgers equation*, J. Differential Equations **90** (1991), 238–287.

7*. ______, *The dissipation of nonlinear dispersive waves: the case of asymptotically weak nonlinearity*, Comm. Partial Differential Equations **17** (1992), 1665–1693.

(*) Provided by the author during the translation.

8*. C. J. van Duyn and L. A. Peletier, *Asymptotic behavior of solutions of a nonlinear diffusion equation*, Arch. Rational Mech. Anal. **65** (1977), 363–377.

9*. M. Escobedo and E. Zuazua, *Large time behavior for convection-diffusion equation in* $\mathbb{R}^5$, J. Funct. Anal. **100** (1991), 119–161.

10*. J. Ginibre and G. Velo, *On a class of nonlinear Schrödinger equations with nonlocal interaction*, Math. Z. **170** (1980), 109–136.

11*. M. Grillakis, J. Shatah, and W. A. Strauss, *Stability theory of solitary waves in the presence of symmetry*. II, J. Funct. Anal. **94** (1990), 308–348.

12*. N. Hayashi and T. Ozawa, *Smoothing effect for some Schrödinger equations*, J. Funct. Anal. **85** (1989), 307–348.

13*. S. Kamin and L. A. Peletier, *Large time behavior of the porous media equation with absorption*, Israel J. Math. **55** (1986), 129–146.

14*. T. Kato, *On nonlinear Schrödinger equations*, Ann. Inst. H. Poincaré Phys. Théor. **46** (1987), 113–129.

15*. S. Klainerman, *Long-time behavior of solutions to nonlinear evolution equations*, Arch. Rational Mech. Anal. **78** (1982), 73–98.

16*. E. Mitidieri, *Estimates from below for the solutions to a class of second order evolution equations*, Differential Integral Equations **3** (1990), 1101–1111.

17*. G. Ponce and L. Vega, *Nonlinear small data scattering for the generalized Korteweg-de Vries equation*, J. Funct. Anal. **90** (1990), 445–457.

18*. M. E. Schonbek, *Uniform decay rates for parabolic conservation laws*, Nonlinear Anal. **10** (1986), 943–956.

19*. W. A. Strauss, *Dispersion of low-energy waves for two conservative equations*, Arch. Rational Mech. Anal. **55** (1974), 86–92.

Recent Titles in This Series

(Continued from the front of this publication)

97 **Itiro Tamura,** Topology of foliations: An introduction, 1992
96 **A. I. Markushevich,** Introduction to the classical theory of Abelian functions, 1992
95 **Guangchang Dong,** Nonlinear partial differential equations of second order, 1991
94 **Yu. S. Il′yashenko,** Finiteness theorems for limit cycles, 1991
93 **A. T. Fomenko and A. A. Tuzhilin,** Elements of the geometry and topology of minimal surfaces in three-dimensional space, 1991
92 **E. M. Nikishin and V. N. Sorokin,** Rational approximations and orthogonality, 1991
91 **Mamoru Mimura and Hirosi Toda,** Topology of Lie groups, I and II, 1991
90 **S. L. Sobolev,** Some applications of functional analysis in mathematical physics, third edition, 1991
89 **Valeriĭ V. Kozlov and Dmitriĭ V. Treshchëv,** Billiards: A genetic introduction to the dynamics of systems with impacts, 1991
88 **A. G. Khovanskiĭ,** Fewnomials, 1991
87 **Aleksandr Robertovich Kemer,** Ideals of identities of associative algebras, 1991
86 **V. M. Kadets and M. I. Kadets,** Rearrangements of series in Banach spaces, 1991
85 **Mikio Ise and Masaru Takeuchi,** Lie groups I, II, 1991
84 **Đáo Trọng Thi and A. T. Fomenko,** Minimal surfaces, stratified multivarifolds, and the Plateau problem, 1991
83 **N. I. Portenko,** Generalized diffusion processes, 1990
82 **Yasutaka Sibuya,** Linear differential equations in the complex domain: Problems of analytic continuation, 1990
81 **I. M. Gelfand and S. G. Gindikin, Editors,** Mathematical problems of tomography, 1990
80 **Junjiro Noguchi and Takushiro Ochiai,** Geometric function theory in several complex variables, 1990
79 **N. I. Akhiezer,** Elements of the theory of elliptic functions, 1990
78 **A. V. Skorokhod,** Asymptotic methods of the theory of stochastic differential equations, 1989
77 **V. M. Filippov,** Variational principles for nonpotential operators, 1989
76 **Phillip A. Griffiths,** Introduction to algebraic curves, 1989
75 **B. S. Kashin and A. A. Saakyan,** Orthogonal series, 1989
74 **V. I. Yudovich,** The linearization method in hydrodynamical stability theory, 1989
73 **Yu. G. Reshetnyak,** Space mappings with bounded distortion, 1989
72 **A. V. Pogorelev,** Bendings of surfaces and stability of shells, 1988
71 **A. S. Markus,** Introduction to the spectral theory of polynomial operator pencils, 1988
70 **N. I. Akhiezer,** Lectures on integral transforms, 1988
69 **V. N. Saliĭ,** Lattices with unique complements, 1988
68 **A. G. Postnikov,** Introduction to analytic number theory, 1988
67 **A. G. Dragalin,** Mathematical intuitionism: Introduction to proof theory, 1988
66 **Ye Yan-Qian,** Theory of limit cycles, 1986
65 **V. M. Zolotarev,** One-dimensional stable distributions, 1986
64 **M. M. Lavrent′ev, V. G. Romanov, and S. P. Shishat·skiĭ,** Ill-posed problems of mathematical physics and analysis, 1986
63 **Yu. M. Berezanskiĭ,** Selfadjoint operators in spaces of functions of infinitely many variables, 1986

(See the AMS catalog for earlier titles)